ALGEBRA in the Early Grades

STUDIES IN MATHEMATICAL THINKING AND LEARNING
Alan H. Schoenfeld, Series Editor

Artzt/Armour-Thomas • *Becoming a Reflective Mathematics Teacher: A Guide for Observation and Self-Assessment*
Baroody/Dowker (Eds.) • *The Development of Arithmetic Concepts and Skills: Constructing Adaptive Expertise*
Boaler • *Experiencing School Mathematics: Traditional and Reform Approaches to Teaching and Their Impact on Student Learning*
Carpenter/Fennema/Romberg (Eds.) • *Rational Numbers: An Integration of Research*
Cobb/Bauersfeld (Eds.) • *The Emergence of Mathematical Meaning: Interaction in Classroom Cultures*
Cohen • *Teachers' Professional Development and the Elementary Mathematics Classroom: Bringing Understandings to Light*
Clements/Sarama/DiBiase (Eds.) • *Engaging Young Children in Mathematics: Standards for Early Childhood Mathematics Education*
English (Ed.) • *Mathematical and Analogical Reasoning of Young Learners*
English (Ed.) • *Mathematical Reasoning: Analogies, Metaphors, and Images*
Fennema/Nelson (Eds.) • *Mathematics Teachers in Transition*
Fennema/Romberg (Eds.) • *Mathematics Classrooms That Promote Understanding*
Fernandez/Yoshida • *Lesson Study: A Japanese Approach to Improving Mathematics Teaching and Learning*
Kaput/Carraher/Blanton (Eds.) • *Algebra in the Early Grades*
Lajoie • *Reflections on Statistics: Learning, Teaching, and Assessment in Grades K-12*
Lehrer/Chazan (Eds.) • *Designing Learning Environments for Developing Understanding of Geometry and Space*
Ma • *Knowing and Teaching Elementary Mathematics: Teachers' Understanding of Fundamental Mathematics in China and the United States*
Martin • *Mathematics Success and Failure Among African-American Youth: The Roles of Sociohistorical Context, Community Forces, School Influence, and Individual Agency*
Reed • *Word Problems: Research and Curriculum Reform*
Romberg/Fennema/Carpenter (Eds.) • *Integrating Research on the Graphical Representation of Functions*
Romberg/Carpenter/Dremock (Eds.) • *Understanding Mathematics and Science Matters*
Schliemann/Carraher/Brizuela (Eds.) • *Bringing Out the Algebraic Character of Arithmetic: From Children's Ideas to Classroom Practice*
Schoenfeld (Ed.) • *Mathematical Thinking and Problem Solving*
Senk/Thompson (Eds.) • *Standards-Based School Mathematics Curricula: What Are They? What Do Students Learn?*
Sophian • *The Origins of Mathematical Knowledge in Childhood*
Sternberg/Ben-Zeev (Eds.) • *The Nature of Mathematical Thinking*
Watson • *Statistical Literacy at School: Growth and Goals*
Watson/Mason • *Mathematics as a Constructive Activity: Learners Generating Examples*
Wilcox/Lanier (Eds.) • *Using Assessment to Reshape Mathematics Teaching: A Casebook for Teachers and Teacher Educators, Curriculum and Staff Development Specialists*
Wood/Nelson/Warfield (Eds.) • *Beyond Classical Pedagogy: Teaching Elementary School Mathematics*
Zaskis/Campbell (Eds.) • *Number Theory in Mathematics Education: Perspectives and Prospects*

For additional information on titles in the Studies in Mathematical Thinking and Learning series visit www.routledge.com

ALGEBRA in the Early Grades

Edited by

James J. Kaput • David W. Carraher • Maria L. Blanton

Lawrence Erlbaum Associates
Taylor & Francis Group
270 Madison Avenue
New York, NY 10016

Lawrence Erlbaum Associates
Taylor & Francis Group
2 Park Square
Milton Park, Abingdon
Oxon OX14 4RN

Lawrence Erlbaum Associates is an imprint of Taylor & Francis Group, an Informa business

Printed in the United States of America on acid-free paper
10 9 8 7 6 5 4 3

International Standard Book Number-13: 978-0-8058-5473-2 (Softcover) 978-0-8058-5472-5 (Hardcover)

Visit the Taylor & Francis Web site at
http://www.taylorandfrancis.com

Dedication

It is a bittersweet experience for us to put the finishing touches on this manuscript without our dear friend and colleague, Jim Kaput. Jim originated the volume and labored long and hard to turn the promise of early algebra into reality. He found it rewarding to play a pivotal role in a book that reflects the depth of children's thinking while recognizing the challenges for educators and the importance of grounding the work in mathematical foundations.

We offer this volume as evidence of his vision and leadership.

—David Carraher and Maria Blanton

Contents

Preface

This book is the result of approximately 15 years of thinking, research, and collective discussion about the role of algebra in school mathematics. It originated in the context of the Algebra Working Group (AWG), an early online discussion group led by the first editor and supported by the National Center for Research in Mathematical Sciences Education (NCRMSE) based at the University of Wisconsin Center for Education Research. For 4 years, this group—which included approximately 40 researchers and leaders in mathematics education, some of whom were members of the National Council of Teachers of Mathematics *Standards* development committees—held vigorous discussions about the nature and role of algebra in school mathematics. The question "What is algebraic reasoning?" was raised regularly, but it was never fully answered. Nonetheless, many of the issues explored eventually became the basis of research reported in the present volume, including our characterizations of algebraic reasoning.

As the AWG continued to hold small conferences, produce discussion documents, and sponsor symposia and papers at research conferences to further the discussions of algebra learning, it became clear to an AWG subgroup that a long-term solution to the challenges of school algebra required reconceptions of school mathematics in the early grades. This, in turn, would require a major research and development initiative. In the early to mid-1990s, a smaller group known as the Early Algebra Research Group (EARG) began to meet to focus on research issues concerning algebraic reasoning in the elementary grades. The National Center for Student Learning and Achievement (NCISLA), a successor to NCRMSE, supported EARG for the next several years as it began to investigate these issues and share the results with the math education community. Some of these appear, in revised form, as chapters in this book.

Indeed, the present volume contains echoes of previous conversations from the algebra working groups, and we have tried to recognize here the origins of ideas and findings in prior work as much as possible. Even so, researchers today are still defining and mapping the space of possibilities for developing algebraic reasoning in the elementary grades. Replication and data refinement represent ongoing and future work. Because the important effects of early algebra are based on long-term longitudinal change ranging across elementary grades to middle school and beyond, this volume represents a starting point of a story that will be a long time in the telling.

STRUCTURE AND ORGANIZATION OF THE BOOK

Our goal in this volume is to contribute to a comprehensive, research-based "early algebra story" that helps answer such questions as: What is algebra, mathematically and pedagogically? What is its curricular place in school mathematics, especially elementary grades mathematics? How can it be fruitfully taught and learned in the early grades? What are the important cognitive and symbolization differences between arithmetic and algebra? Are there developmental constraints to learning algebra? Which aspects of algebra are especially learnable? Are some approaches to teaching algebra to young learners more promising than others? How can teachers build algebraic reasoning, especially if their own experience with the subject has been deficient, and how can the institutional setting in which they work support this?

The early algebra story offered in this book is only a beginning, and, in fact, is the beginning of several stories with several starting points, alternative plots, and plausible happy endings. The contributions to the present volume address different aspects of the story and offer diverse scenarios of how it may play out. Some illustrate the potential of building algebraic reasoning in the context of generalizing properties of numbers and operations. Some emphasize the role of reasoning with physical quantities, including quantities described initially with letters. At least one contribution envisions geometry taking a leading role in early algebra developments. Other chapters explore theoretical underpinnings of early algebra or issues concerning implementing early algebra in districts, schools, and classrooms.

Part I takes a foundational approach by examining the nature of algebra and its roots in naturally occurring human powers. Chapter 1 defines what we term the *algebra problem* and its historical roots and tackles the problem of what we mean by *algebra* and *algebraic reasoning*. In particular, it portrays algebra as both an inherited body of knowledge and as activity that people do. Further, it offers a content analysis, consistent with most analyses of the past decade, of the key aspects and strands of school algebra that can serve as a common reference across chapters. Chapter 2 attempts to deepen our analysis of what we mean by algebra by focusing on how core aspects of algebra arise through symbolization processes. Chapter 3 examines how children's capacity for algebraic reasoning is deeply connected to innate powers of generalizing as a human activity.

Finally, chapter 4 discusses the implications of using physical quantities and quantitative relationships as a springboard for algebraic reasoning, and chapter 5 uses elementary teacher professional development as a context for lifting out the role of argumentation in establishing personal and social mathematical certainty. All chapters in Part I are intended to orient the reader toward some of the foundational ideas that underpin the early algebra enterprise.

By looking specifically at children as they reason algebraically, Part II brings a classroom-grounded interpretation to the foundational ideas described in Part I. This section, which some may view as the heart of the book, evokes the question: Why aren't similar mathematical discussions and behaviors being promoted in classrooms more widely? or Why didn't I get to learn algebra this way? Although its chapters may be regarded as optimistic, it is an optimism grounded in concrete, documented classroom experience. Chapter 6 looks at classroom stories of children engaged in generalized arithmetic, particularly as they notice and generalize about properties of numbers, whereas chapter 7 explores children's reasoning about changes over time and the generalizations inherent in concepts of speed, distance, and time. Chapter 8 looks at children's work with arithmetic number sentences and the relationship between the recognition and production of these sentences. In a departure from number and quantity, chapter 9 argues how visualization inherent in geometry can be used to leverage children's algebraic thinking. Chapters 10 and 11 use functions as a context for pushing children's algebraic thinking. In particular, chapter 10 uses children's work with functions to argue that early algebra is not algebra early, whereas chapter 11 examines children's use of multiple representations to coordinate information about a function expressed in multiple ways. Finally, in chapter 12, the authors examine the mutually supportive relationship between negative numbers and algebraic thinking and how negative numbers can be meaningfully taught within a curriculum that integrates algebraic thinking.

The early algebra story is more than a story about foundational ideas or student thinking and learning because it inevitably must confront the matter of implementation. Who can enact the story and how? Can we get there from here with the available resources, or must we find ways to build capacity based on those resources, particularly the human resources of teachers in elementary schools today? The optimism contained in this volume regarding what children can do is tempered by the realizations that large-scale implementation and professional development is complex and the research basis for all aspects of early algebra, including its implementation, is still in its formative stages. Part III begins to address this by examining issues of implementation involving teacher practices, professional development, and curriculum materials that support the kinds of mathematical work seen in Part II. It offers examples of the conditions and materials that make algebraic reasoning in elementary grades possible. In particular, chapter 13 examines student thinking as a way to leverage teacher professional development in early algebra, whereas chapter 14 looks more broadly at how the institutional setting in which teachers work constrains or supports their early algebra practice. In chapter 15, the author uses insights from children's work to discuss challenges of curriculum and implementation in an early algebra approach that uses mathematical generality (rather than the particular of numbers) as a starting point for developing children's understanding of structure. Chapter 16 brings into relief issues of curriculum design, building the case that curricula should explicitly build on the generality inherent in arithmetic. Finally, chapter 17 examines the algebraic trajectory through which students evolve as they work in a particular curriculum, *Math Workshop.*

INTENDED AUDIENCES OF THE BOOK

The book deliberately encompasses diverse points of view that address different aspects of the early algebra story. Because it is intended for a diverse audience, we did not enforce a single style of exposition or scholarly detail in order that different readers might approach the book according to their own interests and purposes. Those looking for an overall theoretical framework for algebraic thinking may find Part I of particular interest. Teachers and teacher educators may find Parts II and III especially useful. Policymakers are likely to find Part III of particular interest. Overall, our intent was that the book be useful and accessible to anyone who would be a reader of the NCTM *Principles and Standards* and its support material, as well as typical state curriculum frameworks.

Acknowledgments

Many avenues of support converged to make this volume possible. Tom Romberg and Tom Carpenter, through their leadership of NCRMSE and its successor, NCISLA, provided sustained support of both the Algebra Working Group (AWG) and the Early Algebra Research Group (EARG) through Grants R117G10002 and R305A60007 of the U.S. Department of Education Office of Educational Research and Improvement. Their support was constant, even when the conceptual foundations of early algebra research were yet to be built and the work that made these foundations appear reasonable was yet to be done. Several meetings of the Early Algebra Research Group were held at the University of Massachusetts Dartmouth thanks to their support.

In 2001, the National Science Foundation supported the participation of several contributors in the 12th ICMI Study Group in Melbourne, Australia, through the grant "An International Perspective on the Role of Algebra in Elementary School Mathematics" (NSF REC-0140104). Jim Kaput co-chaired the multiday symposium on early algebra, during which precursors of several of the present chapters were first introduced.

We would like to thank Naomi Silverman of Lawrence Erlbaum Associates for her assistance throughout the development of this volume. We would also like to thank the reviewers for the reactions and advice they offered on earlier drafts: Eric Knuth, of the University of Wisconsin, Madison; and one anonymous reviewer. Our colleague, Alan Schoenfeld, has been an enthusiastic supporter and series editor of this project. We also thank him for writing the closing chapter for the book.

Special thanks to Becky Moniz, at the University of Massachusetts–Dartmouth, for her unflagging help, throughout the project, in managing the logistics for the meetings and manuscripts.

Finally, we thank our families for bearing with us and providing encouragement during the long gestation of this book.

—James Kaput, David Carraher, and Maria Blanton

About the Contributors

Virginia Bastable
Summer Math for Teachers
Mount Holyoke College
South Hadley, Massachusetts

Dan Battey
College of Education, Division of Curriculum & Instruction
University of Arizona
Tempe, Arizona

Maria L. Blanton
Department of Mathematics
University of Massachusetts
Dartmouth, Massachusetts

Timothy Boester
Department of Mathematics
Department of Educational Psychology
University of Wisconsin
Madison, Wisconsin

Bárbara M. Brizuela
Department of Education
Tufts University
Medford, Massachusetts

Thomas P. Carpenter
Wisconsin Center for Education Research
University of Wisconsin
Madison, Wisconsin

David W. Carraher
TERC
Cambridge, Massachusetts

Barbara Dougherty
School of Education
University of Mississippi
University, Mississippi

Darrell Earnest
TERC
Cambridge, Massachusetts

Megan Loef Franke
Department of Education
University of California
Los Angeles, California

E. Paul Goldenberg
Education Development Center, Inc.
Newton, Massachusetts

James J. Kaput
Department of Mathematics
University of Massachusetts
Dartmouth, Massachusetts

Richard Lehrer
Department of Teaching and Instruction
Peabody College Vanderbilt University
Nashville, Tennessee

Nitza Mark-Zigdon
School of Education
Tel-Aviv University
Tel-Aviv, Israel

John Mason
Centre for Mathematics Education
The Open University
Milton Keynes, U.K.

Stephen Monk
Department of Mathematics
University of Washington
Seattle, Washington

Luis Moreno
Center for Research
and Advanced Studies
Mexico City, Mexico

Irit Peled
Department of Teaching
University of Haifa
Haifai, Israel

Susan Jo Russell
TERC
Cambridge, Massachusetts

Deborah Schifter
Education Development Center
Newton, Massachusetts

Analúcia D. Schliemann
Department of Education
Tufts University
Medford, Massachusetts

Alan H. Schoenfeld
University of California
Berkeley, California

Judah L. Schwartz
Departments of Education and
Physics and Astronomy
Tufts University
Medford, Massachusetts

Nina Shteingold
Education Development
Center
Newton, Massachusetts

Erick Smith
Cayuga Pure Organics
Brooktondale, New York

John P. (Jack) Smith III
Department of Counseling,
Educational Psychology
and Special Education
Michigan State University
East Lansing, Michigan

Patrick W. Thompson
Department of Mathematics
and Statistics
Arizona State University
Tempe, Arizona

Cornelia Tierney
TERC
Cambridge, Massachusetts

Dina Tirosh
School of Education
Tel-Aviv University
Tel-Aviv, Israel

A Skeptic's Guide

A SKEPTIC'S GUIDE TO ALGEBRA IN THE EARLY GRADES

Some readers may wonder whether it is wise to introduce advanced mathematical concepts and methods to young learners. They may doubt whether young children are capable of learning algebra. They may question whether the mathematics problems presented in the subsequent chapters are truly about algebra. Some may consider it unrealistic to expect teachers to fit algebra into an already bulging curriculum.

Here, we address several such doubts about early algebra.[1] Hopefully, this discussion helps the reader understand why the idea of introducing algebra in the early grades often evokes strong feelings among educators and parents. Indeed, the issues inherent to early algebra are complex, and our understanding of these issues is evolving as we explore in more detail what young children can do. We leave it to the reader to weigh the evidence provided in this book regarding these matters.

ARE YOUNG CHILDREN READY FOR ALGEBRA?

Developmental readiness is the notion that people can only learn certain ideas and concepts when requisite mental structures or concepts are in place. Clearly there is some validity to this idea. It seems obvious that 1-year-old infants will not learn to read and write no matter how hard we try to teach them. There is simply too much knowledge about spoken language skills that children need to acquire before one can expect them to master reading and writing. Similarly, one would expect that calculus, at least as we now know it, would be out of reach for very young students. Many ideas from algebra, especially those in the realm of abstract algebra, would also appear to be inaccessible to young learners. We leave aside the question of whether the ideas are inherently too abstract for young minds or whether they simply rest on a knowledge base too extensive to be acquired in a short time

[1]See chapter 1 for a more formal treatment of what we mean by early algebra.

span. However, we recognize that it is reasonable to assume that many subdomains of algebra are going to be inappropriate for young learners.

Instead of asking whether young children are ready for algebra, perhaps we need to ask whether there are any algebraic concepts, ideas, and methods within reach of young learners—at least enough to justify the systematic inclusion of algebra in early mathematics curricula. Said another way: What kinds of algebraic concepts can children learn in instructional settings that support algebraic thinking? These are important questions that highlight differing psychological perspectives on teaching and learning. But, perhaps more importantly, these questions need to be considered in light of the kind of algebra we want young children to experience.

Some people hold that students should only be taught things after they have expressed interest in learning about them. As someone once commented in a mathematics education conference: "Shouldn't we wait [i.e., hold off algebra instruction] until students express a need for *n*?"

Despite its romantic appeal, much of education involves the appropriation of information, techniques, and concepts that evolved over long periods of time. Representational systems in mathematics—place-value notation, graphs, tables, and algebraic script—took decades, if not centuries, to evolve. Students can no more invent the tools of science and mathematics than they can invent spoken language. Students will surely make discoveries while they acquire existing knowledge, but on their own they will not reinvent representational systems such as algebraic script.

Much of the work described in this book is consistent with the Vygotskian view that learning precedes and facilitates development. In the present context, this stance corresponds to the view that instruction can promote the development of children's understanding of early algebra. As a result, much of the empirical evidence presented here is gathered from classrooms where the central aim is to understand what kinds of algebra children can do from instruction that attends to algebraic thinking.

SHOULDN'T KIDS LEARN BASIC SKILLS FIRST?

Some people believe that, even if young children are not developmentally hampered from learning algebra, algebra instruction should wait until elementary content such as arithmetic is well mastered. Their point makes sense if one assumes that algebra necessarily follows arithmetic. No one doubts that algebra-as-we-were-taught-it follows arithmetic-as-we-were-taught-it. The question is whether this is the best way to parse mathematics. A premise of early algebra research is that arithmetic and, more generally, early grades mathematics have been approached in ways that downplay generality. Proponents of early algebra question whether it necessarily needs to be that way. They argue that children may be able to think about structure and relationships even before they have been instructed in the use of literal symbols. For example, when children are asked to determine whether the sum of two (very large) odd counting numbers will be even or odd, they can make predictions and justify them in terms of algebraic properties of numbers without the use of literal symbols. Empirical studies such as those presented here can help us determine how young learners come to grasp such ideas.

In emphasizing the importance of higher order thinking skills such as generalization, it is not necessary to diminish the importance of routine skills such as computational proficiency. Computational fluency is important, but it may be possible to master basic skills while developing more advanced skills. Algebraic activities provide rich, meaningful contexts in which children can practice computational fluency and even enjoy doing so. Children exploring a version of the Handshake Problem that requires generalization (How many handshakes are possible among a group of n people if each pair shakes hands once?) encounter opportunities to practice basic skills such as multiplication and addition, as well as sophisticated ones such as combinatorial reasoning.

CAN TEACHERS ADD ALGEBRA TO AN ALREADY BULGING EARLY MATHEMATICS CURRICULUM?

This question would appear to be rhetorical, based on the following reasoning:

- The curriculum is presently full; there is no room for anything else.
- To add another topic, some existing topics must either be dropped or covered in less depth.
- It is impossible to eliminate existing topics; it is inadvisable to cover the present topics in less depth.

Thus, algebra cannot be added to the early grades mathematics curriculum.

However, early algebra does not necessarily make the elementary school curriculum bigger. Instead, it tends to treat existing topics more deeply, in ways highlighting generalization. On the face of it, this would seem to require more time. However there are good reasons for believing that algebra can unify much of the existing mathematics curriculum under a smaller number of broad concepts. One might even argue that the list of key topics in an "algebrafied" curriculum would actually be shorter than that of existing curricula. In this sense, early algebra has the potential to make the existing curriculum—sometimes criticized as a "mile wide and an inch deep"—into a more connected mathematical experience for children. Yet, although this seems promising, we recognize that there are real challenges to bringing about the necessary types of curricular, instructional, and institutional changes to support this.

TO BE TRULY ALGEBRAIC, SHOULDN'T THE ACTIVITY INVOLVE THE SYNTACTIC RULES OF ALGEBRA?

By and large, the algebra in early algebra tends to overlap with (rudimentary) algebra recognized by the mathematics community. However, some mathematicians and mathematics educators will undoubtedly argue that syntactical rules for manipulating symbolic expressions are not given enough attention in early algebra. As a French mathematician once remarked after reading an early algebra paper, "I liked it, but where's the algebra?" It happens that the 9-year-old students reported on in the paper did not write out algebraic expressions and derive other expressions from them. By that mathematician's standards, the students were not doing algebra.

Early algebra researchers tend to take a broad view of symbolic reasoning. For them, symbolic reasoning includes, but is not restricted to, reasoning with algebraic script. Consider a student who states: "For whatever temperature you give me in degrees Celsius, I can find the temperature in degrees Fahrenheit by multiplying by 5/9 and adding 32." This student is expressing, in the vernacular, a function that can be expressed as $f(n) = n(5/9) + 32$, where n is the temperature in degrees Celsius. Both representations, the one in algebraic script as well as the one in natural language, are symbolic. Both express generalizations. Without a doubt, algebraic script has some distinct virtues over spoken language—it is succinct and well suited to further analysis and derivations. However, the expression in natural language has its own merits: It conveys information about the thermal context that is missing in the algebraic expression. In fact, natural language typically serves as an important starting point for children to learn algebra because it allows them to begin to make sense of and describe algebraic concepts using a known language.

The early algebra researcher is also likely to pay special attention to two other systems of symbolic representation: tabular representations and graphs. To be fair, all four of these symbolic systems (natural language, algebraic script, tabular representations, and graphs) are recognized as legitimate among research mathematicians as well as mathematics educators. Nonetheless, there are sometimes striking differences in the relative weight given to the systems. Early algebra researchers sometimes introduce algebraic script only after students have become well versed in using the other representational systems. Whereas some experts may see this as ill advised or even wrong, the underlying goal of early algebra is for children to learn to see and express generality in mathematics. Initially, the language of generalization is often broadly defined and evolves toward more specialized forms such as algebraic script and Cartesian graphs.

IS ALGEBRA THE SAME AS GENERALIZATION? IF SO, ISN'T IT EVERYWHERE?

People have sometimes criticized inclusive views of algebraic reasoning on the grounds that it becomes difficult to distinguish *thinking algebraically* from *thinking mathematically* or (just plain) *thinking*. Certainly, mathematics educators need to be clear about what they mean by algebraic thinking in cases where students do not use algebraic notation. But it seems unreasonable to restrict the expression *algebraic thinking* to occasions on which students use notation such as "$f(n) = 3n - 7$." A narrow conception of algebraic activity obscures the relations between algebra and early mathematical thinking, leaving mathematics educators with little insight into how the learning of algebraic notation can build on existing skills.

Mason (chap. 3) argues that children bring natural powers of generalizing to the elementary classroom that have long allowed them to discriminate, select, and generalize, in many (nonmathematical) contexts, as a *human activity*. This is seen in how young children learn to differentiate among species, or in the increasingly specialized verbal utterances they use in the acquisition of language. Thus, rather than trying to parse precise boundaries of algebraic generalization, the goal of elementary grades mathematics should be to harness these innate powers in the particular context of number, quantity, visualization, and so forth, so that children engage in generalizing as a *mathematical activity*.

We do not claim that there are easy answers to the previous questions. What we hope to provide through this volume is evidence of children's mathematical thinking that will propel us beyond the arguments to craft workable mathematical, curricular, and instructional solutions to what we describe in chapter 1 as the *algebra problem*. Sometimes—often—this challenges us to reexamine deeply held beliefs about children's mathematical powers.

I

THE NATURE OF EARLY ALGEBRA

This part leads off the early algebra story with examinations of what algebra and algebraic thinking are, how they relate to more general symbolization processes, and their foundation in naturally occurring human powers that appear at the very earliest stages of human development. Chapter 1 chronicles the history of what we term the *algebra problem* and defines, from a content point of view, what we mean by *algebra* and *algebraic reasoning*. Recognizing that algebra is both an inherited body of knowledge and something that people do, it offers an explicit description of two core aspects and three strands of school algebra that appear in virtually all later chapters of the book.

Chapter 2 describes where these essential aspects come from and how they arise as two related types of more universal symbolization processes. In particular, using examples from elementary mathematics, it discusses how symbols and referents come into being as separate entities in our experience, a deep and subtle constructive process of recording, reflecting, and revising. The authors discuss the common phenomenon of attention switching, where thinking sometimes is guided by what the symbols are taken to stand for, and sometimes by the forms of the symbols themselves (including syntax). They also offer a distinction between non-algebraic and algebraic reasoning based on the purposeful expression of and reasoning with generality, and a distinction between proto-algebraic reasoning and (mature) algebraic reasoning based on the latter's use of conventional symbol systems.

The two symbolization processes, and especially the first, can be seen throughout the episodes described in the book. However, they come into particular relief in chapter 5, where E. Smith, using the slightly different language of *representational thinking* and *symbolic thinking* for the two core aspects, applies them primarily in the context of Strands 2 (Function and Variation) and 3 (Modeling). His classroom episodes are taken from professional development work with elementary teachers and are intended to suggest approaches to mathematics and algebraic thinking that are common across age and grade levels. Of special note is his attention to the important and often neglected role of argumentation and the establishing of certainty, both at the social and the personal levels. Given the importance of generalization and the expression of generality in our characterization of algebraic reasoning, the fact that he regards establishing mathematical certainty as a driver of representational thinking adds a major ingredient to our characterization. In this chapter, E. Smith focuses explicitly on the role of argumentation in establishing generality, but the reader will see argumentation playing this role across many chapters as both a form of student mathematical activity and as a pedagogical goal in teacher practice. Indeed, across virtually all cases, we see the importance of active classroom discussion, where students can test the validity of generalizations, the assumptions behind them, and their appropriate range of applicability.

In chapter 3, Mason argues that children arrive at school in possession of all the powers needed to learn how to think algebraically. He sees these as including imagining and expressing, focusing and de-focusing, specializing and generalizing, conjecturing and convincing, as well as classifying and characterizing objects and processes. They have their roots in the earliest recognition of patterns by infants and in their earliest stages of language development. They continue to develop into the broad cognitive and linguistic competencies that children possess when they arrive at school, especially (but not exclusively) children's ability to form, argue for, and express generalizations using natural language. He suggests that we would greatly improve our mathematics education enterprise if we were to take these powers seriously as resources for mathematics teaching and learning. A key goal of his chapter is to convince us that these student powers are indeed the same ones, in different contexts and in nascent forms, perhaps, that are used in serious mathematics learning. He closes with a consideration of some of the reasons that these powers are not well tapped in elementary school, including the common assumptions behind the rush to achieve certain topic coverage and procedural skill development and, even more constraining, the low expectations regarding young students' abilities to engage productively in mathematical activities that we would describe as algebraic.

In chapter 4, J. Smith and Thompson take the point of view that, in order for students to be able to learn and use algebraic statements, these statements need to be experienced as being about something. They suggest that whereas many researchers treat reasoning with and about *numerical* relationships and operations as a basis for algebraic reasoning, an alternative is reasoning with and about *physical quantities* and *quantitative* relationships—quantities such as lengths, weights, times, areas, speeds, and so on, where measurement and units come into play—as opposed to reasoning about numbers, operations, and their properties. By examining a variety of problem situations, they show the depth and richness of extended experience with quantitative reasoning across Grades K–8, how it is of intrinsic value as a way of making sense of many different kinds of situations and phenomena, how it is different from algebraic reasoning, and how algebraic reasoning can be built on a foundation of quantitative reasoning. Relative to the last point, they provide examples illustrating the generality of quantitative reasoning and how it draws the student toward ways of expressing that generality. Hence, it makes the power of algebra immediately apparent once the stage is set. They point out that the traditional elementary mathematics curriculum seldom puts students in this position, where algebra really pays off.

A goal of the chapters in this section is to orient the reader to the whole enterprise of building algebraic reasoning in the contexts of elementary grades mathematics. Although we see much commonality across the chapters, especially in the functional respect given to children's ways of making sense of their world and the central place given to active student expression of their ideas and processes, we also see that differences in the context, reflected in different strands of algebraic reasoning, yield real differences in how algebraic reasoning emerges.

1

What Is Algebra? What Is Algebraic Reasoning?

James J. Kaput
University of Massachusetts, Dartmouth

This introductory chapter provides a shared road map of algebra in the elementary grades and an historical perspective on why we might need such a road map. Because the landscape is quite varied, it is useful to know where we are in relation to the larger territory and what has brought us to this point. We begin with an account of what we have termed the algebra problem and of the evolution in the research base on learning algebra.

THE ALGEBRA PROBLEM

The scholars in this book address what for at least 10 years we have termed the algebra problem. It is especially acute in the United States where, historically, the introduction of school algebra has awaited the completion of a 6- to 8-year computational arithmetic curriculum that has been implemented as independent of algebra. At the turn of the 20th century, with universal elementary schooling already in place, shopkeeper arithmetic was expected for all. Algebra was reserved for the elite, which amounted to a scant 3% to 5% of the population who completed secondary school at that time (National Center for Education Statistics, 1994). Thus, the arithmetic-then-algebra curriculum structure was already well in place as U.S. society evolved toward universal secondary education during the 20th century. The resulting late and abrupt approach to the introduction of algebra was deeply institutionalized and regarded as the natural order of things (Fey, 1984). But, by the dawn of the 21st century, this highly dysfunctional result of the computational approach to school arithmetic and an accompanying isolated and superficial approach to algebra had led to both teacher alienation and high student failure and dropout, especially among economically and socially less advantaged populations. This is the result of the convergence of two unprecedented forces: (a) in response to societal needs for a deeper mathematical literacy, all students are now expected to learn algebra under the "algebra for all" banner; and (b) this expectation is legally codified in high-stakes accountability measures that define academic success in terms of success in algebra.

Political agendas aside, there are some salutary reasons for rethinking and reworking algebra in early grades mathematics. Solving the algebra problem serves four major goals:

1. To add a degree of coherence, depth, and power typically missing in K–8 mathematics.
2. To ameliorate, if not eliminate, the most pernicious and alienating curricular element of today's school mathematics: late, abrupt, isolated, and superficial high school algebra courses.
3. To democratize access to powerful ideas by transforming algebra from an inadvertent engine of inequity to a deliberate engine of mathematical power.
4. To build conceptual and institutional capacity and open curricular space for new 21st-century mathematics desperately needed at the secondary level, space locked up by the 19th-century high school curriculum now in place.

Changes of this magnitude require deep rethinking of the core algebra enterprise and will not be achieved by minor adjustments such as attempting to fix a first algebra course, starting it a year earlier, or legislating 2 years of algebra for all. As Carraher and colleagues argue, early algebra is decidedly not (traditional) algebra early.

Solving the algebra problem involves deep curriculum restructuring, changes in classroom practice and assessment, and changes in teacher education—each a major task. Further, each must be achieved within the capacity constraints of the teaching population, within the limited time and resources available for in-service and preservice teacher development, and within the constraints of widely used instructional materials. Steps in this direction are underway on several fronts. The *Principles and Standards of School Mathematics* (National Council of Teachers of Mathematics, 2000) has advocated an increasingly longitudinal view of algebra, that is, a view of algebra not as an isolated course or two, but rather as a strand of thinking and problem solving beginning in elementary school and extending throughout mathematics education. This perspective was reflected in the original *Curriculum and Evaluation Standards for School Mathematics* (National Council of Teachers of Mathematics, 1989) and now is expressed in curriculum and professional development projects and documents that exemplify the principles and standards highlighting the development of algebraic reasoning in the earlier grades (Cuevas & Yeatts, 2001; Greenes, Cavanagh, Dacey, Findell, & Small, 2001). Some of this groundbreaking work is described in the present volume.

THE STATE OF RESEARCH AND ITS EVOLUTION OVER THE PAST 30 YEARS

The algebra problem has been brought to the fore in our thinking by research that spans several decades. Through the 1980s, research in algebraic thinking and learning focused on student errors and constraints on their learning, especially developmental constraints. A large body of evidence was accumulated that showed students tended to have fragile understanding of the syntax of algebra (e.g., Matz, 1982; Wenger, 1987) or that they had difficulty in interpreting algebraic symbols (e.g., Clement, 1982; Clement, Lochhead, & Monk, 1981; Kaput & Sims-Knight, 1983) or even coordinate graphs (Clement, 1989). Perhaps we should not be surprised, given the curriculum that students experienced and the fact that for many years most research either measured the given curriculum's affect on students under traditional classroom circumstances or took the form of brief interventions aimed at teaching symbol manipulation techniques based on the same narrow syntactical view of algebra that defined the dominant curriculum (e.g., Lewis, 1981; Sleeman, 1984, 1985, 1986). However, recent research on the status of student knowledge based in the traditional arithmetic-then-algebra regime has pointed to specific obstacles to algebra learning that computational arithmetic creates for the learning of algebra. For example, limited approaches to equality and the "=" sign in arithmetic as separator of procedure from result (Kieran, 1992) are now known to interfere with later learning in algebra (Fujii, 2003; McGregor & Stacey, 1997).

Davis's work constitutes an early and noteworthy exception to the forms and results of this early research. His work in the 1960s and 1970s involved sustained and tailored interventions with students of various ages, including elementary age students (Davis, 1975; Davis & McNight, 1978), and directly inspired some of the work reported in this volume (e.g., Franke, Carpenter, & Battey, chap. 13). His work anticipated and set the stage for the optimistic stance on learning algebra that appeared years later and that is reflected in large part in the present volume.

CHALLENGE IN DEFINING ALGEBRA AND ALGEBRAIC REASONING

Recognizing the algebra problem and crafting a solution that spans grades K–12 introduces another complexity, namely, defining "algebra" and "algebraic reasoning," especially as an object of thinking in the elementary grades. What we think algebra is has a huge bearing on how we approach it—as teachers, administrators, researchers, teacher-educators, curriculum developers, framework writers, instructional material evaluators, assessment writers, policymakers, and so forth. Although the views of algebra offered here differ in focus across the chapters, they stress algebra's breadth, richness, and organic relation to naturally occurring human cognitive and communicative powers. Indeed, the kind of narrow view of algebra that has dominated school algebra for years in many countries as primarily syntactically guided, symbolic manipulations not only grossly understates the multiple sides of algebra historically as mathematics, it is also an inadequate foundation for reconsideration of algebra's place in school mathematics. The editors share the view that we need a broader and deeper view of algebra that can provide school mathematics with the same depth and power that the several aspects and strands of algebra have provided mathematics as a theoretical and practical discipline historically. Only by expanding our views of algebra can we hope to use algebra to integrate mathematics across all grades and all topics.

Mathematics as Something We Acquire Versus Something We Do

Lee (2001) has identified several ways to approach answering the two questions in this chapter's title. We begin with two different ways of describing mathematics in general and algebra in particular. First, mathematics is a cultural artifact, something we receive as part of our cultural heritage. This cultural artifact, particularly algebra, is embedded in education systems across the world in very different ways, especially in terms of when algebra is introduced and how tightly it is integrated with other mathematical topics (Kendal & Stacey, 2004). Mathematics is also a set of activities, something people do. For instance, it involves the production of representations to express generalizations, and it involves transforming those representations. It also uses these two kinds of activities in other mathematical activities such as modeling (Bednarz, Kieran, & Lee, 1996; Kieran, 1996, 2004).

The first question in the chapter's title, "What is algebra?", highlights algebra as a self-standing body of knowledge—as a cultural artifact. The second, "What is algebraic reasoning?", highlights algebra as human activity. These differences in how we think about algebra show up in many ways. For example, those who think of algebra as reasoning are inclined to consider students' ways of doing, thinking, and talking about mathematics as fundamental. For them, algebra emerges from human activity; it depends on human beings for its existence, not just historically, but also in the present. Those who think of algebra as an inherited subject matter are comfortable talking about it without thinking about people. They might refer to the commutative law of addition, for example, without having to establish how the law came to be or how students come to learn it (or not). For them, commutativity is a property of mathematics itself. Each view is useful, depending on our purposes, and we shall use both.

A Symbolization Perspective on Algebra and Algebraic Reasoning

In order to provide some structure to the complex web of ideas, notations, and activities that are involved in algebra and algebraic reasoning, we take a symbolization perspective. We take the view that the heart of algebraic reasoning is comprised of complex symbolization processes that serve purposeful generalization and reasoning with generalizations. Although he did not take a broad symbolization view in his examination of school algebra, Usiskin (1988) sought to characterize algebraic reasoning in terms of the familiar uses of letters (as unknowns, variables, parameters, etc.). Our goal, pursued further in the next chapter, is to understand not only these kinds of traditional representational phenomena, but the phenomena that occur as algebraic reasoning is being developed in the context of elementary grades mathematics.

Another complication in describing algebra in a way that respects its richness and subtlety is the fact that it is not a static body of knowledge. It evolves as a cultural artifact in terms of the symbol systems it embodies (most recently due to electronic technologies), and it evolves as a human activity as students learn and develop. Hence, algebra needs to be described both through a snapshot of its structure and function in mathematics today and in mathematically mature individuals, and through a dynamic picture of its evolution historically and developmentally. Detailed descriptions of either type are not possible in a reasonably short book, let alone a single preliminary chapter. Nonetheless, we include reference to the evolving nature of algebra as artifact as we proceed. The next chapter deals with how it evolves within individuals as a human activity.

Most attempts to define algebra historically, blending both a cultural artifact and action perspective, tend to be oriented toward progress in solving equations, where the origin of the equations might be problem situations or simply assertions about numbers or measurement quantities, often surprisingly similar across the millennia (e.g., Katz, 1995). They then turn to the solving of equations in the 16th and 17th centuries apart from their status as models, and simply as mathematical objects of intrinsic intellectual interest (Kline, 1972).[1] And, from a modern perspective (18th century and later), definitions of algebra refer to the use of literal symbols as a central feature of the activity. Indeed, the earliest versions of "equation solving," appearing almost 4,000 years ago in the Rhind Papyrus, took place in stylized natural language rather than in the specialized symbols of what has been termed "rhetorical algebra" (Harper, 1987; Kline, 1972). Relatedly, another feature of algebra that is often taken as a defining one (Usiskin, 1988) is its use of literal symbols, letters. This practice evolved over many centuries and stages and did not stabilize until early in the 18th century (Boyer, 1968).

The next chapter takes a wider, symbolization view of algebra, beyond the uses of letters, in order to accommodate both the variations in symbolization activities that occur in early grades and the increasing variety of symbol systems, including graphical ones, which extend what the traditional systems once did for us. For now, we offer an initial content analysis of algebra intended to respect its richness and diversity across uses and age levels.[2]

A CONTENT ANALYSIS OF ALGEBRA

Two Core Aspects of Algebra

We regard one core aspect of algebraic reasoning to be generalization and the expression of generalizations in increasingly systematic, conventional symbol systems (Core Aspect A). The second core aspect of algebraic reasoning is syntactically guided action on symbols within organized systems of symbols (Core Aspect B). Each of these core aspects of algebra appears in some form across all three strands of algebra as they are defined next (see also Table 1.1). Our characterization merges algebra's two identities as a cultural artifact expressed mainly as conventional symbol systems and as certain kinds of human activities.

We now expand on these aspects and strands from the perspective of school mathematics, with special focus on elementary mathematics. Core Aspect B is typically thought to develop later than Core Aspect A because

---

[1]We limit our general historical references to widely accessible accounts. The reader can substitute other standard histories of mathematics as convenient.

[2]Unless we note otherwise, when we use the word "algebra," we mean both senses of algebra.

Table 1.1
Core Aspects and Strands

The Two Core Aspects
(A) Algebra as systematically symbolizing generalizations of regularities and constraints.
(B) Algebra as syntactically guided reasoning and actions on generalizations expressed in conventional symbol systems.
Core Aspects A & B Are Embodied in Three Strands
1. Algebra as the study of structures and systems astracted from computations and relations, including those arising in arithmetic (algebra as generalized arithmetic) and in quantitative reasoning.
2. Algebra as the study of functions, relations, and joint variation.
3. Algebra as the application of a cluster of modeling languages both inside and outside of mathematics.

rule-based actions on symbols depend on knowing what the allowable combinations of symbols are and on knowing how they relate to each other, especially in terms of which combinations are equivalent to others. The work by Carraher, Schliemann, and Schwartz (chap. 10, this volume) and Brizuela and Earnest (chap. 11, this volume) suggests that we look more closely at how students build understanding of syntax in algebra, and that perhaps we should also look at actions on symbols from a more linguistic perspective, attending to interactions between how symbols are parsed visually and how they are constructed mathematically (Kirshner & Awtry, 2004).

There is considerable diversity of opinion about the roles of the two Core Aspects in early algebra learning. Mathematicians and mathematics educators differ in their views of which of the two core aspects of algebra is more central to defining *algebra*. Some treat rule-based actions on symbols (Core Aspect B) as the hallmark of algebraic reasoning, whether or not these actions serve generalization or modeling, and hence do not regard much of the activity described in this book as truly algebraic. Others, cautious about what Piaget (1964) referred to as *premature formalism*, downplay conventional syntax in favor of the deliberate expression of generalizations and models of situations through whatever means are available, but especially natural language and drawings (Resnick, 1982). Only after students are greatly experienced in expressing in these forms would they be introduced to algebraic notation, graphs, and other classes of conventional representations.

Still others hold the view that Core Aspect B deserves attention fairly early in algebra learning. At the very beginning of algebra instruction, students are encouraged to note regularities and make generalizations (Core Aspect A) using their own resources; but they are soon encouraged to make conventional representational forms their own. The underlying premise is that conventional forms (algebraic notation, graphs and number lines, tables, and natural language forms) can not only express, but also enrich and deepen algebraic reasoning in students. The chapters by Carraher et al. (chap. 10, this volume) and Brizuela and Earnest (chap. 11, this volume), examine how a growing familiarity with the syntax of algebra contributes to students' learning and reasoning. A similar argument is advanced by Boester and Lehrer (chap. 9, this volume), who remark that "spatial structure serves as a potentially important springboard to algebraic reasoning, but also that algebraic reasoning supports coming to 'see' lines and other geometric elements in new lights."

The next chapter looks more deeply at the Core Aspects A and B as symbolization processes, their connections to the nature of mathematics, how algebra evolved historically and how algebra comes into existence for individuals.

Three Strands of Algebra

This section looks at each of the three strands and how the two core aspects are embodied in each of them.

Strand 1. Building generalizations from arithmetic and quantitative reasoning is taken by many educators and researchers as the primary route into algebra. It includes generalizing arithmetic operations and their properties and reasoning about more general relationships and their forms (e.g., properties of zero, commutativity, inverse relationships, etc.). This is the heart of algebra as generalized arithmetic. It includes building the syntactic aspect of algebra from the structure of arithmetic—building the basic idea that one can replace one expression by an equivalent one. It involves looking at arithmetic expressions in a new way, in terms of their form rather than their value when computed.

This strand also includes building generalizations about particular number properties or relationships, for example, a sum of two odds is even, properties of sums of three consecutive numbers, finding and expressing regularities in the 100s table or the addition and multiplication tables, examining why multiplying by 10 or 100 has the effect of appending zeros to the number multiplied, and so on. This activity typically uses the generalizations of generalized arithmetic, although not always explicitly.

A third basic activity in this strand is the explicit expression of computation strategies (both conventional and student-invented), such as compensation strategies wherein we add one to an addend and then subtract it from the total to facilitate a computation. These activities often occur in mixed sequences and overlap. For example, when we use the fact that addition is commutative to simplify the mental computation of 3 plus 18 to 18 plus 3, or when we use commutativity of multiplication to reduce the number of multiplication facts to be learned, we are invoking this strand of algebra. More accurately, our activity can be termed *algebraic* when we are stating these properties explicitly and examining their generality—not when we are using them tacitly.

All of the previous involve the critical step from arithmetic to algebra of expanding the notions of equivalence associated with the "=" sign from that of separator of operation and result to general equivalence (Kieran, 1981, 1994; MacGregor & Stacey, 1997).

Alternative approaches exist within this strand that are not based on the arithmetic of numbers but rather are based on reasoning with physical quantities (see Dougherty, chap. 15, this volume; Smith & Thompson, chap. 4, this volume; and, to some extent, Carraher et al., chap. 10, this volume). All of these approaches, based on number or on quantity, share the central role of generalizing and expressing those generalizations explicitly and systematically. The Dougherty approach is adapted from earlier Soviet work (Davydov, 1975, 1982) and uses a fundamentally different style of initial symbolization that avoids numerals and instead uses letters to denote quantities, thereby embodying generality in the symbolic expression of specific (but unmeasured) cases involving, say, comparisons of lengths or heights or volumes.

Strand 2. The second strand involves generalizing of a fairly particular kind, basically toward the idea of function, where expressing the generalization can be thought of as describing systematic variation of instances across some domain. The syntactic aspect of algebra is usually applied to change the form of expressions denoting regularities, in comparing different expressions of a pattern to determine whether they are equivalent, or in determining when functions take on particular values (e.g., roots) or whether they satisfy various constraints (building and solving equations). This strand is often taken to begin with elementary patterning activities that are often thought to be a necessary precursor to other forms of mathematical generalization[3] (Cuevas & Yeatts, 2001; Greenes et al., 2001), although patterning does not play a significant role in the work described

[3]One-dimensional patterning has the special quality of hiding the variable on which the pattern depends (usually the number of the item in the sequence) and hence focusing strictly on relations between consecutive items in a sequence. Hence, its structure as a function is especially well hidden.

in this book. Sometimes the patterns are generated geometrically, such as in work with triangular or figurate numbers, patterns in areas of figures, and so on. This strand has blurry boundaries in the sense that ideas bound up with the idea of function are both very rich and wide-ranging. They include, for example, the various kinds of change and hence involve the ideas of linearity, rate, and so on. This strand also makes regular use of a wide range of symbol systems beyond the usual character-string based systems, including tables, graphs, and various pedagogical systems such as "function machines."

Strand 3. Modeling as an algebraic activity seems to be of three basic types based on how the two core aspects of algebra are employed. A first type of modeling is number-or quantity-specific, without pretense that a general class of situations is being modeled. In effect, it is an arithmetic problem that requires using the syntactic aspect of algebra to solve (see Bednarz & Janvier, 1996). It typically takes the form of the statement of a constraint, usually in the form of an equation, which then requires the use of the syntactic aspect of algebra to yield a solution. Here the variable is regarded as an unknown rather than as a variable representing a class of situations.

A second type of modeling uses the first core aspect in generalizing and expressing patterns and regularities in situations or phenomena, arising either outside mathematics or from within mathematics (e.g., geometric patterns). Here, the domain of generalization is the situation being modeled, and often the expression of the generalization takes the form of using one or more variables that may then express a function or class of functions. Of course, working with such expressions to gain insight into the situation being modeled usually involves the syntactic aspect of algebra.

A third type of algebraic modeling involves generalizing from solutions to single-answer modeling situations of either the first type above or from pure arithmetic word problems that did not require algebraic maneuvers to solve (what Blanton & Kaput, chap. 14, this volume, 2004, refer to as *algebrafying* an arithmetic problem). Here the algebra enters as one relaxes the constraints of the given problem to explore its more general form, scope, and deeper relationships—including comparisons with other models and other situations. In this type of generalization modeling, the introduction of variables expressing the generality of the situation often takes the form of parameters.

Notes on the Aspects and Strands

Strand 1 at more advanced levels leads to abstract algebra, but can also be regarded as including more elementary activities such as clock arithmetic, working with strings of letters or other symbols according to specified rules, and so on. Strand 2, from the point of view of algebra defined as a mathematical discipline, is not strictly algebra, but analysis. At its more advanced levels, it leads to calculus and analysis. However, we treat it as algebra because its content is a major part of school algebra, especially from middle school onward, and it usually requires using the syntactic aspect of algebra. Some (e.g., Schwartz, 1990) have argued that school algebra be entirely constituted in terms of the idea of function. Strand 3 often involves Strand 2, and its different forms are distinguished on the respective roles of the core aspects of algebra that in turn are reflected in whether variables are treated as unknowns, variables, or parameters. Both Strands 2 and 3 make use of multiple representation types.

The core aspects and strands emphasize algebra's deep, but varied, connections with all of mathematics. It is this web of connections that enables algebra to play the key role across K–12 mathematics that we and others suggest (Kaput, 1999). To this end, this content analysis is consistent with that provided by the National Council of Teachers of Mathematics Algebra Working Group and that appears in various reform documents (e.g., National Council of Teachers of Mathematics, 2000; see also Kieran, 1996).

The next chapter delves more deeply into how the core aspects of algebra arise through symbolization: Core Aspect A through successive processes of record making and actions on those records, and Core Aspect B through the efficiency-driven "lifting of repeated actions on symbols" into syntax. Finally, the next chapter offers a two-stage characterization of algebraic reasoning that respects both core aspects, including the special role of historically developed conventional symbol systems.

REFERENCES

Bednarz, N., & Janvier, B. (1996). Emergence and development of algebra as a problem-solving tool: Continuities and discontinuities with arithmetic. In N. Bednarz, C. Kieran, & L. Lee (Eds.), *Approaches to algebra: Perspectives for research and teaching* (pp. 115–136). Dordrecht, the Netherlands: Kluwer Academic.

Bednarz, N., Kieran, C., & Lee, L. (Eds.). (1996). *Approaches to algebra: Perspectives for research and teaching*. Dordrecht, the Netherlands: Kluwer Academic.

Blanton, M., & Kaput, J. (2004). Design principles for instructional contexts that support students' transition from arithmetic to algebraic reasoning: Elements of task and culture. In R. Nemirovsky, B. Warren, A. Rosebery, & J. Solomon (Eds.), *Everyday matters in science and mathematics* (pp. 211–234). Mahwah, NJ: Lawrence Erlbaum Associates.

Boyer, C. (1968). *The history of mathematics*. New York: Wiley.

Clement, J. (1982). Algebra word problem solutions: Thought processes underlying a common misconception. *Journal for Research in Mathematics Education, 13*(1), 16–30.

Clement, J. (1989). The concept of variation and misconceptions in Cartesian graphing. *Focus on Learning Problems in Mathematics, 2*(1–2), 77–87.

Clement, J., Lochhead, J., & Monk, G. (1981). Translation difficulties in learning mathematics. *American Mathematical Monthly, 88*, 286–290.

Cuevas, G. J., & Yeatts, K. (2001). *Navigating through algebra in grades 3–5*. Reston, VA: National Council for Teachers of Mathematics.

Davis, R. (1975). Cognitive processes involved in solving simple algebraic equations. *Journal of Children's Mathematical Behavior, 1*(3), 7–35.

Davis, R., & McKnight, C. (1980). The influence of semantic content on algorithmic behavior. *Journal of Children's Mathematical Behavior, 3*(1), 39–87.

Davydov, V. (1975). The psychological characteristics of the "prenumerical" period of mathematics instruction. In L. P. Steffe (Ed.), *Soviet studies in the psychology of learning and teaching mathematics* (Vol. 7, pp. 109–206). Chicago: University of Chicago Press.

Davydov, V. (1982). The psychological characteristics of the formation of elementary mathematical operations in children. In T. Carpenter, J. Moser, & T. Romberg (Eds.), *Addition and subtraction: A cognitive perspective* (pp. 224–238). Hillsdale, NJ: Lawrence Erlbaum Associates.

Fey, J. T. (1984). *Computing and mathematics: The impact on secondary school curricula.* Reston, VA: National Council of Teachers of Mathematics.

Fujii, T. (2003). Probing students' understanding of variables through cognitive conflict problems: Is the concept of a variable so difficult for students to understand? In N. Pateman, B. Dougherty, & J. Zilliox (Eds.), *Proceedings of the 27th Conference of the International Group for the Psychology of Mathematics Education* (Vol. 1, pp. 49–66). Honolulu, HI: University of Hawaii.

Greenes, C., Cavanagh, M., Dacey, L., Findell, C., & Small, M. (2001). *Navigating through algebra in prekindergarten–grade 2*. Reston, VA: National Council for Teachers of Mathematics.

Harper, E. (1987). Ghosts of Diophantus. *Educational Studies in Mathematics, 18*, 75–90.

Kaput, J. (1999). Teaching and learning a new algebra. In E. Fennema & T. Romberg (Eds.), *Mathematics classrooms that promote understanding* (pp. 133–155). Mahwah, NJ: Lawrence Erlbaum Associates.

Kaput, J., & Sims-Knight, J. E. (1983). Errors in translations to algebraic equations: Roots and implications. *Focus on Learning Problems in Mathematics, 5*(3), 63–78.

Katz, V. (1995). The development of algebra and algebra education. In C. Lacampagne, W. Blair, & J. Kaput (Eds.), *The algebra initiative colloquium* (Vol. 1, pp. 15–32). Washington, DC: U.S. Department of Education, Office of Educational Research and Improvement.

Kendal, M., & Stacey, K. (2004). Algebra: A world of difference. In K. Stacey, H. Chick, & M. Kendal (Eds.), *The future of teaching and learning of algebra: The 12th ICMI study* (chap. 13). Dordrecht, the Netherlands: Kluwer Academic.

Kieran, C. (1981). Concepts associated with the equality symbol. *Educational Studies in Mathematics, 12*, 317–326.

Kieran, C. (1992). The learning and teaching of school algebra. In D. Grouws (Ed.), *Handbook of research on mathematics teaching and learning* (pp. 390–419). New York: Macmillan.

Kieran, C. (1996). The changing face of school algebra. In C. Alsina, J. Alvarez, B. Hodgson, C. Laborde, & A. Pérez (Eds.), *Proceedings of the 8th International Congress on Mathematical Education: Selected lectures* (pp. 271–290). Sevilla, Spain: S.A.E.M. Thales.

Kieran, C. (2004). The core of algebra: Reflections on its main activities. In K. Stacey, H. Chick, & M. Kendal (Eds.), *The future of teaching and learning of algebra: The 12th ICMI study* (chap. 2). Dordrecht, the Netherlands: Kluwer Academic.

Kirshner, D., & Awtry, T. (2004). Visual salience of algebraic transformations. *Journal for Research in Mathematics Education, 35*(4), 224–257.

Kline, M. (1972). *Mathematical thought from ancient to modern times*. New York: Oxford University Press.

Lee, L. (2001). Early algebra—But which algebra? In H. Chick, K. Stacey, J. Vincent, & J. Vincent (Eds.), *The future of the teaching and learning of algebra* (pp. 392–399). Melbourne, Australia: University of Melbourne.

Lewis, C. (1981). Skill in algebra. In J. R. Anderson (Ed.), *Cognitive skills and their acquisition* (pp. 85–110). Hillsdale, NJ: Lawrence Erlbaum Associates.

MacGregor, M., & Stacey, K. (1997). Students' understanding of algebraic notation: 11–15. *Educational Studies in Mathematics, 33*, 1–19.

Matz, M. (1982). Towards a process model for high school algebra errors. In D. H. Sleeman & J. S. Brown (Eds.), *Intelligent tutoring systems* (pp. 25–50). New York: Academic Press.

National Center for Education Statistics. (1994). *A profile of the American high school sophomore in 1990* (NCES No. 95-086). Washington, DC: Author.

National Council of Teachers of Mathematics. (1989). *Curriculum and evaluation standards for school mathematics*. Reston, VA: Author.

National Council of Teachers of Mathematics. (2000). *Principle and standards for school mathematics*. Washington, DC: Author.

Piaget, J. (1964). *Development and learning. Conference on cognitive studies and curriculum development*. Ithaca, NY: Cornell University Press.

Resnick, L. B. (1982). Syntax and semantics in learning to subtract. In T. P. Carpenter, J. M. Moser, & T. A. Romberg (Eds.), *Addition and subtraction: A cognitive perspective* (pp. 136–155). Hillsdale, NJ: Lawrence Erlbaum Associates.

Schwartz, J. (1990). Getting students to function in and with algebra. In G. Harel & E. Dubinsky (Eds.), *The concept of function: Aspects of epistemology and pedagogy* (pp. 261–289). Washington, DC: Mathematics Associations of America.

Sleeman, D. (1984). An attempt to understand students' understanding of basic algebra. *Cognitive Science, 8*, 387–412.

Sleeman, D. (1985). Basic algebra revisited: A study with 14-year olds. *International Journal of Man–Machine Studies, 22*, 127–149.

Sleeman, D. (1986). Introductory algebra: A case study of misconceptions. *Journal of Mathematical Behavior, 5*(1), 25–52.

Usiskin, Z. (1988). Conceptions of school algebra and uses of variables. In A. Coxford (Ed.), *The ideas of algebra, K–12* (1988 yearbook, pp. 8–19). Reston, VA: National Council of Teachers of Mathematics.

Wenger, R. (1987). Some cognitive science perspectives concerning algebra and elementary functions. In A. Schoenfeld (Ed.), *Cognitive science and mathematics education* (pp. 217–252). Hillsdale, NJ: Lawrence Erlbaum Associates.

2

Algebra From a Symbolization Point of View

James J. Kaput
Maria L. Blanton
University of Massachusetts, Dartmouth

Luis Moreno
Center for Research and Advanced Studies (CINVESTAV)

We use the context of equation solving to examine the process of building structured ways of acting on symbols—the basis for a syntax and the second core aspect of algebraic reasoning. We see this as a process of "lifting out" of repeated equivalence statements. This is an efficient way to substitute statements that are repeatedly seen as giving essentially equivalent information. Through repetition and being made explicit (which amounts to yet another process of generalization, but on symbol-substitution actions), these gain status as a system of substitution rules—a syntax. Because the syntax arises from previously accepted equivalences in the reference system for the symbols, it is guaranteed to yield correct results when used appropriately. We examine certain pedagogical issues associated with learning syntactically defined procedures with and without conceptualizations based in reference fields.

We close with a characterization, based on the centrality of generalization, of which kinds of symbolizations deserve to be called fully algebraic (those that use conventional symbol systems) and those that are quasi-algebraic (those with the same generalization and reasoning purposes as algebraic but that use less conventional symbolizations—thereby extending the notion of *quasi-variables* defined by Fujii, 2003). This characterization thus distinguishes those symbolization activities that are not algebraic—those not used in the service of generalization and reasoning with generalizations. The characterization is applied to the three strands of algebra—generalization of arithmetic and quantitative reasoning, functions, relations and variation, and modeling.

ALGEBRA FROM A SYMBOLIZATION POINT OF VIEW

We regard generalization and symbolization as being at the heart of algebraic reasoning. Why are generalization and symbolization so tightly linked? The answer is utterly basic. The only way a person can make a single statement that applies to multiple instances (i.e., a generalization), without making a repetitive statement about each instance, is to refer to multiple instances through some sort of unifying expression that refers to all of them in some unitary way, in a single statement. But, the unifying expression requires some kind of symbolic form, some way to unify the multiplicity. Generalizing is the act of creating that symbolic object. This is where symbolization in the service of generalization—and algebra—starts, both within individuals and historically. Once a symbolization is achieved, it becomes a new platform on which to express and reason with generality, including further symbolization.[1] This chapter attempts to flesh out these symbolization ideas in a way that will help make sense of the development of algebraic reasoning phenomena reported throughout this book, and in so doing, help clarify more deeply what we mean by "algebra" and how it develops.[2]

Our discussion skirts on the edge of profound issues concerning the nature of and relations among mathematics, language, and thought. This is not the place to plumb them to great depth, but it is important to know that they are close by.

---

[1]We are oversimplifying here, of course, but particularly in the sense that in order to serve as a platform for further generalization, the initial generalization must become part of a coherent system of generalization expressions.

[2]The word *algebra* means both senses of algebra defined in the previous chapter, as a cultural artifact and as something people do.

THE BIG PICTURE FROM A HISTORICAL AND SYMBOLIZATION PERSPECTIVE

A Deep Mathematical Duality: Generalizing and Using Symbols to Reason With Generalizations

Algebraic reasoning shares dual aspects with most of mathematics: Mathematics is about generalizing and expressing generalizations, and it is about using specialized systems of symbols to reason with the generalizations.[3] We regard these as the two underlying aspects of algebra that are at the heart of all algebraic reasoning. The phrase "reason with generalizations" includes both deducing and inducing structure from the configurations of symbols whether or not one is actively and physically transforming them. This, therefore, covers cases where one studies the structure of a single expression or sequence of such, as well as cases where one actively transforms them using rules for transforming symbol complexes. This is a slight elaboration of what was called the second core aspect of algebra in the previous chapter, which focused on syntactically guided transformation of symbolic expressions. Our interest is in understanding how these symbolization processes and the capacity to engage in them come to exist, particularly in individuals.

Although we have much to learn about the process of symbolization, we do know that it cannot be separated from conceptualization. Ideas, especially generalizations, grow out of our attempts to express them to ourselves and others, and our attempts to express them give rise to symbolizations that in turn help build and fill out the ideas, folding back into those ideas so that conceptualization and symbolization become inseparable. As Sfard (2000) put it, mathematical discourse and mathematical objects create each other. Indeed, she argues, as do we, that this process occurs at both the individual and classroom level as well as historically. Moreover, it is a continuous process, not something that one does and completes. We want to look more closely at it, with particular interest in how symbol and referent come to be experienced sometimes as separate and sometimes the same, and how actions on symbols become possible.

One can produce symbolizations of many kinds and with many different purposes. One dimension along which they can vary is the extent to which they are shared in some community, from being the entirely private

[3]We say *most* because there are branches of mathematics where computation or counting takes center stage—as in combinatorics, although even there the objects are often abstract and the processes become subject to generalization, which means they typically call on algebra.

construction of an individual dealing with a single situation to being fully conventional in the sense that they are shared by a wide community of people who are regarded as mathematically educated. The base 10 number system, the standard algebraic symbol system, two-column tables and coordinate graphs are all examples of the latter. And, of course, the main reason that they have become conventional is that they are very useful across wide varieties of situations. Each is powerful in its own way. Each is a highly efficient way of symbolizing, the end result of an historic process of refinement—contributing to algebra's identity as a cultural artifact.

But, once some symbolization of a generalization is established, conventional or not, then the symbols themselves can help further the reasoning process because they have packed within themselves some features of the generalizations in a crystallized, materially stable and usually more compact form. This enables us to compare generalizations by comparing their respective symbolizations, to examine their range of applicability, to deduce inferences by examining their form, and, most importantly for algebra, by acting directly on the symbols themselves to systematically manipulate their configurations into different forms. This action treats the symbols as objects in their own right, without immediate regard for what they might stand for. As we were told by Whitehead (1929) many years ago, "Civilisation advances by extending the number of important operations we can perform without thinking about them" (p. 59).

The evolution of the symbol system that we now take for granted as the most visible, material aspect of algebra as a cultural artifact took place over millennia, including the idea of unknown, the explicit expression of it using special words and then symbols, the idea of representing numerical variables as letters, the idea of representing general classes of mathematical statements using literal coefficients, writing equivalences (which eventually became equations), the idea that it is possible to enumerate sets of special canonical forms that can be handled by established methods, and most especially the substitution rules that constitute a syntax for acting on symbols (Boyer & Merzbach, 1989; Kline, 1972). As we illustrate in the next section, these rules capture and crystallize mental actions as rules for operating on physical symbols. It may be worth remembering that Descartes titled his famous book, in which he lays out what he referred to as his own algebra, *Rules for the Operation of the Mind* (written in the first half of the 17th century and published posthumously in 1701; see Puig & Rojano, 2004, for a detailed discussion).

Bochner (1966), a mathematician and historian, wrote about the profound importance of the emergence of algebra in the 17th century, the compact and systematic use of symbols to represent generalizations and abstractions:

> Not only was this algebra a characteristic of the century, but a certain feature of it, namely the "symbolization" inherent to it, became a profoundly distinguishing mark of all mathematics to follow. ... (T)his feature of algebra has become an attribute of the essence of mathematics, of its foundations, and of the nature of its abstractness on the uppermost level of the "ideation" a la Plato. (pp. 38–39)

In a deep, long-term sense, the process of symbolization allows a kind of lift-off from the bounds of concretely referenced thinking even more powerfully than did the invention of writing (Donald, 1991), because not only does it enable us to overcome the memory and processing limits of raw human cognition (if, indeed, such has ever existed), it introduces new possibilities for abstraction away from the specifics of experience and, most importantly, the creation of new forms of symbol-mediated experience. Historically, in mathematics, this happened many times and in many ways. For example, arithmetic freed itself from its geometric referents in Alexandria (Kline, 1972) and, over an extended period, algebra gradually did the same, completing the separation in the years following Descartes when the operations on numbers became free of the constraints of geometric dimensionality.

Not only can we make symbolized generalizations, but the resulting symbolizations enable us to treat these symbolized generalizations as objects and relations in their own right that can serve as ingredients of further generalizations in an upward spiral of abstraction and mathematical power that was not possible previously. In this way, the algebraic system itself gave birth to new entities such as negative and what we now know as complex numbers. Symbolization in connection with generalization, as pointed out by Mason in chapter 3, begins at a very early age. Indeed, as argued by Werner and Kaplan (1962), these symbolization processes are an integral part of human development. But, when the symbolizations become algebraic, new mathematical worlds become possible. This is the epistemological basis for algebra often being referred to as a gateway to higher mathematics. Next, we turn to the second aspect of algebra to account for—its operational aspect.

The Idea of Acting on Physical Symbols—Operative Symbol Systems

Bochner (1966) also identified a critical new ingredient of modern mathematical symbolism:

> Various types of "equalities," "equivalences," "congruences," "homeomorphisms," etc. between objects of mathematics must be discerned, and strictly

> adhered to. However this is not enough. In mathematics there is the second requirement that one must know how to "operate" with mathematical objects, that is, to produce new objects out of given ones. (p. 313)

Mahoney (1980) points out that this development made possible an entirely new mode of thought "characterized by the use of an operant symbolism, that is, a symbolism that not only abbreviates words but represents the workings of the combinatory operations, or, in other words, a symbolism with which one operates" (p. 142).

This helps us pin down further the second core aspect of algebra, the syntactically guided (rule-based) transformation of symbols while holding in abeyance their potential interpretation. This aspect is both a source of mathematics' power and a source of difficulty for teachers and learners. In a fairly narrow form as manipulating expressions and solving equations, it has dominated school algebra since entering the U.S. secondary school curriculum during the 19th century. Surprisingly, perhaps, much of the practical content of the algebra texts was set by Euler in the mid 1700s (Kline, 1972) and has survived.[4]

The algebraic system was not the first operative system, and, indeed, the idea of a symbolic *calculus* of reasoning goes back at least to Descartes and Leibniz (Kline, 1972). The standard base 10 placeholder number system supports an operative symbolic system in the form of all the usual algorithms for multidigit addition, subtraction, multiplication, and long division that were invented in the 14th and 15th centuries (Swetz, 1987). When one is using, say, the usual algorithm for multiplying multidigit numbers (think of the vertical algorithm for the product of 376 and 1,287), one is acting on the marks on paper according to their position and shape, and not on specific meanings for the numerals involved. This is different from non-rule-based writing interspersed with thinking as when one sketches a diagram while developing an idea or building an argument or presentation. The latter involves alternating actions on physical inscriptions and interpreting their meanings, but the actions are not determined by strict syntactically defined rules.

But operating on physical numerals as a way of number-specific calculating is different from rule-based operating on strings of algebraic literals. The numeric calculation is directed toward producing a single number result represented unambiguously by a string of digits. A rule-based algebraic

[4]This may help explain why the idea of function has not been prominent in school algebra. It was not invented and consolidated as a concept until well into the 19th century, although it had been present implicitly well before then, and the word *function* was actually used in a modern covariation sense by Leibniz (Boyer, 1959).

operation (factor, simplify, combine like terms, solve, etc.) typically involves changing the visible form of the symbols under some equivalence constraint, where the starting configuration may or may not be representing a generalization and where the ending configuration is usually some visible form that is more convenient to the purpose at hand (although in equation solving it may also yield a single numerical result).[5]

A CLOSER LOOK AT SYMBOLIZATION AS A PROCESS

Looking at Versus Looking Through Symbols

As you read this chapter, you are undoubtedly looking through the marks on the page to build in your mind some version of what we are attempting to express. You were not noticing the page itself or the specific marks on it. But, you could just as well stare at the marks on the page and notice their visuographical properties—the horizontal blank spaces, the little hooks that reach down into these spaces, the occasional extra large open spaces with odd shapes, and so forth (squinting helps). Try to look at the marks on your windshield while driving. You are likely to feel a bit endangered as you, quite literally, lose sight of the road. Don't do this in traffic!

As the windshield experience suggests, it is probably impossible to look at and look through symbols simultaneously. This perceptual difference is analogous to a deep epistemological distinction between mathematics as an object of study in its own right versus mathematics as an intellectual tool, as a means of seeing, organizing, and reasoning about experience, including the highly structured experience that takes the form of science and the ever-widening areas of human endeavor where mathematics is applied. This difference is sometimes stated as the difference between mathematics as something to think about versus mathematics as something to think with. This distinction is especially interesting in algebra, because algebra is often used as something to think with, but where the thing that algebra is used to think about is frequently mathematics itself. A simple example is using algebraic symbols to represent general properties of arithmetic, such as $a + b = b + a$, for commutativity of addition of (some class of) numbers a and b.

---

[5]For a long-term evolutionary perspective on the place of operative symbol systems and their connection to the invention of modern computation in the form of operations autonomously executable without the direct engagement of a human partner (computer programs), see Shaffer and Kaput (1999) and Kaput and Shaffer (2002).

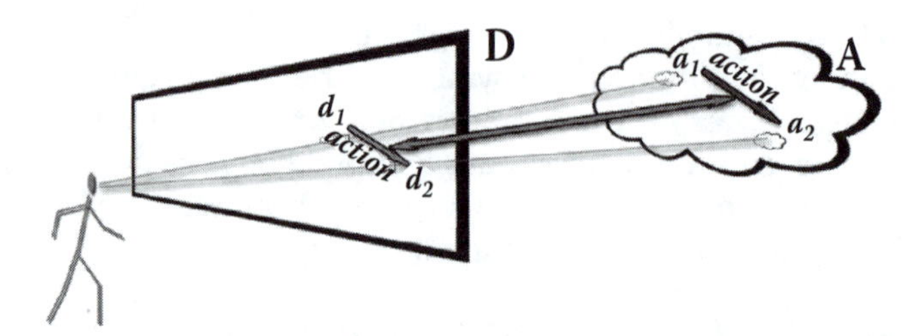

FIGURE 2.1. "Looking at versus through symbolization"—Attentional focus.

The Separation of Symbol From Referent

We begin with a simple diagram, intended to illustrate the difference between *looking at* versus *looking through* symbols that in turn lead us into a closer look at what we mean by *symbol* and *referent*.[6] At times, we are working with the symbols as opaque objects in their own right, perhaps inspecting them for patterns or, in the case of arithmetic algorithms or algebraic character strings, operating on them according to rules that apply to the symbols themselves. Or, in computer environments, we might be acting on spreadsheet cells, simulation controls, or graphical objects—coordinate graphs or geometric constructions. These are depicted in Figure. 2.1 as *actions* in symbol system *D* that relate in some way to *actions* in what we are taking as its reference field *A* (much more on this later). By action, we mean both physical and mental actions. Now, at some points, we might find it useful to look beyond the symbol *D* to what they might "mean"[7]—what they and actions on them might stand for in *A*, the situation at hand.

A word of clarification on what we refer to as *attention lines* emanating from the stickfigure's head: These are not intended as sight lines, as vision lines as Aristotle used them, or even as they are used in optics. Rather, they are a notational element intended to help us talk about the role of *attention*

[6]Many readers might recognize the similarity of this perspective to the classic sense and denotation framework developed by Frege (1984) to describe the process of interpreting and making meaning with symbols (see the excerpts and account by McGuinness, 1984). Although compatible in a general way, ours focuses more intently on actions, and the continuing constructive processes involved—as applied to the work of elementary school.

[7]The reader may notice that the word *meaning* makes relatively few appearances in this chapter. This is quite deliberate because quite often meaning is taken as a primitive, undefined root term, whereas here we are attempting to explore how meaning is built.

and how we switch attentional focus from one kind of thing to another. Nonetheless, the visual metaphor is useful, but, as with all metaphors, we need to be careful in how we use it—and be certain that it does not (ab)use us!

But this figure presumes a *separation* between symbol and referent—what the symbol is taken to stand for. Where did that separation and referential relation come from? To begin to answer this question, we need to take a closer look at the process of symbolization. We then return to diagrams of this type to review from a symbolization perspective some commonly occurring situations in elementary mathematics that are bases for the development of algebraic reasoning.

Setting the Stage—The Classroom-Based Starting Point

We begin our analysis of symbolization with Figure 2.2. The goal is to capture essential features of what we have seen across many instances of the development of algebraic reasoning, including those described in several chapters of this volume, and that reflect many others' accounts of symbolization processes (e.g., Cobb, Gravemeijer, Yackel, McClain, & Whitenack, 1997; Gravemiejer, Cobb, Bowers, & Whitenack, 2000). Whereas the analysis is framed quite broadly in terms of symbolizing activity, we are most interested in those cases where the symbolization is in the direction of the kinds of symbols and uses usually taken as "algebraic." We should note that the majority, but not all such symbolizations described in the literature, turn out to be character string-based symbol systems. But, as seen in Brizuela and Earnest (chap. 11, this volume) and Carraher, Schilemann, and Schwartz (chap. 10, this volume), algebraic symbolization can also include coordinate graphical representations, among others.

We begin with a classroom situation *A* that the students come to be engaged in, to explore, describe, and discuss mathematically. You are invited to refer to any of many classroom episodes throughout this book to find this kind of activity underway. All initial engagement with *A* is already and inevitably mediated by the students' cultural, linguistic, bodily (e.g., kinesthetic, haptic, tactile) resources, their social capacities and inclinations to share experiences, and any prior experience with situations of that type. Importantly, the situation might already be mathematical in the sense that it is about, for example, numbers and operations, sequences of numbers, geometric arrays of dots, or a verbally described sequence of actions on numbers. Alternatively, it might be at least partially mathematized as, for example, being about measured heights of classmates or properties of regular geometric figures or counts of embedded figures. Or, *A* might be a situation that has yet to be mathematized, such as a toy car rolling down a ramp, a cliff eroding, a quilt being sewn, the story of a (real or imaginary) school store's operations, groups of people shaking

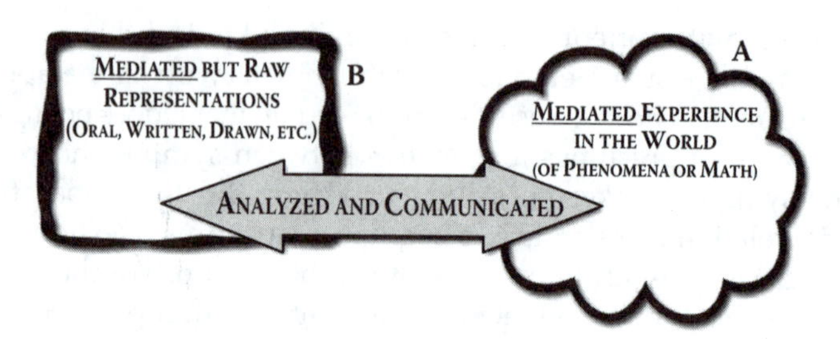

FIGURE 2.2. Beginning the process of symbolization.

hands, or a vase being filled with water. *A* may be presented in more or less mathematical terms—a table, a diagram, or perhaps simply in words, orally or written, or even by enactment, as with handshakes. As is clear, this depiction intends to encompass a wide variety of situations.

We assume that the students are in the kind of classroom that appears across most chapters of this book, a classroom where mathematical discussion is nurtured and supported. (We take the Vygotskian point of view that, later, a student working alone will be able to utilize important aspects of the socially constituted symbolization processes that were internalized during the social process.) Hence, when mathematically confronted with and engaged in *A*, students begin to build oral, written, and drawn descriptions of the situation—records of those aspects of the situation that are accessible to them at the time. These are again mediated by what students bring to the situation, including what they bring by way of symbolization resources, which of course will vary a great deal across grades and students. In this way, students build informal descriptions as depicted by *B*.[8] Because every situation has a history of some kind in the experience of students and teacher, at least in the sense of triggering recall of other situations and analyses, we could have added other *A*s and *B*s that feed into the bidirectional "Analyzed and Communicated" arrow connecting *A* and *B*. (To keep things schematically simple at the outset, we didn't.)

B might initially consist of rough pictures of the situation, with much extraneous detail or sketchiness that misses critical factors, or both (Lesh & Doerr, 2003; Lesh, Landau, & Hamilton, 1983). It might consist of measurements or counts recorded in preliminary, informal ways. Or, it might initially simply take the form of informal discussion (including gesture,

[8]We have oriented our diagram right to left, in part, to be consistent with our first diagram and, in part, to give a sense of pulling back or pulling out of, rather than "processing through" along a left-to-right timeline.

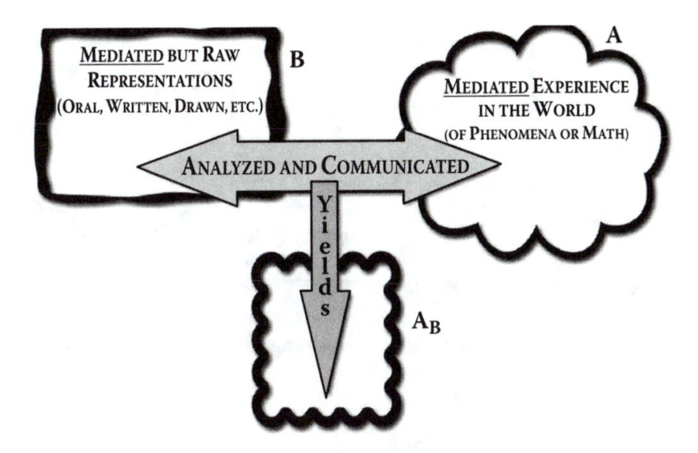

FIGURE 2.3. Building newly mediated conceptualization of A.

mimicry, etc.) in reference to students' coming-to-be-shared view of *A*. Indeed, this is the most likely first symbolization step because natural language is already present as a symbolization resource and, as Mason points out (chap. 3, this volume), it is already well developed as a vehicle for expressing generality.

In all cases, the physical[9] symbolization *B* is built from and tested against observations about *A* in the social context of discussion and interaction. *B* is in the process of being constructed as a new chunk of experience that is separate from *A*. But, as referenced to Sfard earlier (and as classically described by Vygotsky, 1965), the symbolization process is helping the students create a new *conceptualization* of *A*. They are coming to think of *A* in a somewhat new way. Let's call it A_B because it depends on *B*. So we have a newly mediated conceptualization of *A*, related to *B*, that can support the continuation of the symbolization process (see Fig. 2.3).

[9]Note that speech and gesture, although not yielding permanent inscriptions, are nonetheless intended to be included here as physical inscriptions. Note also that we use the term *inscription* when we deliberately want to emphasize that we are treating a symbol as a physical object independent of how or whether it may be interpreted, and independent of its potential participation in a larger system of symbols. We would use the term *notation* if we were regarding it as part of a larger system of symbols, but at the same time independent of issues of interpretation and reference. We use the term *symbol* when we are discussing the use of a notation in a referential context—where the notation may stand for something. Whereas there is no real standard set of terms for these usages, the ones we have chosen seem close to the norm as we have seen it. We have deliberately minimized our use of the term *meaning* because its usage is so broad and so variably defined.

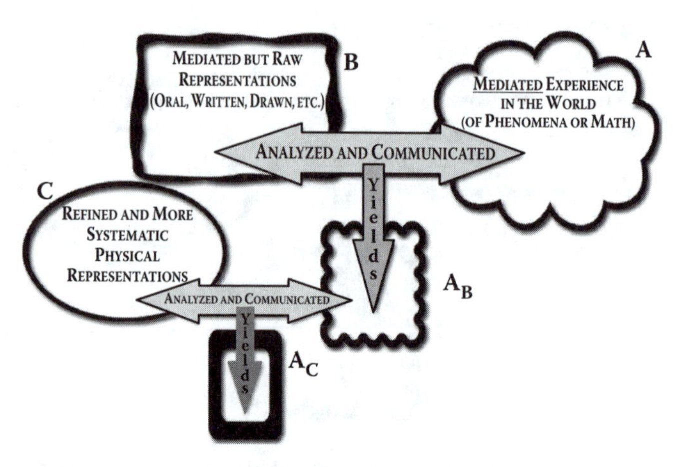

FIGURE 2.4. Continuing the process of symbolization.

So, to continue our description of the dynamic symbolization process, we will add a new snapshot (see Fig. 2.4, which includes a refined symbolization *C*, which in turn yields a refined conceptualization of *A*, denoted A_C in Fig. 2.4). This process has been described in terms of developing a "chain of signification" (Cobb et al., 1997) and, as illustrated by Cobb and colleagues, it is a powerful design principle for organizing curriculum.

The process continues toward, and perhaps reaches, a symbolization *D* that is conventional (at least for the class) and compact and that supports a conceptualization A_D. Figure 2.5 depicts a nested description of the conceptualizations to emphasize that a record, or trace, of all the previous conceptualizations (taken individually or socially, depending on the perspective) and their associated symbolizations remain (depending on how much overwriting has taken place physically—in the case of spoken symbolizations, this is a matter of memory). Such records themselves often act as resources in further symbolization, either within a given episode or across episodes.

We do not delve here into the level of detail needed to analyze these kinds of symbolization processes in depth. However, we note that systematic unpacking of the process through intermediate notations will reveal the kinds of stages found by other researchers, especially the icon-index-symbol stages formulated initially by Peirce (1955), used by Bruner (1973), and examined in detail by Goodman (1968/1972) (iconicity especially), by Deacon (1998) in his account of the evolution of language, by McGowen and Davis (2001) in their Deacon-referenced analysis of

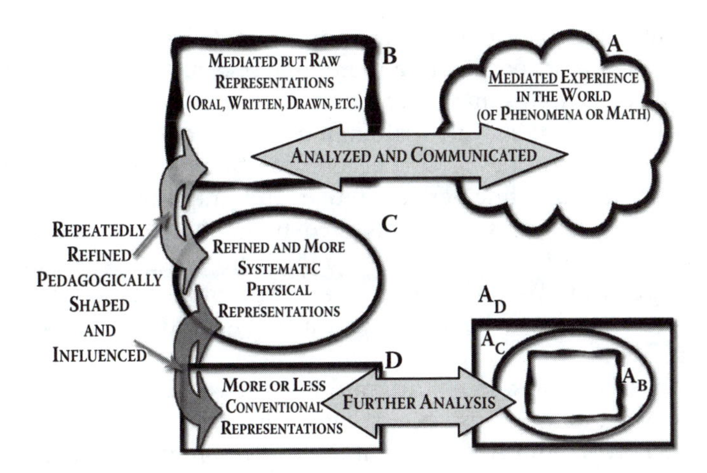

FIGURE 2.5. Nested nature of symbolization and its records.

algebraic symbol interpretation and use, and the broadly eclectic study of mathematical symbol interpretation offered by Drouhard and Teppo (2004). We simply suggest that ours is a common visual framework for reading such analyses.[10]

Another reason for including the nested set of conceptualizations is to emphasize the constructive, additive nature of the process of symbolization, as opposed to the view often taken that it is abstractive and subtractive. We see the abstractive/subtractive view as resulting from a confusion of the visible product, which is often an idealized schematic and simplified description of *A*, with the constructive and often very creative process by which the symbolization product is built. Indeed, a more accurate view of the reflexive and recursive nature of the process would be provided by a nesting of the entire process, whereby each new stage envelops all the prior ones, and the reader is invited to think of the diagrams in this way. This view of symbolization is also very consistent with the notion of concrete abstraction as a constructive, additive process described by Wilensky (1991) and Noss and Hoyles (1996), among others.

One other point regarding the symbolizations: Although we have concentrated on early symbolization and the processes of building a conventional notation system, it is usually the case in students' development

[10]Note that the reconceptualizations, such as $A_{B,}$ can be regarded at the "interpretant" in Peirce's tripartite analysis, but where our emphasis is how the interpretant is actively constructed.

that several notation systems are already present. Thus, the student may be in the position of coordinating more than one such system—as occurs in those episodes, for example, where students are using physical manipulatives in coordination with written and spoken notations for numbers and operations, or in the case of certain of E. Smith's examples (chap. 5, this volume), coordinating verbal, tabular, and graphical notations. From the notational perspective being offered here, students' conceptualizations can be regarded as being elaborated and enriched as additional notations are incorporated into their available symbolization resources. In effect, a preexisting *B* is being brought into play and elaborated to help make sense of *A*.

Of course, in a classroom, with multiple active participants, there are multiple conceptualizations[11] occurring simultaneously. Over the course of the socially mediated conversation based on perceptually shared phenomena and symbolizations that is orchestrated by the attention-directing activity of the teacher, these tend to converge—or, if you prefer—emerge. We see evidence of both the convergence/emergence, as well as the irregularity and unpredictability, of the symbolization process in the many classroom episodes described in this book. Often, the variation across students is a source of progress toward more effective symbolizations as the insight or symbolization of a more advanced student provides the means by which other students come to see the situation in a more productive way and appropriate the symbolization, particularly under the deliberate influence of the teacher (van Oers, 2000). The classroom episodes (especially in Bastable & Schifter, chap. 6, this volume; Schifter, Monk, Russell, & Bastable, chap. 16, this volume; van Oers, 2000) also reveal how the teacher's agenda plays out in the orchestration of the symbolization process. This includes the degree to which the teacher wants to shape the students' symbolizations toward conventional ones, where the symbolizations might only be conventional relative to previously established classroom language (see the "turn-arounds" in Schifter et al., chap. 16, this volume, and later).

Finally, one can see how argumentation associated with justifying and refining the scope of generalizations interacts with and drives the

[11]By use of the term *conceptualization*, we do not mean to imply a cognitivist view of the situation. Instead, we prefer to leave this part of the analysis open to multiple perspectives, including social constructivist and other perspectives because we feel that our framework for analyzing symbolization is consistent with more than one (see Cobb & Bowers, 1999; Cobb, 1994). In effect, the conceptualization may be a distributed entity, distributed across participants and their symbolizations (this is actually our preferred view).

symbolization process, including the chapters by Franke, Carpenter, and Battey (chap. 13, this volume), Blanton and Kaput (chap. 14, this volume), Schifter et al., (chap. 16, this volume), and especially the chapter by E. Smith (chap. 5, this volume), where argumentation toward the development of experienced certainty is shown to be central. Discourse and its social context pull toward explicitness regarding the range of application of an asserted generalization—if it holds for this, does it hold for that? In effect (again, from a Vygotskian perspective), the social context supplies the doubters and those needing convincing more readily than situations where the students working alone must supply these themselves. It also triggers the engagement of the social and conversational resources that are typically readily available to children.

Returning to the Matter of Reference: So Where Did the Separation of Symbol From Referent Come From?

Our general descriptions of symbolization processes are intended to help answer the earlier question: Where did the separation of symbol from referent come from (see Fig. 2.1)? It is the result of a chain-of-signification symbolization process driven by communication that reformulates the initial situation. At each stage, a process of externalization and re-expression are simultaneously underway that creates a separate physical entity. This material product is typically a set of physical inscriptions depicted in the window that the symbolizer metaphorically looks at or through depending on the needs of the situation. Less metaphorically, it is a matter of attentional focus, to the physical inscriptions or their conceptual referents. But the window also acts as a mirror because any reconstituted formulation of the situation reflects back to guide actions, both mental (interpretive actions) and physical (elaborations) on the inscriptions.

A Note on Our Symbolization: Actor Versus Observer, the Flow of Time, the Limits of Static Diagrams, and the Relative Nature of Reference. The astute reader has long realized that the authors are involved in a symbolization process in this chapter. We have introduced two kinds of diagrams, the window type in Figure 2.1 and the process type depicting the process of symbolization in Figures 2.2 through 2.5. In the former, we have a place for the actor, or agent, and emphasize the shifts in attention as well as action, where time is mainly implicit. In the latter diagrams, the flow of time is expressed via the introduction of sequences of new symbolizations.

More important, as a static two-dimensional diagram, each type under-represents both the transformative effect of symbolization and its continual nature by maintaining the identity of A as a reference field when, in fact, as a conceptualization, it changes as it is symbolized—hence the nesting in

Figure 2.5. This process of transformation is especially underrepresented in the window diagrams that we use to track symbolization and symbol use, so the reader is forewarned to take this into account. A better representation would be dynamic, where each symbolization stage would change the reference field in some visible way and do so in a less discrete, less unidirectional and more recursive way.

In our choices of letters, we have deliberately straddled the boundary between being entirely arbitrary in our choices of letters and repeatedly using the same letters for reference field and endpoint symbolizations. We should emphasize that the nature of reference is relative at its core, and what is referent in one description is the result of a prior symbolization, and so where we take up the description has a lot to do with what is symbol and what is referent. It is inevitable that when we undertake the description of a continual process, we make choices about where to begin, which in turn imply choices regarding what is assumed to be in place. This is discussed further later when we examine symbolization associated with algebra.

Remember that these kinds of diagrams only make sense from an external observer's point of view. From an actor's point of view, the window/mirror and the conceptualization are experienced as aspects of the same event. It is in this actor point-of-view sense that Nemirovsky and colleagues (Cobb, 2000; Nemirovsky, 1994; Nemirovsky & Monk, 2000) refer to as *fusion* of symbol and referent, which they deliberately contrast with the *transparency* view that is implicit in Figure 2.1.

Finally, the framework is deliberately intended to be applicable at multiple time scales, ranging from the minute-to-minute changes of an individual or small group at work solving a problem, to the lesson-long changes of a classroom, to the longer term changes of students involving sequences of classes within a grade level (Cobb et al., 1997; Gravemeijer, 2000), to student developmental-level changes, to the development of ideas in the context of research laboratories (Roth, 2003), to long-term historical time-scale processes to help frame accounts of the evolution of symbol systems and their related mathematical concepts and practices (Puig & Rojano, 2004; Sfard, 2000).

TRANSITIONS FROM ARITHMETIC TO ALGEBRA FROM A SYMBOLIZATION PERSPECTIVE

We now look more closely at those situations *A* that are already mathematical and where the symbolization is directly in the service of generalization. This is often where algebraic symbolization is learned initially and where most instances in the book appear. Because of the central place of arithmetic in elementary school, we look more closely at those situations where *A* involves arithmetic statements or number patterns. In this case, *A* already utilizes the results of prior mathematical symbolization (which is one reason the word *mediated* appears in the earlier figures). We make repeated heuristic use of the attentional focus diagrams based on the window metaphor because they provide a simple notation (perhaps too simple for strongly theoretical purposes) for tracking symbolization phenomena that seem to occur across many familiar situations. The reader is invited to imagine a process diagram associated with constructing each new layer of a window diagram: a sequence of receding windows.

Generalizing Properties in Arithmetic: Variations in Attentional Focus

Simple examples of generalizing come from arithmetic as children build concepts of number and operations and isolate their properties: When students notice that they get the same result no matter the order in which they add two natural numbers, they are free to switch their order. Prior to this, as they move from counting all to counting on, they realize, as a matter of efficiency, that it is easier to count on from the bigger of the two numbers, so they switch order to accommodate this strategy. It is this switching that symbolically comes to be instantiated as a substitution. (From an actor's perspective) I can replace 3 + 5 with 5 + 3 if I desire. Thinking in terms of the *A* and *D* in the earlier diagrams, where *A* is regarded as a numeric reference field for the symbol system of numerals and operation signs *D*, this is an action on symbols, in *D*. If it is guided by my understanding that both sums are 8 as numbers, then I am treating the numerals transparently, looking through *D* to act in my conceptualization of *A*, guided by the properties that I already have constructed for them in *A*. This action in *A* might also be supported with actions on another notation system, for example, counting manipulatives of the sort depicted as a *B* and *C* in Figure 2.4—reflecting the fact that prior symbolizations are still available (see Mason, chap. 3, this volume, for a discussion of this case). We depict in Figure 2.6 this kind of typical arithmetic situation in elementary school mathematics. The action is guided by attention to and conceptualization of *A*, not *D*. But its results can be recorded in a physical inscription, in a *D*, which could be oral or written (as shown in Figure 2.6). And, in general, such a record might vary in character, from iconic to indexical to symbolic (see McGowen & Davis, 2001).

In contrast, I may already have generalized this switchability property in the system of numerals and operations and expressed it symbolically as a property of the *symbols*, replacing 3 + 5 by 5 + 3. Here I am guided by my knowing that I can always switch the order of the addends, treating it

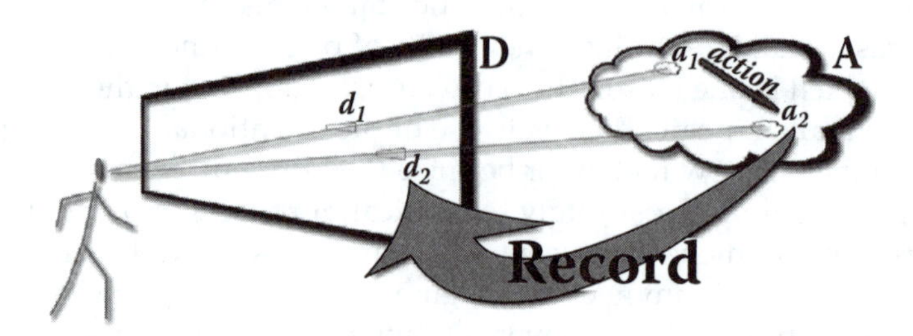

FIGURE 2.6. Action with attentional focus in A, recorded in D.

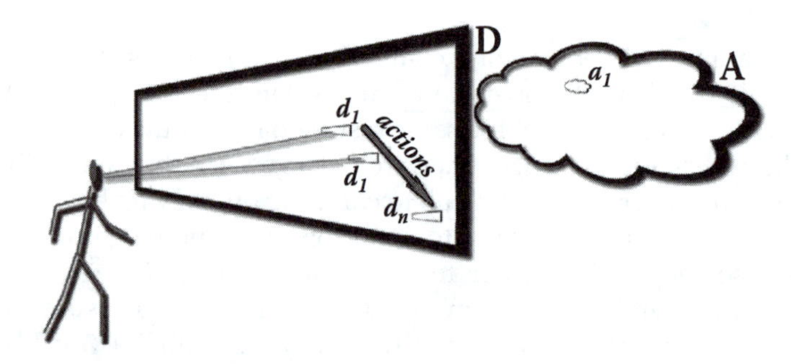

FIGURE 2.7. Actions based on attention to a symbolization D of A.

as a figural symmetry about the "+" sign. That is to say, it is guided directly by my knowledge of the configurations of inscriptions in *D*, not by my knowledge of the size, decomposability, and so on, of the numbers, which has already been compacted into my knowledge of *D*. This attention to *D*, rather than A, is depicted in Figure 2.7.

Outwardly, the difference between action driven by knowledge of and attention to *A* versus knowledge of and attention to *D* may not be obvious to an observer, especially in a simple case such as that mentioned previously. It might be more obvious if, say, a student who needed to add 11 and 9, aligned 11 and 9 vertically and talked about "carrying the 1" to get 20. Here we are in the position of Figure 2.7. This would be clearly different from an approach where the student spoke of "subtracting one from the 11 to add it to the 9 to get 2 10s." This would be depicted by Figure 2.6.

Much of what are often referred to as *student-invented algorithms* are based in a reference field for the numerals (in an *A*), but come, through symbolization, to be expressed in terms of operations on numerals (in a *D*),

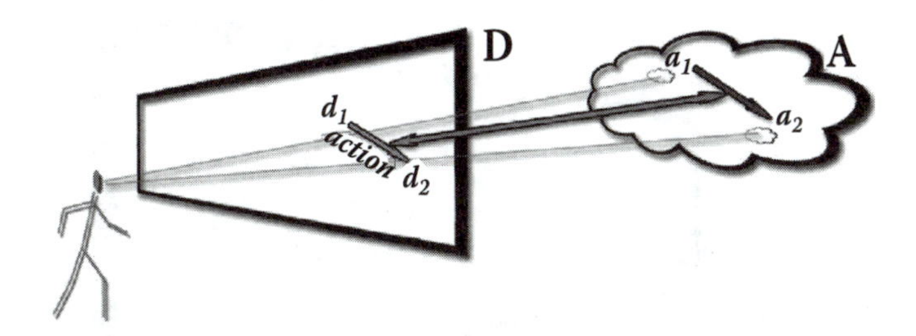

FIGURE 2.8. Symbolization in D of actions based in a conceptualization of A.

or, more often, in an oral (rather than written) *D*. These would be depicted by Figure 2.8, where attention alternates between *D* and *A*, coordinating the two systems of thinking. Indeed, there is considerable pedagogical leverage in symbolizing such strategies (e.g., compensation strategies), because, as with most physical inscriptions, this makes them explicit, discussable, and sharable. Such might take place in multiple stages, beginning first with numerical statements and then statements using literals: For example statements such as $19 + 5 = (19 + 1) + (5 - 1) = 20 + 4 = 24$ would be written before a more general statement such as $a + b = (a + 1) + (b - 1)$. Note that a very similar compensation strategy saves the "correction subtraction to the last step (ignoring associativity for the moment): $19 + 5 = (19 + 1) + 5 - 1 = 20 + 5 - 1 = 25 - 1 = 24$. In each case, the strategy is based on reasoning about numbers and operations that is then recorded in symbols.

By contrast, traditional algorithms learned only as procedures in *D* without active symbolization based in *A* lack the referential connection to *A* and hence cannot support the kinds of attentional coordination afforded by a notational system that is the product of active symbolization. Hence, there is a difference between actions in *D*, where *D* is a result of active symbolization, which allow the potential for coordination with their referential roots, versus actions in a *D*, which is not the result of an active symbolization and where there is no ready access to a referential field. In this latter case, actions in the notation system must be guided strictly by the rules of that notation system, unguided by the previously learned structure of the reference field. See Figure 2.9 where no *A* is present.

In effect, there has been no reconceptualization of *A* in terms of *D*, what was termed A_D earlier (see Fig. 2.3). Indeed, if we were to include a conceptualization it would need to be described as D_D. The lack of conceptual links to the structure renders such a D_D very fragile. This provides another perspective on the difference between purely procedural knowledge and conceptual knowledge as described by Hiebert (1986) and Silver

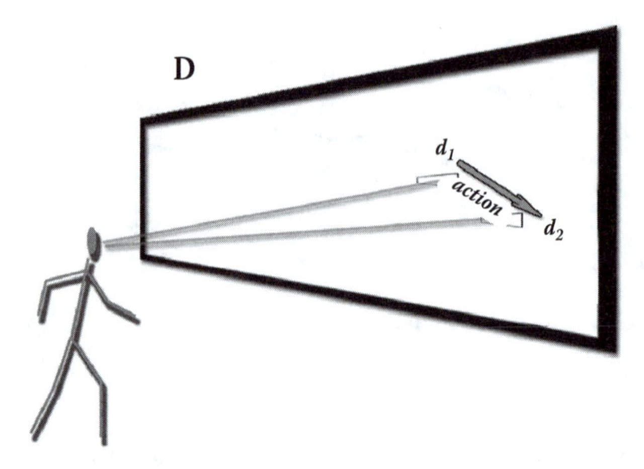

FIGURE 2.9. Operating in a notationally and conceptually isolated system.

(1986), where it is argued that conceptual knowledge is more connected and hence more stable. A similar point is made by Carpenter and Lehrer (1999). E. Smith (chap. 5, this vol.) provides an example of a student (actually, an elementary teacher in a summer course) whose pattern development work in a tabular *D* is disconnected from its blocks-based reference field *A* and so is unable to make sense of irregularities in the table.

The Origins of Algebraic Reasoning Part I: Expressing a Generalization Using an Explicit Variable

Symbolizing Generalizations Beginning in Mathematics: Generalizing Arithmetic or Quantitative Reasoning. Now, what does the emergence of algebraic reasoning look like in terms of our framework? Let us examine a typical case of a student (or classroom) generalizing and expressing a generalization based in arithmetic. We join the action after the students have already developed an ability to write and discuss numbers and operations using the usual numeral symbols and words, and at least some of them can write number sentences in some form. This situation reflects many episodes throughout this book and published elsewhere (e.g., Bastable & Schifter, chap. 6, this volume; Brizuela & Earnest, chap. 11, this volume; Carpenter, Franke, & Levi, 2003; Carraher et al., chap. 10, this volume; Schifter et al., chap. 16, this volume). The students have a multiplicity of specific experiences subject to generalization, but they need a compact way to express them as a unity, as a general statement that covers the multiplicity of instances; indeed, it's not only a multiplicity, it's an infinity of instances. They are in need of a symbolization.

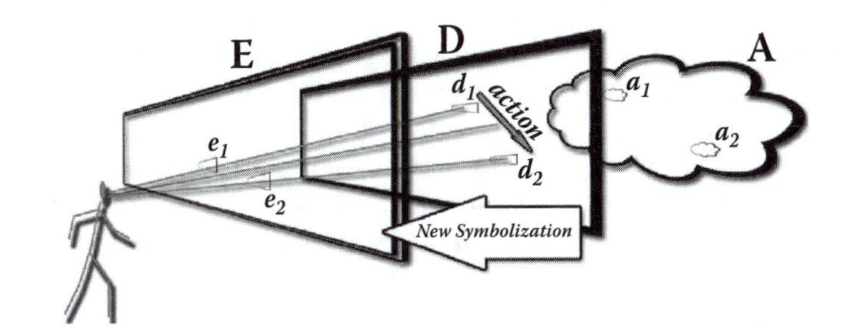

FIGURE 2.10. An algebraic move—using literal variables to express generalizations over some situation.

In the simple case already mentioned, commutativity of addition, the students have already written and discussed sums and perhaps have a name for the process of reversing their order (see e.g., Schifter et al., chap. 16, this volume). These verbalizations could count as an intermediate symbol system as discussed earlier, reflected in Figure 2.2. They have also recorded such reversals in their number sentence symbol system *D*, as depicted in Figure 2.6.

But this now provides the opportunity for another round of symbolization, one that is clearly more algebraic in cultural artifact terms. In particular, now that they have a generalization to express—actually, to re-express—they are ready, and in most cases more than willing, to write a generalization using literals rather than numerals, a statement such as "$a + b = b + a$" that carries the visual structure of their previous numerical statements. In effect, with the likely guidance of the teacher (we must recall how much effort this step took historically), the students are (re)creating a new symbol system *E* based on actions and equivalences experienced in the numerical statements of *D*, likely in concert with verbalizations. The situation is depicted in Figure 2.10.

They are now using regularities in the shape of the symbols in *D*, the visual symmetry of symbols such as $5 + 6$ and $6 + 5$ across the "+" sign in this case, based in prior numerical experience in *A*. They are likely also applying their experience inherited from the culture at large in using letters as shorthand abbreviations and as labels, as well as to denote indeterminate things.[12] The result is some sort of conventional expression of a generalization. Work by Fujii (2003) indicates that appropriate arithmetic

[12]Needless to say, this is a different kind of use of letters—as variables—and eventually the distinction will need to be made clear. Otherwise, a variety of difficulties can be expected, for example, the Students-Professors Problem reversal error (Clement, 1982; Kaput & Sims-Knight, 1983b).

experiences can make this symbolization a natural transition, but that if such experiences are not provided, then students' understanding of literals as variables is limited (see later where we extend his notion of quasi-variables).

This symbolization migrates existing *D* inscriptions such as the operation sign "+" and the equal sign to subtly different roles in *E* (Kieran, 1981). For this to happen as described, the symbols need to have been used in a mutually compatible way in arithmetic, especially in unexecuted number sentences, or as shown in J. Smith and Thompson (chap. 4, this volume), in explicit expression of quantitative relationships. It is a clear message of much prior research that a strictly procedural use of these signs does not provide a foundation for this critical symbolization step, but appropriately designed and executed instruction does indeed support such a step (Carpenter et al., 2003; Carraher, Schliemann, & Brizuela, 2001; Fujii, 2003; Fujii & Stephens, 2001; Kieran, 1992; MacGregor-Stacey, 1997). We should also note that students' conceptualization of addition is enriched and viewable from a more abstract level by this new symbolization in *E* in a kind of retroactive way. Further, it sets the stage for potential generalizations to other number types or to other operations because they can now be stated succinctly in *E*. Finally, it depends on an appropriate symbolization activity in *D*, which is the building of a series of written expressions that can serve the semiotic move of replacing the numerals by literals in analogous expressions. This activity is in support of the second core aspect of algebra, reasoning from the forms of inscriptions. The fact of visual analogues between the arithmetic and the algebraic systems suggests that, rather than being a discrete or discontinuous symbolization jump, the process is better understood as a continuous, analogic process of the substitution of visually similar forms.

For young students whose symbolic resources are limited, the generalization may need to be expressed using their available techniques of discourse based in natural language, including intonation and gesture. In this case, students express generality in the form of particular instances intended to be seen as typical of a range of possibilities where the actual range of the generalization is often tacit. Such a particular instance might also reference physical materials, which can embody generality only if used in this way. See Mason (chap. 3, this volume), Bastable and Schifter (chap. 6, this volume), and Blanton and Kaput (2003) for examples of this type. Argumentation associated with clarifying the range and nature of the generalization pushes students toward more explicit expression of generality and hence the symbolization that makes this possible (E. Smith, chap. 5, this volume).

Symbolizing Generalizations That Begin Outside of Mathematics. The situation here is structurally similar except for the role of *A*, where *A* is not fully mathematized as it was in the previous situation. In the current case, one or more intermediate representations (*B*, *C*, ...) typically need to be constructed before the key generalization step to some representation *E* that captures the essential generalization(s) needed to solve the problem at hand. These will involve counting, measuring, organizing, and applying perhaps some previously developed way of symbolizing, such as a two-column table or perhaps sequences of number sentences. Note that *E* need not be a version of the character string algebraic system but could instead be a coordinate graph or, depending on the level of the students, it could be a two-column table from which generalizations are made orally and tested numerically. The factor that makes it algebraic is that it is a deliberate expression of a purposeful generalization to serve further reasoning. It embodies the first core aspect of algebra.

As was the case in our commutativity example, it pays to build sufficient experience in one symbolization before moving to another symbolic expression of the generalization. Consider, for example, an algebrafied version of the well-studied Handshake Problem, which asks a question of the type: If I have 50 people at a party, how many handshakes are needed if everyone is to shake everyone else's hand exactly once? We can regard the problem situation as an *A* in the previous descriptions. In this form, the 50 amounts to a number being treated algebraically, because it is very difficult to achieve a numerical answer without creating a general functionlike rule for determining the number of handshakes for a variable number of people. It pulls us to create a model that embodies a generalization using another system of representation *B*. One style for achieving this goal is to generate a set of data on the total number of handshakes for a manageable number of people (perhaps organized in a table) and determine a function that outputs the number of handshakes given the number of people (either in efficient closed form, or in far less efficient recursive form and requiring much more labor to yield an answer). This style combines the generalization core aspect and the function strand (2) to form *B* in which to compute a solution.

Another style, far less common but more algebraic in its treatment of the numbers and operations involved, more directly applies Core Aspect 2, reasoning from the form of expressions. This approach expresses the number of handshakes for a given number of people as an unexecuted number sentence that yields the sum of handshakes of all the people up to and including the last person added to the group. Such a sequence, based on repeatedly incrementing the group by 1 and noticing that each new person must shake hands with all the people who previously shook hands, takes the form of 0 for 1 person, 1 for 2 people, $1 + 2$ for 3 people, $1 + 2 + 3$ for 4 people, $1 + 2 + 3 + 4$ for 5 people, and so on. This then suggests that the total for 50 people is the sum of positive integers up to $50 - 1 = 49$. Here the form of the computation more directly reflects the structure of the situation than the prior approach. The B produced here is of a very different kind from that in the other table function style approach, which actually obscures the computational structure. The actual computation is the same as the recursive one, of course, but is expressed somewhat more explicitly (Kaput & Blanton, in press). Indeed, the approach is entirely the same for the symbolically general case with "50 – 1" replaced by "$n - 1$"—so that the step to a literal-based expression is quite direct (an E in the previous description). The power and range of such reasoning with sums is nicely illustrated in Bezuszka and Kenney (2004).

Transitions to Algebra II: A "Lifting Out" of Schematized Actions on Symbols to Build an Operative Notation System: A Symbolization Example Involving Equations

Next, consider the second core aspect of algebra, reasoning from the syntactically defined forms of symbols, especially syntactically guided transformations of symbols. We have previously discussed the historical development of two primary operative notation systems, the numerical algorithm system founded on the base 10 placeholder system, and the algebraic system that uses literals to stand for variables. Let us now take a classroom-learning look from a symbolization perspective at one aspect of the algebraic system, one that involves equation solving.

Consider a situation where students have been working with open number sentences such as 8 + _ = 13 or, after introducing literals representing variables as discussed earlier, using a literal, $8 + x = 13$. After solving and discussing some number of these kinds of sentences, it is noticed that the answer always seems to be of the form 13 – 8, that is, in verbal terms, "you subtract the left-hand number from the right-hand number to get the answer."[13] The students can be thought of as working within a number-sentence symbol system D, depending on their arithmetic prowess, perhaps making occasional use of other systems such as counting chips, fingers, or marks on paper. In this way, they are in the process of building a rule, a generalization that applies to a parallel set of additive number sentences written in the D system. It almost always passes through a "theorem-in-action" stage (Vergnaud, 1982), where the action can be done, but not articulated (see also Mason, chap. 3, this vol.). It also occurs in the form of students being able to predict the next number in a pattern without being able to explicitly articulate the pattern as a formula of some kind.

[13]When the number sentence problems are examined as equations using literals for the unknowns (e.g., as 6 + x = 14 or x + 5 = 12), the action of subtraction from both sides is more easily represented (Usiskin, 1988; see also Peled & Carraher, chap. 12, this volume).

This is another example of the symbolization process as already described, where children's intermediate step could be in form of the verbal version of the rule as given—a *B* in the sense of Figures 2.3 through 2.5. Mathematically, we could regard this as a generalization over a subset of the expressions writable in *D*. As is discussed further later, arguing the range of the generalization is a main driver in building a new symbolization.

At some point, again as the result of a combination of discussion and perhaps the teacher-led cataloging and recording of cases, the rule gets extended to cover cases where the unknown is in the first position, as in "_ + 6 = 15." To ensure that the rule covers all such cases and will extend to more cases in the future, the teacher suggests that they think of it as "subtracting the same number from both sides (of the equation)." Although it need not be written in what we would recognize as algebraic form, this new verbally described operation on the number sentence objects of *D* is another, and major, contribution to building a new symbol system which consists of expressions of generalizations about actions on number sentences in D. It is a distinct representation of general actions in *D*, and as such is part of a new operative symbol system *E* being lifted out of *D* to serve as a new, more general way of thinking about and operating on the number sentence objects of *D*.

This is an algebraic move, as reflected in Figure 2.10, but it is a different kind of move. Whereas the previously described move involved expressing variation across statements in *D*, the new one expresses actions on the inscription objects of *D*. These two kinds of symbolizations embody the two core aspects of algebra both as a cultural artifact and as mathematics that people do, one to express generalization and the other to operate on those generalizations symbolically (the key aspect of algebra emphasized by Bochner earlier).

A common next step in the algebraic symbolization of equation solving involves the action of transposition of symbols as a shortcut for "subtracting the same thing from both sides." At some point in working with such simple additive number sentences, and solving those with the subtraction technique, someone inevitably says that this is the same as moving the known number from the left side to the right side and either subtracting it or changing its sign (so $x = 14 - 6$ or $x = 12 - 5$ in the previous examples). In effect, this introduces another operation in the symbol system for equations.[14] Indeed, it is usually quickly extended to broader kinds of equations, particularly linear equations.

---

[14]We should note, of course, that unless one is using a particular kind of manipulative or software, one does not move symbols, one rewrites them in what amounts to a substitution—in this case replacing a few "subtract-from-both-sides" steps with a rewrite of the result of those steps.

This is an important step for two reasons. First, it is pragmatic in that the procedure makes very efficient use of computational knowledge (subtraction), and second, it simplifies the equation solving by reducing cognitive load to the management of simple actions on a small number of symbols (in the spirit of the Whitehead quotation). And, importantly, this new action is further removed from the referent field *D* because it is no longer expressed in language that applied to the original number sentence objects of *D*. This then adds some error risk, however, due to the potential for misapplication of the operation, which is now more isolated from previously symbolized and conceptualized actions.

The symbolization process of moving from a study of number sentences in symbol system *D*, whose referents are numbers and operations on them with some cognitive reference field A_D, to equation solving of the sort described yields a new symbol system *E*, which supports actions that can be carried out independently of *D* and its conceptualizations A_D. This could be illustrated in a diagram similar to Figure 2.10.

REFLECTIONS ON THE SYMBOLIZATION FRAMEWORK: GENERALIZING AND EXTENDING OUR EXAMPLES

The Two Kinds of Algebraic Symbolization: Generalizing and Action Lifting

Expressing generalizations in compact and stable external form using some kind of notation for variables and then lifting frequently used and wide-scope actions out of symbol systems are the two central ingredients in the development of algebraic reasoning. They are repeated across many contexts and many levels. As we saw in our equation examples, they interact in a deep way. Actions on symbolizations are not possible or fruitful until the generalizations they express are compacted and crystallized by those symbolizations. In an interesting way, the symbolization results in both a separation of symbol from referent and a close connection between the two. Frye (1987) points out that this is an essential feature of symbolization in general. He uses the words *displace* and *condense* and identifies other philosophers of symbolization who discuss this same issue. Productive use of the separation and connection turns out to be a critical skill in algebra, especially when actions on symbols are involved.

But not just any actions are worth lifting out. They must make good use of existing mental function, provide conceptual and linguistic economy, serve a useful purpose across a wide range of cases, and, most importantly, they must be logically coherent with respect to the existing structure of the system on which they are applied. Indeed, this is the essence of being "syntactical." The action needs to produce objects that fit with existing ones in some acceptable way. But, if the symbol system is well built from its reference field, this coherence is built in. This is one aspect of what Wigner (1960) calls the *unreasonable effectiveness of mathematics*. It is also the reason why we can ignore the reference field as we manipulate symbols according to syntactical rules and be assured that the result will be both consistent with the configuration of symbols that we started with and a logical basis for thinking about the reference field. Of course, if the reference field is nonmathematical as in the modeling context, then the new configuration may or may not fit the realities of the reference field. The judgment at that point is an empirical matter and is usually the central concern of the activity.

Some of the lifted actions based in arithmetic can be represented directly in terms of the structure of the system, such as the distributive law of multiplication over addition in the usual number systems, which allows the substitution of $a * (b + c)$ by $a * b + a * c$ or vice versa. The action is an equivalence-preserving substitution, which has parallels in the other basic properties of operations as well as substitution actions such as factoring and expanding polynomials that are built directly on them. Other substitution actions, such as the various laws for exponents or adding fractions, are more complicated but follow from the basic ones. Other actions are those less directly related to basic structure, such as the operations on equations we introduced earlier, or more complicated ones that depend on some special starting configuration, such as the quadratic formula or operations such as addition or multiplication of rational expressions.

Some actions are less explicitly specifiable and, as anyone who has worked on defining a computer algebra system will attest, embody some ambiguity, such as simplify or combine like terms. But all of these actions require attention to the form, the configuration, of the symbols—the *sense* in Frege's terms. It is here where adjustments to approaches to arithmetic are especially important because, as pointed out by Fujii (2003), Livneh and Linchevski (1999), in our own work (Blanton & Kaput, 2004), and by J. Smith and Thompson (chap. 4, this volume), most arithmetic statements are read as instructions to compute and are indeed executed as such, typically leading to a numerical result. Because in algebra they need to be thought about in a fundamentally different way (as illustrated in our earlier Handshake Problem discussion), students have particular difficulty making the transition unless their work with arithmetic treats arithmetic statements in a more algebraic way. In particular, it is important for students to work with unexecuted arithmetic expressions, building sequences of them, transforming them, comparing them with others, and so on. Fujii (2003) refers to substituting different numbers into sequences of unexecuted number sentences as using the numbers as quasi-variables.

This is another aspect of the process of bringing out the algebraic character of arithmetic as suggested by Carraher et al. (chap. 10, this volume).

An Essential Pedagogical Tension

The increased risk for error as one moves away from the original reference fields is repeated across many mathematical contexts and levels. It is one way of viewing the inevitable pedagogical tension between the increase in mathematical power deriving from cycles of symbolization that pull ever farther from concrete reference fields and the resulting decrease in learnability for students.

Our earlier comments about introducing the numeric algorithm system without a symbolization process to ground it in numeric experience apply *a fortiori* here in algebra. If the algebraic system is not introduced via a well-grounded symbolization process in the context of both expressing generality and the lifting out of previously established actions, then it is based only in rules about itself and is symbolically and conceptually isolated from foundations in what the student knows and can do (as in Fig. 2.9).

There is, we feel, an important lesson in the equation example. The action on the numerical-expression-objects of *D* to be lifted out to become an action in a new system *E* is useful across a wide range in *D*, it is important, and it is often repeated. As noted earlier, it is representational economy and efficiency, measured in cognitive[15] terms, and communicative power measured in terms of argument and justification that together drive the symbolization process. One can recognize most of these factors at work in the previous example (although for brevity's sake we did not use classroom data) and in virtually every classroom episode described in this book. But when argument and justification are not present, and the symbolization process is cut short as is so often the case, one can expect the kinds of algebra learning difficulties that were once the staple of research in algebra (e.g., Booth, 1984; Clement, Lochhead, & Monk, 1981; Hart, 1984; Matz, 1982; Wenger, 1987).

Historical Symbolization in Terms of the Framework

As is commonly known (see most any history of mathematics book covering the period from 1400–1800), there have been many episodes across history where actions regarded as legitimate in a symbol system *D* yielded results that violated expectations in the accepted reference field *A* that in

[15]We say *cognitive*, but there is a linguistic aspect as well, especially in terms of visual parsing of written statements. This is one of the messages of recent work by Kirshner and Awtry (2004).

turn, provided a rationale for enlarging or modifying *A*. For example, this occurred when equation solution methods yielded negative roots, or when simply squaring an unknown, adding one, and setting the result equal to zero, one can find oneself trying to take the square root of a negative number.[16] To make things more problematic, the arithmetic of such imaginary numbers can be made consistent with the previously known arithmetic of rational numbers (for an accessible account, see Jones, 1954, reprinted in Swetz, 1994). The same is, of course, true of negative numbers. However, there may be a well-developed notation system *D* and an accepted reference field *A* containing objects that, at one point, are not represented in *D*, but where *D* is extended in a natural way (in hindsight) to represent them, as Stevin extended the base 10 placeholder system for whole numbers to decimals by allowing digits representing negative powers of 10 (Kline, 1972; Moreno & Waldegg, 2000). A similar situation occurred when the idea of fractional exponent, which was seen simply as another, more computationally efficient way to write powers of roots, was extended to arbitrary real numbers. This, of course, required a limit process and helped generate new issues of continuity of the real numbers (Boyer, 1959).

These kinds of historical phenomena remind us that the process of symbolization in school mathematics has a deeply transformative effect on what is known—ideas of numbers and ideas about operations on numbers are transformed during the process, a process that is almost never a matter of re-representing what was known. Noble, Nemirovsky, Dimattia, and Wright (2004) draw an analogy with getting to know a person; as one gets to see individuals in new circumstances, our knowledge of those people change, and for us, they are experienced differently from the first meeting. Our knowledge is richer, deeper, and more generative. We can predict future behaviors across wide varieties of circumstances, relate those individuals to other people we know, and so on. It is in this sense that the process of symbolization, particularly in the service of generalization in the development of algebraic reasoning, can deepen, enrich, and make coherent the experience of elementary school mathematics.

Errors in Terms of the Framework

There are two broad classes of errors. One involves mistakes in using the rules of a symbol system within a symbol system, as when one mishandles a carrying across zero in the standard multidigit addition algorithm or makes erroneous manipulations of algebraic expressions (Matz, 1982).

---

[16]For an amusing quotation by William Frend in his 1796 algebra textbook bashing the idea of negative numbers and roots of negative numbers, see Kenner (1958), reprinted in Swetz (1994).

The other involves cross-system interpretation errors, as when a correct action is performed in one system and is then misinterpreted in another (Wenger, 1987). The latter is most likely to occur in the context of modeling nonmathematical situations where the interpretation in either direction is subject to error (Booth, 1984; Clement, 1982; Kaput & Sims-Knight, 1983a, 1983b). The table blocks symbol-referent breakdown described by E. Smith (chap. 5, this volume) and mentioned earlier is of this type.

The Actor Versus Observer Perspectives: Fusion Versus Transparency

The type of notation system that is involved in symbol use is likely to be a factor in the experience of fusion, or the extent to which the symbolizer experiences the notation as something separate from what is being represented. In particular, a key difference may result from the difference between analogue and iconic notations on the one hand and character-based notations on the other (Goodman, 1968/1972). In the former case, the notation embodies physical features that can be mapped directly onto its presumed referent, especially when discussed in a language that applies both to the situation/phenomenon and the symbols used to describe it. A coordinate graph whose height may be increasing over an interval being used to describe some phenomenon, which is taken to have some quantitative aspect that is likewise increasing is an example of this duality phenomenon. In this and many cases (Roth, 2003), the features of the symbol and those of its referent are experienced and discussed univocally, as essentially the same thing by sophisticated users such as scientists.

This is a different kind of notation–referent relationship and is more iconic than that which occurs when the notation involves an arbitrary visual shape as with alphanumeric characters (e.g., algebraic expressions), where the relation between the configuration of the notation and the phenomenon is far less direct and in need of much more interpretation (i.e., symbolic as mentioned earlier). Knowing for which values of the domain that a polynomial is increasing can involve some very fancy mathematics compared to knowing, based on visual inspection, where its graph is pointing upward or downward. It is our sense that the process of working through the behavior of a highly indirect symbol system, and hence living in it, focuses attention on the symbols and hence away from the represented phenomena. As the studies by Roth (2003) of scientists working to represent complex phenomena have shown, this would tend to make fusion harder but not impossible. An important feature of the cases Roth studies is the deep engagement of the scientists in the process of symbolizing the phenomena that they are studying; their representations of the phenomena and their understandings of those phenomena co-evolve, as was the case of the elementary school children studied by Lehrer and Pritchard (2002) building understanding of two-dimensional space, and Lehrer, Schauble, Carpenter, and Penner (2000) building understanding of various science concepts.

We feel that it is useful to have both the actor and the observer perspectives available when describing symbolization. We take the view that versions of this symbolization process need to be repeated all across mathematical schooling in order to build both a rich web of representational resources and the habits of mind that enable students to continue to extend and apply these resources.

So Which Symbolization Processes Are Algebraic and Which Are Quasi-Algebraic?

The previous chapter offered a content analysis of algebra: the two core aspects, expressing generalizations and operating on the resulting symbolizations, and three strands, generalizing from arithmetic and quantitative operations and their structures and properties, functions/variation, and modeling. This chapter describes how this content arises from a symbolization perspective. But symbolization occurs across all mathematical activity ranging from primitive counting and recording to the most esoteric advanced mathematical research. Operating on and reasoning with symbols, including common arithmetic, are likewise ubiquitous. So which symbolizations among the three strands deserve to be called *algebraic*? Whereas some arbitrariness in choices is unavoidable, we have a sufficient framework in place to offer a principled characterization of which kinds of symbolization activity are algebraic, which are quasi-algebraic (defined in the next section) and which are not algebraic.

Deliberate Generalization Is at the Core. We regard a symbolization activity as *algebraic* if it involves symbolization in the service of expressing generalizations or in the systematic reasoning with symbolized generalizations using conventional algebraic symbol systems (including more recent graphical and dynamic systems). Extending Fujii's (2003) idea of quasi-variables, we define an activity as *quasi-algebraic* if it satisfies the same conditions except that it may use any symbols, including traditional arithmetic ones, informal ones (including oral speech and physical manipulatives), or idiosyncratic ones. Thus, the use, in support of deliberate generalization of quasi-variables in this sense, as well as the algebraic use of numbers, both qualify as quasi-algebraic activity. Indeed, most of the classroom-based activities depicted in this book qualify as quasi-algebraic. Note that *algebraic use of numbers* refers to engaging students in reasoning with numerical statements that are being analyzed not for purposes of computation but for their structure, as when a teacher, attempting either to get across or diagnose an understanding of some general property about numbers, asks whether the sum of two odd numbers is even by asking about the sum of 327 and 459, or whether 327 + 459 is the same as 459 + 327, where the computation is beyond the current capacity of her students.

In contrast, manipulation of conventional algebraic symbols apart from acts of generalization (e.g., multiplying polynomials for the sake of practice) does not qualify as algebraic activity according to our characterization, despite the fact that it may be of educational value—in the same way that playing the scales does not qualify as musical activity, despite the fact that it may be necessary in order to be musical later. Also, arithmetic problem solving in pursuit of one-answer arithmetic solutions is neither algebraic nor quasi-algebraic. On the other hand, doing a sequence of arithmetic computations in pursuit of a pattern or generalization is quasi-algebraic.

Hence, we see that use of conventional symbol systems plays an important role in distinguishing algebraic from quasi-algebraic symbolizing activity. However, our characterization involves a more subtle matter concerning the mental activity of the symbolizer, one that may make the distinction difficult to determine in certain cases. For example, consider a student building a traditional algebraic expression or equation while solving a one-answer word problem. Is the student engaged in an activity that we would term *algebraic*? We suggest that this depends on whether the student is writing the expression or equation deliberately to express a general statement about a range of possibilities generated by substituting specific values for the variables in the expression or equation or, alternatively, simply repeating a rehearsed procedure that leads to an answer for the given kind of problem. In the former case, we would regard the activity as algebraic and, in the latter, lacking the crucial ingredient of the intent to express generality, we would deem it as nonalgebraic activity.

The use of conventional symbol systems is a necessary condition for an activity to be algebraic, but it is certainly not sufficient. And, in the case of quasi-algebraic activity, the diagnosis may be even more subtle because, for instance, a student may be attempting to state the general through the use of a specific case, as Mason points out, and where the intent might be evident only in the student's vocal intonation. The key here is not to be able to diagnose every student activity to determine which category it falls within, but rather to outline some guidelines that help distinguish between those kinds of tasks and those kinds of responses to tasks that are on the path to algebraic reasoning from those that are not.

Quasi-Algebraic Reasoning Across the Three Strands of School Algebra. In our content analysis of the three strands of school algebra in chapter 1 (this volume), the role of generalization is most salient in the first strand, which centers on building generalizations from arithmetic and quantitative reasoning. We wish to emphasize that this strand is meant to encompass higher order algebraic symbolizations—generalizations on systems of symbols. This includes abstract algebra. But in its quasi-algebraic forms, it could involve the properties of clock arithmetic, for example.

In the case of the Joint Variation and Functions strand, again, quasi-algebraic activity is marked by building generalizations of specific quantitative relationships across a wide variety of notations, including physical concrete materials such as sequences of blocks or colored rods. Good examples can be found in chapters 5, 10, and 11 (this volume), particularly in the analyses by E. Smith (chap. 5, this volume). In effect, the move from quasi-algebraic to algebraic activity is determined by the use of conventional notation systems for the representation of variables and functional relationships (both character-string and graphical), where the generalization that renders it quasi-algebraic or algebraic is a statement that embodies a joint-variation relationship.

In the Modeling strand, the situation is a bit more complex because certain kinds of modeling situations force the use of some sort of (conventional) algebraic approach, whereas others are more sensitive to the background of the problem solver and might readily be the subject of quasi-algebraic approaches. In particular, explorations, especially systematic ones, using sequences of arithmetic statements usually mark a quasi-algebraic approach. See Bednarz and Janvier (1996) for a discussion of this issue framed in terms of the difference between arithmetic and algebraic problems.

Much more could be said regarding the details in each strand, including the interactions between the use of conventional notation systems and the thinking that is involved in each strand. For example, see Usiskin (1988), where the different uses of notations for variables are outlined. However, this chapter is intended to introduce rather than exhaustively examine symbolization in algebra.

REFERENCES

Bednarz, N., & Janvier, B. (1996). Emergence and development of algebra as a problem-solving tool: Continuities and discontinuities with arithmetic. In N. Bednarz, C. Kieran, & L. Lee (Eds.), *Approaches to algebra: Perspectives for research and teaching* (pp. 115–136). Dordrecht, the Netherlands: Kluwer Academic.

Bezuszka, S., & Kenney, M. (2004). That ubiquitous sum: 1 + 2 + 3 + ... + *n*. *The Mathematics Teacher, 98*(5), 316–321.

Blanton, M., & Kaput, J. (2003). Developing elementary teachers' algebra eyes and ears. *Teaching Children Mathematics, 10*(2), 70–77.

Blanton, M., & Kaput, J. (2004). Design principles for instructional contexts that support students' transition from arithmetic to algebraic reasoning: Elements of task and culture. In R. Nemirovsky, B. Warren, A. Rosebery, & J. Solomon (Eds.), *Everyday matters in science and mathematics.* (pp. 211–234) Mahwah, NJ: Lawrence Erlbaum Associates.

Bochner, S. (1966). *The role of mathematics in the rise of science.* Princeton, NJ: Princeton University Press.

Booth, L. R. (1984). *Algebra: Children's strategies and errors.* Windsor, England: NFER-Nelson.

Boyer, C. (1959). *The history of calculus and its historical development.* New York: Dover.

Boyer, C., & Merzbach, U. (1989). *A history of mathematics* (2nd ed.). New York: Wiley.

Bruner, J. (1973). *Beyond the information given.* New York: Norton.

Carpenter, T., Franke, M. L., & Levi, L. (2003). *Thinking mathematically: Integrating arithmetic and algebra in elementary school.* Portsmouth, NH: Heinemann.

Carpenter, T. P., & Lehrer, R. (1999). Teaching and learning mathematics with understanding. In E. Fennema & T. Romberg (Eds.), *Classrooms that promote understanding* (pp. 19–32). Mahwah, NJ: Lawrence Erlbaum Associates.

Carraher, D. W., Schliemann, A. D., & Brizuela, B. (2001). Can young students operate on unknowns? In M. V. D. Heuvel-Panhuizen (Ed.), *Proceedings of the 25th conference of the International Group for the Psychology of Mathematics Education?* (Vol. 1, pp. 130–140). Utrecht, the Netherlands: Freudenthal Institute, Utrecht University.

Clement, J. (1982). Algebra word problem solutions: Thought processes underlying a common misconception. *Journal for Research in Mathematics Education, 13*(1), 16–30.

Clement, J., Lochhead, J., & Monk, G. (1981). Translation difficulties in learning mathematics. *American Mathematical Monthly, 88*, 286–290.

Cobb, P. (1994). Where is the mind? Constructivist and sociocultural perspectives on mathematical development. *Educational Researcher, 23*(7), 13–20.

Cobb, P. (2000). Conducting teaching experiments in collaboration with teachers. In A. Kelly & R. Lesh (Eds.), *Research design in mathematics and science education* (pp. 307–334). Mahwah, NJ: Lawrence Erlbaum Associates.

Cobb, P., & Bowers, J. (1999). Cognitive and situated learning perspectives in theory and practice. *Educational Researcher, 28*(2), 4–15.

Cobb, P., Gravemeijer, K., Yackel, E., McClain, K., & Whitenack, J. (1997). Mathematizing and symbolizing: The emergence of chains of signification in one first-grade classroom. In D. Kirshner & J. A. Whitson (Eds.), *Situated cognition theory: Social, semiotic, and neurological perspectives* (pp. 151–233). Hillsdale, NJ: Lawrence Erlbaum Associates.

Deacon, T. (1998). *The symbolic species: The co-evolution of language and the brain.* New York: Norton.

Donald, M. (1991). *Origins of the modern mind: Three stages in the evolution of culture and cognition.* Cambridge, MA: Harvard University Press.

Drouhard, J.-P., & Teppo, A. R. (2004). Symbols and language. In K. Stacey, H. Chick, & M. Kendal (Eds.), *The future of teaching and learning of algebra: The 12th ICMI study* (chap. 9). Dordrecht, the Netherlands: Kluwer Academic.

Frege, G. (1984). The thought: A logical investigation. In B. McGuinness (Ed.), *Collected papers on mathematics, logic and philosophy* (pp. 58–77). Oxford, England: Blackwell.

Frye, N. (1987). The symbol as a medium of exchange. In J. Leith (Ed.), *Symbols in life and art* (pp. 3–16). Montreal, Quebec, Canada: McGill-Queen's University Press.

Fujii, T. (2003). Probing students' understanding of variables through cognitive conflict problems: Is the concept of a variable so difficult for students to understand? In N. Pateman, B. Dougherty, & J. Zilliox (Eds.), *Proceedings of the 27th conference of the International Group for the Psychology of Mathematics Education* (Vol. 1, pp. 49–66). Honolulu, HI: University of Hawaii.

Fujii, T., & Stephens, M. (2001). Fostering an understanding of algebraic generalisation through numerical expressions: The role of quasi-variables. In H. Chick, K. Stacey, J. Vincent, & J. Vincent (Eds.), *The future of the teaching and learning of algebra* (Proceedings of the 12th ICMI study conference, pp. 258–264). Melbourne, Australia: University of Melbourne.

Goodman, N. (1968/1972). *Languages of art: An approach to a theory of symbols.* Indianapolis, IN: Bobbs-Merrill.

Gravemeijer, K., Cobb, P., Bowers, J., & Whitenack, J. (2000). Symbolizing, modeling, and instructional design. In P. Cobb, E. Yackel, & K. McClain (Eds.), *Symbolizing and communicating in mathematics classrooms* (pp. 225–274). Mahwah, NJ: Lawrence Erlbaum Associates.

Hart, K. (1984). *Children's understanding of mathematics: 11–16.* London: John Murray.

Hiebert, J. (1986). Conceptual and procedural knowledge in mathematics: An introductory analysis. In J. Hiebert (Ed.), *Conceptual and procedural knowledge: The case of mathematics* (pp. 1–27). Hillsdale, NJ: Lawrence Erlbaum Associates.

Kaput, J., & Blanton, M. (2005). Algebrafying the elementary mathematics experience in a teacher-centered, systemic way. In T. A. Romberg, T. P. Carpenter, & F. Dremock (Eds.), *Understanding mathematics and science matters.* (pp. 99–125) Mahwah, NJ: Lawrence Erlbaum Associates.

Kaput, J., & Shaffer, D. (2002). On the development of human representational competence from an evolutionary point of view: From episodic to virtual culture. In K. Gravemeijer, R. Lehrer, B. van Oers, & L. Verschaffel (Eds.), *Symbolizing, modeling and tool use in mathematics education* (pp. 277–293). London: Kluwer Academic.

Kaput, J., & Sims-Knight, J. E. (1983a). Errors in translations to algebraic equations: Roots and implications. *Focus on Learning Problems in Mathematics, 5*(3), 63–78.

Kaput, J., & Sims-Knight, J. E. (1983b). Misconceptions of algebraic symbols: Representations and component processes. In H. Helm & J. D. Novak (Eds.), *Proceedings of the international seminar: Misconceptions in science and mathematics.* (pp. 63–78) Ithaca, NY: Cornell University.

Kieran, C. (1981). Concepts associated with the equality symbol. *Educational Studies in Mathematics, 12,* 317–326.

Kieran, C. (1992). The learning and teaching of school algebra. In D. A. Grouws (Ed.), *The handbook of research on mathematics teaching and learning* (pp. 390–419). New York: Macmillan.

Kirshner, D., & Awtry, T. (2004). Visual salience of algebraic transformations. *Journal for Research in Mathematics Education, 35*(4), 224–257.

Kline, M. (1972). *Mathematical thought from ancient to modern times.* New York: Oxford University Press.

Lehrer, R., & Pritchard, C. (2002). Symbolizing space into being. In K. Gravemeijer, R. Lehrer, B.van Oers, & L. Verschaffel (Eds.), *Symbolizing, modeling and tool use in mathematics education* (pp. 59–86). London: Kluwer Academic.

Lehrer, R., Schauble, L., Carpenter, S., & Penner, D. (2000). The interrelated development of inscriptions and conceptual understanding. In P. Cobb, E. Yackel, & K. McClain (Eds.), *Symbolizing and communicating in mathematics classrooms* (pp. 325–360). Mahwah, NJ: Lawrence Erlbaum Associates.

Lesh, R., & Doerr, H. (2003). Foundations of a models and modeling perspective. In R. Lesh & H. Doerr (Eds.), *Beyond constructivism: A models and modeling perspective on mathematics teaching, learning, and problem solving* (pp. 3–33). Mahwah, NJ: Lawrence Erlbaum Associates.

Lesh, R., Landau, M., & Hamilton, E. (1983). Conceptual models in applied mathematical problem solving research. In R. Lesh & M. Landau (Eds.), *Acquisition of mathematics concepts and processes* (pp. 263–343). New York: Academic Press.

Livneh, D., & Linchevski, L. (1999). Structure sense: The relationship between algebraic and numerical contexts. *Educational Studies in Mathematics, 40,* 173–196.

MacGregor, M., & Stacey, K. (1997). Students' understanding of algebraic notation: 11–15. *Educational Studies in Mathematics, 33,* 1–19.

Mahoney, M. (1980). The beginnings of algebraic thought in the seventeenth century. In S. Gankroger (Ed.), *Descartes: Philosophy, mathematics and physics.* (141–155) Sussex, England: Harvester.

Matz, M. (1982). Towards a process model for high school algebra errors. In D. H. Sleeman & J. S. Brown (Eds.), *Intelligent tutoring systems* (pp. 25–50). New York: Academic Press.

McGowen, M., & Davis, G. (2001). Changing pre-service elementary teachers' attitudes to algebra. In H. Chick, K. Stacey, J. Vincent, & J. Vincent (Eds.), *The future of the teaching and learning of algebra* (Proceedings of the 12th ICMI study conference). Melbourne, Australia: University of Melbourne.

McGuinness, B. (Ed.). (1984). *Collected papers on mathematics, logic, and philosophy.* Oxford, England: Blackwell

Moreno, A. L., & Waldegg, G. (2000). An epistemological history of number and variation. In V. Katz (Ed.), *Using history to teach mathematics: An international perspective* (pp. 183–190). Washington, DC: Mathematical Association of America.

Nemirovsky, R. (1994). On ways of symbolizing: The case of Laura and the velocity sign. *Journal of Mathematical Behavior, 13,* 389–422.

Nemirovsky, R., & Monk, S. (2000). "If you look at it the other way ... " An exploration into the nature of symbolizing. In P. Cobb, E. Yackel, & K. McClain (Eds.), *Symbolizing and communicating in mathematics classrooms* (pp. 177–221). Mahwah, NJ: Lawrence Erlbaum Associates.

Noble, T., Nemirovsky, R., Dimattia, C., & Wright, T. (2004). Learning to see: Making sense of the mathematics of change in the middle school. *International Journal of Computers for Mathematics Learning, 9*(2), 109–167.

Noss, R., & Hoyles, C. (1996). *Windows on mathematical meaning: Learning cultures and computers*. Dordrecht, the Netherlands: Kluwer Academic.

Peirce, C. S. (1955). *The philosophical writings of Peirce* (J. Buchler, Ed.). New York: Dover.

Puig, L., & Rojano, T. (2004). The history of algebra in mathematics. In K. Stacey, H. Chick, & M. Kendal (Eds.), *The future of teaching and learning of algebra: The 12th ICMI study* (chap. 8) (pp. 189–224). Dordrecht, the Netherlands: Kluwer Academic.

Roth, W.-M. (2003). *Toward an anthropology of graphing*. Dordrecht, the Netherlands: Kluwer Academic.

Sfard, A. (2000). Symbolizing mathematical reality into being: Or how mathematical discourse and mathematical objects create each other. In P. Cobb, E. Yackel, & K. McClain (Eds.), *Symbolizing and communicating in mathematics classrooms* (pp. 37–98). Mahwah, NJ: Lawrence Erlbaum Associates.

Shaffer, D., & Kaput, J. (1999). Mathematics and virtual culture: An evolutionary perspective on technology and mathematics education. *Educational Studies in Mathematics, 37*, 97–119.

Silver, E. A. (1986). Using conceptual and procedural knowledge: A focus on relationships. In J. Hiebert (Ed.), *Conceptual and procedural knowledge: The case of mathematics* (pp. 181–198). Hillsdale, NJ: Lawrence Erlbaum Associates.

Swetz, F. (1987). *Capitalism and arithmetic: The new math of the 15th century*. Chicago, IL: La Salle.

Swetz, F. (1994). *Five fingers to infinity: A journey through the history of mathematics*. Chicago, IL: Open Court.

Usiskin, Z. (1988). Conceptions of school algebra and uses of variables. In A. Coxford (Ed.), *The ideas of algebra, K–12: 1988 yearbook* (pp. 8–19). Reston, VA: National Council of Teachers of Mathematics.

van Oers, B. (2000). The appropriation of mathematical symbols: A psycho-semiotic approach to mathematics learning. In P. Cobb, E. Yackel, & K. McClain (Eds.), *Symbolizing and communicating in mathematics classrooms* (pp. 133–176). Mahwah, NJ: Lawrence Erlbaum Associates.

Vergnaud, G. (1982). A classification of cognitive tasks and operations of thought involved in addition and subtraction problems. In T. P. Carpenter, J. M. Moser, & T. A. Romberg (Eds.), *Addition and subtraction: A cognitive perspective* (pp. 39–59). Hillsdale, NJ: Lawrence Erlbaum Associates.

Vygotsky, L. (1965). *Thought and language* (E. Hanfmann & G. Vakar, Trans.). Cambridge, MA: MIT Press.

Wenger, R. (1987). Some cognitive science perspectives concerning algebra and elementary functions. In A. Schoenfeld (Ed.), *Cognitive science and mathematics education* (pp. 217–252). Hillsdale, NJ: Lawrence Erlbaum Associates.

Werner, H., & Kaplan, B. (1962). *Symbol formation*. New York: Wiley.

Whitehead, A. N. (1929). *The aims of education and other essays*. New York: MacMillan.

Wigner, E. (1960). The unreasonable effectiveness of mathematics in natural sciences. *Communications in Pure and Applied Mathematics, 13*(1), 1–14.

Wilensky, U. (1991). Abstract meditations on the concrete and concrete implications for mathematics education. In I. Harel & S. Papert (Eds.), *Constructionism* (pp. 193–203). Norwood, NJ: Ablex.

3

Making Use of Children's Powers to Produce Algebraic Thinking

John Mason
The Open University

By the time children get to school, they have already displayed enormous powers for making sense of the worlds they inhabit: the material world of things, the mental world of images, the symbolic world of labels and words, and the social world of practices. They know what food to expect at different meals, they know about getting dressed, and they know about different rules of behavior applying in different situations. The 7-year-old who, after multiplying 45 by 8 in her head through repeated doubling, suddenly observes "Hey dad, every number is half another number; (pause) and every number and every number and a half is half of another number" (Sue Johnston-Wilder, personal communication, June 3, 2002); the 4-year-old who observes "if the canal floods we will get wet"; the 3-year-old who, sometime after being told he was in the outskirts of Paris, asked, "Are we in the out-trousers yet?" (Noticeboard, 2004); and the 2-year-old who responds to the sound of a car in the drive in the evening with "daddy!" beautifully display pattern-generated expectation, as does the child who is learning to walk, crawl, or roll over, or the newborn recognizing mother from smell and feel as well as sight, despite her different perfumes, clothes, and hairdo.

The claims made in this chapter are that the central problem of teaching is to get learners to make use of those powers and to develop them, and that algebraic thinking is what happens when those powers are used in the context of number and relationships. When textbook authors and teachers are tempted to do for learners what the learners could already do for themselves by using their own powers, they increase rather than decrease learner dependency. They actually contribute to the gradual atrophy of those powers, at least within the context of mathematics, and this in turn contributes to growing disaffection and disinterest in school mathematics. Nowhere is the need for those powers more evident than in the teaching of algebra, and, as is argued here, this means in arithmetic as well.

Algebra is and has been for a long time, the mathematical watershed for most adults, in the sense that it is when they meet algebra that they decide that mathematics is not for them and so cut themselves off from a vital part of their cultural heritage and their mathematical potential. Brandford (1908) observed that "the radical mistake of algebraical teaching for many generations was in passing by a jump from Particular Arithmetic to purely Symbolic Algebra, and thereby omitting a sufficient training in Generalized Arithmetic ... the simplest type of significant symbolic algebra" (p. 253).

When does algebraic thinking begin? A traditional answer based on current curricula would suggest that it begins in secondary school with courses labeled *algebra*. My answer, building on suggestions of Gattegno (1973, 1987), is that the powers necessary for algebraic thinking are being used by children as soon as they leave the womb, if not before. The child lying in his or her crib making sounds is not simply rehearsing the tonalities of phrases and sentences before using words. Babies are experiencing pausing and hence associating (the basis for associativity), ordering (the basis for commutativity), and perhaps even distributing ("ah" pause "mm bb" and "ah mm" pause "ah bb"). Instead of seeing arithmetic as the necessary prerequisite for algebra as generalized arithmetic, it is possible, and pedagogically effective, to see both as arising from the use of the same powers.

Put another way, to learn arithmetic beyond memorizing a counting poem and a few number-name facts, it is actually necessary to think (pre-)algebraically. Both algebra and arithmetic are natural outcomes of the application of human powers to counting and calculating. Algebraic thinking is required in order to make sense of arithmetic, rather than just performing arithmetic instrumentally. It is also necessary whenever some calculation is to be repeated many times, especially when there might be variations, such as when an entrepreneur is developing pricing policies.

To substantiate these claims, the chapter begins with a quick summary of some of the most important powers possessed by all children who can walk and talk. The use of these powers is briefly illustrated in the domain of arithmetic, before illustrating how algebra emerges from the use of these powers, ending with consideration of what those unaccustomed to making use of learners' powers might put forward as objections or obstacles to this perspective.

POWERS

Human beings make sense of their experience. The word *sense* is pertinent, because as philosophers have proposed and debated for centuries, the senses are the basis for our contact with the material world (Buchdal, 1961; Whitehead, 1911/1948). Although words connected with sight dominate English idioms ("I see what you mean," "a little foresight would have helped"), many people reject the frozen metaphor of "understanding is seeing." They prefer to use metaphors involving sound ("I hear what you are saying"), the visceral ("I don't feel you are listening to me."), and touch, both in its physical and its emotional sense ("I was touched by your saying . . .").

Sense is made by using natural powers to collect, classify, assimilate, accommodate, and even reject sensations, whether physical or imagined, remembered or constructed, literal or metaphoric. Thinking in terms of natural powers is certainly not new. According to Dewey (1897), "The child's own instincts and powers furnish the material and give the starting point for all education" (p. 77).

The following powers are some that are relevant to mathematics, and particularly to algebraic thinking.

Imagining and Expressing

Everyone has the power to imagine, where *imagine* and *imagery* are taken in the broadest sense, encompassing any or all of the senses. Recall some particular detail of a scene on a recent holiday, and really be there in your mind. You can do this even while reading or driving, and the vividness of your recall assists you in describing vividly to others. It also provides access to details that don't come immediately to mind at first. Imagery and imagination are the means by which people direct the energies produced by their emotions. For example, several of the 3,000-year-old Upanishads use the metaphor of the human psyche as a chariot with horses, driver, and owner (Rhadakrishnan, 1953): Mental imagery (the reins) is the means for the intellect (the driver) to control and direct the emotions (the horses) that pull (provide the motive power for) the chariot (the body).

Watching a baby in the cot moving and gurgling, or an older child playing house, it is clear that children have a rich world of imagination. Klein (Torretti, 1978) described the faculty of forming images as naive intuition" proposing that physical experience develops into mechanical (material) intuition and optical experience develops into projective, geometric intuition. To imagine what is not physically present, and to dwell in that world while remaining physically present in the material world is a fundamental power, which is crucial to the development of mathematical thinking. It is, after all, the basis for modeling: moving from the material (or other world) through the imagined to a symbolic world, resolving the problem there, then reinterpreting results via the world of imagery back into the original world (Open University, 1978).

Charles Saunders Peirce found it difficult to decide between mathematics as "the science of necessary conclusions" (quoting his father Benjamin Peirce, 1870) and as "the study of the hypothetical states of things" (Peirce, 1902/1956, p. 1778). By augmenting the imagination, mathematizing these images, and deducing necessary conclusions, he emphasised how valuable it is to try out conjectures imaginatively or symbolically through mathematics, rather than performing dangerous or catastrophic experiments in the material world. Young children are adept at using their power to imagine as part of modeling, using old boxes and cartons to stand in for cars and trains, caves and tents, secret chests, and make-believe presents.

Young children love to have someone read to them. Being read to is an important way of receiving adult attention, and it has the effect of stimulating powers of imagination. In the safety of adult presence, children can let themselves go into a world of fantasy. Later, they will be called on to express the products of their imagination in speech, sound, pictures, movement, and writing, as well as in mathematical symbols. Learning to control mental imagery is an important precursor to all forms of thinking, whether in images, diagrams, or symbols. It is not enough to acknowledge that learners possess these and similar powers: The power to imagine and to express needs to be invoked in the classroom, frequently and effectively, by encouraging students to express perceived generalities, relationships, connections, properties, and so on.

Everyone also has the power to express themselves, by which is meant to express something of their mental world. The urge to express is clear: Children running, jumping, and shouting is just one manifestation. Different people prefer different media but, as many people are currently pointing out (Bruner, 1991, 1996; Chamberlin, 2003; Donald, 1991), human beings are narrative animals: They have a deep-seated need to tell (portray, display, act out) stories that account for their experiences and their history, and a strong need to recount these to others as the basis for social interaction. Not all young children express themselves coherently at first (Freudenthal, 1991). To learn to tell stories, to learn to write essays, they need to become explicitly aware of the fact that other people's inner worlds are different from their own, and that the contents of their own inner world are not directly accessible to others. They need to learn to describe explicitly what they are seeing, hearing, feeling, or touching to someone who does not share their world. Consciousness emerges through learning to tell stories about oneself.

Focusing and De-Focusing

Very young children somehow learn to discern details in what they are seeing and hearing. By focusing their attention, they actually create objects out of their sensations through stressing some features as foreground, and ignoring others as background. This power enables them to distinguish between different objects, and to detect relationships between objects or between features of objects. They could not learn language without this discernment, nor could they distinguish food from non-food, or parents from themselves. They teach themselves how to focus their attention on some detail, to shift from one detail to another, and also to defocus in order to take in a larger whole. For example, at the end of this sentence, pause and become aware of the individual alphabetic characters, perhaps comparing and contrasting the *m* and the *w*; then de-focus until you are aware of the page as a whole and the flow of white spaces down the page between the words. This power is possessed by everyone (see Kaput, Blanton, & Moreno, chap. 2, this volume, who describe it in terms of *attentional focus*).

Speaking of powers possessed by children, Gattegno (1970) suggested that "a child brings the ability to notice differences and to assimilate similarities" (pp. 25–26). If people did not constantly discern some details while ignoring others, then experience would either be flat and dull through being undifferentiated, or intensely and even unbearably sharp and particular through being distinguished in every detail, as C. S. Lewis (1945) suggested in *The Great Divorce: A Dream*. It is not sufficient to discern detail, however. Children naturally detect relationships between details, and they shift their attention from specific elements to relationships between elements, and from relationships between specific details to relationships as properties that objects may or may not have. Pierre and Dina van Hiele used different language for essentially the same observation (van Hiele, 1986), although they tend to treat discerning, relating, and property making as levels of understanding and types of reasoning rather than as moment-by-moment shifts in the structure of human attention. Pirie and Kieren (1989) use similar ideas expressed differently in their onion-layer model of understanding.

The issue for teaching lies in getting learners to discern details and recognize relations that the teacher knows are fruitful to discern and to recognize. Shifting to seeing relationships as possible properties of objects is nontrivial but often expected and assumed in mathematics classrooms; going further and declaring a definition or assuming a property as an axiom from which to make deductions is another important but nontrivial shift. For example, recognizing that subtracting 1 from 7 and from 16 gives a multiple of 3 in both cases is one thing; identifying "subtracting one and getting a multiple of 3" as a property is quite another; focusing attention only on numbers that are 1 more than a multiple of 3 and deducing facts about all of them involves a further significant shift in how one attends, and to what.

It has been suggested that one of the major contributions made by Isaac Newton is the dual perspective of, on the one hand, using axioms from which to deduce explanations of natural phenomena as in his *Principia*, while at the same time basing his optics deductions on experimental evidence (Buchdahl, 1961).

Learning to ignore is also an important part of learning to focus, but not easy to work on explicitly. The learner desperately wanting to work with a particular number rather than with an as-yet-unspecified number needs to let go of the particular. This comes with time and experience, and with having attention drawn away from particulars. Similarly, the child fixated by the miraculous act of making a mark with a pencil cannot attend to drawing "some thing" and, if pestered by adults asking "What is it?", is likely to lose the opportunity to focus on mark making. The young child entranced by building towers out of rods and blocks intended for number work is unlikely to appreciate another "use," and so may simply not hear or make sense of what the teacher is trying to demonstrate or get the child to do. The adult stuck in a pattern of using a mnemonic to remember some formula has attention diverted to the mnemonic that could be more usefully directed toward the problem at hand. These all illustrate the hazards of attention focused on one thing while the teacher wants the learner's attention on something else. They demonstrate the inappropriate direction of attention and use of powers by the teacher. Each fixation can play an important role for a time, but later has to be subordinated or transcended to enable more sophisticated development.

Specializing and Generalizing

Just to make sense of language, children have to generalize because language is essentially general, as Vygotsky (1965) and others suggested:

> At any age, a concept embodied in a word represents an act of generalization. But word meanings evolve. When a new word has been learned by the child, its development is barely starting; the word at first is a generalization of

> the most primitive type; as the child's intellect develops, it is replaced by generalizations of a higher and higher type—a process that leads in the end to the formation of true concepts. ... (p. 83)

Nouns such as *cup* and *chair*, which are used in a wide variety of situations and for a variety of objects depending on the circumstances, are inherently general. To specify a particular cup or chair, it is necessary to modify with adjectives, or to point physically. Numbers, such as 3 or –3.7 seem concrete to an adult, but they too are abstracted generalizations. Counting numbers arise from stressing the count and ignoring what is being counted, generalizing across contexts. Negatives and decimals are generalized extensions of counting in order to extend arithmetic operations such as subtraction and division so that all possible questions using those operations can be answered (or, as with division by zero, are outlawed).

Focusing on one aspect is a form of stressing. Stressing some feature necessarily implies an ignoring of other details or aspects and, as Gattegno (1970) pointed out so forcefully, this is the basis of generalization and abstraction, which are in turn the basis for language: "Without stressing and ignoring, we cannot see anything. We could not operate at all. And what is stressing and ignoring if not abstraction? We come with this power and use it all the time" (p. 11).

Generalizing also lies at the very heart of mathematics: "Another characteristic of mathematical thought is that it can have no success where it cannot generalize" (Peirce, 1902/1956, p. 1778). But, hand in hand with generalizing goes the opposite, particularizing, or "specializing," as Polya (1957, 1962) referred to it: "We need to adopt the inductive attitude [which] requires a ready ascent from observations to generalizations, and a ready descent from the highest generalizations to the most concrete observations" (1957, p. 7).

For example, to make sense of a statement such as "the sum of an even number of consecutive odd numbers is divisible by double the number of numbers," it is necessary to specialize to some confidence-inspiring objects: These may be particular numbers, or they may be generalities expressed in symbols. But, the purpose of the specializing is to make sense, to enable a re-construction of the general, expressed in a more familiar language and a more manipulable symbolism.

When children discern a particular object, they have to "ascend," to "see through the particular to the general" in order to be able to speak about it. Doing this explicitly as part of an overt practice is part of what it means to think mathematically. As Whitehead (1911/1948) observed, "To see what is general in what is particular and what is permanent in what is transitory is the aim of scientific thought" (p. 4). For example, from the fact that 5 + 7 and 9 + 11 are both divisible by 4, and that 5 + 7 + 9 + 11 is divisible by 8, and choosing to see these facts as generic rather than particular, gives rise to conjectured generalizations that can then be tested and generalized.

When children hear someone else speaking, they have to be able to "see the particular in the general"—that is, to specialize to particular instances. But then that is what everyone has to do in order to make sense. The sentences you are reading are necessarily general, being cast in language, and you are probably trying to instantiate them from your experience, trying to locate specific or particular examples and counterexamples. This too is a natural power every child possesses. Exercising a little imagination at the same time leads to seeking extreme or special cases, which might challenge or test a general assertion in a particular case, leading to a conjecture or counterexample. Looking for counterexamples is a practice, which develops with encouragement, although parents may not always appreciate this behavior at home. Looking intentionally for examples and counterexamples are practices that are vital to mathematical thinking and need constant encouragement so that learners can refine their techniques.

We know that young children learn from particular instances through generalizing; what does not seem to be very common is to call on those powers explicitly when we try to instruct young people. For example, despite young children readily contradicting an assertion such as that "all cats are striped" by referring to a counterexample (O'Connor, 1998), this power is rarely seen being used in school, presumably because of social pressures and implicit contracts. Rather, as teachers we often try to do the work for them: We provide particular cases, we display methods, and we provide worked examples. We then expect them to generalize, yet rarely do we explicitly and intentionally prompt them to use their powers to generalize, nor display that power being used. As Branford (1908) put it: "Beware, my fellow teacher, lest you unconsciously and incautiously supply the children with a generalization, which they have not as yet of themselves perceived and reached ... such a blunder I find myself repeatedly committing, so easy a trap is it to fall into" (p. 43). The issue, then, is not whether learners can generalize, but what teachers can do to stimulate them to use and develop that power effectively and appropriately.

Conjecturing and Convincing

Piaget (1971) promulgated the notion that children are essentially explorers, constructing meaning from their experiences. They do this by making conjectures and modifying these as and when they run into counterexamples. Lakatos (1976) analyzed mathematicians going through similar processes historically. Of course, none of this is overt for the young child, but the challenge for teachers is to create an atmosphere in which this natural power is called on and developed, because it is essential to the proper functioning of mathematical thinking, in general, and algebraic thinking, in particular.

Vygotsky promoted the view that higher psychological processes are acquired through participation in and exposure to practices of relative experts who display those processes. For example, exposure to others who overtly simplify in order to understand and then re-complicate, who specialize in order to re-generalize, who construct examples and nonexamples to illustrate and exemplify, is much more likely to influence learners than lack of such exposure. In other words, it helps if teachers are mathematical with, and in front of, the people they teach. This includes displaying their own use of the powers that learners are expected to employ, not to "try to do the work for them," but exposing learners to mathematical practices.

An important feature of each classroom is the atmosphere and ethos arising from the ways of working valued and promoted by the teacher. Things that are said and done that seem to work all too readily become assertions and habits. In ordinary life, these turn into the stories that people tell themselves in order to justify their actions. Mathematics is more rigorous about its stories: They have to convince other people. Mathematicians work best in a conjecturing atmosphere in which conjectures are articulated in order to try them out, see how they sound and feel, to test them and so to see how to modify them as and when necessary. Mathematicians try to convince colleagues, who try to find counterexamples to their assertions. Everyone recognizes that things are said in order to "get them out," to use the attempt to articulate in order to stop ideas and possibilities from tumbling around like clothes in a drier, getting more and more tangled. Those who disagree offer counterexamples and suggestions for modifications, rather than acting in ways that could be taken as criticism of the person making the conjecture. Mathematical thinking is deepened by periods of individual work to specialize and generalize, to rehearse expressions of what is being imagined, and by periods of collective negotiation and search for ways to modify utterances so as to block counterexamples.

This is the sort of atmosphere in which mathematics thrives, and it can be established in any classroom at any age. Legrand (1993) describes similar practices under the heading of *scientific debate* (also summarized in Mason, 2001a). There is an important feature of conjecturing that is perhaps not as strongly developed in most of us as it might be: Having articulated a conjecture, it is wise not to believe it (Polya, 1957).

Classifying and Characterizing

When was the last time you sorted or organized a collection of objects?: putting dishes away, putting laundry away, dealing with mail, weeding the garden, organizing papers. What did you have to do in order to achieve the sorting? Did you get a sense of pleasure when it was finished? We sort and organize all the time, and although it may sometimes be seen as a chore to start with, there is often a sense of satisfaction when things have been sorted. Each act of sorting involves stressing some (relevant) features and ignoring others, which in turn requires us to be able to discriminate those features.

Sorting and organizing objects is typical of challenging tasks that young children seem to enjoy. For example, sorting beads by color and by size, sorting attribute blocks by color, by size, by thickness, by color and size, by color and thickness, and so on, occupy young children for hours. Such activity is extremely valuable because it exercises and develops fine-motor control at the same time as leading them to discriminate according to particular attributes while ignoring others, and experiencing generalizations associated with properties. For example, Freudenthal reports Bastiaan at age 4 years, 2 months playing with rings of plastic on a piece of wire, then pulling a bottle cap out of his pocket and saying "now I must do something with this"; Freudenthal (1978) suggests making a hole, and suddenly Bastiaan says, "a screw-nut has a hole where the screw fits in" and pulls a nut from his pocket to put on the wire (pp. 194–195): A connection was made, as was a generalization. Young children often display a desire to organize their own environment. For example, two young girls were observed sorting and organizing the books in a bookshelf, by size, which of course included books of a series being together. The desire to impose order is manifested very early. Every experience is classified (unconsciously) in order to assimilate it into current schema and so "make sense of it." If it resists classification, then it is either rejected out of hand, or schemas are altered in order to accommodate it (Piaget, 1971).

For example, number names are quickly classified into even and odd according to the last digit, but seeing that this is the same as being divisible by two requires some explicit work; recognizing and naming polygonal shapes requires discerning the presence of vertices and straight sides, recognizing the relationship between the number of vertices and the number of edges, and using this number to characterize the shape. In fact, most shapes children are shown in mathematics, at first, are in fact regular. It is no wonder, therefore, that they develop what Fischbein (1993) called *figural concepts* based on perceived properties that may not be what experts discern and stress.

It is logical that children will not sort collections of objects on the basis of features that they are not distinguishing. It is also reasonable that tasks, which involve sorting according to characteristics that are not at the forefront of their attention, may prompt them to discriminate other features (although they may not immediately respond as expected). This is a good example of the kind of sensitivity that directs effective teaching: noticing what learners distinguish and what they do not, and challenging them appropriately to go beyond what they can currently do with ease. By being aware of the powers children are using, it is possible to expose explicitly the criteria that are deemed important, and to work against learners making inappropriate or undesired generalizations.

For example, when children first start playing the game of 20 questions, they find it hard not to ask very specific, specialized questions. If told it's an animal, they ask immediately if it is a specific animal such as a dog, and then proceed to try other animals. However, I have never encountered any child asking if it was a specific dog, such as their own pet, which suggests that by the time 20 questions seems a reasonable game, children have already encountered some degree of stressing of some features while ignoring others and employ specializing without being aware of it. By participating in the game with older children, they discover that there are more general questions like "does it have four legs?" that cut down the possibilities. In this way, they are enculturated into ways of classifying. Very young children sometimes find it difficult to accept a familiar class (crows) as a subclass of another (birds), just as learners find it confusing that a square can also be a rectangle, a triangle is a special case of a trapezium, and an integer can also be seen as a fraction. Although the power is available, it sometimes has to be brought to the surface and learners must be allowed to experience its use before they integrate it into their thinking.

Children as Powerful Agents

The fact that every child has displayed these powers does not mean that they are automatically thinking mathematically. Rather, in order to be thinking mathematically, these powers need to be exercised, developed, and for the most part brought to the surface so that they can be used intentionally. The issue is not whether these powers are available, but that people sometimes fail to make use of them. One reason for this is that these powers may have been downplayed or ignored in school. Where teachers are induced to downplay reasons, to bypass convincing learners by appealing to their powers to reveal and express mathematical structure, where teachers and texts proceed by worked examples and repetitious but unstructured, unprincipled practice, and where time is not devoted to reconstructive reflection on the underlying structure of exercises, children's powers and energies are not just being wasted but abused.

One of the telling features of the use (and abuse) of children's powers is that young children, immersed in the use of their powers, are not put off by only partially comprehending what is going on around them. They make what sense they can, and they work away at it. The very young child lying in the crib can be heard making sounds that rehearse the tone patterns of sentences, before they have words. They are immersed in language, but they are not disheartened by not understanding.

Once children have been to school for a time, they begin to show petulance and disinterest when things get too hard. Why is it that not understanding becomes an issue when children are in school, but not understanding is of no concern when they are working for themselves with their own powers? For example, children practicing skateboarding or gymnastics will display perseverance that is never witnessed in classrooms. Could it be that in the classroom the use of their powers has been usurped, that they have begun to be enculturated into a climate of non-use of many of their own powers, in order to do just what the teacher wants?

How, then, can children's natural powers be harnessed to enrich the emergence of algebraic thinking?

ARITHMETICAL THINKING AS PRE-ALGEBRA

Looked at from the point of view of the use of powers, even arithmetic, commonly seen as the ground on which algebra is based and from which algebra is derived, cannot actually be grasped fully without algebraic thin-king, that is, without the use of the powers associated with generalizing. "In order to do arithmetic you already need to do algebra!" (Hewitt, 1998, p. 19).

In the following examples, which are just a few of many others, attention is directed to instances in which learners' powers are used spontaneously, and that can be exploited to support increasingly sophisticated algebraic thinking.

Rhythm and Sequence

Young children are sometimes invited to thread beads on a string, or to place objects in a line or in a tower. One covert or inner aim (Mason, 1992; Tahta, 1980, 1981) of such tasks is to put the children in a situation in which they can work at refining and integrating their fine-motor control. The aim is not to get beads threaded, but to subordinate the actions so that they no longer need to pay so much attention to getting the needle into the hole, or the blocks stacked so they don't fall over. Because expert functioning requires little or no attention paid to details of the action, to become expert requires removing attention from performing that action. Consequently, being given lots of practice in which full attention can be given to performing the action is more likely to develop a habit of requiring full attention than actually helping the person integrate through subordination of attention (Gattegno, 1987). It makes more sense, therefore, to engage learners in a macro, or outer, task, which attracts attention to a larger aim (e.g., trying to make a sequence) while at the same time the learner is practicing the action on specific examples, preferably exercising their own choice and powers to create relevant examples for themselves rather than working through someone else's script (Hewitt, 1996).

In the case of beads and towers, a task might be to design a repetitive sequence as simple as:

blue, red, blue, red, blue, red, ...

Or, it may be as complex as:

blue, red, blue, blue, red, blue, blue, blue, red, ...

This can be used as the basis for clapping instructions (hard on blue, soft on red, etc.).

Of course, it is much more interesting to make up your own sequences, thereby exposing the dimensions of possible variation and the corresponding ranges of permissible change of which you are aware (Marton & Booth, 1997; Mason & Johnston-Wilder, 2004; Watson & Mason, 2004). The point is to make it sufficiently challenging so that attention is drawn away from the physical doing and into what cognition is good at, which is monitoring and directing.

As Johnson (1987) has pointed out, the fundamental means for making sense of abstract ideas is the metaphor of position or location in space, which we experience through the senses. The groundwork for appreciating and employing such metaphors lies in early connections, which children make between the physical-visceral and the cognitive-mental. For example, poems with strong rhymes and rhythms are popular with young children. The rhythm appeals to their body-based pattern recognation and generalization. At the same time, they are making links between body and mind, between enaction and cognition, which is a fundamental form of abstraction. They develop the physical coordination to clap a rhythm of hard and soft claps (e.g., hard soft soft, hard soft soft, ...), replacing these with claps and non-claps. Furthermore, extending a sequence beyond any diagrammatic score from which to read the clapping, such as the aforementioned blue and red sequences, encourages learners to move inward and make use of their power to imagine beyond what is physically present. They can also count out loud at the same time, and then use a number chart to select and name the numbers corresponding to hard claps. There are many different ways to carry out such activity, with the aim of viscerally relating regular number counts to rhythm.

More challenging is making the count silent, clapping loudly on some numbers and quietly on the others. In the United Kingdom, Fizz Buzz is a game with various rules, such as people count in turn; if your number is divisible by 3, you say "fizz" as many times as 3 divides your number; and, if it is divisible by 5, you say "buzz" as many times as 5 divides your number. Simpler versions have everyone counting out loud but becoming silent if there is a digit 3 or 5 in the numeral. There is great delight when most are silent, but a few say the number out loud inappropriately. Whereas the outer task is simply to engage in the game, the inner task is to anticipate and think about the number before it is said, thus calling on the powers of imagining and focusing.

Dealing With Not Knowing

At first, having two counts (e.g., 3 cows in one field and 5 cows in another) is a state of not knowing how many animals there are altogether in both fields, or how many there would be if they were all put into one field. Notice how mental imagery is ever present. The cows in fields could, of course, be beans in bags, and many other contexts; ultimately, learners are expected to recognize that there is a method that involves attending to the numbers and ignoring the objects. Notice too that moving to "3 cows and 5 horses makes how many animals?" requires a shift in what you are attending to: You have to let go of the particular animals, and re-see them just as animals, which then permits a focus on the numbers.

Much has been made by researchers of learners' propensity for getting an answer no matter what the problem asks, such as in the well-studied *l'âge du capitaine* (Baruk, 1985; Freudenthal, 1991; Merseth, 1993). It is typical of a wide class of situations in which learners display use of a method they have been taught implicitly, namely, ignore the objects and just add (deal with) the numbers. They are being very mathematical and show that they learned: It's just that they have learned too general a method. One of the many mistakes in early algebra ($3x + 5y = 8$ or $8xy$) (Booth, 1988; Kücheman, 1981) could be accounted for in the same way. Similarly, the classic error 0.3×0.3 = "oh point nine" based on 0.4×0.4 = "oh point sixteen" and 0.5×0.5 = "oh point twenty-five" is a brilliant conjecture based on inappropriate data!

It is common to find "hidden answer" tasks in textbooks, of the form

$$\square + 3 = 8 \text{ and } 5 + \_ = 8.$$

These are presumably intended by authors to expose learners to algebraic ideas but are not always seen by children as anything to do with unknowns but rather as arcane arithmetic tasks (Carpenter & Fennema, 1992; Franke, Carpenter, & Battey, chap. 13, this volume). An alternative situation that is closer to children's experience is to publicly count a number of objects, cover them with a cloth, and ask the children if they know how many objects are under the cloth (Floyd, Burton, James, & Mason, 1981). There is an immediate opportunity to work on the difference between not knowing when you look because they are covered up, and knowing even though they are out of sight (mental imagery). Contrasts can be made with covering up an uncounted number of objects. (Note connections with Piagetian preservation tasks; see Donaldson, 1978.) Then someone reaches under the cloth and extracts some objects, which are seen and counted. How many objects remain? Now the answer is not known (at first), but perhaps there is a method for finding out just from what is currently known (without peeking). Notice an instance here of moving to working with properties.

Box equations are but a drop in the ocean of potential for thinking algebraically within arithmetic, so concentration on them as a dominant style of task almost certainly ensures that learners learn to do them mechanically, arithmetically, avoiding or circumventing any algebraic thinking, any sense of the box as an as-yet-unknown to be discovered.

For example, it is common to set sequences of tasks such as $4 + 6 = ?$; $3 + 7 = ?$; $2 + 8 = ?$, $1 + 9 = ?$ These are known as number bonds to 10, and it is a curricular aim to have children integrate and automate these pairs. But, there is a much more important generalization that learners are expected to pick up later when doing subtractions. Stated in full generality, it would be $(x + d) + (y - d) = x + y$. There are of course intermediate generalizations achieved by specializing $x + y$ to a particular invariant value (especially 10 as the base of our number naming system) and then varying d (as with the previous examples), or specializing d and varying x and y (as in mental arithmetic methods). (See Carpenter, Franke, & Levi, 2001, for an exploration of this case; and Schifter, Monk, Russell, & Bastable, chap. 16, this volume, for others.) The important pedagogical questions ask whether these generalizations are left implicit, whether learners are called on to use their own powers to make these various generalizations for themselves but are supported and encouraged to do so, or whether the generalizations are imposed on the learners before they have come to them themselves.

There is an unfortunate clash in most curricula, because learners are immersed for some time in a culture in which not knowing answers (to arithmetic calculations) is treated negatively, and then suddenly introduced to algebra in which not knowing is treated positively as an opportunity to use symbols, as a way of working with not knowing. In the 19th century, Mary Boole suggested that algebra arose when someone

> perhaps a woman, said "How stupid we've been! We have been dealing logically with all the facts we knew about this problem, except the most important fact of all, the fact of our own ignorance. Let us include that among the facts we have to be logical about, and see where we get to then ... let us agree to call it *x*, and let us always remember that *x* stands for the Unknown." (Tahta, 1972, p. 55)

Acknowledging ignorance, not knowing, makes it possible to denote it, and then manipulate it as if it were known. No wonder children who have been trained in dependency and in getting single numerical answers find algebra confusing!

Methods as Implicit Generalization

No one teaching arithmetic would expect learners to memorize all possible additions of two two- or three-digit numbers, or the products of all pairs of two-digit numbers. The whole point of arithmetic is to learn methods for determining them, as Eliot (1855/1990) observes in an essay about Thomas Carlyle: "It has been well said that the highest aim in education is analogous to the highest aim in mathematics, namely, not to obtain *results* but *powers*, not particular solutions, but the means by which endless solutions may be wrought" (p. 343).

Unfortunately, mathematics is not always seen or taught in this way. De Morgan (1865), writing at about the same time, railed against teaching by rote:

> Mathematics is becoming too much of a machinery; and this is more especially the case with reference to the elementary students. They put the data of the problems into a mill and expect the result to come out ready ground at the other end. An operation which bears a close resemblance to that of putting in hemp seed at one end of a machine and taking out ruffled shirts ready for use at the other end. This mode is undoubtedly exceedingly effective in producing results, but it is certainly not soaked in teaching the mind and in exercising thought. (p. 2)

If arithmetic is about learning methods, as well as some elementary facts, it is curious that learners are subjected to particular calculations all the time. Of course, saying in words how you add two three-digit numbers, or do two- more or digit multiplications is extraordinarily difficult: To see this for yourself, try developing a spreadsheet to do these things with the constraint that a cell may only ever hold a single digit! So we embark on getting learners to learn methods that can't even be spoken completely, much less written down. We do this by subjecting learners to many particular cases, with explanations, while at the same time overlooking a multitude of other opportunities to encourage children to use those same powers overtly. The fact that so many children do as well as they do is a testament to their use of their powers without explicit intention.

Whenever a learner solves a problem, there is available the question "What is the method that was used?", which is intimately tied up with the question "What can be changed about the problem and still the same technique or method will work?" or "What is the class of problems which can be solved similarly?" S. Brown and M. Walter (1981) suggest asking "What if ... something changed?" or "What if not ... ?" Watson and Mason advocate explicitly asking learners to consider what *dimensions of possible variation* and corresponding *ranges of permissible change* they are aware of (Mason & Johnston-Wilder, 2004; Watson & Mason, 2004) as stimulus to becoming aware of, and even expressing features of, the general class of problem of which the ones considered are representative. This is essentially the same as what Blanton and Kaput (chap. 14, this volume; in press) refer to as *algebrafying arithmetic tasks*. Learners aware of general classes are much more likely to recognize the type of a problem on a test. Learners who have constructed and solved their own problems of a given type—perhaps responding to "construct a problem of this type which is easy," "... which is hard in some way," "... which would challenge a friend or the teacher," "... which would be good for learners next year to work on," "... which demonstrates that you know how to do problems of this type"—are again much more likely to be able to deal with examination questions, for they have exercised creativity and choice, as well as powers of focussing and de-focusing, generalizing, and so on.

Tracking Arithmetic

Children are often exposed to arithmetic as the study of producing a single number answer to sometimes quite complex number situations. But arithmetic, as has been suggested already, is much more. Instead of completing calculations, tracking one or more numbers through a calculation displays the role played by the number—what some refer to as the *algebraic use of numbers* (see Zaskis, 2001): "As one example, consider the ever-popular Think of a Number games: Think Of A Number; add 2; multiply by 3; add 4; add twice the number you first thought of; divide by 5; subtract 2; you now have the number you first thought of." There are, of course, many variants (see also Goldenberg & Shteingold, chap. 17, this volume). But adolescents and even younger students seem to get great pleasure from the apparent mystery of how the prediction can be made. To find out how, it is useless to try an example by doing all the calculations because then the generality is obscured. It is vital to resist the temptation to do calculations. Thus, starting with 7 (the least popular number between 1 and 10) gives

$$7; 7 + 2; (7 + 2) \times 3 = 7 \times 3 + 6; 7 \times 3 + 6 + 4; 7 \times 3 + 10 + 7 \times 2 = 7 \times 5 + 10;$$
$$7 + 2; 7$$

The mystery is solved because you can look through the 7, seeing it as a placeholder, a slot, a variable into which any other number could be dropped. Learners are experiencing that something is invariant (the structure of the arithmetic computation) despite the fact that the actual number used is allowed to change. This theme of invariance in the midst of change pervades mathematics. Indeed, much of algebra at school and beyond can be seen as the study of invariance and invariants.

Zazkis (2001) has found that using very big and unwieldy numbers can assist learners who are reluctant to let go of particulars and to contemplate general or as-yet-unknown numbers. By replacing the 7 in the Think of a Number by something enormous, such as 987654321, learners' attention is diverted away from the particular and onto the method of the calculation.

If, in a problem situation, you can check whether a proposed answer is correct, then in principle you can set out the constraints imposed by the problem situation in algebra. All you have to do is to track the arithmetic of a proposed solution, then treat it as a slot, substitute a letter, and see what happens, for example:

> A Father at his Death left his three Sons all his Money in this manner: to the Eldest he gave half of it, wanting 44 Pounds; to the Second he gave one third of it, and 14 Pounds more; to the Youngest he gave the Remainder, which was less than the Share of the second Son by 82 Pounds; What was each Son's Share? (Ward, 1713, Question 31, p. 224)

To check a proposed solution of, say 100 pounds, the eldest received *100*/2 – 44, the second received *100*/3 + 14, and the youngest received *100*/3 + 14 – 82. These three must then add up to the total of *100*. Replacing the proposed solution by a letter f leads to $(f/2 - 44) + (f/3 + 14) + (f/3 + 14 - 82) = f$. So $7f/6 - 98 = f$, so f was $6 \times 98/7 = 84$ Pounds. Learners with an eye on general methods for a class of similar problems might track all the numbers to find that:

$$f(1/2 + 1/3 + 1/3) + 2 \times 14 - 82 = f$$

$$f = (82 - 2 \times 14)/(1 - 1/2 - 1/3 - 1/3)$$

Now any of the initial data can be changed and the bulk of the thinking is done: All that remains is some arithmetic.

Of course, students of arithmetic are working problems to practice their arithmetic. Their attention is probably fully caught up in the complexities of adding fractions, subtracting from 1, and dividing. They may find that holding all the arithmetic without compacting the numbers by doing calculations is a strain on their attention. But, once they can suspend the desire to do computations, they have what is tantamount to a formula for a whole class of problems of the same type, by treating each datum as something, which could be varied.

Newton, writing in 1669, thought it a simple matter to learn to set up equations, even before knowing how to solve them:

> After the novice has exercised himself some little while in algebraic computation, . . . I judge it not unfitting that he test his intellectual powers in reducing easier problems to an equation, even though perhaps he may not yet have attained their resolution. Indeed, when he is moderately well versed in this subject and conceives he has some degree of skill in the art of eliciting from the circumstances of a question as many equations as suffice to implement fully all conditions and knows how to reduce all those equations (should there be several) to a final one which satisfies the question, then will he with greater profit and enjoyment contemplate the nature and properties of equations and learn their algebraic, geometrical and arithmetical resolutions.
>
> But when some problem has been proposed, the practitioner's skill is particularly demanded when it comes to designating all its conditions by an equal number of equations. To do this, let him in the first place examine whether the propositions or phrases by which it is enunciated are all fit to be denoted in algebraic terms in the same way that we express our concepts in Greek or Latin characters. Should this happen (as usually is the case in questions which relate to numbers or abstract quantities) let him then set names on the unknowns and also on the quantities if need be, denoting the sense of the question algebraically, and its circumstances thus translated into algebraic terms will yield as many equations as suffice to resolve it. (Whiteside, 1972, pp. 428–429)

Newton's near contemporary Ward (1713) realized that it was not always quite so easy:

> When any Problem or Question is proposed to be Analytically Resolved; it is very requisite that the true design or meaning thereof be fully and clearly comprehended (in all its parts) that so it may be truly abstracted from such ambiguous words as Questions of this type are often disguised with; otherwise it will be very difficult, if not impossible to state the Question right in its various substituted Letters, and ever to bring to an Æquation, by such various Methods of ordering those Letters as the Nature of the question may require.

> Now the knowledge of this difficult part of the work, is only to be obtained by Practice, and a careful minding of the Solution of such Leading Questions as are in themselves very easie. And for that Reason I have inserted a collection of several Questions; wherein there is a great variety. (pp. 175–176)

Mathematicians turned their attention instead to the solution of equations, and modern algebra began. That is perhaps why, since then, word problems slink out of most algebra texts and migrate to arithmetic and school algebra texts.

Implicit Awareness of Arithmetic Structure

Children's awarenesses of how numbers can be manipulated are often intuitive and implicit, making use of theorems in action (Vergnaud, 1981), rather than having explicit and conscious awareness of rules. It is in bringing these awarenesses—these functionings—to the surface and becoming aware of them that algebra emerges. For example, the young child who when asked "3 + 5?" then "5 + 3?" announces after a pause that "anything plus anything is anything plus anything" is not talking nonsense. That child is struggling to control the essential generality of language to express an awareness about how numbers work. For 3 + 5 is equal to 5 + 3 NOT because they are both equal to 8, but because adding is based on the physical action of combining, and combining does not depend on order. Algebraic thinking is the thinking that results in general statements about number (in school) and about structure (in university). Put another way, when objects are labeled and the labels are manipulated, you have algebra (see Kaput et al., chap. 2, this volume), or as Tahta (1989) put it, "The geometry that can be told is not geometry [but algebra]."

Children behave with numbers as if they know that addition and multiplication are order independent. They usually operate as if subtraction and division do depend on order, although in the midst of a three-digit subtraction they may be tempted to take smaller from larger as part of the method or algorithm that they have constructed from past experiences (S. Brown & van Lehn, 1980). Children also act as if multiplication distributes over addition in some circumstances. They use these rules as theorems in action, particularly when stimulated to discuss and exchange different ways of doing arithmetic mentally (Anghileri, 2000, 2001).

When teachers decide to introduce negative numbers, then fractions, and then decimals, they are recapitulating developments, which took years, even centuries to be adopted fluently and without cavil. Instead of telling learners rules, or trying to provide pseudocontexts in which the rules are evident, it is possible to get learners to determine what the rules should be by appealing to their sense of structure and awareness of how simpler numbers work. The guiding principle is that whatever is decided must extend what has gone before. An excellent way to do this is to run patterns backward:

$$2 \times 4 = 8 \qquad 2 \times 3 = 6 \qquad 2 \times 2 = 4 \qquad 2 \times 1 = 2 \qquad 2 \times 0 = ?$$

$$2 \times (?) = ?$$

Something as simple as this can be worked on by looking for what is changing and how it is changing at each stage, and demanding continuation. One or two experiences like this are unlikely to be sufficient. What will be convincing is over and over meeting the same rule arising from extending different patterns. This idea is developed further later under the label of Tunja Sequences.

ALGEBRAIC THINKING

As Bednarz, Kieran, and Lee (1996), among others, point out, arithmetic proceeds from the known to the unknown. At each step, something fresh and previously unknown is calculated. Algebraic thinking allows people to solve much more complicated problems, because it starts with acknowledging ignorance of the unknown, as Mary Boole put it so poetically, denoting that ignorance, and then doing calculations with it as if it were known. Thus, the unknown is found from the known by means of the unknown. The psychological impact and significance of this switch of tactic is not always acknowledged in traditional presentations of algebra.

It is common to think in terms of *algebra* as the use of letters to stand for numbers and even to use the phrase *generalized arithmetic* for this clerical activity. Indeed, there is a long history going back to the 16th century of introductory algebra books treating algebra in this way. I am proposing that it is much more fruitful to think of generalized arithmetic as meaning the result of learners generalizing their experience with numbers, and expressing generality about properties of numbers arising from a variety of situations in the material, mental, and symbolic worlds, culminating in expressing generality about the rules of arithmetic and then taking these to be the rules for algebraic manipulation. Thus, algebra is most usefully seen not as symbol manipulation (or worse, study of the 24th letter of the alphabet), not as arithmetic with letters, not even as the language of equations, but as a succinct and manipulable language in which to express generality and constraints on that generality.

Proceeding to Algebra Without Going Via Arithmetic

Davydov (1972), working within a Vygotskian framework, argues that very young children can work with relationships without going via quantities at all. His ideas are being developed in a longitudinal study by Dougherty and colleagues (Dougherty, 2003; chap. 15, this volume). See also Freudenthal (1974, 1978) and J. Smith and Thompson (chap. 4, this volume). Children are introduced to qualitative comparisons of area, length, and mass before they have numbers with which to quantify. They demonstrate authority and control of symbolic notation for comparison ($A > B$) and for manipulating these in order to express intuitive, perceptual connections. Although the symbols are necessarily labels for objects rather than quantified measures, there is a noticeable development from label-for-object to label-for-measure-of-quality, even though these are not translated into numbers at the beginning. Davydov and colleagues also showed that young learners could tackle word problems with letters rather than numbers: Sometimes the presence of particular numbers attracts thinking into particulars rather than into structures (Freudenthal, 1974, 1978).

One thing to be learned from this work is that young children possess the power to think and articulate abstractions and generalities independent of numbers. Therefore, waiting to introduce symbols until arithmetic has been mastered may fail to make use of children's powers when they are available. There is strong resonance with Gattegno's view, for he saw algebra as stressing operations (combining, distributing, dividing, permuting) and consequently ignoring the objects (numbers), which are being treated in generalized ways and denoted by letters. This led him to his definition of algebra as the "study of the dynamics of the mind." Generalized arithmetic is then only part of algebra, for there are other relationships between objects, which can usefully be expressed and generalized, as the Davydov-inspired work has shown.

One consequence of distinguishing between arithmetic and algebra in relation to problem solving is that people sometimes try to classify problems as either arithmetic (can be done using only arithmetic) or algebraic (requires the use of symbols and their manipulation). However, it is very difficult to find a problem, which cannot be done by arithmetic if you are clever enough. For example, consider the following two problems:

> Cistern: One tap fills a cistern in 1 hour, another in 2 hours; how long does it take when the taps are used together?
> Couriers: One courier sets out from city *A* to take a message to city *B* a specified distance apart, traveling at a specified speed; some specified time later, a courier sets out from city *B* to city *A* traveling at a different specified speed to the first. When and where do they meet?

Freudenthal (1991) describes how children faced with the cistern-filling problem looked at the proportion of water contributed by each tap and quickly calculated the answer. Of course, using algebra can reveal internal structure, for, as he goes on to point out, people rarely recognize the courier problems as having the same structure as cistern problems, at least until they solve both problems in a general form.

What matters is not whether a problem can or cannot be done in one way or another, but whether learners have the flexibility and confidence to choose to use arithmetic or algebra according to the situation. It is the thinking that is algebraic, or not, not the task.

One of the consequences of failing to make use of children's powers over many generations is that those who have not succeeded with algebra question its value, thinking that for them, arithmetic is sufficient. After all, they were successful without algebra, so why impose it on everyone? Reflection on the presence of arithmetic and algebra in everyday life soon reveals that customers want numbers: They want the price they are to pay. By contrast, entrepreneurs need policies, which are essentially algebraic formulae or at least general procedures for working out prices to be charged, taxes, and so on. Sometimes they get them a bit wrong: In supermarkets in the United Kingdom, the following is a remarkably common phenomenon: "Hot Cross Buns sell in packets of 6 for 90p. A special deal offers you a second packet for half price. On a particular day packets approaching their sell-by date are reduced to 40p each. How much do I pay for two of these packets?" (Alan Parr, e-mail February 11, 2003)

The answer turns out to be 35, because the computer is programmed to subtract 45p for the second one, so it becomes cheaper to buy two and throw one away than to buy just one. This illustrates just how important it is for the entrepreneur to think about generality so as to get their formulae, their algebra, correct. It affords an opportunity in class to generalize by looking for dimensions of possible variation such as treating the individual numbers as variables, and by working out the price to be charged for 3, 4, . . . n packets.

Algebra should not be taught because some people who become entrepreneurs will need to think that way. Algebra should be taught to everyone because it is the natural outcome of the use and development of the relevant powers. Put another way, no one should be confined to arithmetic calculations as their basic numeracy because to do so stunts or even blocks their access to the kind of thinking that is essential for participation in a democratic society: recognition and use of general methods, testing and challenging of generalizations, and questioning the modeling assumptions on which such generalizations are often made.

From what has been said so far, it is clear that the position being advocated is that algebraic thinking is required for success in arithmetic, so the two are intertwined. Some difficulties with arithmetic may even arise because algebraic thinking, and especially, learners' own powers, are not being activated. Similarly, some difficulties learners have with algebra may be due to the absence of use of the powers and particularly of algebraic thinking when learning arithmetic.

MAKING USE OF POWERS TO PRODUCE ALGEBRA

The claim has been made that algebra, like arithmetic, emerges from the use of children's natural powers. There is room here to give only a few examples of how this can happen, but other chapters in this book can be read through the lens of learners' powers. See also South Notts Project (undated 1970s) and Malara and Navarra (2003).

Awareness of and Expressing Generality

Whenever a teacher becomes aware of an implicit generality in some particular instances, there is an opportunity to make a choice: to pause for a moment and prompt learners to try to express that generality before continuing, or to keep going. Whenever the teachers catch themselves uttering a generality, or hear a learner expressing a generality, there is an opportunity to make a choice: to pause for a moment and prompt learners to specialize (construct a particular example) or to keep going. As Malara and Navarra (2003) have demonstrated in their ArA1 project, learners often begin by babbling in a new language, whether it is the language they hear around them or the symbolic language of generalization in mathematics. Indeed, if they are not immersed in the language, and finding reasons to want to use it in order to express themselves, then it is no wonder if they pick up at best a few fragments. There is no need to rush learners from pictures, which can be read generically, to single-letter expressions. Abbreviating arises perfectly naturally when learners find themselves wanting to write down complicated expressions repeatedly.

Picture-pattern sequences (Mason, 1988b; Mason, Graham, Pimm, & Gowar, 1985; South Notts, n.d.) provide just one context for generalizing. For example, for each of these sequences (Fig. 3.1), learners can be invited to specify a method of describing how to draw more and more members of the same sequence. They then use that method to justify a way to count how many segments and how many corners there will be in some picture much farther along the sequence. Both simpler and more complicated sequences can be generated by learners, so they can challenge themselves and each other.

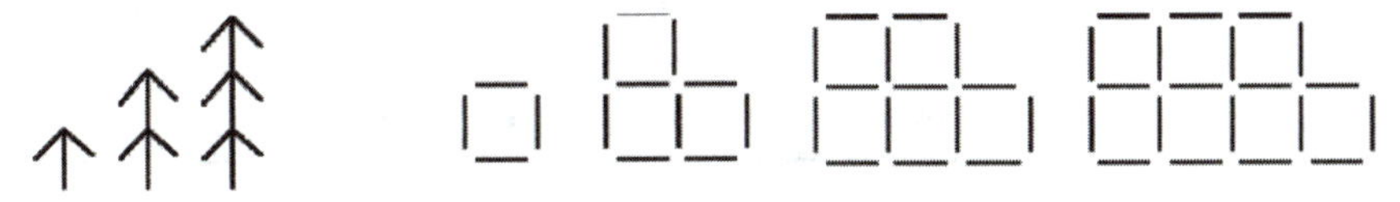

FIGURE 3.1 Picture-pattern sequences.

Number properties such as that the sum of three consecutive numbers is divisible by 3, that the product of two consecutive numbers is divisible by 2, and that one more than the product of two numbers differing by 2 is a perfect square can also be generalized extensively.

Sequences of number relationships, such as either of the following two columns, can be extended downward following the pattern emerging due to the counting numbers:

$$2 \times (1 + 3) = 2 \times 1 + 2 \times 3 \qquad (2 - 1) \times (2 + 2) = 2^2 + 2 - 2$$

$$3 \times (2 + 3) = 3 \times 2 + 3 \times 3 \qquad (3 - 1) \times (3 + 2) = 3^2 + 3 - 2$$

$$4 \times (3 + 3) = 4 \times 2 + 4 \times 3 \qquad (4 - 1) \times (4 + 2) = 4^2 + 4 - 2$$

The presence of terms from a familiar sequence, such as the counting numbers, makes it almost automatic to fill in the next and succeeding lines. The brain automatically detects both systematic change, and invariance, all manifested in "the almost visceral desire to continue." Watson (2000) described this as "going with the grain," making a useful contrast with the overt, mathematical sense making of "going across the grain," which means pausing to address what is the same and what is different about each statement (see also Freudenthal, 1991). In the first case, one observation is that multiplication "distributes over the addition of 3," inviting further generalization of the invariant 3s. In the second case, the expression $x^2 + x - 2$ always factors, so factoring a quadratic is seen as the expression of a generalization of a collection of arithmetic facts.

Thinking in terms of with versus across the grain adds substance to Polya's (1957) still overlooked phase of "looking back." Learners experience "going across the grain" when they interpret a generalization achieved by going with the grain. It is the going across the grain, the interpreting, that is the important part of the task, not merely the completion of the sequence!

Equation sequences can also be extended upward, revealing through the desire for the pattern to continue, the necessary properties of multiplication by negative numbers. In Mason (2001b), I called these Tunja

Table 3.1

Portion of Infinite Grid of Similar Cells

	$2 \times (1 + 3) = 2 \times 1 + 2 \times 3$	
$1 \times (2 + 3) = 1 \times 2 + 1 \times 3$	$2 \times (2 + 3) = 2 \times 2 + 2 \times 3$	
		$3 \times (3 + 3) = 3 \times 3 + 3 \times 3$

sequences, because teachers in the city of Tunja, Colombia, asked me how to teach factoring quadratics to learners who were struggling with negative numbers. Basic structural properties of arithmetic can be exploited in a similar manner long before factoring is encountered and reinforced while learning about factoring. Simple patterns, and more complex ones, can even be put into a grid so that two different dimensions can be varied at the same time (see Table 3.1). Learners can be invited to extend the grid downward and upward as well as right and left, justifying their proposals in terms of structure exhibited in patterns. Because learners sometimes experience considerable difficulty generalizing two things at once (Mason, 1996), using grids in this way offers support for and experience of multiple generalizations. These ideas are exploited in Flash (Mason, 2006).

The learners' rules for the arithmetic of negative numbers, fractions, and even decimals can be experienced and formulated by learners on the basis of the use of their own powers, thus helping them make the transition for mathematical warrants from authority to reason and logic.

From the simple but powerful awareness that adding something to one number and subtracting it from another leaves the sum invariant, or adding the same number to both leaves the difference invariant, to the more complex task of classifying all numbers that are both square and triangular, the same mathematical possibilities are present: specializing in order to make sense of a generality through reconstructing it for yourself, by discerning what is changing and what is invariant.

As learners become used to being expected to express generalities, and to treat such expressions as conjectures to be tested with particular and special cases, and then to attempt a justification, most will internalize the practice so that the teacher need only prompt occasionally. The outstanding conjecture that needs to be tested systematically is that such learners will be much more efficient and effective at learning mathematics, because they are behaving as natural mathematicians, because they are using their own natural powers. Anecdotal evidence and reason suggest that this conjecture is well founded.

Working With the As-Yet-Unknown

Whenever learners have solved a problem, it is possible to turn the question around and ask for other problems of the same type, which would give the same answer. This does more than illustrate the pervasive and creativity-demanding mathematical theme of "doing and undoing" (Mason, 1988a; SMP, 1984) or inverses (Groetsch, 1999; Melzak, 1983). It does more than engaging the natural power and desire to try to classify objects (all problems giving the same answer). It also engages learners immediately in thinking about as-yet-unknown values, similar to those they started with, which will also give the same answer.

For example, we know that 5 + 5 = 10, but what other pairs will give the same sum? Having discovered perhaps that 29 leaves a remainder of 2 on dividing by 3, what are all the numbers, which have the same property? Variations include what are all the numbers that when divided into 29 leave a remainder of 2? Or, put in a form more accessible to younger children, what numbers can you start with and count by threes to get to 29? A similar question is embedded in the analysis of the 31 game: How can you guarantee a win as the first player if players alternately choose numbers from 1 to 5, and the winner is the first to get the cumulative total to 31 (see e.g., Brousseau, 1997; Freudenthal, 1991). Trying particular cases while attending to the structure—to the choices available—is a productive way of approaching the task and its generalizations. Transforming arithmetic tasks in this way by converting a *doing* calculation into an *undoing* encourages learners to make choices and to act creatively. It provides them with experience of multiplicity in the as-yet-unknown, in contrast to tasks that seek single, arithmetically calculated correct answers.

Challenging learners to specify a method to be used by someone else (perhaps at the end of a telephone or e-mail) on data that they have not yet collected, uses learners' mental imagery to place them in a position of not knowing what numbers are to be used, and so seeking for some way to refer to numbers that are as-yet-unknown.

Freedom and Constraint

The mathematical notion of a variable arises naturally both in the context of expressing generality and in referring to as-yet-unknown numbers. These joint aspects of what mathematicians call variables are present at a very early age, and they are united by seeing situations in terms of freedom and constraint. I am thinking of a number (freedom is complete within the meaning intended of *number*); oh yes, and it is between 2 and 3 (a constraint is added); and its decimal name does not use the digits 5 or 9 (another constraint); how close can it be to 5/2? Here the sense of freedom diminishes with constraints. A simpler version is obtained by reversing (doing and undoing again) the traditional arithmetic question of 3 + 4 = ? to give 7 = ? + ?, where the question marks denote numbers to be filled in. Thinking for the moment in terms of positive whole numbers, how much freedom is there for your first choice? And when you have made that choice, how much freedom remains? Sometimes your first choice renders the second choice impossible. Altering the range of permissible change for numbers gives access to a different sense of freedom. Indeed, all schoolbook problems can be seen as examples of situations in which you start with complete, or at least very great freedom, and impose constraints. The problem is to construct an object, even all-possible objects, satisfying the constraints (Watson & Mason, 2004). Often it makes it easier to solve a problem if you impose the constraints one at a time. The 7 = ? + ? has many different dimensions of possible variation, including changing the number of question marks and changing the operation(s). Tasks like these provide learners with direct experience of working with the as-yet-unknown as an expression generality with constraint (Floyd et al., 1981).

Toward Manipulation: Multiple Expressions for the Same Thing

One of the delightful features of getting learners to express generalities, such as how many objects will be required to make a pattern or to satisfy some constraint, is that very often there are many different ways to express the same thing. For example, the number of chairs needed when tables that seat four people are put together in a long row can be expressed in several different ways (see Table 3.2).

Table 3.2

Different Ways to Express the Same Generality

2 + 2 × number of tables
number of tables + number of tables + 2
4 × number of tables – 2 × (number of tables – 1)
1 + 2 × number of tables + 1

Each expression asserts the same thing yet each looks different. Indeed, each displays a different way of seeing. Learning to see the situation through the structure of an expression strengthens the learners' sense of there being different ways to express the same thing, strengthens their experience of interpreting symbolic expressions, and also leads to the desire to manipulate expressions: Surely, there must be a way to see that these expressions always give the same answers, without having to go back to the tables and chairs! The desire to manipulate algebraic expressions can arise naturally and spontaneously as a result of learners using their powers. At the same time, of course, students are building their accompanying notion of equivalence of symbolic expressions.

Guessing and Testing

The child who can search for and find a glass or a fork when asked for one has demonstrated the power to specialize (particularize). The child who can offer a contrary example to an assertion has demonstrated the power to specialize with a purpose, which involves maintaining contact between the general and the particular. *Guess & Test* is not only good mathematics, but it can develop over time into more sophisticated versions: *Try & Improve,* in which the guess is modified according to some principle rather than being essentially random; *Spot & Check* when an answer is tried and found to be correct (it is often difficult just from marks on paper to distinguish between a lucky guess and an insightful spotting of the correct value); *Using Structure,* where the values tried are built up in some way using structural features of the problem (e.g., when given the perimeter of an isosceles triangle with sides twice the base, some evident use of 5 as a multiplier or divisor); and finally, denoting the as-yet-unknown by a manipulable label.

The methods of *false position* and *double false position,* which dominated the medieval mathematics curriculum in Europe, were devised very early on by Egyptian, Indian, and Chinese mathematicians to deal with problems involving proportions and linear relationships, respectively. The method was based on trying a convenient value and then using the incorrect answer to modify the trial value to obtain the correct answer. Isaac Newton then developed an iterative process to solve equations based on a form of *Try & Improve,* using fluxions to structure the improvement. Although not for a moment suggesting that this ontogeny would be usefully recapitulated by modern learners, it shows the depth and significance of developing *Guess & Test* into a successful general method that both spurred and used the development of algebra.

Summarizing

Getting learners to make use of their powers is not simply an approach to algebra or even an approach to mathematics. It is mathematics. It is the evocation of mathematical thinking. Algebra is seen as a powerful language in which to express relationships as generalities, enabling those relationships to be seen as properties and hence as the basis for deductions (see Kaput et al., chap. 2, this volume, for further amplification of this point). This includes and subsumes the use of algebra to solve word problems, indeed to solve any problems in which what is unknown is not readily deducible from what is known. By treating algebraic thinking as a natural consequence of the use of learners' powers, algebra can be released from its gatekeeper role as some kind of intelligence test and as an intellectual obstacle of adolescence. It is a natural birthright of all human beings.

POSSIBLE OBSTACLES

Any approach to teaching encounters obstacles, precisely because the teacher is only able to provide fodder on which learners can act, and prompts to direct learner attention. Most obstacles to learning algebra that emerge from research can be accounted for on the basis of failure to make use of expressing generality as the foundation stone for children's experience of arithmetic, much less algebra.

Stacey and MacGregor (2001) question the efficacy of an approach to algebra based on expressing generality in the context of picture patterns and tables of consecutive values of functions, for three reasons: first, because there is no research evidence that it is more, or even as, effective as introducing letters to stand for as-yet-unknown numbers; second, because research on learners' responses to pattern formulating tasks shows low facility; and third, because picture patterns stress relationships between terms and hence induce a recursive or inductive specification of rules (you multiply by two and add one for the next), rather than functional relationships (the nth term is ...). The difficulty with demanding research evidence in advance is that expressing generality is not a strategy to be used and then tested, but a holistic approach to mathematics. The whole approach needs to be tried properly and fully to test it as an approach. Testing learners on isolated items with which they have little familiarity, and certainly little breadth of experience, demonstrates only that learners do best on tasks that closely match their training. Stressing the expression of generality is working to educate awareness, not just train behavior. This takes time. It is not a substitute for some topic, but integral to every lesson. Indeed, a lesson without the opportunity for learners to generalize cannot be considered to be a mathematics lesson. Stacey and MacGregor do point to success with the use of spreadsheets for the use of symbols for unknown quantities (Rojano & Sutherland, 1993), and they make suggestions about improving a pattern-based approach, without perhaps going as far as is advocated here.

Evidence for the importance, value, and effectiveness of stressing and exploiting expressing generality from the earliest age is available, however, in the sense that those people who choose to exploit their mathematical powers show evidence of behaving in the ways being promoted here (Krutetskii, 1976). These are the learners who quickly pick up ways of thinking, ways of using their powers that constitute mathematical thinking. Most learners require more explicit encouragement and prompts to get them thinking algebraically in mathematics lessons.

Although it is often thought only *high attainers* can *do algebra*, there is mounting evidence to the contrary, including in several chapters of this book. Learners classified as *low attainers* can display the use of their powers to think mathematically when they are treated appropriately (Ahmed & Williams, 1992; Boaler, 1997; Ollerton & Watson, 2001; Watson, De Geest, & Prestage, 2004). It seems daft, therefore, to construct curricula along lines that are not compatible with the exploitation and use of children's natural powers. Charting routes that oversimplify for learners, break down topics into tiny discrete steps, present generalizations to them rather than provoking them into making them for themselves, and focus on training behavior without educating awareness (Gattegno, 1987), cannot be making effective or efficient use of learners' powers. Approaches, which provide plenty of particular examples and plenty of practice yet never prompt for generalization, and in which the examples provided are not structured so as to enable learners to discern relevant dimensions of possible variation, cannot be considered pedagogically effective. Each of these instructional moves attempts to reduce or remove the fundamental tension identified by Brousseau (1997) and expressed succinctly as: "The more clearly and particularly the teacher specifies the behaviour sought from learners, the easier and more likely it is that learners will produce that behaviour without actually generating it from themselves, and hence, without learning from the doing." This tension is endemic and inescapable. Teaching becomes ineffective when attempts are made to reduce it by going to extremes; flexibility and variation around the mean is much more effective.

Timing

There is a complex issue regarding expressing generality and timing. There does not appear to be time in an already overcrowded curriculum for learners to take the time to express and re-express, challenge, and modify conjectures. I take the opposite view: It is a singular waste of human energy and time not to promote learners' use and development of their natural powers. Worse, failure to do so produces disaffection and disruption by some, which makes learning difficult for others.

Furthermore, if learners develop the habit if using their own powers, then later teaching can be much more efficient. Learners can be expected to use those powers, and so will require much less time going over old topics and being trained in new techniques in new topics. They will be learning how to learn more efficiently and effectively.

If learners are always left to themselves to generalize from ill-structured tasks, then many will never work out that that is what they are supposed to do (which partly accounts for the present parlous state of algebra learning). If learners are offered situations and experiences, which are structured so as to suggest and invite generalization, then more learners may experience inner generalization, but may not integrate it into their functioning. If learners are periodically explicitly invited not only to generalize but also to express those generalities, and to negotiate the articulation of those generalities with colleagues as well as with a teacher, then there is a much greater chance that more learners will experience the pleasure of exercising their own power to generalize.

If learners are always given a set of exercises that starts simply and gradually becomes more complicated, then they are likely to form the impression that this is how mathematics is. If, instead, they are sometimes given complex or general statements to consider, to modify if necessary, and to convince themselves, then a friend, then a sceptic (see Mason, Burton, & Stacey, 1982) whether it is always true, they experience problem solving in its richest and most valuable form. If they are supported in simplifying through specializing, trying out particular cases not just to establish a pattern of numbers by going with the grain, but rather through which to become aware of structure that can then be articulated as a generalization by going across the grain, then they are likely to get pleasure not only from making choices as to which particular cases to consider (using their power to specialize), but also from appreciating the generalization they reconstruct and perhaps even extend.

If learners are only ever asked to work with what is present to them (physical objects, diagrams already drawn, symbolic expressions already formed), then they are likely to feel disempowered. If learners are invited to imagine what is not present, to discern details, to seek relationships, to identify properties, and most of all, go beyond what is present to them in the material world, then their powers of imagining will be used fruitfully rather than trivially.

If learners' attention is drawn to different dimensions of possible variation, then they are more likely to form a rich web of meaning for a concept. If learners are invited to manipulate familiar objects not as the core of a task, but as part of trying to see for themselves what is going on as they try to articulate for themselves a generality, then they can use and strengthen their mathematical powers and take algebra in their stride.

The notion of *scaffolding and fading* (S. Brown, Collins, & Duguid, 1989) is vital here. If learners have a steady and constant diet in terms of the format in which tasks are presented and developed by teachers, then they will become inured to and dependent on that format. As with all human beings, invariants soon disappear from attention unless there is some contrasting change going on. The picture on the wall is noticed for a few weeks and then tends to disappear. If, however, there is a varied diet, if sometimes the teacher starts from the general and invites and supports learners in simplifying in order to make sense and then regeneralize for themselves, if the teacher sometimes explicitly works on getting learners to express and negotiate conjectured generalities, but at other times makes—at best—indirect reference to those opportunities, and sometimes acts as if assuming that learners will do this for themselves, then learners are much more likely to take initiative and to internalize the strategies. They are also more likely to develop their powers in a positive ascending spiral of pleasure and self-confidence, instead of a descending spiral of don't, can't, won't as the opportunities for using their powers decrease.

CONCLUSIONS

People of all ages have the necessary powers to make sense of their experience and, especially, to make sense mathematically. The issue is not whether or not these powers exist, but how to make use of them by not obstructing or obscuring the issue, not trying to do the work for them so that they park their powers at the classroom door. Using your own powers is motivating; using your own powers creatively is exhilarating. Education is not about training learners to be dependent on experts, but about fostering independence and creativity. The principal interpersonal attribute required is trust: trusting children to use their powers and trusting the use of those powers to be effective and productive. This is not a recipe for hands-off discovery learning. On the contrary, it is extremely demanding on teachers because they have to be aware not only of their learners' current state, but of appropriate potential mathematical and sociocultural developments. This they do through enhancing their own awareness (Mason, 1998).

Algebra, in particular, is vital to all citizens. As Gattegno (1970) put it:

> ... how can we deny that children are already the masters of abstraction, specifically the algebra of classes, as soon as they use concepts, as soon as

> they use language, and that they of course bring this mastery and the algebra of classes with them when they come to school.
> ... the central point is this: the algebra is an attribute, a fundamental power, of the mind. Not of mathematics only.
> Without algebra we would be dead, or if we have survived so far, it is partly thanks to algebra—to our understanding of classes, transformations, and the rest. (pp. 24–25)

Why then does further mathematics prove to be so difficult for so many? My claim is that education systems militate against supporting teachers to get learners to make choices and to use their own powers. There is so much desire to control, to standardize, sometimes masked by equity rhetoric, but always making the same mistake: failing to support and encourage teachers to use their own powers and to make choices creatively in order to promote the same for their learners.

ACKNOWLEDGMENTS

I would like to express my appreciation and gratitude to Jim Kaput for his continuing enthusiasm for, and interest in, the expression of these ideas, and to Maria Blanton for her support, encouragement, and helpful suggestions, which improved the presentation and content of the chapter. The ideas expressed here were developed in a book *Developing Thinking in Algebra* for an Open University Course in 2005.

REFERENCES

Ahmed, A., & Williams, H. (1992). *Raising achievement in Mathematics Project report*. London: WSIHE/HMSO.

Anghileri, J. (2000). *Teaching number sense*. London: Continuum.

Anghileri, J. (2001). *Principles and practices in arithmetic teaching: Innovative approaches for the primary classroom*. Buckingham, England: Open University Press.

Baruk, S. (1985). The captain's age: On errors in mathematics. *L'âge du capitaine: De l'erreur en mathématiques*. Seuil: Seuil Point Sciences.

Bednarz, N., Kieran, C., & Lee, L. (Eds.). (1996). *Approaches to algebra: Perspectives for research and teaching*. Dordrecht, the Netherlands: Kluwer Academic.

Blanton, M., & Kaput, J. (in press). Design principles for instructional contexts that support students' transition from arithmetic to algebraic reasoning: Elements of task and culture. In R. Nemirovsky, B. Warren, A. Rosebery, & J. Solomon (Eds.), *Everyday matters in science and mathematics*. Mahwah, NJ: Lawrence Erlbaum Associates.

Boaler, J. (1997). *Experiencing school mathematics: Teaching styles, sex and setting*. Buckingham, England: Open University Press.

Booth, L. (1988). Children's difficulties in beginning algebra. In A. Coxford & A. Schulte (Eds.), *The ideas of algebra, K–12, NCTM 1988 yearbook* (pp. 20–32). Reston, VA: National Council for Teachers of Mathematics.

Branford, B. (1908). *A study of mathematical education*. Oxford, England: Oxford University Press.

Brousseau, G. (1997). *Theory of didactical situations in mathematics: Didactiques des mathématiques, 1970–1990* (N. Balacheff, M. Cooper, R. Sutherland, & V. Warfield, Trans.). Dordrecht, the Netherlands: Kluwer Academic.

Brown, J., & van Lehn, K. (1980). Repair theory: A generative theory of bugs in procedural skills. *Cognitive Science, 4*, 379–426.

Brown, S., Collins, A., & Duguid, P. (1989). Situated cognition and the culture of learning. *Educational Researcher, 18*(1), 32–41.

Brown, S. I., & Walter, M. I. (1982). *The art of problem posing*. Philadelphia: Franklin Institute Press.

Bruner, J. (1991). The narrative construction of reality. In M. Ammanti & D. Stern (Eds.), *Rappresentazioni e Narrazioni* (pp. 17–42). Roma-Bari: Laterze.

Bruner, J. (1996). *The culture of education*. Cambridge, MA: Harvard University Press.

Buchdahl, G. (1961). *The image of Newton and Locke in the age of reason* (Newman History and Philosophy of Science Series). London: Sheed & Ward.

Carpenter, T. P., & Fennema, E. (1992). Cognitively guided instruction: Building on the knowledge of students and teachers. In W. Secada (Ed.), *Curriculum reform: The case of mathematics education in the U.S.* [Special issue]. *International Journal of Educational Research, 17*(5), 457–470.

Carpenter, T. P., Franke, M. L., & Levi, L. (2001). *Thinking mathematically: Integrating arithmetic and algebra in elementary school*. Portsmouth, NH: Heinemann.

Chamberlin, J. (2003). *If this is your land, where are your stories?: Finding common ground*. Toronto, Canada: Knopff Canada.

Davydov, V. (1972). *Types of generalization in instruction: Logical and psychological problems in the structuring of school curricula* (Soviet Studies in Mathematics Education, Vol. 2). Reston, VA: National Council for Teachers of Mathematics.

de Morgan, A. (1865). A speech of Professor de Morgan, president, at the first meeting of the London Mathematical Society. In *Proceedings of the London Mathematical Society* (pp. 1–9). London: London Mathematical Society.

Dewey, J. (1897). My pedagogic creed by Dewey. *The School Journal, 54*(3), 77–80.

Donald, M. (1991). *Origins of the modern mind: Three stages in the evolution of culture and cognition*. Cambridge, MA: Harvard University Press.

Donaldson, M. (1978). *Children's minds*. New York: Norton.

Dougherty, B. (2003). Voyaging from theory to practice: Measuring up. In N. Pateman, B. Dougherty, & J. Zilliox (Eds.), *Proceedings of the 27th conference of the International Group for the Psychology of Mathematics Education held jointly with the 25th conference of the PME-NA* (Vol. 1, pp. 17–30). Honolulu, HI: University of Hawaii.

Eliot, G. (1990). Thomas Carlyle. In G. Eliot (Ed.), *Selected essays, poems and other writings* (pp. 343–348). Harmondsworth: Penguin. (Original work published 1855)

Fischbein, E. (1993). Figural concepts. *Educational Studies in Mathematics, 24*(2), 139–162.

Floyd, A., Burton, L., James, N., & Mason, J. (1981). *Developing mathematical thinking in Algebra* [Open University course]. Milton Keynes, England: Open University Press.
Freudenthal, H. (1974). Soviet research on teaching algebra at the lower grades of the elementary school. *Educational Studies in Mathematics, 5*, 391–412.
Freudenthal, H. (1978). *Weeding and sowing: Preface to a science of mathematical education.* Dordrecht, the Netherlands: Reidel.
Freudenthal, H. (1991). *Revisiting mathematics education.* Dordrecht, the Netherlands: Kluwer Academic.
Gattegno, C. (1970). *What we owe children: The subordination of teaching to learning.* London: Routledge & Kegan Paul.
Gattegno, C. (1973). *In the beginning there were no words: The universe of babies.* New York: Educational Solutions.
Gattegno, C. (1987). *The science of education: Part I. Theoretical considerations.* New York: Educational Solutions.
Groetsch, C. (1999). *Inverse problems: Activities for undergraduates.* Washington, DC: Mathematical Association of America.
Hewitt, D. (1996). Mathematical fluency: The nature of practice and the role of subordination. *For the Learning of Mathematics, 16*(2), 28–35.
Hewitt, D. (1998). Approaching arithmetic algebraically. *Mathematics Teaching, 163*, 19–29.
Johnson, M. (1987). *The body in the mind: The bodily basis of meaning, imagination, and reason.* Chicago: University of Chicago Press.
Krutetskii, V. A. (1976). *The psychology of mathematics abilities in school children* (J. Teller, Trans.). Chicago: University of Chicago Press.
Küchemann, D. (1981). Algebra. In K. Hart (Ed.), *Children's understanding of mathematics 11–16* (pp. 102–119). London: Murray.
Lakatos, I. (1976). *Proofs and refutations: The logic of mathematical discovery.* Cambridge, England: Cambridge University Press.
Legrand, M. (1993). *Débate scientifique en cour de mathématiques* (Repéres IREM, No. 10, Topiques ed.). Paris: Topiques.
Lewis, C. S. (1945). *The great divorce: A dream.* London: Geoffrey Bles.
Malara, N., & Navarra, G. (2003). *ArAl: A project for an early approach to algebraic thinking.* Retrieved August 2004 from http://www.math.unipa.it/~grim/SiMalara Navarra.PDF
Marton, F., & Booth, S. (1997). *Learning and awareness.* Mahwah, NJ: Lawrence Erlbaum Associates.
Mason, J. (1988a). *Doing and undoing* [Project update]. Milton Keynes, UK: Open University Press.
Mason, J. (1988b). *Expressing generality* [Project update]. Milton Keynes, UK: Open University Press.
Mason, J. (1992). Doing and construing mathematics in screen space. In B. Southwell, B. Perry, & K. Owens (Eds.), *Space—The first and final frontier: Proceedings of the 15th annual conference of the Mathematics Education Research Group of Australasia* (MERGA-15, pp. 1–17). Retrieved from http://cme.open.ac.uk/article_jhm_merga.htm
Mason, J. (1996). Expressing generality and roots of algebra. In N. Bednarz, C. Kieran, & L. Lee (Eds.), *Approaches to algebra: Perspectives for research and teaching* (pp. 65–86). Dordrecht, the Netherlands: Kluwer Academic.

Mason J. (1998). Enabling teachers to be real teachers: Necessary levels of awareness and structure of attention. *Journal of Mathematics Teacher Education, 1*(3), 243–267.

Mason, J. (2001a). Mathematical teaching practices at tertiary level: Working group report. In D. Holton (Ed.), *The teaching and learning of mathematics at university level: An ICMI study* (pp. 71–86). Dordrecht, the Netherlands: Kluwer Academic.

Mason, J. (2001b). Tunja sequences as examples of employing students' powers to generalize. *Mathematics Teacher, 94*(3), 164–169.

Mason, J. (2006). *Structured variation grids.* Retrieved November 2006 from mcsmail/open.ac.uk/jhm3

Mason, J., Burton L., & Stacey K. (1982). *Thinking mathematically.* London: Addison-Wesley.

Mason, J., & Johnston-Wilder, S. (2004). *Designing and using mathematical tasks.* Milton Keynes, England: Open University Press.

Mason, J., with Johnston-Wilder, S., & Graham, A. (2005). *Developing thinking in algebra.* London: Sage

Mason, J., Graham, A., Pimm, D., & Gowar, N. (1985). *Routes to, roots of algebra.* Milton Keynes, England: Open University Press.

Melzak, Z. (1983). *Bypasses: A simple approach to complexity.* New York: Wiley.

Merseth, K. (1993, March). How old is the shepherd?: An essay about mathematics education. *Phi Delta Kappan, 74,* 548–554.

Newman, J. R. (Ed.). (1956). *The world of mathematics.* New York: Simon & Schuster.

Noticeboard. (2004). *Mathematics teaching 186.* Accessed from www.atm.org.uk/mt/archive/met186files/TM-MT186-20-21.pdf

O'Connor, M. (1998). Language socialization in the mathematics classroom: Discourse practices and mathematical thinking. In M. Lampert & M. Blonk (Eds.), *Talking mathematics in school: Studies of teaching and learning* (pp.17–55). Cambridge, England: Cambridge University Press.

Ollerton, M., & Watson, A. (2001). *Inclusive mathematics: 11–18.* London: Sage.

Open University. (1978). *M101: Mathematics foundation course.* Milton Keynes, England: Author.

Peirce, C. S. (1956). The essence of mathematics. In J. Newman (Ed.), *The world of mathematics* (Vol. 3, pp. 1773–1783). London: Allen & Unwin. (Original work published 1902)

Piaget, J. (1971). *Biology and knowledge.* Chicago: University of Chicago Press.

Pirie, S., & Kieren, T. (1989). A recursive theory of mathematical understanding. *For the Learning of Mathematics, 9*(4), 7–11.

Polya, G. (1957). *How to solve it.* New York: Anchor.

Polya, G. (1962). *Mathematical discovery: On understanding, learning, and teaching problem solving* (combined ed.). New York: Wiley.

Rhadakrishnan S. (1953). *The principal upanishads.* London: Allen & Unwin.

Rojano, T., & Sutherland, R. (1993). Towards and algebraic approach: The role of spreadsheets. In I. Hirayabashi, N. Nohoda, K. Shigematsu, & Fou-Lai Lin (Eds.), *Proceedings of the 17th International Conference for the Psychology of Mathematics Education* (pp. 189–196). Japan: University of Tsukuba.

SMP 11-16. (1984). *Doing and undoing*. Cambridge, England: Cambridge University Press.

South Notts Project. (n.d.). *Material for secondary mathematics*. Nottingham, England: Shell Centre, University of Nottingham.

Stacey, K., & MacGregor, M. (2000). Curriculum reform and approaches to algebra. In R. Sutherland, T. Rojano, A. Bell, & R. Lins (Eds.), *Perspectives on school algebra*, (pp. 141–153). Dordrecht, the Netherlands: Kluwer Academic.

Tahta, D. (1972). *A Boolean anthology: Selected writings of Mary Boole on mathematics education*. Derby: Association of Teachers of Mathematics.

Tahta, D. (1980). About geometry. *For the Learning of Mathematics, 1*(1), 2–9.

Tahta, D. (1981). Some thoughts arising from the new Nicolet film. *Mathematics Teaching, 94*, 25–29.

Tahta, D. (1989). Is there a geometrical imperative? *Mathematics Teaching, 129*, 20–29.

Torretti, R. F. (1978). *Philosophy of geometry from Riemann to Poincare*. Dordrecht, the Netherlands: Reidel.

van Hiele, P. (1986). *Structure and insight: A theory of mathematics education* (Developmental Psychology Series). London: Academic Press.

Vergnaud, G. (1981). Quelques orientations théoriques et méthodologiques des recherches françaises en didactique des mathématiques. In *Some theoretical orientations and French research methodologies in teaching mathematics* (Vol. 2, pp. 7–17). Grenoble: Edition IMAG.

Vygotsky, L. (1965). *Thought and language* (E. Hanfmann & G. Vakar, Trans.). Cambridge, MA: MIT Press.

Ward, J. (1713). *The young mathematician's guide: Being a plain and easie introduction to the mathematicks in five parts* (2nd ed.). London: Tho. Horne.

Watson, A. (2000). Going across the grain: Mathematical generalization in a group of low attainers. *Nordisk Matematikk Didaktikk (Nordic Studies in Mathematics Education), 8*(1), 7–22.

Watson, A., De Geest, E., & Prestage, S. (2004). *Deep progress in mathematics: The improving attainment in mathematics project*. Oxford, England: Oxford University Press.

Watson, A., & Mason, J. (2004). *Mathematics as a constructive activity: The role of learner-generated examples*. Mahwah, NJ: Lawrence Erlbaum Associates.

Whitehead, A. (1948). *An introduction to mathematics*. London: Oxford University Press. (Original work published 1911)

Whiteside, D. (Ed.). (1972). *The mathematical papers of Isaac Newton: Vol. 5. 1683–1684*. Cambridge, England: Cambridge University Press.

Zazkis, R. (2001). From arithmetic to algebra via big numbers. In H. Chick, K. Stacey, J. Vincent, & J. Vincent (Eds.), *Proceedings of the 12th ICMI study conference: The future of the teaching and learning of algebra* (pp. 676–681). Melbourne, Australia: University of Melbourne Press.

4

Quantitative Reasoning and the Development of Algebraic Reasoning

John P. (Jack) Smith III
Michigan State University

Patrick W. Thompson
Arizona State University

> No doubt, it is difficult for a teacher to teach something which does not satisfy him entirely, but the satisfaction of the teacher is not the unique goal of teaching; one has at first to take care of what is the mind of the student and what one wants it to become. (Poincaré, 1904, p. 255)

In keeping with the theme of this book—the early development of algebraic knowledge and reasoning—we describe how students might develop knowledge and ways of thinking in elementary and middle school that support their learning of algebra. When we use the term *algebra*, we are not referring to the content of the algebra I course currently taught in most American middle and high schools. We mean the expression, manipulation, and formalization of mathematical concepts and structures mediated by explicit, rule-governed notational systems. As such, the content of algebra, to us, depends on ideas of coherence, representation, generalization, and abstraction. To address the development of algebraic reasoning, especially a meaningful and useful algebra, we must first address a more fundamental problem in mathematics teaching and learning.

For too many students and teachers, mathematics bears little useful relationship to their world. It is first a world of numbers and numerical procedures (arithmetic), and later a world of symbols and symbolic procedures (algebra). What is often missing is any linkage between numbers and symbols and the situations, problems, and ideas that they help us think about. Preparing students for algebra should not mean importing parts of an algebra I course into the earlier grades. Rather, it should involve changing elementary and middle school curricula and teaching so that students come to use symbolic notation to represent, communicate, and generalize their reasoning.

The opening quote from Poincaré highlights our central point. As we design an early algebra program for elementary and middle-school students, we must avoid the temptation to make it resemble the algebra familiar to us as adults. There are too many problems with traditional algebra I in the United States—perhaps the most serious of which is students' inability to find meaning and purpose in it—to use it as a model for our efforts (Silver, 1997). For this reason, a fresh approach is needed. Indeed, we must craft our expectations so that students build a kind of algebraic competence that is rich, generative, and multipurpose.

For most of us, *algebra* means the content of traditional algebra I and the courses that follow it. For authors of algebra I textbooks, algebra is a tightly integrated system of symbolic procedures, each of which is closely connected with a particular problem type. The procedures are often introduced as the mathematical means to solve specific types of problems, but the focus quickly becomes learning how to manipulate symbolic expressions. These procedures are then practiced extensively and later applied to specific problem situations (i.e., word problems). Teaching this content involves helping students to interpret various commands—solve, reduce, factor, simplify—as calls to apply memorized procedures that have little meaning beyond the immediate context. For many students, this reduces algebra to a set of rituals involving strings of symbols and rules for rewriting them instead of being a useful or powerful way to reason about situations and questions that matter to them. Consequently, many students limit their engagement with algebra and stop trying to understand its nature and purpose. In many cases, this marks more or less the end of their mathematical growth.

Many mathematics educators have recognized the deep problems of content and impact of algebra I and have made introductory algebra a major site in curricular reform efforts (Chazan, 2000; Dossey, 1998; Edwards, 1990; Fey, 1989; Heid, 1995; Phillips & Lappan, 1998). In one class of proposals, algebra is presented as a set of tools for analyzing realistic problems that outstrip students' arithmetic capabilities. In contrast to algebra I, problem situations involving related quantities serve as the true source and ground for the development of algebraic methods, rather than mere pretext (Chazan, 2000; Lobato, Gamoran, & Magidson, 1993; Phillips & Lappan, 1998). These introductions to algebra aim to develop students' abilities to use verbal rules, tables of values, graphs, and algebraic expressions to analyze the mathematical functions embedded in the problem situations, and centrally involve computer-based tools and graphing calculators to achieve these goals (Confrey, 1991; Demana & Waits, 1990; Heid, 1995; Schwartz & Yerushalmy, 1992).

Other proposals have emphasized the abstract and formal aspects of mathematical practice, suggesting that introductory algebra should develop students' abilities to identify and analyze abstract mathematical objects and systems. For example, Cuoco (1993, 1995) characterized algebra as the study of numerical and symbolic calculations and, through the development of a theory of calculation, the study of operations, relations among them (e.g., distributivity), and mathematical systems structured by those operations. Cuoco's proposal reflects mathematicians' interest in the study of increasing abstract and general algebraic systems.

Two working groups, directed to chart algebra reform K–12, have proposed a more pluralistic approach (National Council of Teachers of Mathematics Algebra Task Force, 1993; National Council of Teachers of Mathematics Algebra Working Group, 1997). They identified four basic conceptual themes in current algebra reform proposals—functions and relations, modeling, structure, and representation and language—which in turn can be explored in various mathematical contexts, such as growth and change, number, pattern and regularity (National Council of Teachers of Mathematics Algebra Working Group, 1997). Rather than mandating one best introductory algebra, these educators anticipated different courses that emphasize different themes and draw from different contexts (see also Bednarz, Kieran, & Lee, 1996). Kaput's (1995) characterization of algebra and algebra reform was similarly pluralistic, identifying five major strands of algebraic thinking, which alternately focus on mathematical process (generalization, formalization, manipulation), content (structures, functions), and language.

Given this proliferation of alternatives, one approach to early algebra would pick one view of algebra and develop scaled-down introductory versions in earlier grades. For example, the current pre-algebra course common to middle schools is a scaled down version of algebra I. We suggest this would be a mistake. We believe it is possible to prepare children for different views of algebra—algebra as modeling, as pattern finding, as the study of structure—by having them build ways of knowing and reasoning that make those mathematical practices different aspects of a more central and fundamental way of thinking. Alternative views of what algebra is can be seen as different emergent aspects of making sense of one's world quantitatively. To invoke a biological metaphor, the development of quantitative reasoning can serve as the conceptual rootstalk for many different approaches to algebra. Because the stalk can support multiple branches, wedding early algebra to one or another approach is unnecessary and limiting.

We advocate an early emphasis on developing children's ability to conceive of, reason about, and manipulate complex ideas and relationships, as an equal complement to numerical reasoning and computation. Children who develop a rich capacity for reasoning about general relationships among quantities will possess the conceptual foundation for learning and making sense of different programs and views of algebra. This chapter describes a conceptual orientation toward what is going on in complex quantitative situations, showing how teachers can help students make mathematical sense of those situations. The key claim in our argument is: If students are eventually to use algebraic notation and techniques to express their ideas and reasoning productively, then their ideas and reasoning must become sufficiently sophisticated to warrant such tools.

There is a reciprocal relationship between the long-term development of students' algebraic abilities and the long-term development of their reasoning from which these abilities emerge. If algebra, meaning the use of representational practices that employ systematic use of symbols to express quantitative and structural relationships, is to become students' means of expressing and supporting their thinking, then they must have experiences from whence the thinking that those practices support emerges. Likewise, if they are to develop thinking that calls for representational practices that employ systematic use of symbols to express quantitative and structural relationships, then the roots of those practices must be present in their early activities.

ALGEBRA, SITUATIONS, AND QUANTITIES

For the mathematically sophisticated, the best approach to complex mathematical problems is to use the tools of algebra to help manage that complexity.[1] We move quickly away from the problem situations themselves,

[1]Mathematics educators use the term *problem* in various different ways. In this chapter, our problems are mathematical tasks suggested by verbal descriptions of situations constituted by interrelated quantities. Such problem situations can generate many problems. In students' terms, our problems are always word problems. In addition, we follow Schoenfeld's (1985) view that problems, in contrast to exercises, cannot be solved by routine, well-practiced methods. They require thinking.

with all their complex relationships, toward the formality of algebraic and numerical expressions and manipulations. In appealing to algebraic methods to solve complex problems, we elect not to use less formal approaches that are tied more closely to the situation. We also tend to devalue this informal reasoning in comparison to algebraic methods. In this devaluation, we miss an essential connection between the two kinds of thinking: That more concrete, intuitive, and situation-specific patterns of reasoning, appropriately supported and nurtured over a period of years, can foster students' development of the algebraic reasoning we value so highly. If our goal is for students to understand and use algebra, then the success of an early algebra program will depend on supporting the development of formal reasoning from an informal foundation.

To illustrate the common separation of formal, algebraic reasoning and informal reasoning, compare a traditional algebraic solution to the following problem to one that more directly involves the quantities and relationships in the problem situation:

> Problem 1. I walk from home to school in 30 minutes, and my brother takes 40 minutes. My brother left 6 minutes before I did. In how many minutes will I overtake him? (Krutetski, 1976, p. 160)

A typical algebraic solution to this problem involves assigning variables, writing algebraic expressions, and eventually stating and solving an equation. If t represents the number of minutes I have walked, then whenever I have walked t minutes my brother will have walked $t + 6$ minutes. If d stands for the number of miles from home to school, and if my brother and I travel at constant speeds, then my walking speed is $d/30$ miles per minute and my brother's is $d/40$ miles per minute. Using the general relationship that "rate multiplied by time equals distance" ($d = r * t$), these expressions can be stated in the equation, $(t + 6)\ d/40 = t\ d/30$, which can be easily solved from its equivalent form, $(t + 6)/40 = t/30$. Related motion problems like Problem 1 are common in algebra textbooks because algebraic methods are presumed necessary to solve them. As long as no difficulties arise, this algebraic approach makes few explicit references to how speeds, times, and distances are related in the situation. The basic idea is to move out of the situation and its constituent quantities and into the world of symbolic expressions and equations.

But this problem can also be solved by reasoning about the relationships among distances, walking rates, and times of travel without the support of variable assignments or algebraic expressions. Here is one example of this approach, which we will call *quantitative reasoning*:

- I imagine myself walking behind my brother, seeing him ahead of me. What matters in catching up with him is the distance between us and how long it takes for that distance to become zero.
- The distance between us shrinks at a speed that is the difference of our walking speeds.
- I take 3/4 as long as brother to walk to school, so I walk 4/3 as fast as brother.
- Since I walk 4/3 as fast as brother, the distance between us shrinks at the rate of 1/3 of brother's speed.
- The time required for the distance between us to vanish will therefore be three times as long as it took brother to walk it in the first place (6 minutes).
- Therefore, I will overtake brother in 18 minutes.

Like the algebraic solution, this reasoning is quite sophisticated, requiring a rich understanding of how times, speeds, and distances are related, how those relationships can be used to draw inferences, and how numerical values can be inferred from those that are given. It also has the same level of potential generality as the algebraic solution. If different initial numerical values were given, the calculations in the solution might become more cumbersome, but the logic of the reasoning would not change. The two solutions differ most visibly in their use of algebraic symbols. However, they differ more deeply in the former's focus on translating relationships into symbols and the latter's focus on expressing and working directly with those relationships.

In proposing quantitative reasoning as a root for algebraic thinking, we acknowledge that the former does not develop easily or quickly. In fact, the student who produced this solution achieved his proficiency from a wide variety of experiences over several years. Our thesis is that students' quantitative reasoning is worth years of attention and development, both because it increases the likelihood of success with algebra and because it makes arithmetic and algebraic knowledge more meaningful and productive.

Algebraic reasoning is characterized by its generality and by the role that symbolic expressions play in stating general relationships, comparing and manipulating them, and facilitating many numerical evaluations. Quantitative reasoning, when developed throughout children's elementary and middle school years, develops mathematical ideas of similar generality that students will eventually find sensible to express in algebraic notation. Put simply, quantitative reasoning provides conceptual content for powerful forms of representation and manipulation in algebra.

Before we proceed, it is important to emphasize that we are not using the terms *quantity* and *quantitative reasoning* as synonyms for *number* and *numerical reasoning*. Indeed, our central purpose here is to show how the elementary years can be used to support the development of students' quantitative reasoning by focusing their attention away from thinking strictly about numbers and numerical operations. In our view, conceiving of and reasoning about quantities in situations does not require knowing their numerical value (e.g., how many there are, how long or wide they are, etc.). Quantities are attributes of objects or phenomena that are measurable; it is our *capacity* to measure them—whether we have carried out those measurements or not—that makes them quantities (Thompson, 1989, 1993, 1994). In this sense, we follow Piaget's (1952, 1970) meaning of quantity and quantification. But, as we do, we also acknowledge that other analysts draw much closer associations between quantity and number (e.g., Fey, 1990; Fuson et al., 1997).

RELATIONSHIPS BETWEEN QUANTITATIVE REASONING AND ALGEBRAIC REASONING

The two prior solutions to Problem 1 suggest a stark contrast. One translated the relationships in the situation into traditional algebraic expressions and looked like algebra; the other directly manipulated the relationships among the quantities in the situation—elapsed times, walking speeds, and walking distances. Whereas this contrast exemplifies the character of quantitative reasoning as a distinct form of mathematical thinking, we stress the connections between quantitative reasoning and algebra, as well as their differences. Before we consider how such sophisticated quantitative reasoning can be nurtured over the years, we return to Problem 1 and examine three solutions in greater detail. These solutions (the two already given and one more) show how quantitative reasoning can underlie and motivate reasoning with symbols.

The traditional algebraic solution involved generating and solving the equation,

$$(6+t)\frac{d}{40} = t\frac{d}{30}.$$

But, what sort of thinking could motivate the initial variable assignments and the symbolic expressions for times and speeds found in that equation? If equation writers understood the problem situation (rather than memorized a script for this problem type), their reasoning might have had some of the following character:[2]

[2]We do not claim that all cases of reasoning sensibly with algebraic symbols follow these exact steps. This is only one example of sensible algebraic reasoning on this problem.

- Since we both begin from home, I will catch my brother when we both have walked the same distance from home.[3] We both walk any distance by traveling at some speed for some amount of time.
- I do not know how far it is from home to school, but I can think of it as some number of miles, which I will designate by *d*. My brother walks *d* miles in 40 minutes, so his speed is (Eq. 2) $\frac{d}{40}$ miles per minute. I walk *d* miles in 30 minutes, so my walking speed is $\frac{d}{30}$ (Eq. 3) miles per minute.
- At any moment in my walk, I have walked $t\frac{d}{30}$ (Eq. 4) miles in t minutes. After I start, brother will have walked for 6 minutes longer than I, so when I have walked t minutes, brother will have walked (t + 6) minutes. Therefore, he will have walked $(t+6)\frac{d}{40}$ (Eq. 5) miles when I have walked t minutes.
- I will catch brother when he and I have walked exactly the same distance from home. At that moment, our two distances will be the same, so the formula for his distance, $(t+6)\frac{d}{40}$ (Eq. 6), and the formula for my distance, $t\frac{d}{30}$ (Eq. 7), will have the same value. So, I am looking for values of *t* that make the sentence $(6+t)\frac{d}{40}=t\frac{d}{30}$ (Eq. 8) true for any value of *d*.

We consider this reasoning a good example of using algebra with understanding. The main content of that understanding is a solid conceptual grasp of how the quantities of walking speeds, times traveled, and distanced traveled from home are interrelated. This elaborated algebraic solution differs from the first bare-bones one in that it restates the problem in terms of the reasoner's own experience with relative motion, sketches the logic for transforming walking times into speeds, sees the expressions for distance as complete and continuous descriptions of motion, and generates an equation to determine where those distance expressions produce the same value. Indeed, every step in the solution expresses some conceptual relationship between two or more quantities in the situation, and it is these relationships that motivate and justify the various algebraic expressions. One role for quantitative reasoning in complex problem solving is therefore to provide the content for algebraic expressions so that the power of that notation can be exploited.

Another role of quantitative reasoning is to support reasoning that is flexible and general in character but does not necessarily rely on symbolic expressions. We return to the nonalgebraic, quantitative solution of

[3]The reasoner assumes that the walkers follow the same path.

Problem 1 and unpack it to show how such sophisticated reasoning might grow and how it shares the generality that characterizes algebraic reasoning:

- I imagine myself and my brother walking. What matters in catching up with him is the distance between us and how long it takes for that distance to vanish.

The reasoner projects herself into the problem situation, adopting the perspective of actually looking at her brother walking ahead of her. From this perspective, her distance and brother's distance from home are irrelevant, and the only thing that matters is the distance between them. It is this step of imagining oneself into the problem situation that so often eludes students:

- The distance between us shrinks at a rate that is the difference of our walking speeds.

This is quite a sophisticated inference. When two quantities change at constant rates, in the same direction, and we consider how rapidly their measures move apart, we are asking at what rate the difference between them changes. If we consider each one changing for a unit of time, then the added difference will be the difference of their rates. Thus, the rate at which the excess of one over the other changes is the difference of the two quantities' rates. In terms of distance and speed, the rate at which the distance between the walkers' changes is the difference of their walking speeds. We hasten to add that by *difference* we do not mean the result of subtracting. Rather, we mean the distance that is created by comparing how much one distance exceeds or falls short of the other:

- I take 3/4 as long as brother to walk to school, so I walk 4/3 as fast as brother.

This is another sophisticated inference. It exploits a general understanding of speed as a rate of change of distance with respect to changes in time. Because we walk the same distance, our walking speeds are different only because our travel times are different. Longer travel times mean slower walking speeds. If I walk the same distance as brother in three fourths the time, then I would walk one third again as far as brother in the same amount of time (Fig. 4.1). So I walk four thirds as fast as brother because I walk four one thirds (4/3) as far as brother in the same amount of time:

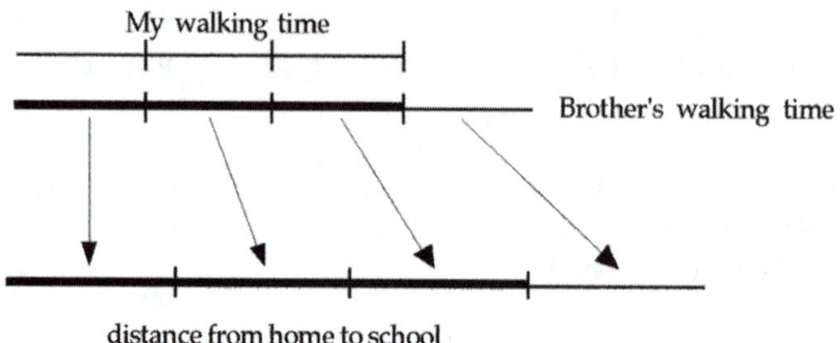

I walk 1/3 of the way to school in each 1/4 of Brother's time, so I walk 4/3 of the way to school in 4/4 of Brother's time. Therefore, I walk 4/3 as fast as Brother, because I go 4/3 as far as Brother in the same amount of time.

FIGURE 4.1. Relating my speed to brother's speed based on the relationship of my time to his.

- Since I walk 4/3 as fast as brother, the distance between us shrinks at the rate of 1/3 of brother's speed.

Walking 4/3 as fast as brother means that my speed is 1/3 greater than my brother's speed. So, in a given amount of time, I not only walk as far as brother, I walk an extra one third of the distance he has walked. Therefore, the distance between us shrinks at the rate of one third of brother's speed:

- The time required for the distance between us to vanish will therefore be three times as long as it took brother to walk it in the first place (which was 6 minutes).

If brother took some amount of time to walk some distance, and another person walked at one third of brother's speed for the same amount of time, then that person will walk one third of the distance brother walked. So, in 6 minutes, the amount of time brother used to get ahead of me, the distance between us will shrink by one third of the distance he walked in 6 minutes. I need to shrink this distance three times, so it takes me 3 times 6 minutes, or 18 minutes, to catch brother:

- Therefore, I will overtake brother in 18 minutes.

This solution illustrates some important features of quantitative reasoning and its origins. First, quantitative reasoning draws heavily on everyday experience. The basic approach—studying how the distance between the walkers decreases—depends on the reasoner projecting herself into the situation and invoking the visual imagery of catching up. Once framed in that way, the solution proceeds by drawing on relationships among speeds, times, and distances. The change in distance-between-walkers is cast as a rate of change and expressed in terms of the brothers' walking speeds, and numerous quantitative manipulations of the speed–time–distance relationship support numerical inferences about the value of various quantities. Finally, although it is grounded in everyday experience, it is difficult to imagine how students could develop this level of facility without focused instruction, which draws on and stretches their abilities to state general relationships and make inferences from them.

We do not offer this solution as paradigmatic of quantitative reasoning. Indeed, quantitative reasoning does not typically follow any standard pattern or routine like the variable assignment and equation solving in traditional algebraic problem solving. Quantitatively oriented solutions tend to vary more widely than algebraic solutions to the same problem, primarily because they are grounded in how students conceive of situations, and there is tremendous range in these conceptions. To illustrate this variety, we present another quantitative solution to Problem 1 produced by a less mathematically mature student. This student's reasoning was less general but grounded in a solid, concrete understanding of constant speed as a rate of change:

- Imagine the distance from home to school cut up into 30 pieces. Each piece is how far I walk in 1 minute.
- Imagine the distance from home to school also cut up into 40 pieces. Each of these pieces is how far brother walks in 1 minute.
- Brother's 1-minute-distance piece will be 3/4 the length of my 1-minute-distance piece.

This statement directly compares the length of the 1-minute-distance to brother's 1-minute-distance:

- In 6 minutes, Brother will travel six [of his 1-minute distances], so he will travel 18/4 of my 1-minute-distance piece. That is how far ahead of me brother is.

Six iterations of 3/4 of my 1-minute-distance piece is 18/4 of my 1-minute-piece:

- When I start walking, I will move closer to brother by 1/4 of my 1-minute-distance piece each minute.

If for every 1-minute-distance piece I move brother moves 3/4 of that 1-minute-distance piece, then I am gaining by the difference each minute:

- I will make up 18/4 (eighteen one fourths) of my 1-minute-distance piece in 18 minutes when I gain on brother at the rate of 1/4 1-minute-distance piece each minute.

Several features of this solution are worth highlighting. Although it may appear entirely arithmetical and concrete, it involves quantities whose actual values are unknown (distance from home to school, "my" 1-minute-distance, brother's 1-minute-distance), yet from which the reasoner derives essential information ("in 1 minute, brother travels 3/4 the distance I do"). Her study of 1-minute-distances and how they build up over time also coordinates distances and times without the appeal to speed-as-rate as a mediator, whereas a rate conception of speed was central to the previous quantitative solution. This contrast underscores the inappropriateness of expecting particular statements of conceptual relationships in quantitative solutions. Quantitative relationships—especially complicated, multiplicative ones—can be expressed in many ways. This reasoning describes in verbal terms what was expressed in symbols in the bare-bones algebraic solution. Brother's 1-minute-distance piece is equivalent to the formula $d/40$ if d were used to represent the distance from home to school. Finally, ideas of functional covariation—how one quantity varies in relation to the variation of another—are central to this student's reasoning. Segmenting the total distance into 1-minute-distance intervals provides a framework for comparing and coordinating distances and eventually to quantify how much she is gaining.

But reasoning of this sort, even if less sophisticated than the previous solution, does not spring forth either quickly or spontaneously. It must be carefully nurtured over many years. On the one hand, it requires positive support from curricula and pedagogy that extend children's existing abilities. The mental operations used in quantitative reasoning must be built in many contexts and over relatively long periods of time. On the other, it means avoiding the classroom orientation that quickly shifts the focus away from making sense of situations and toward calculation (A. G. Thompson, Philipp, Thompson, & Boyd, 1994). It opposes the prevailing view that mathematics is about getting "the answer," in numerical or symbolic form. Senseless patterns of thinking emerge for students once meaning and purpose in mathematics disappear and students' expectation of making sense is therefore difficult to restore. The next two sections attempt to illustrate how richer capacities for quantitative reasoning can be nurtured in the elementary and middle school years, beginning with additive situations.

QUANTITATIVE REASONING AND ARITHMETIC REASONING

We now look more closely at what we mean by quantities and relationships between them. Just as we contrasted the emphasis on symbolic procedures with reasoning about quantities and relationships in Problem 1, we emphasize here the difference between reasoning about numbers and calculations and reasoning about quantities in a problem situation typically seen as arithmetic:

Problem 2. At some time in the future John will be 38 years old. At that time he will be three times as old as his daughter Sally. Sally is now 7 years old. How old is John now? (Adapted from A. G. Thompson et al., 1994)

A Numerical/Computational Solution

From one perspective, Problem 2 is a three-step arithmetic word problem. To solve it, students must first divide 38 by 3 to determine that Sally's age in the future is 12 2/3 years; then subtract 7 from 12 2/3 to determine that the difference between her age then and now is 5 2/3 years; and finally subtract 5 2/3 from 38 to determine John is 32 1/3 years old now. Most middle grades students know that there are three numbers they must use (38, 3, and 7) and four operations (addition, subtraction, multiplication, and division) to choose from. They know they must find the right sequence of operations on the right pairs of numbers (including intermediate results like 12 2/3) to produce the correct final answer. From this perspective, classroom discussions usually center on those issues—which numbers, which operations, and in what order? Although some attention might be given to justifying the operations (e.g., noting that "three times as old" is a clue to multiply), the primary focus is numerical and computational.

A Quantitative/Conceptual Solution

A different approach is also possible—one that centers on what many people call *understanding the problem* (see Riley, Greeno, & Heller, 1983, for a detailed example of this general view). From this perspective, the

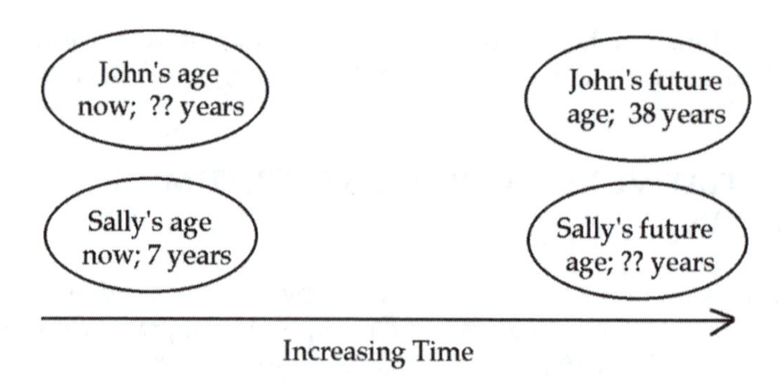

FIGURE 4.2. Everyone ages as time passes

problem concerns quantities, their properties, and relationships among them, and its solution involves reasoning about those relationships and eventually linking them to numerical operations. This perspective focuses on helping students conceptualize situations irrespective of the numerical information with which they are presented and the calculations they can produce.

To illustrate this approach, we first identify four quantities in Problem 2: "John's age at some future time," "Sally's age at that same future time," "Sally's age now," and "John's age now." We can think about these ages and many relationships between them (e.g., one person is older, their ages change at the same rate) without knowing their numerical values. The fact that we happen to know the values of two of them—"John's age at some future time; 38 years" and "Sally's age now; 7 years"—is incidental to our ability to think about their ages changing over time.

These four quantities by themselves do not represent what is going on in Problem 2. To comprehend fully, we need to recognize three important relationships that integrate the quantities into a coherent structure. First, there is the temporal relationship that time moves on from now to then, which relates John's ages and Sally's ages (Fig. 4.2).[4]

Second, we need to recognize that John's ages (now and in the future) and Sally's ages (now and in the future) stand in a specific quantitative relationship to each other, commonly called a difference. The difference between John's two ages, for example, is the amount of time by which his age in the future exceeds his age now. Although we cannot immediately

[4]We use ovals to represent quantities and place inside those ovals all relevant information about those quantities—their name, their units of measure, and any numerical value or expression that is given or can be inferred—here and in all subsequent figures. This is a convenient notation, but there is nothing mathematically or cognitively unique or essential about it.

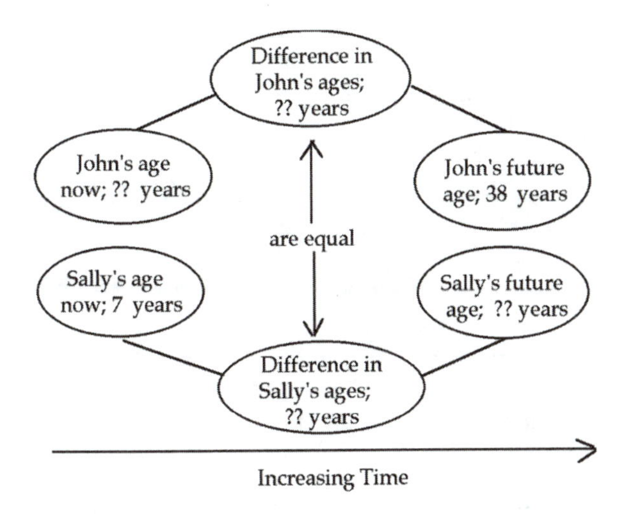

FIGURE 4.3. Time passes in equal amounts for everyone.

determine the value of this difference, we can imagine it. We also know that however much Sally grows older, John (and every other person) will grow older by the same amount, so the difference between Sally's present and future ages will be the same as the difference in John's present and future ages. We represent these relationships as new quantities (the differences in Fig. 4.3) linked to the age quantities.

A third relationship provides a crucial link between John's and Sally's ages. At some point in the future, John's age will be three times as great as Sally's. This relationship is typically called a *ratio*. We use that term to indicate a quantity that expresses a multiplicative comparison of two other quantities. A ratio's measure describes how many times as great the measure of one quantity is as the measure of the other. As with differences, we refer to ratios as quantities. Both are born of comparison and therefore have dual existences: They express relationships between two quantities (either a multiplicative [ratio] or additive [difference] comparison), and they are quantities in their own right (as a measurable attribute of such comparisons).

When we include the ratio between John's and Sally's future ages to the structure presented in Figure 4.3, we have identified, analyzed, and related the quantitative information relevant to answering the question of how old John is now. In traditional terms, we have built an understanding of the problem, which will support and justify our arithmetical reasoning (Fig. 4.4).

With such an understanding, it is easy to decide (and justify) what calculations are needed. When John's future age is three times as large as Sally's future age, Sally's age at that moment will be 1/3 as large as

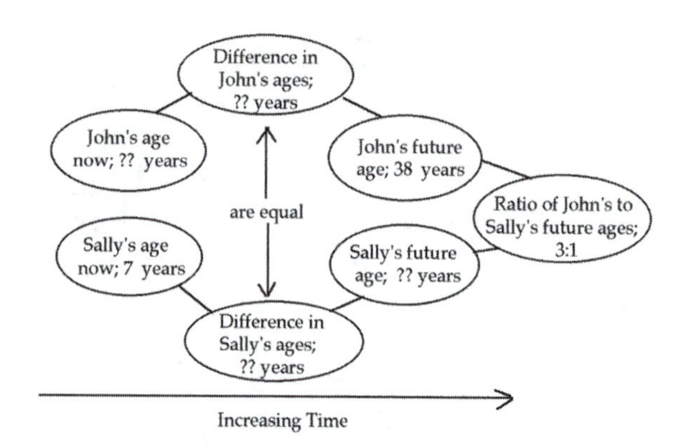

FIGURE 4.4. Relationships among quantities.

John's. Thus, it makes sense to divide "John's age in the future" by 3 to determine the value of "Sally's age in the future" (Fig. 4.5). Knowing the values of Sally's current and future ages, we can determine by how much older she grows, which is 38/3 – 7, or 17/3, years. Because we know that John grows older by the same amount as Sally (17/3 years), and that in 17/3 years he will be 38, we can make our final computation to find that John's age now is 32 1/3 years (Figure 4.5).

So, What's the Point?

The purpose of these extended examples is to emphasize the richness that mathematical reasoning can have when we focus on quantities and relationships among them instead of on numbers and arithmetic operations. Sowder (1988) showed that when students do not attend to quantities and relationships, their problem solving quickly becomes a matter of ungrounded debate about choosing numbers and operations. An emphasis on the quantitative aspects of situations reorients students' mathematical focus in three important ways, affecting both the development of their arithmetic reasoning and their future prospects in algebra.

First, the quantitative/conceptual approach makes thinking about the quantities and their relationships a central and explicit focus of solving the problem. The resulting conceptual structure that we have represented in a series of figures can be used to explain and justify both quantitative inferences and numerical computations. Such figures create public frameworks that students and teachers can use to think about situations, making it less likely that students will see numerical computations as materializing from nowhere. They also create contexts for examining mathematical issues that are unlikely to arise in the numerical/computation approach. For example, a teacher could raise the question of whether the

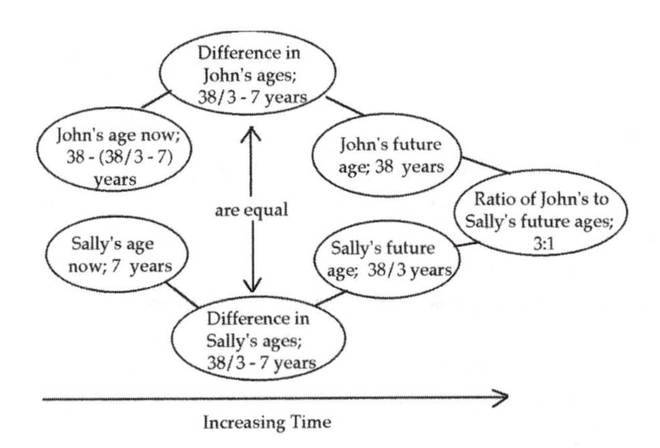

FIGURE 4.5. Computing within relationships among quantities.

ratio between John's and Sally's ages remains constant as they get older. Questioning how ratios of people's ages change over time could lead to an examination of how ratios change as the related quantities increase in equal increments and how those increments themselves must change for ratios to remain constant.

Second, this focus on thinking about and representing general relationships between quantities (i.e., the relationships inherent in the quantities themselves, not the specific numerical values they take on) supports the kind of conceptual development that will eventually make algebra a sensible tool for thinking and problem solving. We say this because quantities are inherently indeterminate. We can imagine comparing two heights without knowing their specific measures. Heights are quantities that we understand as being measurable, but knowing their measures does not add to conceptualizing the comparison. Rather, knowing their measures simply adds information about the comparison. Algebraic notation and methods are powerful tools for stating, analyzing, and manipulating general relationships, but without ideas of substantial generality to express, students will find little sense in and little use for algebra.

Third, the quantitative/conceptual approach also suggests an early route to algebraic symbols in its focus on representing the general numerical relationships, rather than specific computations. If students write their calculations in open form (e.g., "38/3 – 7" instead of "5 2/3") and focus on the nature of that calculation rather than its result, they can adjust more easily to using expressions in place of computed values. The purpose of writing open expressions is to record a chain of reasoning clearly—a rationale that is useful in arithmetic as well as algebraic reasoning. Open expressions make it much easier for students to think about the effects of changing a given numerical value in the situation and therefore support the shift in focus from particular to general relationships (Mason, chap. 3, this volume). Also, if students regularly use expressions to represent values, it is a much smaller step to using formulas to represent values.[5] Of course, it is incumbent on teachers to draw children's attention consistently to the nature and purpose of their activity—that they are, in fact, representing a number without actually having to compute it and that they are reasoning about many similar problems all at one time.

In emphasizing the importance of open numerical expressions as an entry point to algebraic formalism, we recognize that they are a standard early topic in the traditional algebra I and pre-algebra curricula. However, with students' prior experiences dominated by numerical computation, most of them do not understand what purpose open expressions serve. After all, if you can complete a computation, then why state it in incomplete terms? When the focus is on grasping, stating, and exploring general relationships between quantities, open expressions serve a clearer purpose: to connect general relationships (like differences) to specific situations, quantities, and numerical values.

The comparison of the two solutions to Problem 2 also illustrates other features of quantitative reasoning that we attempt to elaborate in the balance of the chapter, specifically:

- Some quantities arise directly from measuring things; others, like differences and ratios, arise from quantitative operations—operations on other quantities. Quantitative operations (e.g., multiplicative comparison) are not the same as numerical operations (e.g., multiplication) despite the frequent similarity in terminology.
- Quantities that result from quantitative operations exist in two different senses, as quantities in their own right and as relationships between the two quantities. It can be conceptually demanding to reason and communicate about such quantities because we must distinguish and coordinate these two senses and, when necessary, shift between them.
- Quantitative reasoning produces essential non-numerical inferences about quantities and how they relate in the problem situation. It is often the glue that holds arithmetic reasoning and algebraic reasoning together.

[5]With just a slight change in information, children could engage in the same pattern of inferences as depicted in Figure 4.5, with a symbol to represent a numerical value. Activities designed with this end in mind—to generalize a *pattern of inferences* instead of a pattern of numbers—lead much more naturally to generating symbolic expressions as models of quantitative relationships.

- Cognitive resources other than spoken language are useful (and often necessary) in managing quantitative reasoning in complex problem settings. One class of resources is diagrams that represent relationships in sensible, public ways.
- Mathematicians and educators lack a standardized, accepted terminology for quantities. Some terms used in this chapter, like difference, represent relatively standard usage; others, for example, ratio and rate, are less standard (see P. W. Thompson, 1994). But the names used to designate types of quantities are less important than the way people think about them.

QUANTITATIVE REASONING IN COMPLEX ADDITIVE SITUATIONS

Situations that involve complex additive relationships (i.e., more than three related quantities) can be an important site for the early development of students' quantitative reasoning. Developing young students' abilities to reason with additive situations prepares them for algebra in multiple ways. It provides occasions for them to think about situations systematically, initially ignoring matters of calculation and instead focusing on what is going on. When additive situations include large numbers of interrelated quantities, this complexity presses students' abilities to understand, represent, and express those relationships and therefore develops their non-numerical mathematics skills. If we want students to learn and use algebra as a sensible tool for expressing their thinking and solving problems, then work with complex problems must come first.

Students who have mastered addition and subtraction as operations on numbers may have much more to learn about additive relationships among quantities. For example, consider the following two problems:

Problem 3. Thomas has 38 baseball cards and 13 more than his friend Alex, How many baseball cards does Alex have?

Problem 4. Jim, Sue, and Tom played marbles. Sue won 6 marbles from Jim and 5 from Tom. Jim won 3 marbles from Tom and 4 from Sue. Tom won 12 marbles from Jim and 2 from Sue. (Compare Tom's number of marbles before and after the game.) (Adapted from P. W. Thompson, 1993)

By Grade 3, Problem 3 is not difficult for most students, despite that more could be mistaken as a cue to add 38 and 13. But, even older students struggle with Problem 4 because they must manage many quantities (collections of marbles and changes in those collections), construct relationships among those quantities, and reason about those relationships, instead of simply choosing numbers from the problem statement and computing. (We placed parentheses around the last sentence in Problem 4 to emphasize that situations can be the instructional focus without becoming problems—that is, without asking for a quantity's value.)

Problem 4 illustrates some of the quantitative issues that can receive attention when the main goal is to help students understand what is going on in situations. A teacher might ask, "Is it possible for Jim to win 3 marbles from Tom if Tom won 12 marbles from Jim?" and then make their understanding of that possibility the focus of her lesson. The ensuing discussion, when oriented in this way, could move in the direction of considering how the changes in the various collections of marbles interrelate (see e.g., P. W. Thompson, 1993).

Developing students' ability to reason with complex additive relationships means rethinking the notion of problem in elementary mathematics. We cheat our students if our problems are always requests for calculations. The next few sections illustrate a different sense of problems—that of situations where students conceptualize and reason about relationships between quantities. If students' experience of problems changes, so can the kinds of questions that teachers can ask, the kinds of assessment that make sense, and the character of students' capabilities. We must guard against underestimating students' quantitative reasoning abilities. They come to school fully capable of learning to reason with simple additive relationships and that reasoning can develop in impressive ways with thoughtful instruction (Carpenter & Moser, 1984; Carpenter, Moser, & Romberg, 1982; Fennema, Carpenter, & Peterson, 1989; Fuson et al., 1997). A major task of the elementary curriculum should be to build on that competence.

Sources of Differences

This section describes how differences—a key component of additive problem situations—can arise in a variety of ways. We also examine the kinds of questions that can be posed in complex additive situations; and discuss some general aspects of students' quantitative reasoning in these situations and how teachers can support and extend it. This section does not even begin to sketch out a K–8 curriculum in additive reasoning, but it does provide examples of how traditional word problems can be adapted to support the development of quantitative reasoning.

Many complex additive situations centrally involve one or more differences, quantities that measure how much one quantity exceeds or falls short of another. Differences arise in situations in at least three ways. In some cases, they emerge when actions physically change quantities in the situation. For example, in Problem 4, Sue's act of winning 6 marbles from Jim created the difference, "the 'new' marbles in Sue's collection."[6] Differences also result from comparisons of two quantities that remain wholly intact in the situation, for example, between two people's height (how much taller?), between size of classes (how many more children?), between two people's driving speed (how much faster?). Third, differences can emerge when quantities change over time but without any physical act of transfer. For example, we can conceive of the daily fluctuations in the price of some commodity relative to its "base" price as a sequence of differences, one for each day.

The Basic Case. In the conceptually simplest case, a difference compares two quantities that are not themselves the result of other quantitative operations (i.e., they are not differences, ratios, or rates). Conceptualizing a difference means thinking about three quantities in relationship to each other.

The situations in Problems 2–4 included examples of basic differences, such as the difference between John's age now and Sally's age now (Problem 2) and the difference between Sue's marbles before and after playing Jim (Problem 4). In the first example, the values of the difference and one quantity were known, so the value of the third quantity could easily be calculated. In the second, only the value of the difference was known so there were many possibilities for the number of marbles in Sue's collection before and after playing Jim. It is not necessary—and this is a crucial point—to know the values of either quantity to conceptualize their difference and reason about it (Fig. 4.6). Conceptualizing a difference only requires thinking about the excess (or deficit) of one quantity over another. As Carraher, Schliemann, and Brizuela (2000) have shown, even primary age children can conceptualize differences and reason about them in relation to their constituent quantities.

Operations on Differences. In some situations, two or more basic differences are present, and reasoning about those situations can involve additive comparisons or combinations of those differences.[7] Conceptualizing a

---

[6]Alternatively, the same act can be seen as creating the difference, "the marbles that are no longer in Jim's collection." The character of the difference depends on how the reasoner conceives the situation.

[7]We could just as easily consider differences of ratios or any other quantities that are themselves the result of a quantitative operation. Because the context is additive situations, we only discuss a difference of differences.

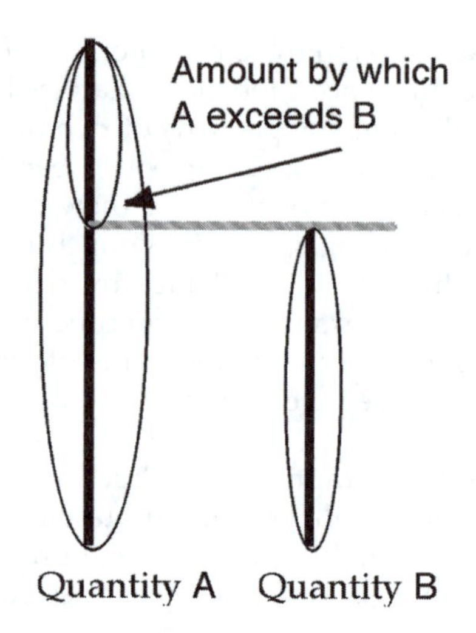

FIGURE 4.6. Difference of two quantities.

difference of differences or a combination of differences means coordinating the relationships among seven quantities (although not necessarily simultaneously; Fig. 4.7).

In the example "Allen is 27 years older than Alva and Alva is 13 years younger than Denise," we can think about the relationship between Allen's and Denise's ages as a comparison: How much older than Alva is Allen than Denise? Relative to Alva's age, Allen is "more older" than Denise, by exactly 14 years. In, "the price of regular unleaded gas at Sam's station increased 5.9¢ one month, decreased the next month, and then was 3.7¢ lower at the end of those two months," the change during the second month can be thought of as a combination of differences (Fig. 4.8).

As in the basic case, conceptualizing a difference or a combination of differences, however, does not require or necessarily follow from knowing the specific values of the differences that are compared. In the statement, "Toni compared her height to her brother's height and Melissa compared her height to her brother's height. They found that Toni is taller than her brother by more than Melissa is taller than hers," we know neither the specific value of any of the three differences nor the specific values of the four individuals' heights. Yet, we can imagine a comparison of Toni's and Melissa's differences and think about that difference as a quantity. Even if the value of the difference of height differences were

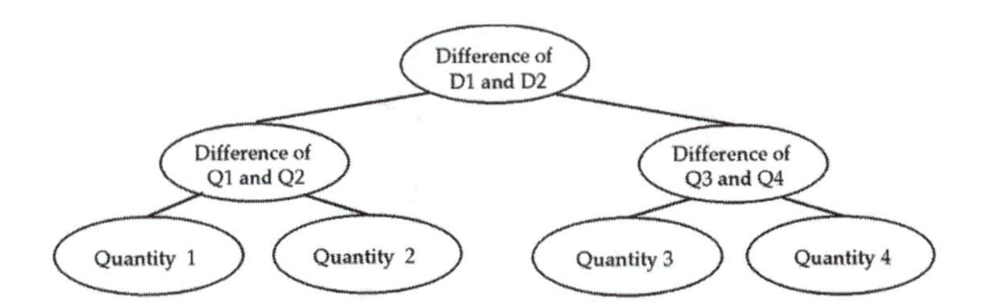

FIGURE 4.7. Difference of differences.

known, many different heights and height differences could produce it. For this reason, situations involving differences of differences are often complicated to sort out, make sense of, and speak about. There are more quantities to keep track of, they participate in multiple relationships, and assigning or changing their numerical value may have complicated effects.

Patterns of Differences. Other situations involve additive comparisons of two quantities, which change many times (even continuously) in the situation. Repeated or continuous additive change creates the possibility of conceiving a pattern of differences. Conceptualizing a pattern of differences means grasping the *collection* of changes as an object of consideration. This approximates thinking about differences as a function of two variable quantities.[8] In many cases, changes take place over time, so differences can be arranged and thought about as a temporal sequence. For example, the performance of a business is often conceptualized as a series of differences between revenues and expenses over some units of time, say, months (Fig. 4.9).

Curriculum: Problems and Problem Situations

In developing a curriculum to support students' additive reasoning capabilities, it is important to consider the kind and complexity of the situations presented. Because the overall goal is to prepare students to use mathematics to think about their world, it is sensible to choose and/or

[8]If one quantity did not vary, then the pattern of differences would be similar to a function of one variable. If both quantities vary, then the differences would a function of two variables.

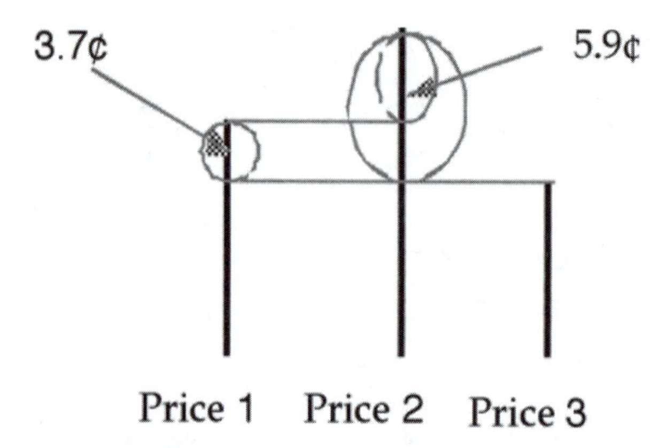

FIGURE 4.8. The difference between the 2nd and 3rd months' prices is seen as a combination of two differences.

develop situations and quantities for which they have rich, everyday experience. This general principle does not imply a strict realism (e.g., "kids can only think about quantities they have directly experienced"). Rather, it means that the situations where students can draw on broadest reservoirs of personal experience, thinking, and talk (e.g., motion, growth, physical characteristics) are dependable places to start. They are contexts where students' reasoning may be initially the most developed. But other features also contribute to the complexity and difficulty of situations. In general, additive situations become more complex as the total number of quantities increases, the level of interrelation among quantities increases, and the number of quantities with known values is decreases.[9]

Students' abilities to explain and reason are clearly dependent on and strongly influenced by the kind of questions teachers pose. In most elementary classrooms, students are asked to find the value of one quantity from the given values of other quantities, often in conceptually simple situations. A curriculum of quantitative reasoning must include different sorts of questions about more complex situations. Sometimes these questions will be calls to find a value of a quantity (or, perhaps more fruitfully, a range of values; see later); at other times, they will be questions

[9]These considerations should be treated as general guidelines for preparing and selecting situations for the classroom, not as hard and fast rules. Children and classrooms differ in their reactions to situations and teachers must be prepared to experiment with and, if necessary, adjust their problem situations in response to their students.

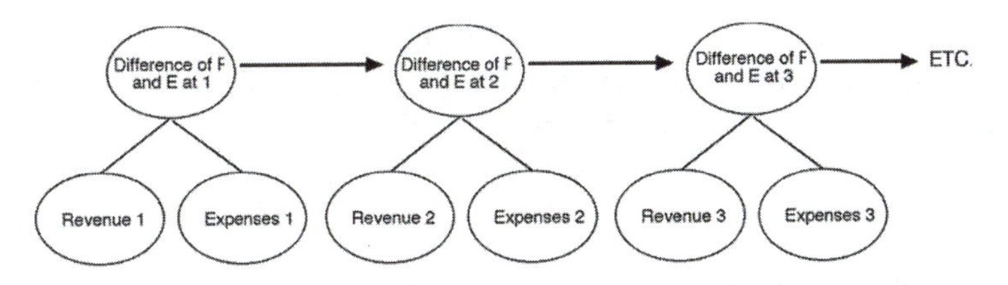

FIGURE 4.9. Chain of differences.

about the nature or behavior of a quantity. Problems 5 and 6 provide two examples:

> Problem 5. Sam and Joseph each had a shorter sister, and they argued about who was more taller than his sister. Sam won the argument by 14 centimeters. He was 186 cm tall; his sister was 87 cm; and Joseph was 193 cm tall. How tall was Joseph's sister? (Adapted from P. W. Thompson, 1993)

In this case, although the right series of three numerical subtractions will produce the "answer," the complexity of Problem 5 makes it important to think about and make sense of the situation by representing and sorting out the quantities and their relationships. It makes reference to four people's heights, two brother–sister height differences, and a difference between those differences. Without making sense of the situation as a set of related relationships, the connection between the difference of differences (whose value is 14 centimeter) and Sam's family difference (whose value is 99 centimeters) can be quite mysterious. So one strategy for developing students' quantitative reasoning is to pose problems that are too complex for them to apply an over learned strategy for solving simpler problems.

A second curricular design strategy to highlight quantitative structure is to include problems that permit many numerical answers. For example, Problem 5 can be adapted as follows:

> Problem 6. Sam and Joseph each had a shorter sister, and they argued about who stood taller over their sister. Sam won the argument by 14 centimeters. He was __ cm tall; his sister was __ cm; Joseph was __ cm tall; and his sister was __ cm tall. What numbers can you put in the blanks so that everything works out? (Adapted from P. W. Thompson, 1993)

Because only one quantity has a given value, this form of the situation shifts attention further toward the structure of the additive relationships.

Six of the seven quantities can take on different values, and students must coordinate their choices of height values to generate appropriate values of the three differences. This strategy of asking students to reason from a difference rather than to a difference can be applied to generate a wide range of interesting additive problems.

Student Reasoning and Pedagogical Considerations

All students have the capacity, both existing and potential, to develop significantly better quantitative reasoning skills than is currently likely in most elementary mathematics classrooms (Carpenter et al., 1982; Nemirovsky, Tierney, & Ogonowski, 1993; Vergnaud, 1982, 1983). But that development depends on students' committing to the goal of describing situations, quantities, and relationships as clearly as possible as they see them. Students, particularly those who only see mathematics as finding numerical answers by arithmetic, are unlikely to accept that goal if clear reasoning is not valued and rewarded, appropriate support for clearer thinking and communication is not provided, and multiple perspectives on situations are not welcomed. Each of the requirements makes specific demands on teachers and the norms they maintain in their classrooms (Wood, Cobb, Yackel, & Dillon, 1993).

In comprehending and communicating quantitative situations in their own terms, students have two primary means of expression: verbal descriptions and various sorts of external diagrams.[10] Producing a conceptually clear verbal description does not require using the "right" terminology; it is a matter of taking the task of clear description for others seriously, listening to others' reactions, and clarifying and refining when necessary. When students' verbal descriptions lose clarity and do not communicate well—either to teachers or their peers—they can be asked to show their thinking in a diagram. As with verbal descriptions, the nature of drawings will vary with the situation and the student, but here is a representative example. To manage the quantitative complexity of Problem 6, one fifth-grade student used pencils placed side-by-side (Fig. 4.10) to represent heights and basic differences (see P. W. Thompson, 1993, for a more complete account).

Such difference diagrams can provide useful grounding for discussions of problems like Problem 6, where the value of a difference is known but

[10]We also acknowledge the role of gesture in mathematical reasoning. We focus our discussion on verbal descriptions and diagrams because they are usually more permanent forms of expression (e.g., when words or phrases are written on the board) and therefore more accessible foci of group discussion and understanding.

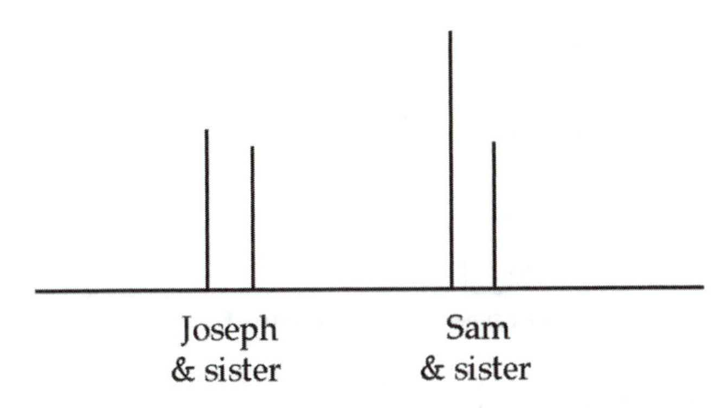

FIGURE 4.10. Concrete comparison of differences.

the values of the other quantities are not. More broadly, and perhaps more algebraically, they can help students to think about quantitative relationships at a level of generality beyond the immediate problem. With some experience and practice they can also easily be adjusted and annotated to fit a wider range of problems. For example, diagrams may make it easier for students to explore whether the winner in Problem 6 could be the shorter boy, when he stands taller over his sister by more than the other boy (Fig. 4.11).

Teacher Questions and Classroom Discussion

Teaching quantitative reasoning involves two main components: choosing a sequence of situations and providing appropriate support for students' reasoning. Teachers can prepare in advance for the help they provide their students. For each situation, they should first decide for themselves what quantities are involved, how they are related, and how they would describe the situation quantitatively. Then they should imagine how their students might describe the situation differently and what conceptual difficulties might be lodged in their descriptions (e.g., the common event of confusing the value of a difference with one of its constituent quantities). Although students' interpretations cannot be completely predicted in advance (in fact, presuming to know exactly which errors are coming can lead to terrible problems), teachers do not need to wait for the discussion to start thinking about how they might respond.

Because the central goal is to focus on quantities and how they relate in situations, and because this represents a major mathematical change of focus for many students, it is important to open discussions with

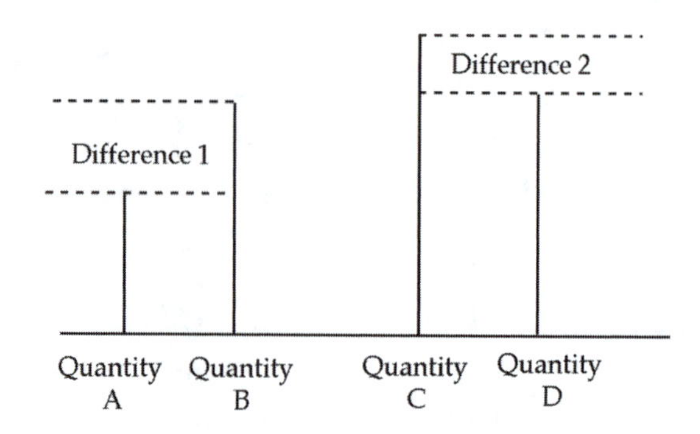

FIGURE 4.11. A general scheme of additive comparisons.

questions that lead to discussions of quantities, not numbers. A useful opening question can be a general one, like what is going on here? The goal is to get students to describe situations as they see them. In supporting quantitative discussions, the most central skill is careful listening. Using their own knowledge of the situations, teachers can listen for which quantities are mentioned, which are central for particular students, and how students see relationships between those quantities. For situations that express additive relationships, teachers should especially listen for how students are discriminating differences from other quantities in the situation.

In teaching fifth graders about complex additive situations, one of us opened the discussion of Problem 5 as follows (P. W. Thompson, 1993):

Teacher (T):	What are they doing?
Several students:	Arguing.
T:	What are they arguing about?
S1 & S2:	Who's taller.
S3:	Who's taller than Sam's sister and Joseph's sister.
T:	Are they arguing about who's taller?
Several students:	No.
S4:	Who was taller than their sisters.
T:	Who was taller than their sister?
Several students:	Yeah.
T:	Are they both taller than their sisters?
Several students:	Yes.
S1:	Who was *more* taller.
T:	Who was *more* taller? What does that mean?

(Students move quickly to modeling the situation using four pencils; they also try adjusting their model, at the teacher's request, so that the shorter brother is actually the winner.)

This is one example of students reasoning quantitatively (note the absence of references to numbers) with appropriate teacher support. The opening question led to increasingly accurate descriptions of the comparisons in the situation. The teacher was able to point to where students needed to revise their views without being explicit about how they should change or doing it for them. Students were still "in charge" of their thinking.[11]

Perhaps the most challenging task in supporting quantitative reasoning is to listen for and respect alternative descriptions of the quantities in the situation while pressing all participants for clarity. This is challenging work when two or more quite different views of the situation have been expressed. For example, in Problem 5, one student might emphasize the differences and another the relationships between the constituent quantities (the heights). When quite different perspectives arise, teachers can help their students by highlighting the contrasts between them and then managing when each view gets worked on. Leading the discussion to elaborate only one view at a time may help the class come to see different descriptions as simply different perspectives of the same situation.

TWO FINAL EXAMPLES: DEVELOPING QUANTITATIVE FOCUS

We have stressed that the current K–8 mathematics curriculum, with its emphasis on numbers and arithmetic, falls well short of adequately developing students' quantitative reasoning as a foundation for algebra. To achieve a better curricular balance between quantitative/situational reasoning and numerical reasoning, many additive situations currently expressed as numerical word problems can be adapted to focus on quantities and relationships. This section wraps up with two examples that illustrate how simple this process can be.

---

[11]It is worthwhile to note that, in contexts like this, we recommend against moving too quickly to simplify the situation by asking children to work with numbers. The point of this example is to illustrate how a teacher can support students' thinking about the situation without drawing their attention to calculating an answer.

A Single Difference

Many word problems in the primary curriculum are situated in quantitative change, the gains and losses of everyday objects, for example:

Problem 7. Tony had 11 marbles but he lost 4 marbles to Marguerite in a game. How many marbles did Tony have after the game?

Instead of following up Problem 7 with many others with the same structure, Problem 8 can be given:

Problem 8. Sharon lost 6 marbles to Philip in a game. What can we say about the number of Sharon's marbles before and after the game?

Problem 8 asks students to reason from a difference, rather than to combine the values of a difference and one constituent quantity to find the missing value of the other. Although students may initially object that no answer exists, once five or six pairs of values have been generated for Sharon's marbles, they can explore the properties of those pairs. For example, they can determine that the smallest number of marbles Sharon could start the game with was 6. Of course, if this problem is presented in isolation, students will likely not accept it as a problem or understand its role in the development of their thinking. However, when it is one small piece in a more deliberate, multiyear quantitative curriculum that challenges and extends their reasoning capacities, students' reaction may be quite different.

Coordinating Two (or More) Differences

The upper elementary curriculum contains many complex, multistep additive word problems, like Problem 9:

Problem 9. An elementary school has two fourth-grade classes, Room 5 and 7, and two fifth-grade classes, Room 6 and 8. There are 52 fourth graders in the school, which is 7 more students than are in the fifth grade. There are 22 students in Room 6. How many students are in Room 8?

Whereas this situation is certainly more complex than Problems 7 and 8, the reasoning required of students is minimized by the structure of the quantities and their known values. The value of the difference between the combined sizes of fourth-grade classes and fifth-grade classes determines

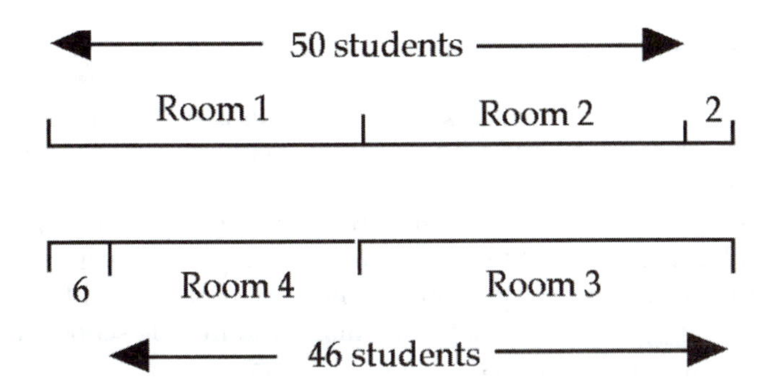

FIGURE 4.12. Complex arrangement of differences and combinations.

the total number of fifth graders (52 – 7 = 45), and the given number of students in Room 6 determines the number of students in Room 8 (45 – 22 = 23). Only one difference appears in this situation; it compares the sizes of the two combined classes and has a value of 7 children.

Problem 10 describes a similar situation with a slightly different structure to provoke greater attention to quantitative relationships:

> Problem 10. The same elementary school has two first-grade classes, Rooms 1 and 2, and two second-grade classes, Rooms 3 and 4. Rooms 1 and 2 together have 50 students, and Rooms 3 and 4 together have 46 students. Room 1 has 6 more students than Room 4 and Room 2 has 2 fewer students than Room 3. How many students can be in each room? Is there only one possible size of each class?

The reasoning generated by Problem 10 will likely be quite different for a number of reasons. Because the size of the individual classes is only restricted by the size of grades (total first and total second graders), many different values will "work." There is also the interesting and important relationship between the differences between classes (e.g., Room 1 has 6 more children than Room 4) and the difference between grades (the first grade has 4 more children than the second grade). If students do not raise the issue, then teachers can ask if the relationship between three differences ("6 more" combined with "2 less" is equivalent to "4 more" overall) is coincidence or not. Diagrams like Figure 4.12 can help support students' analysis of this situation. Also, as suggested before, they provide teachers with a convenient structure for generating new problems by varying the given numerical values and helping students to trace the impact of these changes and look for generality in the situation.

CONCLUSIONS

We have attempted to sketch out one proposal for early algebra: an approach to elementary and middle school mathematics that both readjusts the current K–8 focus on arithmetic (numbers and operations) and supports the development of algebraic reasoning. We question the content of algebra I as the presumed standard for early algebra development and with it, the presumed developmental linkage between arithmetic and algebra. We suggest instead that elementary and middle school curricula be reconceptualized in terms of students' quantitative, arithmetic, and algebraic reasoning. In contrast, we recognize that we have not outlined a curriculum in quantitative reasoning. Although that is a pressing and important task, it is not one that can be addressed in a single chapter. Instead, we have tried to show how simple adaptations of current curricula, when taught with a different emphasis, can make students' mathematical experiences much richer quantitatively.

Quantitative Reasoning Is a Central Dimension of Students' Mathematical Development. It is related to, but in an important sense independent of, both arithmetic and algebra. It is also foundational for both arithmetic and algebra, providing content and meaning for numerical and symbolic expression and computation. The common *arithmetic to algebra* framework is too limiting and narrow; discussions of early algebra should be framed in terms of connections among quantitative reasoning, arithmetic reasoning, and emergent algebraic reasoning (Fig. 4.13). In each area, we should consider reasonable goals for students' learning, available curricula, useful teaching tools, and research on students' capabilities.

The Current Emphasis on Numerical and Symbolic Expression and Manipulation Is Fundamentally Flawed. It is flawed largely because it fails to substantively connect mathematics to students' experiential world. The implications of this disconnection are varied, profound, and negative for too many students by the middle school years. For students to learn mathematics that is powerful and productive, more attention must be given to the development of quantitative reasoning. Developing students' abilities to conceptualize and reason about situations in quantitative terms is no less important that developing their abilities to compute.

A Rich Program of Quantitative Reasoning Spurs the Development of Students' Conceptual and Representational Capacities as It Connects Mathematics to the World of Objects and Situations, Measurement, and Change. It pushes students to examine, articulate, and represent general relationships among

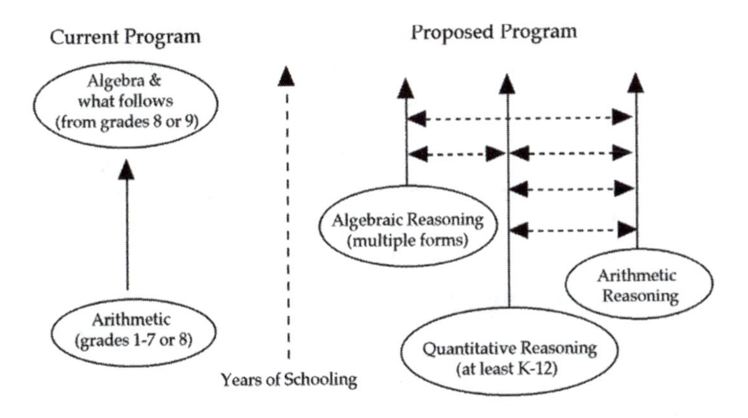

FIGURE 4.13. Two views of the introduction of algebra.

and between quantities. Without the understanding of such general, conceptual relationships, students find little need or sense in learning the tools of algebra. If, on the other hand, students develop mathematical ideas of sufficient complexity—among them complex quantities and relationships between quantities—their expression, manipulation, and further abstraction in algebraic notation can become a more meaningful and sensible activity.

Building Sophisticated Quantitative Reasoning Skills for the Majority of Students Is Not a 1- or 2-Year Program; It Requires Development Throughout the Elementary and Middle School Years. Students often come to school with substantial quantitative competence in additive relationships and build that competence, in and outside of classrooms. However, their development of skilled quantitative reasoning will depend on instructional programs that recognize and extend students' existing abilities. These programs will require work on more complex additive situations and relationships and, even more centrally, on developing students' abilities to conceptualize and reason about multiplicative quantities and relationships (Harel & Confrey, 1994; Vergnaud, 1983, 1988).

Some important components of such a program include: suitably chosen situations and problems, greater instructional focus on making sense of quantities and relationships in those situations than on finding answers, emphasis on reasoning and expression that is neither numerical or symbolic (in the sense of traditional algebraic symbols), and support for representing and communicating reasoning clearly and publicly in diagrams and open-form expressions.

In closing, we address two potential objections to our position: one concerning content balance in the K–8 curriculum and another concerning the claim that quantitative reasoning prepares students for diverse approaches to algebra.

Throughout, this chapter has argued for better curricular balance between teaching and learning about number/operation and quantity/quantitative reasoning. But, although quantity and number are central categories of elementary mathematics, they do not comprise a comprehensive K–8 curriculum. We believe that students should also experience extended work with metric and nonmetric geometry, data and statistics, as well as introductory probability. Each can provide important ideas to represent and reason about in greater generality with algebraic symbols and methods (see Boester & Lehrer, chap. 9, this volume, for examples from geometry). We acknowledge but cannot resolve the fundamental dilemma that there is more worthwhile mathematics to learn than there is space in the school curriculum to teach and learn it. Competition between ideas and subfields of mathematics is a necessary, perhaps not wholly negative, result. Our push to shift from a number-only orientation (K–8) to a number and quantity orientation reflects our present task of reconceptualizing early algebra as something more than the common view of algebra as generalized arithmetic.

Likewise, we do not believe that a strict conceptual distinction between number and quantity is psychologically defensible or educationally useful. Children can find wonder and engage deeply in the nature of numbers qua numbers and in quantities and situations. Any good mathematics curriculum should recognize and nurture both interests. Moreover, it can be difficult to decide if a person's reasoning is more numerical or more quantitative, particularly when the quantities are numerically specific. In most cases, there is a natural dialogue between the two mathematical dimensions. But softening the distinction between number and quantity does not undermine our fundamental argument that much more attention should be given to thinking about quantities, relationships, and situations.

We argued initially that a K–8 program that gives more attention to quantitative reasoning can support numerous conceptually different and sensible introductions to algebra, as replacements for the current U.S. algebra I course. But our proposal could be seen as supporting, most directly, the more applied view of algebra, algebra as modeling. We have, after all, repeatedly emphasized the importance of making sense of problem situations and that an important kind of mathematical reasoning examines general relationships between quantities. Indeed, we think that introducing algebra as a set of tools for expressing reasoning about complex situations and for generalizing their solutions has substantial promise, especially when the assumption is that all students will take and learn algebra (Silver, 1997). But we also believe that this quantitative, applied introduction can easily support the subsequent shift toward the more formal, structural side of algebra. Algebraic knowledge that has grown from a quantitative root stalk can serve as the basis for moves toward increasing abstraction and focus on abstract structure, making them abstractions from students experience and in students' reasoning about that experience.

Finally, we note that whereas we were familiar with the Davydov and El'konin approach to basing early mathematics instruction on ideas of quantity (Davydov, 1975, 1982; El'konin & Davydov, 1975), we have only recently become aware of curricular materials being developed and researched that bind ideas of quantity with inscriptional practices that mean to provide a bridge between quantitative and symbolic reasoning (Dougherty, chap. 15, this volume). In this approach, children are asked to think about quantities' measures and relationships among them, and to attend to what one can deduce from those relationships. For example, the Davydov and El'konin curriculum intends that children read "$A/B = 5$" as "Quantity A, measured in units of Quantity B, has a measure of 5." It is then a small step to deducing that "A is 5 times as large as B," which introduces the mathematical notion of fraction as reciprocal relationships of relative size (P. W. Thompson & Saldanha, 2003), while respecting the starting point that numbers are measures and symbolic statements capture quantitative relationships. Clearly much work remains to be done.

ACKNOWLEDGMENTS

The authors are grateful for the David Carraher's insightful comments on a previous draft of this chapter. The first author also acknowledges, in memoriam, the hospitality and collegial support of Alba Thompson during his working visit to the Thompson home. Discussions between the authors during that visit played an important role in shaping the content of this chapter.

REFERENCES

Bednarz, N., Kieran, C., & Lee, L. (Eds.). (1996). *Approaches to algebra: Perspectives for research and teaching*. Dordrecht, the Netherlands: Kluwer Academic.

Carpenter, T. P., & Moser, J. M. (1984). The acquisition of addition and subtraction concepts in grade one through three. *Journal for Research in Mathematics Education, 15*, 179–202.

Carpenter, T. P., Moser, J. M., & Romberg, T. A. (Eds.). (1982). *Addition and subtraction: A cognitive perspective*. Hillsdale, NJ: Lawrence Erlbaum Associates.

Carraher, D. W., Schliemann, A. D., & Brizuela, B. (2000, October). *Children's early algebraic concepts.* Plenary address presented at the 22nd annual meeting of the North American Chapter of the International Group for the Psychology of Mathematics Education, Tucson, AZ.

Chazan, D. (2000). *Beyond formulas in mathematics and teaching: Dynamics of the high school algebra classroom.* New York: Teachers College Press.

Confrey, J. (1991). *Function Probe©* [Software]. Ithaca, NY: Cornell University.

Cuoco, A. (1993). Toward one meaning for algebra. In W. D. Blair, C. B. Lacampagne, & J. Kaput (Eds.), *Algebra initiative colloquium* (Vol. 2, pp. 207–218). Washington, DC: Department of Education.

Cuoco, A. (1995). Some worries about mathematics education. *Mathematics Teacher, 88*(3), 186–187.

Davydov, V. V. (1975). The psychological characteristics of the "prenumerical" period of mathematics instruction (A. Bigelow, Trans.). In L. P. Steffe (Ed.), *Soviet studies in the psychology of learning and teaching mathematics* (Vol. 7, pp. 109–205). Chicago: University of Chicago.

Davydov, V. V. (1982). The psychological structure and contents of the learning activity in school children. In T. P. Carpenter, J. M. Moser, & T. A. Romberg (Eds.), *Addition and subtraction: A cognitive perspective* (pp. 224–238). Hillsdale, NJ: Lawrence Erlbaum Associates.

Demana, F. D., & Waits, B. (1990). Instructional strategies and delivery systems. In E. L. Edwards (Ed.), *Algebra for everyone* (pp. 53–61). Reston, VA: National Council of Teachers of Mathematics.

Dossey, J. A. (1998). Making algebra dynamic and motivating: A national challenge. In Center for Science, Mathematics, and Engineering (Ed.), *The nature and role of algebra in the K–14 curriculum* (pp. 17–22). Washington, DC: National Academy Press.

Edwards, E. L. (Ed.). (1990). *Algebra for everyone.* Reston, VA: National Council of Teachers of Mathematics.

El'konin, D. B., & Davydov, V. V. (1975). Learning capacity and age level: Introduction (A. Bigelow, Trans.). In L. P. Steffe (Ed.), *Soviet studies in the psychology of learning and teaching mathematics* (Vol. 7, pp. 1–12). Palo Alto, CA: School Mathematics Study Group and National Council of Teachers of Mathematics.

Fennema, E., Carpenter, T. P., & Peterson, P. L. (1989). Learning mathematics with understanding: Cognitively guided instruction. In J. Brophy (Ed.), *Advances in research on teaching* (pp. 195–221). Greenwich, CT: JAI.

Fey, J. T. (1989). School algebra for the year 2000. In S. Wagner & C. Kieran (Eds.), *Research in the learning and teaching of algebra* (pp. 199–213). Reston, VA: National Council of Teachers of Mathematics.

Fey, J. T. (1990). Quantity. In L.A. Steen (Ed.), *On the shoulders of giants* (pp. 61–94). Washington, DC: National Academy Press.

Fuson, K. C., Wearne, D., Hiebert, J. C., Murray, H. G., Human, P. G., Olivier, A. I., Carpenter, T. P., & Fennema, E. (1997). Children's conceptual structures for multi-digit numbers and methods of multi-digit addition and subtraction. *Journal for Research in Mathematics Education, 28,* 130–162.

Harel, G., & Confrey, J. (Eds.). (1994). *The development of multiplicative reasoning.* Albany, NY: SUNY Press.

Heid, M. K. (1995). *Algebra in a technological world*. Reston, VA: National Council of Teachers of Mathematics.

Kaput, J. (1995, October). *A research base supporting long term algebra reform?* Invited address, annual meeting of the North American Chapter of the International Group for the Psychology of Mathematics Education, Columbus, OH.

Krutetskii, V. A. (1976). *The psychology of mathematics abilities in school children* (J. Teller, Trans.). Chicago: University of Chicago Press.

Lobato, J., Gamoran, M., & Magidson, S. (1993). *Linear functions, technology, and the real world: An algebra 1 replacement unit*. Berkeley, CA: University of California.

National Council of Teachers of Mathematics Algebra Task Force. (1993). *Report from the Algebra Task Force*. Reston, VA: Author.

National Council of Teachers of Mathematics Algebra Working Group. (1997). *A framework for constructing a vision of algebra: A discussion document*. Final report to the Board of Directors. East Lansing, MI: Michigan State University.

Nemirovsky, R., Tierney, C., & Ogonowski, M. (1993). *Children, additive change, and calculus* (Working Paper No. 2-93). Cambridge, MA: TERC.

Phillips, E., & Lappan, G. (1998). Algebra: The first gate. In L. Leutzinger (Ed.), *Mathematics in the middle* (pp. 10–19). Reston, VA: National Council of Teachers of Mathematics.

Piaget, J. (1952). *The child's conception of number*. London: Routledge & Paul.

Piaget, J. (1970). *The child's conception of movement and speed*. New York: Basic Books.

Poincaré, H. (1904). Les definitions en mathematiques. *L'Enseignement des Mathematiques [Definition in mathematics], 6*, 255–283.

Riley, M. S., Greeno, J. G., & Heller, J. I. (1983). Development of children's problem-solving ability in arithmetic. In H. P. Ginsburg (Ed.), *The development of mathematical thinking* (pp. 153–196). New York: Academic Press.

Schoenfeld, A. H. (1985). *Mathematical problem solving*. Orlando, FL: Academic Press.

Schwartz, J., & Yerushalmy, M. (1992). Getting students to function in and with algebra. In G. Harel & E. Dubinsky (Eds.), *The concept of function: Aspects of epistemology and pedagogy* (MAA Notes, Vol. 25, pp. 261–289). Washington, DC: Mathematical Association of America.

Silver, E. A. (1997). "Algebra for all"—increasing student's access to algebraic ideas, not just algebra courses. *Mathematics Teaching in the Middle Grades, 2*, 204–207.

Sowder, L. (1988). Children's solutions of story problems. *Journal of Mathematical Behavior, 7*, 227–238.

Thompson, A. G., Philipp, R. A., Thompson, P. W., & Boyd, B. A. (1994). Calculational and conceptual orientations in teaching mathematics. In D. B. Aichele & A. F. Coxford (Eds.), *Professional development for teachers of mathematics* (1994 NCTM yearbook, pp. 79–92). Reston, VA: National Council of Teachers of Mathematics.

Thompson, P. W. (1989, April). *A cognitive model of quantity-based reasoning in algebra*. Paper presented at the annual meeting of the American Educational Research Association, San Francisco.

Thompson, P. W. (1993). Quantitative reasoning, complexity, and additive structures. *Educational Studies in Mathematics, 25*, 165–208.

Thompson, P. W. (1994). The development of the concept of speed and its relationship to concepts of rate. In G. Harel & J. Confrey (Eds.), *The development of multiplicative reasoning* (pp. 179–234). New York: SUNY Press.

Thompson, P. W., & Saldanha, L. A. (2003). Fractions and multiplicative reasoning. In J. Kilpatrick, G. Martin, & D. Schifter (Eds.), *Research companion to the* Principles and Standards for School Mathematics (pp. 95–114). Reston, VA: National Council of Teachers of Mathematics.

Vergnaud, G. (1982). A classification of cognitive tasks and operations of thought involved in addition and subtraction problems. In T. P. Carpenter, J. M. Moser, & T. A. Romberg (Eds.), *Addition and subtraction: A cognitive perspective* (pp. 39–59). Hillsdale, NJ: Lawrence Erlbaum Associates.

Vergnaud, G. (1983). Multiplicative structures. In R. Lesh & M. Landau (Eds.), *Acquisition of mathematics concepts and processes* (pp. 127–174). New York: Academic Press.

Vergnaud, G. (1988). Multiplicative structures. In J. Hiebert & M. Behr (Eds.), *Number operations and concepts in the middle grades* (pp. 162–181). Hillsdale, NJ: Lawrence Erlbaum Associates.

Wood, T., Cobb, P., Yackel, E., & Dillon, D. (Eds.). (1993). Rethinking elementary school mathematics: Insights and issues. *Journal for Research in Mathematics Education, 6*.

5

Representational Thinking as a Framework for Introducing Functions in the Elementary Curriculum

Erick Smith
Cayuga Pure Organics

> If I had to explain what is algebra to a student, I would say: "think of all that you know about mathematics. Algebra is about making it richer, more connected, more general, and more explicit. . . . " (Ricardo Nemirovsky, Communication to the Algebra Working Group, September 19, 1994)

This chapter has several objectives. First, I want to describe a theoretical framework for thinking about algebra and algebraic thinking. Although I think it is important to think of algebra in the widest possible context in the elementary years, there is still reason to be careful and perhaps even precise in describing what we mean by algebraic thinking. One important distinction is between two kinds of algebraic thinking, which I term *representational thinking* and *symbolic thinking*, respectively. Symbolic thinking is related to the way one understands and uses a symbol system (of which the usual algebraic character string-based one is, of course, a primary example) and its associated rules. In symbolic thinking, the focus is on the symbols themselves, without regard for what they might refer to. On the other hand, representational thinking is reserved to designate the mental processes through which an individual creates referential meaning for some representational system. Kaput refers (chap. 1, this volume) to these in terms of the two core aspects of mathematical thinking.

A second goal of this chapter is to argue that thinking about how individuals create their own mathematical certainty is an important part of representational thinking and a neglected route to making mathematics more connected, more general, and more explicit in Nemirovsky's terms. A third goal is to describe an approach to building student understanding of functions that is applicable in the elementary years that builds on students' sense-making activities around constructing units, working toward an understanding of linear functions reflecting a longitudinal approach to the function strand of algebra. Finally, I argue that occupied context plays an important role in connecting mathematical certainty with functional thinking—that is, in allowing functional thinking to become part of the algebraic process of enriching our mathematical experience.

Throughout, there are cross-references to the two lead-off chapters of the book because the analyses here, although originating several years prior to the writing of chapters 1 and 2, provides strong and specific illustrations of the elements of algebraic reasoning and symbolization outlined in those chapters.

ALGEBRA AS GENERALIZATION: REPRESENTATIONAL THINKING

As pointed out by Kaput (chap. 1, this volume), algebraic reasoning involves many forms and flavors of generalizing activity and representational systems for young children. The importance of keeping a wide view of what counts as algebra is apparent in several examples from Bastable and Schifter (chap.6, this volume). In many of their examples, the algebra takes the form of natural language statements, for example, the conjecture by Knox and Adam that if you take two consecutive numbers, you add the lower number to it's square to the higher number and you get the higher number's square. When children mature, these ways of representing will become differentiated and conventionalized as they come to serve differing purposes and as the individual becomes more aware of conventional classes of symbol systems. Thus, algebra as a support for representational thinking narrows to certain symbol systems, and algebra eventually comes to include symbolic thinking. However, in the elementary grades, our primary interest is in representational thinking, in how children create meaningful representations and in so doing build and express generalizations. Initially, it makes little sense to make an a priori decision about what kinds of representations will count as algebra, but instead to investigate how different representational forms become useful tools in relation to the problems and issues of the students, especially in how they express and argue for generalizations. We will see that representational thinking can cover a broad range of generalizing activities, including some that, at first glance, would be described as arithmetic.

For example, for a 5-year-old, the use of the number 2 to designate a class of sets having two members can be as algebraic as a second grader's use of a table to represent a class of combinations, a third grader's use of a graph, a fourth grader's use of a pattern, or a fifth grader's use of a conventional algebraic equation. From the experiential perspective of the individual, each of these can be a process of generalizing, thus creating meaning for a representational system. On the other hand, as Mason notes in chapter 3 (this volume), expressing distinctions can reify them, contributing to the sense of objectiveness, which we develop from our contact with the material world and generalize to abstract ideas. Thus, number becomes as concrete and objective as chair or spoon. Fractions, decimals, and *x*s can also become equally concrete if the process of reflective abstraction is provoked, supported, and permitted rather than blocked (Mason, chap. 3, this volume; 1996). That is, a fourth grader who has objectified the counting numbers will seldom be engaged in representational thinking when doing computational arithmetic as opposed to when she is using numbers in modeling activity, for example, making tables. Likewise, we might argue that a professional mathematician's use of x (or elements of any symbolic system) would not necessarily involve representational thinking unless the activity involves using these symbols in a process of generalization. That is, symbol manipulation in itself, at any age, does not involve representational thinking because it is symbolic thinking, thinking guided by working knowledge of the symbol system itself.

Thus, we might view the relationship between representational thinking and symbolic thinking as one of leapfrog. Whereas the representational thinking of the 5-year-old involves creating meaning for integers (algebra for the 5-year-old), this representational thinking leads to the objectification of the integers, which the fourth grader can then use as objects that form a pattern. This relates to the symbolization processes, as outlined in chapter 2 by Kaput, Blanton, and Moreno (this volume), whereby the "representing" system becomes, through a new symbolization, a represented system. Ultimately, the idea of categorizing some problems as algebraic and some not algebraic independently of the child mistakenly places the algebraic activity in the problem rather than in the thinking of the child.

Taking an individual perspective is one way of looking or noticing, and does not intend to indicate a view of individuals as isolated sense-makers (or representational thinkers) within their experiential world. That is, one can emphasize an individual perspective within a framework that sees learning as occurring through social interactions within social and cultural contexts. In fact, it is hard to imagine representational thinking in any complex way as occurring without such cultural tools as representational systems that children encounter in or out of school. Cobb, Boufi,

McClain, and Whitenack (1997) have suggested the term *collective reflection* to indicate the process of reflective abstraction that takes place within the context of a reflective discourse. In analyzing an episode from a classroom (to be discussed later) in which several students seemed to be using the classroom discourse as a means to reflect on and objectify the results of a previous activity, they state:

> We speculate that the children were reflecting on and objectifying their prior activity by virtue of their participation in the discourse. In other words, the children did not happen to spontaneously begin reflecting at the same moment. Instead, they were reflecting because they were participating in the discourse. It seems reasonable to talk of *collective discourse* in such instances to stress that it was a communal activity. (p. 7)

So emphasizing an individual perspective does not minimize the importance of the social practice. Rather, it is a deliberate attempt to focus on how the reflecting of the various individuals might take place, what gets objectified for different students, and how this plays a role in their own sense making, recognizing that this takes place within such communal activities.

As a framework for seeking, provoking, and understanding representational thinking in elementary classrooms, the following three aspects of the theoretical framework are all equally important:

1. *Constructivist aspect*: Generalization as a constructive process all children engage in when working on mathematical problems.
2. *Sociocultural aspect*: Algebra as different kinds of representational systems, both conventional mathematical systems and invented/-contextual systems, including the (perhaps tacit) rules for using these systems.
3. *Integrated aspect*: Representational thinking as interweaving the children's construction of mathematical generalizations with a representational system. The next section argues that the construction of mathematical certainty is an important part of these generalizing processes.

This proposed theoretical framework is close to the pragmatic approach advocated by Cobb (1994) and Cobb and Bowers (1999). However, it is also compatible with other approaches advocated in this volume. Mason's discussion of the broad contexts in which children's generalizations take place is in the same spirit. In addition, J. Smith and Thompson (chap. 4, this volume) question the value of a top-down approach, that is, identifying something to be called algebra (e.g., patterns and functions, formal properties, modeling, structure, language, etc.) and scaling it down to be appropriate for elementary school. Instead, they argue:

> We believe it is possible to prepare children for different views of algebra—algebra as modeling, as pattern finding, or as the study of structure—by having them build ways of knowing and reasoning which make those views appear as emergent aspects of a central and fundamental way of thinking. (p. 97)

MATHEMATICAL CERTAINTY

The idea of mathematical certainty has been seriously challenged. From Lakatos's *Proofs and Refutations* (1976) to Ernest's (1991) fallible mathematics, there is a growing consensus that mathematical knowledge, rather than reflecting eternal truths, is a fallible human construction whose truths can and do change across history and circumstance (Davis & Hersh, 1999; Kline, 1980). Although this is not the place for a detailed examination of this complex issue, we can identify another, more local, kind of certainty that is central to learning mathematics—the certainty most adults feel and express when asked about their own mathematical claims (e.g., that $2 + 2 = 4$). In teaching mathematics, we do want our students to feel sure about the answers they propose. However, we also want that certainty to be internal, to be built by students based on their understanding of the problem situation. Yackel (1993) suggests that, in school, certainty often comes from an authority figure and this is often reinforced by classroom dialogue in which the teacher only questions incorrect answers. That is, the students become certain of a correct answer when the teacher does not question it. Teachers attempting to create an inquiry-based classroom often need to confront the previous reinforcement that students have received for relying on external authorities, including the textbook. She describes a classroom interaction where the teacher has posed a problem and a student, DN, has proposed an answer (six). Even though the proposed answer is correct, the teacher continues the dialogue asking: There are six? Alright six. Is that right class? In the ensuing dialogue, the student interprets the teachers questioning to indicate that six is incorrect and successively proposes other answers (Seven? Eight?). After a few minutes, the teacher, Mr. K, asks a different question:

Mr. K: What's your name?
DN: My name is Donna Walters
Mr. K: What's your name?
DN: My name is Donna Walters
Mr. K: If I were to ask you (your name) … again, would you tell me your name is Mary?

DN: No.
Mr. K: Why wouldn't you?
DN: Because my name is not Mary.
Mr. K: And you know your name is— ... If you're not sure you might have said your name is Mary. But you said Donna every time I asked you because what? You what? You know your name is what?
DN: Donna.
Mr. K: Donna. I can't make you say your name is Mary. So you should have said, Mr. K. Six. And I can prove it to you. (Yackel, 1993, p. 5)

Even if we might not want to discourage student conjectures when they lack a proof, we do sympathize with Mr. K's goal that students should construct certainty in their mathematical knowledge that is comparable to their certainty of their name and, hopefully, be able to articulate their reasons for that certainty. However, if their mathematical certainty is not based on eternal truths or on external authority, then we expect the student's experience to play a central role in the construction of this certainty. This section argues first for such an experiential-based mathematical certainty and, second, that the construction of mathematical certainty is a basic driver of representational reasoning.

Constructing Mathematical Certainty

Why do we feel so certain that $2 + 2 = 4$? Although some might answer this from a formalist perspective, this almost certainly plays little role in the creation of the profound certainty most people feel for this claim. From a constructivist perspective, we would see an important role for the experience of combining or putting together two sets of objects, of connecting the combining action with a quantifying (counting or measuring) action, and of reaching the certainty that, regardless of the ways in which the two sets are combined, the outcome is the same. Thus, quantification, combination, and the consistency of the quantitative result of combination are fundamentally based in experience.

However, to make a general claim of certainty about an additive statement requires the coordination of this experiential base with the mediational role of language. To claim with certainty requires one to be able to discern which combinations of sets count and which ones do not count as $2 + 2 = 4$. Language allows one to name sets of objects and their quantification, which ultimately provides the power to identify when the quantification of two combinations of sets is the same. This is clearly a two-way process: It is through our experience that we initially create meaning for language, but this language then becomes the tool that mediates our experience and ultimately allows for general claims of mathematical certainty.

One interesting setting in which to examine this process is in relation to Piaget's nonconservation tasks with young children. Based on various tasks, Piaget concluded that young children do not conserve quantity under certain conditions. A typical task supporting this conjecture is to show a child a row of eggcups, each containing an egg, and then ask whether there are more eggcups or eggs. Most children say there is the same amount of both. The eggs are then spread out over a large space and the eggcups pushed close together. Almost all 4- and 5-year-old children will say there are now more eggs. And, as Papert (1993) states, "They will defend this position even under extensive cross-questioning and even when pressure is placed on them to change their minds ... " (p. 154).

Such students have clearly constructed a certainty about this situation. However, from an adult perspective, we see this certainty as wrong and often interpret this error as originating in a failure to conserve quantity due to an undeveloped sense of logical necessity. Papert (1993), however, suggests another possibility:

> A sensible objection that casts light on what is really being learned is that the children are more likely to have misunderstood the question than to hold the bizarre non-conservationist opinion. They think they are being asked about the space occupied and not about number. In one sense the objection must be true. If the children really understood the question as we do, they would answer as we do. But the objection deepens rather than trivializes Piaget's experiment. There may indeed be a misunderstanding, but it is not a "mere verbal misunderstanding." It reflects something deep about the child's mental world. (pp. 154–155).

Papert (1993) is not arguing that young children do conserve quantity, rather that their use of language may not allow them to distinguish between the language of space and the language of numeric quantity. He continues, "The work being done in the concrete period is that of gradually growing the relevant mental entities and giving them connections so that such distinctions become meaningful. When you or I see six eggs, the sixness is as much part of what we see as the whiteness or the shape of the individual objects" (p. 155). Within Papert's framework, one could argue that these students, in sticking to their position, have constructed a mathematical certainty by connecting their experience of space and quantity with their arguments for more eggs. The claim for more eggs than eggcups is a means of closing the cycle in relation to their own experiences and their understanding of language. From this perspective, they would be, in fact, correct in the claim they are making. Of course, this does not mean that the certainty they have constructed is permanent, rather that it is a certainty within the context of the experiential world of the children. As this experiential world evolves, particularly as their understanding of more grows closer to the taken-as-shared knowledge of adults, this certainty will also evolve. For the student as sense-maker, the certainty will both be infallible within the context of his experiential world and also change as that world grows and changes through experience and communal activity.

In this example, the power of the mediational role of language becomes apparent. Coming to know that the quantity of eggs does not change under rearrangement says less about the structure of the world than it does about the role of language in structuring experience. As adult language is introduced into the experiential world of learners, they must engage in a process of negotiation to produce a fit between language and experience. Thus, we come to an understanding that any action on a set that does change the quantity is to be excluded from those things we call rearranging a set. Likewise, in the case of addition, any action on two sets, one numbering 100 and the other numbering 200, which does not result in a single set numbering 300 is excluded from that which we call addition. Certainty arises out of this negotiating process among experience, conceptual operations, and language. It should be clear, however, that although we construct this certainty, it is very much a certainty. The construction of our mathematical certainty and of our experiential world is intertwined in such a way that they feel very much inseparable. The possibility is unimaginable precisely because that is the part of the world we have made. Goodman (1978) makes a similar point.

Mathematical Certainty and Representational Thinking

This argument views mathematical certainty as emerging from an interweaving of individual experience with language and/or other representational systems. Fundamentally, it is an argument that generalization is not an inductive process based on empirical experience, but rather an interactive process of deciding which experiences count and which ones do not count as cases of a particular representation. It can be regarded as an example of the symbolization process sketched by Kaput et al. in chapter 2 (this volume), where students are building the symbolizations *B* and *C* from their shared reference field *A*.

Drawing from Cobb et al. (1997), and in keeping with Kaput et al. (chap. 2, this volume), this suggests that initially students look through the representation to see particular experiential examples. As the learner becomes more sophisticated (compatible with taken-as-shared cultural knowledge) in deciding which experiences count, the representation comes to take on a life of its own. Because of the selective process of deciding which experiences count, the representation becomes part of the certain mathematical knowing/knowledge of the learner.

Table 5.1
Distribution of Five Monkeys in Two Trees

5	0
2	3
3	2
0	5
4	1
1	4

Cobb et al. (1997) provide an example from a first-grade classroom where the students are asked to decide how many different ways five monkeys could play in two trees (one large, one small). As the students propose possibilities for distributing the monkeys across the two trees, the teacher records them in a table with the small tree drawn on the left side of the table and the large tree on the right. As possibilities are proposed, the teacher records them as shown in Table 5.1.

As Cobb et al. (1997) point out, up to this point the focus of the discourse had been on generating the possible ways the monkeys could be in the trees (p. 263). However, at this point there is an initial shift in the discourse as students attempt to decide whether there can be any more possibilities by checking new suggestions against the record in the table. Cobb et al. continue:

> A further shift in the discourse occurred when the teacher asked:
> Teacher: Is there a way that we could be sure and know that we've gotten all the ways?
> Jordan: [Goes to the overhead screen and points to the two trees and the table as he explains] See, if you had four in this (big) tree and one in this (small) tree in here, and one in this (big) tree and four in this (small) tree, couldn't be that no more. If you had five in this (big) tree and none in this (small) tree, you could do one more. But you already got it right here (points to 5 | 0). And if you get two in this (small) tree and three in that (big) tree, but you can't do that because three in this (small) one and two in that (big) one—there is no more ways, I guess.

Cobb et al. (1997) are interested in the shift in the discourse so that the results of that activity (naming the possibilities) were emerging as explicit objects of discourse that could themselves be related to each other. It is this feature of the episode that leads us to classify it as an example of *reflective discourse.*

In addition, it can be argued that Jordan (and perhaps others) has constructed (or is in the process of constructing) a certainty that there can be no more possible combinations of monkeys in trees. Cobb et al. (1997) suggest that although as observers we separate the signifier (table) from the signified (monkeys in trees), for Jordan this distinction is not so clear. They suggest that Jordan is looking through the table to see the partitions and that, for him, this distinction (between signifier and signified) had collapsed and the table entries *meant* particular partitioning of monkeys (p. 270).

It would seem, then, that Jordan is creating this certainty by interweaving the actions of the class (creating possible combinations) with the representation created by the teacher. In interweaving his understanding of the table with the associated activities of placing monkeys in trees, it seems safe to assume that he is also building an understanding that no other possible entree in the table would count as one of the combinations of interest and any possible division of the monkeys is already in the table. It is precisely because Jordan is in the process of creating these relationships that he is engaging in representational thinking. That is, the shift in discourse noted by Cobb might also be thought of as a shift in the thinking of Jordan and, presumably, other students, from thinking of the particular to representational thinking. This is what Kaput et al. (chap. 2, this volume) refer to as the new conceptualization of the reference field *A* resulting from the symbolization process that began with *A*. As this occurs, the table, for these students, begins to emerge as a representation of a set of combinations rather than only a record of particular actions. However, once Jordan reaches the stage of separating signifier from signified, he makes his argument simply in terms of the symmetry of the number combinations in the table. So this representational thinking has, in a sense, accomplished its task and evolves into predominantly symbolic thinking that is thinking about the possible manipulation of symbols. In Kaput et al.'s terms, the students have moved on to create a symbol system *D* with an integrity and functionality of its own.

In addition, we might imagine Jordan having experiences with other combination problems and eventually coming to objectify the previous table as a representation of any situation involving partitions of five objects into two sets. At this point, he will be able to make arguments about any of these combination problems by reasoning about the table as an object, that is, without the necessity of looking through the table at a particularly contextual situation. Again, representational thinking indicates the processes through which the table becomes objectified as representing certain activities. But, when one moves to working on the object (the table) itself, this, like other kinds of symbol manipulation, involves cognitive processes, which I would call symbolic rather than representational because they are now operating with symbolic elements apart from what they might stand for. In Kaput et al.'s terms, they will have moved on to create a symbol system *E*, which is now functionally independent from the reference field *A* that they started with and is also applicable without reference to the intermediate symbolizations that helped give rise to it.

This hypothetical Jordan has engaged in representational thinking in creating this table as an "algebraic representation" of certain problems and has constructed a mathematical certainty by interweaving his experience in particular situations, which brings its anchoring power and stability, with this cultural tool. That allows him to use the representation and allowable actions on the representation (exchanging rows, etc.) in nonalgebraic ways to make claims about combinations of five items. The importance of the distinction between representational thinking (and its relation to certainty) and symbol manipulation should be, I believe, a central part of algebra teaching and learning. This is precisely the kind of thinking that is underemphasized or ignored in much mathematics instruction, where instead representations are taken as a given and the focus is on the manipulation of them. The unfortunate result is that these symbols never become representational for many students, leaving them without an experiential basis for their mathematical understanding. Again, in Kaput et al.'s (chap. 2, this volume) terms, it is an attempt to build a symbol system without the active, constructive process of symbolization.

A FRAMEWORK FOR FUNCTIONAL THINKING

The remaining part of this chapter extends this framework for representational thinking along a particular line that I call *functional thinking*, which is one of the key strands of algebraic thinking as described by Kaput (chap. 1, this volume). Functional thinking, at least for our purposes here, is representational thinking that focuses on the relationship between two (or more) varying quantities, specifically the kinds of thinking that lead from specific relationships (individual incidences) to generalizations of that relationship across instances. The algebraic reasoning part of functional thinking occurs as children invent or appropriate representational systems to represent a generalization of a relationship among varying quantities.

The following six activities are proposed as underlying functional thinking and thus the construction of functions:

Engaging in a Problematic Within a Functional Situation

1. Engaging in some type of physical or conceptual activity.
2. Identifying two or more quantities that vary in the course of this activity and focusing one's attention on the relationship between these two variables.

Creating a Record

3. Making a record of the corresponding values of these quantities, typically tabular, graphical or iconic.

Seeking Patterns and Mathematical Certainty

4. Identifying patterns in these records.
5. Coordinating the identified patterns with the actions involved in carrying out the activity.
6. Using this coordination to create a representation of the identified pattern in the relationship.

As in the case with Jordan described earlier, the representation in (6) may be physically identical to the record already created. However, the mental activities of interest in functional thinking are the processes by which this record becomes a generalized representation of the relationship and how the individual creates a mathematical certainty about this generalized relationship. Also, these activities may not follow the exact sequence listed here. Creating a record, for example, may be an integral part of focusing on a relationship.

Confrey and Smith (1994) proposed a general framework for teaching functions, based on the use of contextual problems, prototypes, multiple representations, and transformations. In addition, they described a distinction between *covariation* and a *correspondence* approach to functions. Although this approach was aimed primarily at middle and secondary grades, many of its features are relevant to a discussion of elementary students building ideas of function. Indeed, it is intended to provide an illustration of how to apply new approaches originally developed in later grades to earlier grades.

ENGAGING IN A PROBLEMATIC WITHIN A FUNCTIONAL SITUATION

Constructivists have long argued that learning occurs when the individual constructs a problem (Confrey, 1991; Piaget, 1970; von Glasersfeld, 1984). Because constructivists have focused on the individual construction of knowing (Smith, 1995), they have also emphasized the individual nature of such problems. In order to distinguish what the individual is attempting to solve from the social statement of the problem, Confrey (1991) had suggested the use of the word "problematic":

> A problem(atic) is only defined in relation to the solver. A problem is only a problem(atic) to the extent to which and in the manner in which it feels problematic to the solver. When defined this way, as a roadblock to where a student wants to be, the problem(atic) is not given an independent status. The problematic acts as a perturbance, i.e. a call to action. (p. 136, my parentheses)

To initiate the study of functions, we want to create situations where the various problematics for individual students center on the issue of the relationship between two varying quantities. However, as indicated by Confrey, neither a written problem statement nor participation in any particular social activity necessarily creates appropriate problematics. Even though the creation of a problematic is an individual process, it will take place within a social context—that is, through the teacher's suggestion of appropriate problems, combined with classroom discussion and negotiation and engagement in an activity. Activities may involve actual physical processes an individual carries out over time or they may be related to cognitive reflection on imagined activities. In either case, such activities become a problematic when there is something the individual wishes to know in relation to the activity. When this something has to do with the relation between two varying quantities, then the individual begins the process of functional thinking. Thus, the genesis of functional thinking occurs when an individual engages in an activity, chooses to pay attention to two or more varying quantities, and then begins to focus on the relationship between those quantities. It is the focus on a *relationship* that is central to the concept of function. Even though I would argue that the potential for the individual to engage in functional thinking is a matter of the individual construction of an appropriate problematic, teachers, by creating activities, describing variable quantities, and posing appropriate questions and engaging in classroom discourse, provide the opportunity for individuals to engage in functional thinking. As Cobb et al. (1997) state:

> Our rationale for positing an indirect linkage between social and psychological processes is therefore pragmatic and derives from our desire to account for such difference in individual children's activity. As we have noted, this view implies that participation in an activity such as reflective discourse constitutes the conditions for the possibility of learning, but it is the students who actually do the learning. (p. 272)

This is very much in line with the perspective of situated cognition (Kirshner & Whitson, 1997; Lave & Wenger, 1991). Thus, the contextual problems described by Confrey and Smith (1991) are social constructs designed to facilitate this process. Using a phrase initially suggested by Monk (1989), I use the term *functional situation* to describe a situation where one or more students are engaged in a problematic related to the relationship between two variables.

CREATING A RECORD

When students engage in solving problems related to functions, the most common entry point is in the creation of a table (Confrey & Smith, 1994). In Kaput et al.'s terms (chap. 2, this volume), this is the building of a *B* based on a functional situation *A*. Typically, students make two columns and record corresponding entries for values of the variables of interest. Creating such tables often plays an essential role in the development of functional thinking. However, as the student moves from creating a record to reflecting on that record as a representation of a relationship, it may also be important to encourage students to construct additional representations as well (*C*s and *D*s, as described by Kaput et al., chap. 2, this volume). Confrey (1992) has described an epistemology of multiple representations as "a claim that legitimacy of knowledge in mathematics evolves in relation to the multiple forms in which the idea might be displayed. We take the position that it is through the interweaving of our actions and representations that we construct mathematical meaning" (p. 11).

This suggests that encouraging students to create diverse forms of their records may facilitate their engagement in functional thinking, where the diversity occurs both within and across students. The reader likely recognizes the aforementioned as providing an example of the recording activity sketched by Kaput et al. (chap. 2, this volume) in their Figure 2.6.

PATTERNS AND MATHEMATICAL CERTAINTY

If emphasizing a correspondence between the members of two sets were our only criteria, then the creation of a record, particularly of a table, would suffice to construct a function—that is, placing two values in the same row indicates that those two values are the corresponding members of the two sets. However, part of the argument for focusing on functional thinking is the emphasis on constructing a relationship between variables that extends beyond the mere designation of a correspondence. As already discussed, an important part of that process is the construction of a certainty in that relationship. Although potentially tentative, this certainty is an essential part of the feeling of absoluteness and permanence that are part of the dominant cultural view of mathematics. We will explore a few examples that serve also to illustrate the Modeling Strand of algebra identified by Kaput in chapter 1 (this volume) and how it interacts with the Function Strand and the processes of generalization and symbolization.

Two concepts are central to this process. First is the distinction that Confrey and Smith (1991) make between a covariational and a correspondence approach to functions. In a correspondence approach, the emphasis is on the relation between corresponding pairs of variables. For example, in Table 5.2, the focus would be on the relation between x and y, which might be described as twice x plus one or algebraically as: $y = 2x + 1$. This is, of course, the conventional approach taught in most classrooms where algebraic equations are the primary representation.

Table 5.2
Variation Between Two Quantities

x	y
1	3
2	5
3	7
4	9
5	11

In the covariational approach, the focus is on corresponding changes in the individual variables. So the description of the same table might be: When x increases by 1, y increases by 2.

Although it may be important that students eventually construct equations expressing a correspondence relationship, the initial interest when introducing functions is in how students use either of these approaches in constructing a relationship between variables. It has been our experience that the covariation approach is most often chosen by students, particularly when the first variable is created as an indexing variable. In fact, moving down a table is often parallel to the actions one takes in constructing the values of the variables. For example, if one variable is time, one often makes sense of changes (of another variable) over time. We expect that in many cases, students' initial understanding of a functional situation will be constructed through a covariational approach. As they seek patterns in the records they have made, the noticed pattern will typically be in how one moves down a table. Note that this style of reasoning depends on the shape of the inscription and, with graphical inscriptions, the reasoning can take very different forms (e.g., Tierney & Monk, chap. 7, this volume). Here the parallel lists pull the reasoner toward certain styles of thinking rather than others.

Confrey and Smith (1991) also use the idea of prototypic functions to indicate the various classes of functions typically studied ($y = \chi$ for linear; $y = a^{\chi}$ for exponential; $y = \chi^2$ for quadratic, etc.) and also indicate the idea of a characteristic operation or action for different prototypic functions. For certain prototypes, this characteristic action is evident in a covariational approach. For example, with linear functions it is constant addition (constant difference), and with exponential functions it is constant multiplication. Although not as obvious for other prototypes, there are cases where we have observed high school and university students creating quadratics as a constant double summation (constant second difference). For a teacher attempting to create situations, which will become functional situations for her students, there are at least three ways in which she can approach a characteristic operation:

Table 5.3
Changes in Distance From Cliff to House

Year	*Distance*
1990	100
1991	96.5
1992	93
1993	89.5
1994	86

1. She can make it part of the problem statement: In 1990, a house was built 100 inches from the edge of a cliff. If the cliff is eroding toward the house at 3.5 inches per year, when should the owner move out?
2. She can make it part of the analysis of collected data: In 1990, a house was built 100 inches from the edge of a cliff. Over the last 5 years, the owner has made a record of the distance of the cliff from the house, which is shown in Table 5.3. When should the owner move out? (This situation could be altered by using data with a variable annual difference that averaged 3.5 inches/year).
3. She can allow it to evolve in relation to the students' understanding of the functional situation: In 1990, a house was built 100 inches from the edge of a cliff. By 1994, the cliff was only 86 inches from the house. When should the owner move out?

Unfortunately, many problems used in mathematics instruction are of either the first or second type. In the first case, the student is cut off from both finding a pattern and from the context of the activity itself. That is, the student need not deal with erosion as an issue in solving the problem. The description of the context becomes simply a vehicle for carrying (quantitative) information.

The second case does require finding patterns, but is also potentially cut off from the context of the activity itself. The strength of the second alternative is that it may allow for a process of theory building, thus supporting the kinds of science experiments where one collects data, looks for patterns, and then reexamines the context as a process of theory building to connect the two.

The strength of the third alternative is that it can support the building of connections between the conceptual actions one imagines in carrying out the activity with identified (and projected) patterns. In effect, it affords a generalization and symbolization process of the sort described earlier and sketched by Kaput et al. (chap. 2, this volume). In the particular example given, this may be difficult and complex for younger children.

However, in a classroom discussion I had with 10th- and 11th-grade students, conjectures were offered that could support a linear (constant conditions), quadratic (change in the hardness of the rock would lead to constantly increasing, or decreasing, erosion each year), and exponential (increase in the rate of water flow at the base of the cliff would lead to a doubling of the amount of erosion each year) process. Whether or not an expert would agree that these conjectures were realistic, they were offered in the context of conceptualizing possible conditions that lead to erosion, thus offering the potential to connect the activity to the creation of records and to the patterns in those records.

Constructing Mathematical Certainty

A second example of a problem that fits the third alternative is the Stone Path problem[1]: Your neighbor has asked you to build a stone path from her back door to a bird feeder 44 feet away. She has purchased 15 circular stones, each one foot in diameter and wants the space between each stone to be the same. She would like the last stone to touch the bird feeder, but will allow you to place the first stone any distance from the house you wish. Before going to do the work, you need to make a plan for how you will lay out the stones.

When students initially work on their plan and draw diagrams, they typically deal with the whole distance and partition it into stones and equal spaces. Thus, they do not necessarily deal with the problem as a sequential set of actions. However, as teachers, we ask them to label their stones in order and then make a table showing the distance of each stone from the house (or the feeder). Assuming the students do engage in this activity, we have actively encouraged them to change the way they are thinking about the problem to create a functional situation. In doing so, and in making the record of distances, they will typically identify the constant difference as a pattern in the table. In a graph, they initially identify the straight pattern of dots, but often have difficulty coordinating this pattern with the constant difference in the table. One issue is the tendency to see the graph as an iconic representation of the path, thus the equal distances in the path are seen as the equal distances between dots on the graph, rather than the equal vertical distances between dots. However, there is a basis for an ongoing discussion and investigation that should eventually connect this graphical pattern to the table pattern and in turn to the conceptual action of placing the stones at equal distances. It is in this kind of coordination among created representations (table and graph)

---

[1]Jere Confrey, 1994.

with their own conceptual representation of the activity that should support the construction of certainty about the relationship—for example, if the path was extended with the same equal spacing, then the pattern in the table and the graph would also be extended. The important point is that the justification for this extension is not simply an empirical observation of an initial pattern in either the table or graph combined with an extension of that pattern. Rather, it is an actively cognizing process of connecting a regularity in the actions taken in the situation with those patterns. This, in turn, creates the certainty that the continuation of the numerical/graphical pattern is connected to the extension of that regular activity. It is when this relationship is created that the table and graph, which were initially simply records of a local event (placing 15 stones), can be viewed as a generalized representation. This can also support the construction of symbolic expressions, which can be interpreted as statements of a constructed mathematical certainty.

The process described is the basis of what I call *functional thinking*. The emphasis on the creation of one's own conceptual certainty about a relationship is central to this process, because the certainty should play an essential role in creating a function as a representation of a relationship. It is also a strong illustration of the process of symbolization, building and coordinating a *B* and *C* to represent a situation *A*, as outlined by Kaput et al. (chap. 2, this volume).

THE ROLE OF CONCEPTUAL UNITS IN FUNCTIONAL THINKING: THE LINK TO YOUNGER CHILDREN'S REASONING

This chapter is primarily interested in linear functions and proposes that there is opportunity to provide situations for students to construct linear relationships based on their creation of conceptual units at an early age.[2] These conceptual units arise from situations involving repeated addition, which develop from the cognitive structures related to counting. The appeal of introducing linear functions in the elementary years is that their development can be seen as a natural extension of counting.

In his work on counting, Steffe (1987) describes a series of units where each is a result of integrating operations on previously constructed units. The units he describes are: numerical composite, abstract composite, symbolic motor, so many more than, iterable, and measurement. The teaching interviews, which he uses to describe these units all involve activities

[2]These are not measure units, although they may be related in certain situations as in this problem (Kaput & West, 1994).

where a child is asked to, in effect, do a double count. For example, one child is asked how many times he needs to count if he counts 12 blocks by twos. By careful observation of the ways in which the children respond to these questions, Steffe builds his models of the kinds of units they are using. These various units indicate areas that seem to be essential in the development of functional thinking in linear situations. That is, being able to create these units would seem to be essential to building a covariational understanding of a pattern in a linear table and coordinating that with patterns in graphs or other representations. In effect, students are building the units on which the idea of rate of change will be built. And this rate, in turn, is the "m" in $Y = mX + b$, which is the constant difference in the table and the slope in the graph of the relationship.

Confrey (1994) proposed a definition for conceptual unit, which acknowledges the integrating operations described by Steffe (1987), but emphasizes its connection to a repeated action. Confrey and Smith (1994) elaborated on her definition:

> A unit is the invariant relationship between a successor and its predecessor; the unit is (created as the result of) the repeated action involved in numeration. For example, in order to create a *counting* unit of one, a child must first recognize a multitude composed of objects sharing a particular quality. ... The unit which (the child) creates is the result of the operation of carrying out mentally the repeated action. (p. 142)

What seems to separate a unit from an object is that to construct an object, one must, among other things, isolate and coordinate a group of sensory signals to form a more or less discrete visual item or thing (von Glasersfeld, 1989). For Confrey and Smith (1991), a unit is constructed in conjunction with a repeated counting action. A unit arises in the action involved in repeatedly putting together (or taking apart) objects in coordination with this counting sequence.

In functional situations, this coordination becomes more complex. In their discussion, they use the Cliff Problem in the version described earlier, which includes a data table showing the distance from the house to the cliff for each year for 5 years (number 2 earlier), but also include a third difference column showing the constant difference in the distances. They state:

> According to our units analysis, the repeated action in the example is the annual loss of 3.5 inches. Thus we claim, in this example, the unit is –3.5 inches. The decision to segment one's experience into annual intervals sets the stage for a unitizing operation to create a unit of –3.5 inches. Thus, a unit is constructed in relation to an experiential situation which involves both segmentation and an invariance across segments. (Confrey & Smith, 1991, p. 145)

The importance of this analysis is that an essential part of constructing a function is in the way the individual constructs a repeated action in relation to a segmentation of experience, and that process results in the creation of a unit. However, in relation to the functional thinking model described earlier, this analysis needs further elaboration. In functional thinking, an individual needs to identify two or more quantities that vary in the course of this activity. Identifying a quantity that varies is closely related to Confrey et al.'s description of constructing a unit. In the Cliff Problem example, we might assume that the first unit constructed is a year, in effect borrowing a culturally available segmentation of a temporal experience. For whatever reason, the individual has *chosen* to focus on a year as an object and, in counting the passing of years, creates the year as an *initial* unit. In constructing this initial unit and choosing to focus on its relationship to distance, the distance unit, –3.5 inches, is a *derived unit*, that is, it is not created as an intentional segmentation of distance by the individual, rather as a result of the initial construction of the year as a unit.

This distinction is important for two reasons: First, it focuses attention on the intentions of the individual in constructing the initial unit. In the Cliff Problem, for example, this could have been because a year was what the individual cared about (to the closest year, I need to know when the cliff will be within 30 inches of the house). Second, because of the individual's knowledge of the relationship between climate and erosion, she might have chosen a year because it is the unit of time in which she expects the successive variations in erosion will be smallest (e.g., if months were chosen, one might expect several times the amount of erosion in the rainy season as in the dry season). In this case, there would be an interactive relationship between the construction of the initial and derived unit. However, the intention would still be eventually to choose time as the initial unit and then determine the magnitude of the derived distance unit. Conversely, one might choose an initial distance unit, say one foot, then derive a time unit by measuring the time taken to erode that distance.

Being aware of the distinction between the initial and derived units is necessary in developing an understanding of the relationship between variables. However, it is not sufficient. In the Cliff Problem, for example, one might understand why a year was chosen as the initial unit and that distance is a derived unit. But the context of the problem, if limited to the presentation of 5 years of data, still has no connection to the relationship between units. Both years and distance in the table are what Confrey and Smith (1991) called additive units—that is, the operation involved in going from predecessor to successor is one of constant addition. From a student's perspective, the problem context might seem to have little connection to the kind of units involved. That is, it might make equal sense for the distance units to be multiplicative—annual erosion increases by a constant rate, or percent. It might make sense for the distance units to be linear annual erosion increases by a constant additive amount. It might make sense for the distance units to be random—after all, why would one expect to find a pattern? If we do not facilitate ways for the learners' experience of the context to become related to their construction of the operation that creates the unit, then we are using what I would call a "vacant context." The context is simply a convenient setting, playing no real role in the conceptual development of the functional relationship. The alternative is to occupy the contexts we created by encouraging our students to relate their experiences of these contexts to the ways in which units are constructed. As an occupied context, the Cliff Problem is complex, and we might want to be careful about how we use such contexts. However, as already mentioned, students will make hypotheses about this context that support the construction of a variety of units. In making these hypotheses, they bring their previous experience into their construction of units.

The proposed model of functional thinking emphasizes the importance of the development of mathematical certainty. Focusing on how we distinguish between initial and derived units in relation to the intentions of an individual who occupies a context is central to the creation of certainty. It is one's experience within the context (the students' experience of a reference field in Kaput et al.'s (1994) terms that allows one the possibility of creating the conceptual certainty by building the conceptual units: The certainty in the relationship between the units is intrinsically tied to the ways in which one has developed an understanding of the context—in how one occupies the context. It is a certainty rooted in prior experience in quite specific ways. It may well be true, even desirable, that at the time one constructs this certainty, one is able to understand how different occupants might construct different certainties, or that even altering one's own relation to the context might change this certainty. However, the emphasis in developing functional thinking is in how the individual constructs a mathematical certainty as an occupant of a context.

FUNCTIONAL THINKING WITH PATTERN BLOCKS: AN ELEMENTARY EXAMPLE

Finally, we turn briefly to a classroom example. The class was a summer course for elementary teachers taught by a colleague and myself, which focused on various issues related to early algebra. One of the themes of the class was patterns and functions. Early in the class, we had had a discussion of visual patterns where we first developed examples of different kinds of patterns. We then turned to trying to describe why we would call certain visual images a pattern. This led to the inclusion of certain geometric terms: symmetry, rotation, reflection, and translation, and similarity and repetition as properties of patterns.

Table 5.4
John's Data for the Number of Repetitions of a Unit and the Total Number of Exterior Sides

Repetitions	*Sides*
1	9
2	17
3	25
4	33

We introduced pattern blocks to the teachers and initially just asked them to experiment with creating various patterns. Pattern blocks are blocks of various colors, which come in certain geometric shapes: hexagon, square, triangle, parallelogram, trapezoid, and so on. As might be expected, many kinds of patterns were created, some quite complex.

After discussing the various patterns, we asked the teachers to work on creating a particular kind of pattern—one, which if repeated, would stretch in a straight band as far as one wished to expand it. As various patterns were created, people noted that what made a pattern in this context was some initial arrangement that was repeated. Creating such a pattern led to the construction of a unit. As more patterns were created, we posed a question to the group: Can you find a relationship between the number of repetitions (of the unit) and the total number of exterior sides in your pattern?[3] Eventually, most groups created a table showing the number of repetitions and the number of sides. One of these tables is shown in Table 5.4. John, the teacher who wrote it on the board, explained his reasoning in finding a pattern in the table as follows:

I was trying to look at how much each number goes up by. I thought there might be a pattern there but then I saw a better pattern. These numbers here (pointing at the right-hand or ones' digits in the numbers under "Sides") go down by two and these numbers here (pointing at the left-hand or tens' digits in the same column) go up by one. So if we (inaudible) that pattern, I'm going to predict that if we had five repetitions (writes a five under the four), this here would go up by one (writes a 4′ under the first 3 in 33) and this here would go down by two (writes a one under the second 3 in 33, making the number 41 under 33) and then you get 59 (writes 59 under 41) and then you get ... (writes 67 under 59). To check it you would have to go count all those sides which would be kind of tedious.

The first three repetitions of their pattern are shown in Figure 5.1. At this point, the table on the board has been drawn (see Table 5.5). Another

[3]In posing this question, I set the stage for the visual block pattern to be the initial unit and the number of additional sides added each time a pattern unit was appended to be the derived unit.

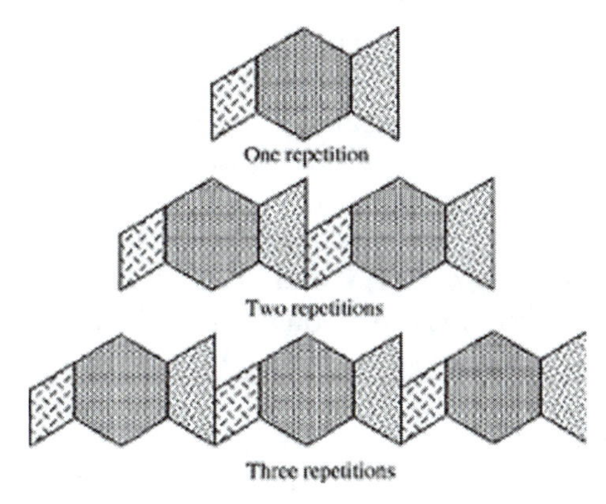

FIGURE 5.1. The first three repetitions of teachers' pattern block pattern.

Table 5.5
John's Continuation of His Data Table

Repetitions	*Sides*
1	9
2	17
3	25
4	33
5	41
	59
	67

teacher suggests that each number in the original table goes up by eight. John replies:

> Well actually we have, no one was eight and the next one was nine. (Turning to his table on the board). I mean the difference between these two is eight (pointing at 9 and 17), but then the difference (pauses uncertainly while pointing at 17 and 25).
> John had apparently miscalculated the difference between 17 and 25 as 9. Other teachers say that the difference is eight also, then say that all the differences in the original table were eight. John asks:
> Are they all eight? Oh, there's a pattern there too then. All these numbers go up by eight. (short pause) I thought that was interesting the way you could—without having to do it you can predict the way it's going to be.

John assumes he has finished and returns to his seat. As the teacher, I hypothesized (to myself) at this point, that John had initially erred in calculating the differences and thus sought and found the pattern of down by two, up by one. However, I was inclined to believe that now knowing that there was a constant difference of 8, he would likely adopt this model. I took his last comment in this light, but posed to the class the following question: *An interesting question is: Do both of those descriptions of the pattern work?*

After a few seconds, several teachers asked what I meant by "work." I rephrased my question asking if both of the descriptions of the pattern in the table could work together. Then John said, "It seems to work as far as I have done it, 6 repetitions. Forty-one plus eight is (pauses). That's not right is it. (Another teacher says it should be 49). It should be 49 if that was the pattern. So I don't think you can use that (emphasizing 'that')."

I continue to feel that John is rejecting his initial hypothesis, but ask: "Which one?" He replies, "You couldn't use that you are going to add 8 to each number to get the next."

> Other teachers suggest that they need to check it against the blocks.
> After a short time, John says, while pointing at the 41 on the board, "I do get 41 for the fifth repetition."

John seems to feel that he is verifying the pattern he suggested. He continues building another repetition. After a minute, he says "Forty-nine" (turns and looks at the board) and "So that doesn't work."

Several teachers then point out that adding eight does work and seem satisfied. However, as the class continues, John sits and looks alternately between his table on the board and the blocks for several minutes.

The importance of this episode is, I believe, in illustrating the tenacity with which John holds to the initial pattern he has found in the table, despite the feeling of the expert in the class (the teacher) that he would naturally turn to the seemingly more obvious constant difference pattern, once it had been pointed out. However, this did not happen. In retrospect, it seems clear that for John the constant difference model carried no more plausibility than his (and perhaps less). John was reluctant to give up on his pattern even at the end of the episode when he seems to have created both a discrepancy for his pattern and at the same time additional evidence for the constant difference model.

Often, even in problem-solving classrooms, there is a tendency for the teacher to assume at this point that the pattern had been established. As the teacher in this situation, I felt strongly that the repeated addition of a repetition would result in a constant addition of sides—thus I felt that I had a basis for feeling some certainty in the situation. However, I was making this assumption from the basis of considerable experience in relating certain kinds of activities to repeated addition models. This was not the case for my students. When I asked how we could be certain that this numeric pattern for the total number of sides would continue indefinitely as we continued to extend the pattern, there were no suggestions.

Within this episode, all involved seem to be operating within a vacant context. Even though the block pattern is used to provide evidence, there is no apparent effort to connect the repeated action with the blocks (append a repetition) with the hypothesized patterns in the table. Even though I knew that this was important, I decided to focus temporarily on other issues in order to provide more experience with the numeric patterns created and to coordinate them with graphical patterns. The tables were entered into a spreadsheet program, which was able to create graphs automatically. There was an extended discussion of the pattern in the graphs (straight lines of discrete points) and of the relationship between the table pattern and the graph pattern. This focused particularly on the question: Where can we find the constant numeric difference in the graph?

Eventually we returned to the blocks, where I posed the problem: You have found a constant difference pattern in your table and a straight-line pattern in your graph. I would like you to work on finding, showing, or creating an explanation of how we might find that constant difference in the patterns you have made with the blocks.

The ensuing discussion was not taped, thus can only be summarized from field notes. However, quotations are relatively intact. There was considerable discussion in John's group about what constituted a side. Even though they had initially counted them (but in a vacant context), they now were trying to coordinate their sense of a side with the action of appending an additional unit of blocks. In the figure, it can be seen that there is initially a long side on the right of the block unit. However, when another unit is appended, this side is cut in half, and one half disappears at the joint. The discussion centered for several minutes around the issue of whether this shortened side should count as one or as one half of a side.

It is noteworthy that this issue was not raised initially. Part of what makes the episode a vacant context is that the counting of the sides was independent of the action of making the sides (appending an additional unit). Later, the class was focused on the particular issue of how the sides changed within the making—that is, within the repeated action of appending a new unit. Eventually, there was a consensus that the shortened half-side actually should count as a complete side. Shortly after that, the group offered its explanation. John again was the spokesman. He begins successively adding block units on the table while explaining what is happening. The discussion is framed in terms of the side that is lost in the action of appending another unit: "Why is it that we are adding 8 when we know that one repetition has 9, the second one only adds 8 so we are losing a side somewhere. We are losing this side (pointing at the short side of the left of the unit being appended). So that's 9, but since we are losing this one side here, it's 9 plus 8."

John then repeats the action of appending a unit with a similar explanation. Other groups eventually offered similar explanations. In the final discussion, where the groups were asked if they could determine how many sides there would be after 25 turns, 50 turns, or any large number of turns, their reasoning reflected this understanding. For example, John's group described how to find the number of sides after 25 repetitions as 25 * 9: take 25 block units of nine, which would be sides. But 24 of these units would lose a side when being appended, thus their calculation was created as 25 * 9 – 24. Eventually they expressed this algebraically as $S = n * 9 - (n - 1)$.

This example is the story of one person's construction of the relationship between the initial unit (adding a unit of blocks to the pattern) and the derived unit (the change in the number of sides). It also connects each of the four goals set forth at the beginning of this chapter. Initially, we see John simply seeking a pattern in his table with no reference to the blocks themselves. In this sense, he is not occupying his context, but rather has abandoned it in favor of the representational system that he established (i.e., his table). Although the pattern he initially finds seems quite powerful at first, when carried back to the context, John finds a contradiction. It is this contradiction that motivates him to revisit the context and then to construct a relationship between the constant action of adding another unit to the blocks (the initial unit), the constant difference in his table, and the constant in the linear equation. This provides a basis for the construction of certainty between the constant action on the blocks and the constant in the equation. Activities such as this can provide a curricular basis that supports students in concretizing the symbolic representation of linear relationships and eventually of more general symbolic representations.

CONCLUSIONS

The approaches to algebra and functions described herein are intended to support the introduction of algebraic and functional concepts throughout the elementary curriculum. At present, these are models intended to support the development of additional appropriate activities and their use with these students. It is essential to develop contexts that allow students to build functional relationships, connect actions in the context with the representational system, and then build mathematical certainty about that relationship through the coordination of repeated actions in the context with the related elements of the representational system—the records of those actions or their consequences. This is an essential aspect of symbolization as described in chapter 2 (this volume).

For linear functions, the key is to focus on the construction of a derived unit of constant addition as being the central concept of linearity—the idea of constant rate of change. It is the focus on functions from a covariational or rate-of-change perspective that allows this central component of linearity to be realized.

In most traditional algebra texts, situations where the students have the opportunity to explore patterns that they find in tables and graphs within the repeated action, which constitutes the initial unit, are often minimized or avoided altogether. This has begun to change with the availability of the NSF-funded "standards-based" curricula (e.g., Investigations: Russell, Tierney, Mokros, & Economopoulos, 2004; Connected Math: Lappan, Fey, Fitzgerald, Friel, & Phillips, 1997; Math in Context: Core Plus Mathematics Project, 1997; Interactive Mathematics Program, IMP, 2002; etc). These curricula tend to place considerable emphasis on connecting compelling contexts and representations and offer a rich environment for creating tables, graphs, and equations that correspond with the given context. Although this approach represents a major improvement over traditional approaches to algebra, it is often not clear how many of these problems support an occupied context where rate of change is fundamental in connecting the action of the context to the creation of the various prototypic functions, especially linear, exponential, and quadratic. The approach developed in this chapter suggests that more work needs to be done in this area. Indeed, several chapters in the next section illustrate the kinds of work needed.

REFERENCES

Cobb, P. (1994). Where is the mind? Constructivist and sociocultural perspectives on mathematical development. *Educational Researcher, 23*(7), 13–20.

Cobb, P., Boufi, A., McClain, K., & Whitenack, J. (1997). Reflective discourse and collective reflection. *Journal for Research in Mathematics Education, 28*(3), 258–277

Cobb, P., & Bowers, J. (1999). Cognitive and situated learning perspectives in theory and practice. *Educational Researcher, 28*(2), 4–15.

Confrey, J. (1991). The concept of exponential functions: A student's perspective. In L. Steffe (Ed.), *Epistemological foundations of mathematical experience* (pp. 124–159). New York: Springer-Verlag.

Confrey, J. (1992). Using computers to promote students' inventions on the function concept. In S. Malcom, L. Roberts, & K. Sheingold (Eds.), *This year in school science 1991: Technology for teaching and learning* (pp. 141–174). Washington, DC: American Association for the Advancement of Science.

Confrey, J. (1994). Splitting, similarity, and rate of change: New approaches to multiplication and exponential functions. In G. Harel & J. Confrey (Eds.), *The development of multiplicative reasoning in the learning of mathematics* (pp. 293–332). Albany, NY: State University of New York Press.

Confrey, J., & Smith, E. (1991). A framework for functions: Prototypes, multiple representations, and transformations. In R. G. Underhill (Ed.), *Proceedings of the 13th annual meeting North American Chapter of the International Group for the Psychology of Mathematics Education* (Vol. 1, pp. 57–63). Blacksburg, VA: Conference Committee.

Confrey, J., & Smith, E. (1994). Exponential functions, rate of change, and the multiplicative unit. *Educational Studies in Mathematics, 26*, 135–164.

Core Plus Mathematics Project. (1997). *Contemporary mathematics in context: A unified approach*. Dedham, MA: Janson Publications.

Davis, P. J., & Hersh, R. (1999). *The mathematical experience*. New York: Mariner Books.

Ernest, P. (1991). *The philosophy of mathematics education*. London: Falmer.

Goodman, N. (1978). *Ways of worldmaking*. Hassocks, Sussex: Harvester Press.

Interactive Mathematics Program (IMP). (2002). Berkeley, CA: Key Curriclum Press.

Kaput, J., & West, M. (1994). Missing value proportional reasoning problems: Factors affecting informal reasoning patterns. In G. Harel & J. Confrey (Ed.), *The development of multiplicative reasoning in the learning of mathematics* (pp. 236–287). Albany, NY: State University of New York Press.

Kirshner, D., & Whitson, J. A. (1997). *Situated cognition: Social, semiotic, and psychological perspectives*. Mahwah, NJ: Lawrence Erlbaum Associates.

Kline, M. (1980). *Mathematics: The loss of certainty*. New York: Oxford University Press.

Lakatos, I. (1976). *Proofs and refutations: The logic of mathematical discovery*. Cambridge, England: Cambridge University Press.

Lappan, G., Fey, J., Fitzgerald, W., Friel, S., & Phillips, E. (1997). *Connected mathematics project*. White Plains, NY: Dale Seymour.

Lave, J., & Wenger, E. (1991). *Situated learning: Legitimate peripheral participation*. New York: Cambridge University Press.

Mason, J. (1996). Expressing generality and roots of algebra. In N. Bednarz, C. Kieran, & L. Lee (Eds.), *Approaches to algebra: Perspectives for research and teaching* (pp. 65–86). Dordrecht, The Netherlands: Kluwer Academic.

Monk, S. (1989, March). *Student understanding of function as a foundation for calculus curriculum development*. Paper presented at the annual meeting of the American Educational Research Association, San Francisco.

Papert, S. (1993). *The children's machine*. New York: Basic Books.

Piaget, J. (1970). *Genetic epistemology*. New York: Norton & Norton.

Russell, S. J., Tierney, C., Mokros, J., & Economopoulos, K. (2004). *Investigations in number, data, and space*. Glenview, IL: Pearson Scott Foresman.

Smith, E. (1995). Where is the mind? Knowing and knowledge in Cobb's constructivist and sociocultural perspectives. *Educational Researcher, 24*(7), 23–24.

Steffe, L. (1987). *Units and their constitutive operations in "multiplicative contexts."* (A report prepared under a grant from the National Science Foundation).

von Glasersfeld, E. (1984). An introduction to radical constructivism. In P. Watzlawick (Ed.), *The invented reality* (pp. 17–40). New York: Norton.

von Glasersfeld, E. (1989). Cognition, construction of knowledge, and teaching. *Synthese, 80*(1), 121–140.

Yackel, E. (1993, July). *The evolution of second grade children's understanding of what constitutes an explanation in a mathematics class*. Paper presented at the International Conference of Mathematics Education, Quebec, Canada.

II

STUDENTS' CAPACITY FOR ALGEBRAIC THINKING

Part II concerns algebra learning among relatively young students. The contributors believe that it is important for students to learn algebra early and as an integral part of mathematics instruction. Because readers may not share their views, they have tried to make their assumptions clear and to provide backing for their claims. This backing comes in the form of theoretical arguments, data, and even stories from classrooms. They have attempted to write persuasively. But, there is also the realization that readers will judge any claims not just in light of the evidence presented, but also based on their own experience in teaching, learning, and interpreting research in mathematics education.

The authors in this part regard generalization as fundamental to learning algebra. They sometimes take differing stances about the nature of generalization. Moreover, they may have different views about the circumstances for promoting the development of generalization and algebraic understanding.

The chapters focus on how conventional representation systems (i.e., graphs, written notation, tables, natural language) capture thinking of a general nature. As Part II progresses, the authors increasingly argue that learning to represent algebraic relations in conventional forms imparts changes in the nature of students' reasoning.

In chapter 6, Bastable and Schifter approach algebraic understanding as a growing awareness of the general properties of numbers. For example, positive integers have a commutative property under the operations of addition and multiplication: The order of the addends (or factors, in the case of multiplication) does not matter. Students come to notice such properties even before they have learned to express them through conventional algebraic notation such as $a + b = b + a$. An additional issue concerns the (implicit or explicit) domain of generalization, by which the authors mean the class of numbers to which a generalization applies. For young students, numbers are the counting numbers (or natural numbers). Only gradually do they extend their view of number to include integers, rational numbers, and real numbers.

In chapter 7, Tierney and Monk take a view consistent with that of Bastable and Schifter. (This should not be surprising, given that they have frequently worked together.) However, the number patterns to which they refer are not the invariant properties of the number system(s) that Bastable and Schifter speak of. Instead, they are properties inherent in a set of data. For example, the changing measures of distance over time convey information about speed for a particular person during a particular segment of a trip. These matters involve generalization to the degree that students realize that changes in distance over time always provide information about speed, for any values of distance and time (as long as the difference in time does not equal zero).

In chapter 8, Mark-Zigdon and Tirosh focus on the written forms of number sentences. They find that the recognition of correct usage precedes the production of correct forms. This finding is consistent with learning in other areas, including language acquisition and the understanding of musical notation.

According to Aleksandrov, Kolmogorov, and Lavrent'ev (1969), "Arithmetic and geometry are the two roots from which has grown the whole of mathematics" (p. 24). Algebra is generally understood as having derived from the arithmetical root. In chapter 9, Boester and Lehrer highlight algebra's indebtedness to the geometric root of mathematics, noting that "spatial structure serves as a potentially important springboard to algebraic reasoning, but also that algebraic reasoning supports coming to 'see' lines and other geometric elements in new lights." Their argument is not historical but rather psychological: "Visualization bootstraps algebraic reasoning and algebraic generalization promotes 'seeing' new spatial structure."

Carraher, Schliemann, and Schwartz (chap. 10) describe how "Early Algebra Is Not the Same as Algebra Early" with respect to: the role of background contexts and quantities, how notation is introduced, and its connectedness to existing topics in the early mathematics curriculum. They argue that algebra requires a shift in focus from the individual case to sets of cases—a kind of generalization for which the concept of function will play a key role. They find that young students can make remarkable progress in this shift when topics of early grades mathematics are reframed in this way.

In chapter 11, Brizuela and Earnest look at students in the process of trying to coordinate diverse representations of the same function. Students are asked to decide which is a better birthday present: (a) receiving $2 for each dollar one already has, or (b) receiving $3 for each dollar one has, but having to return $7 to the giver. Nine-year-old students initially favor one deal over the other. But when they consider a variety of scenarios, they realize that the outcome varies according to the starting amount. The authors focus on the students' work of making and interpreting graphs, and coordinating information in a graph with information in a table.

Peled and Carraher (chap. 12) claim there is a mutually supportive relationship between negative numbers and algebraic thinking: Negative numbers can be meaningfully taught within an "algebrafied" curriculum and, likewise, negative numbers can facilitate the development of concepts important for the growth of algebraic thinking. The concepts in question include functions, equations, and especially the additive structure. The chapter further discusses the extension of the additive part–part–whole structure to deal with more than set inclusion situations.

REFERENCE

Aleksandrov, A. D., Kolmogorov, A. N., Lavrent'ev, M. A. (1969). *Mathematics, its content, methods, and meaning*. Cambridge, MA: MIT Press.

6

Classroom Stories: Examples of Elementary Students Engaged in Early Algebra

Virginia Bastable
Mount Holyoke College

Deborah Schifter
Education Development Center, Inc.

We were active participants in 1990s discussions of the possibilities for the development of algebraic reasoning in elementary grades and devised for ourselves a working definition of *early algebra*. We began with the premise that algebra provides concise notation for expressing generalizations about number systems. Then we chose to look for occasions when children engage with such generalizations—generalizations that they might later express in algebraic language, but that they would now express in whatever means of communication are available to them. That is, we worked with the notion of early algebra as the exploration of generalizations about arithmetic operations.

From our experience in K–6 classrooms, we knew that when one listens to students' discussion of arithmetic ideas, one often hears them engaged at a level of generality that can be considered algebraic. When the arithmetic classroom environment is designed to follow children's thinking and provides elementary students with the opportunity to pursue their own questions, they display interest and ability in formulating and testing generalizations. Although these students do not, of course, use conventional algebraic symbols to express their ideas, the kinds of arguments they pose and the kinds of reasoning they display have parallels in formal algebra.

Once we formulated this conception that early algebra involves making and expressing generalizations, we reviewed our projects' collection of classroom cases through this lens to locate examples of such thinking. This chapter presents classroom episodes from grades one to six that illustrate a shift from thinking in terms of specific arithmetic statements to exploring more general assertions. These examples can be used to analyze the kinds of mathematical thinking involved, to consider the conditions that allow such reasoning to take place, and to determine connections between the mathematics of the K–8 curriculum and the formal algebra program. For each set of episodes, we pose questions that arise as we consider implications for what an early algebra program could be.

The episodes have been drawn from two sources. The first is a set of narratives about classroom events written by participants in the SummerMath for Teachers Program. Most were produced in the context of a project designed to support such writing (Schifter, 1994, 1996a, 1996b). The original narratives are referenced; most appear in the professional literature. Teachers' actual names are used; students are given pseudonyms.

The second source is data collected in Teaching to the Big Ideas (TBI), a 4-year teacher development project conducted collaboratively by the Education Development Center, TERC, and SummerMath for Teachers (Schifter, Russell, & Bastable, 1999). These examples are based on field notes taken by project staff that visited the classrooms of participating teachers or on teachers' written accounts of episodes from their own teaching. In these examples, both teachers and students are referred to pseudonymously.

The teachers from whose classrooms these examples have been drawn have all been working in teacher enhancement projects designed to support the development of a teaching practice that focuses on children's mathematical thinking. Although by looking at these materials we can identify ways in which the children's thinking is algebraic, the teachers themselves actually had no intention to teach algebra. Rather, we claim that, when instruction is designed to build on children's mathematical ideas and to foster children's mathematical curiosity, children are likely to exhibit algebraic ways of thinking by generalizing in the context of lessons in arithmetic, geometry, or measurement.

HOW GENERAL IS GENERAL?

As one develops an ear for children's mathematical conversation, one begins to hear their questions and observations. Children are curious about such generalities as the commutative property and the inverse relationship of addition and subtraction. For example, children talk about backward facts or turnarounds and seem to act as if their generalizations would hold for all whole numbers. However, adult listeners must be careful not to impose their own understandings of generalization onto the children. As the following episodes alert us, we must be careful to ask, what is the domain over which this generality holds for these children?

Episode 1: Noticing the Commutative Property of Multiplication

Virginia Brown (1996), a third-grade teacher, began a unit in multiplication by asking her class to mentally formulate an answer to the following problem:

> Kevin has three pencil cases in his desk with twelve pencils in each case. How many pencils does Kevin have?

After the class agreed on the answer of 36—adding 12 + 12 + 12—she showed them that the problem could be represented as repeated addition and expressed as a multiplication statement: $3 \times 12 = 36$. Although her intention was that her students should then break into small groups to work on a set of similar problems, the lesson at once took an unexpected turn.

Taking a step back from the original problem context, Jeff looked at number patterns and shared his observation that you could break each of the 12s into two 6s, giving you $6 + 6 + 6 + 6 + 6 + 6$, or $6 \times 6 = 36$. Then, Tom suggested that you can break each of the 6s into two 3s to get $12 \times 3 = 36$. And, at this, Anna exclaimed, "Wow, we have found a lot of things that equal 36. Oh. look! This one is the backwards of our first one, 3×12."

The children continued to find ways to break apart and then regroup the numbers to total 36. Looking at the column of twelve 3s, Steve offered that if you circle three 3s, you end up with four groups, giving you $4 \times 9 = 36$. At this, Joe declared, "And so we can add another one to the list because if $4 \times 9 = 36$ then $9 \times 4 = 36$, too."

Now, Anna reacted to this last claim, asking, "Does that always work? I mean, saying each one backwards will you always get the same answer?" Brown responded, "That's an interesting question. What do you think?" Anna replied, "I'm not sure. It seems to, but I can't tell if it would always work. I mean for all numbers."

For homework, Brown asked the class to think about ways to determine the truth of Anna's question. The next day, various children explained their thinking by noting such number pairs as 3×4 and 4×3. Whereas some children used manipulatives to illustrate their examples, Anna was not totally convinced: "But I'm still not sure it would work for all numbers." The teacher decided to table the question but to continue to explore multiplication by introducing arrays.

Two weeks later, Brown reminded the children of Anna's question. She asked if anyone could think of a way to use arrays to prove that the answer to a multiplication equation would be the same no matter which way it was stated? The class thought about this for a while—some alone, others with partners—until Lauren timidly raised her hand. "I think I can prove it." Lauren held up three sticks of 7 Unifix cubes. "See, in this array I have three 7s. Now watch. I take this array," picking up the three 7-sticks, "and put it on top of this array." She turned them ninety degrees and placed them on seven 3-sticks she had previously arranged. "And look, they fit exactly. So 3 × 7 equals 7 × 3 and there's 21 in both. No matter which equation you do it for, it will always fit exactly."

At the end of Lauren's explanation, Jeremy, who had been listening intently, could hardly contain himself. He said that Lauren's demonstration had given him an idea for an even clearer way to prove it. "I'll use the same equation as Lauren, but I'll only need one of the sets of sticks. I'll use this one." He picked up the three sticks of seven. "When you look at it this way," holding the sticks up vertically, "you have three 7s." Then he turned the sticks sideways. "But this way you have seven 3s. See? ... So this one array shows both 7 × 3 or 3 × 7."

Anna nodded her head. Although Lauren and Jeremy had demonstrated with a 3 × 7 array, the representation convinced her of the general claim. "That's a really good way to show it, and so was Lauren's. It would have to work for all numbers."

Once Anna raised the question of "will this work for all numbers?" for the class, not only did Brown take advantage of her students' interest and provide time for them to work on this idea, she also kept the question in mind and asked them to reconsider it 2 weeks later after they had become familiar with arrays and thus had additional tools for representing multiplication.

Episode 2: Does the Order of Addends Matter?

A group of third graders calculated that if they found $1 bill, 2 quarters, 5 dimes, 5 nickels, and 13 pennies in their bank, they would have $2.38. As the children shared their strategies, the following question came up: Does it matter what order you use to add? What might happen if you rearranged the coins and bill? Would you still get the same answer? Their teacher, Elizabeth Parsons, described the class' response:

> Everyone seemed eager to try their own sequence of money. The room started to buzz with excitement. Before they began to work alone with pencil and paper, I asked people to raise a hand if they were thinking we'd

> still get $2.38. About 15 of the 25 kids put hands up. There are always a few indecisive ones with a hand sort of up, then sort of down. However, when I asked about people who thought the amount of money may be a different amount than $2.38, seven or eight hands were raised.

Parsons was surprised that so many of her students hadn't internalized that you can add numbers in any order and maintain the same sum. She decided it would be time well spent to devote one and a half class periods adding up the coins in different ways. At first, the children worked alone, and then discussed their strategies with partners, then in small groups, and finally as a whole class. Several days later, Parsons asked the class to think back on this work: "I asked them to write about ... if they were sure, sort of sure, had no idea, or something in between about the answer turning out the same if we add the coins in other orders."

Among the children's responses (spelling and punctuation have been reproduced):

- I thought that it woulden't and I thought it would. I had a half and half unbalenced maind. the first time I tried it I got $2.38 I tried it again and I got the same exzact ansor. So I thought that $2.38 would be the right ansor. So I stuck with the ansor $2.38 and kept it in the corner of my head and put it up on the board.
- I wan't to know if it troind out to be $2.38 or did it trin out to Be something alise like $2.63 or did it trin out to be $2.00
- I relly had to work on it when I did it, it turnd out the sam as the others $2.38. But when I tryd again I got, $2.33 but when I showd it I had a mastak so I got, $2.38! I think that if any one in the world did this, I think they would get $2.38. Because every one got that, and we think it's right.
- When L came up with his idya I didn't know what it was going to be or if it would be the same answer of 2.38
- I'm pretty sure it would come out to the same answer because there are pitickuler coins

On reading her children's responses, the teacher noted: "My understanding of what goes on in children's thinking has changed.... The ability to see that a sum will stay the same no matter how the order of the addends is switched is not a matter of 'getting it' or not." Reflecting on years of experience with third graders' work on addition, Parsons now viewed her students' process with new insights:

> I now see children reaching a plateau of understanding when they can generalize [that] all pairs of single digit numbers can be reversed. They really KNOW this when they don't have to stop to try it out with each new pair.

> They may reach another plateau when they feel the same is true for two-or three-digit pairs of numbers added in reversed order. It seems natural for their understanding to take longer if there are many more than two addends, many more possible orders, and bigger numbers to add.

Having formulated these new hypotheses about her students' understanding, Parsons can now check them against her students' thinking as they continue to encounter these and similar questions.

Episode 3: Encountering the Inverse Relationship Between Addition and Subtraction

Jill Lester (1996), a second-grade teacher, gave her students the following problem:

> I went for a ride to Vermont which is 54 miles from here. After traveling for 27 miles, I stopped to have a cup of coffee. How much farther did I have to travel to reach Vermont?

As was routine in this classroom, first one child read the problem aloud and then the students had time to think before offering possible answers. When they began to discuss their ideas about the problem, Tom spoke up: "I don't know whether to add or subtract. It seems more like a plus than a minus, but ... " Sam responded quickly. "I think it's both. You can count down from 54 to 27 or you can count up from 27 to 54."

The class continued to discuss the problem, bringing out base 10 blocks to model the calculation. And although they raised many mathematical issues—subtracting 7 from 4, trading ten ones for one ten—the question about adding or subtracting recurred throughout. After Mark solved the problem by laying out two 10-sticks and 34 units and then removing 27 units from the pile, he exclaimed, "Now I'll solve it by adding. I have 27 and I need to know how many more it takes to make 54." As Tom watched Mark, he commented, "Now I get it! I was mixed up with the plus and the minus. I was mixed up about where to start."

Interviewed later, Lester explained that this was not the first time the class had discussed whether you added or subtracted to solve missing addend problems; the question arose quite regularly. Although there were different levels of clarity, most of the children had come to realize that either operation could be used, but they were still intrigued and wanted to test out both ways.

Lester added that in the last several years—since she had begun to change the way she taught to allow her students to voice and follow their own lines of thinking—every group of students had raised and worked through this relationship between addition and subtraction. One year, she said, the first time she presented her class with a missing addend problem, about half the students saw it as addition, the other half as subtraction, and everyone wondered how that could possibly be. Taking advantage of their curiosity, Lester assigned the children to work in pairs—one partner solving the problem by addition, the other by subtraction—and the class spent several days making up and then testing new missing addend problems. After a week they were satisfied that if adding on worked, so would taking away, and vice versa.

Lester was able to recognize that as her students were working on becoming proficient with subtraction computations, they were noticing the interrelationship of the operations of addition and subtraction. She shifted the focus of the lesson so it would include opportunities for her students to examine this more general mathematical principle.

Commentary: How General Is General?

This first set of episodes illustrates children's interest in considering operations as objects of study in themselves (Resnick, 1992). These children are not just enumerating number facts, but are working to express relationships that a set of number facts might exemplify. The episodes also raise questions about precisely what generalizations children are making when they encounter such properties as commutativity or the inverse relationship between addition and subtraction.

In the first example, Virginia Brown's students quickly move beyond the problem context, three 12-pencil cases, becoming curious about the different factor pairs of 36. As they break apart and recombine numbers to create new multiplication facts, they notice that some are "backwards of" others: $3 \times 12 = 36$ and $12 \times 3 = 36$.

Once this observation has been made, some children are inclined to generalize the pattern: "We can add another one to the list, because if $4 \times 9 = 36$ then $9 \times 4 = 36$, too." However, one child, Anna, objects: "Does that always work? I mean, saying each one backwards will you always get the same answer?" And it took another 2 weeks of explorations in multiplication until members of the class, including Anna, became convinced that, in fact, this backward property, commutativity, works when you multiply ANY pair of numbers.

Initially, it seems that Brown's class makes a generalization more far-reaching than Parsons' class. Brown's students make claims about all pairs of numbers independent of any given context, whereas Parsons' students consider a particular problem: the total of a given set of coins and bills.

However, as Parsons reflects on her students' work, she alerts us to the possibility that her students are operating at another level of generalization. Parsons believes that her students already understood that addition is commutative for small natural numbers: "All pairs of single-digit numbers can be reversed." Yet they did not necessarily extend this property to addition of larger numbers or to sums with more than two addends. When confronted with a problem involving the sum of 26 two- and three-digit numbers, her children needed to work with the specific set of numbers to see if order makes a difference. Thus, although Parsons' class was working within the context of a single problem, they (or some of them) may have actually been extending a generalization that they had already made for a more restricted set. In fact, we do not know if Brown's students would still have to do analogous work if they were to encounter, say, 1,344 × 568, or 32 × 54 × 58 × 39.

The question of the scope of generalization arises again as we consider Lester's case. The children discuss how her trip-to-Vermont problem can be solved either by adding or by subtracting: You can start with 54 and take away 27, or you can start with 27 and add on to 54. And Lester recognizes the children's curiosity as common—virtually all of her students, year after year, become intrigued by the notion that any missing addend problem can be solved by subtracting.

However, the generalization is Lester's and ours. On this day, the children in her class talk about adding and subtracting using the numbers 27 and 54. And, Lester tells us, whenever confronted with a missing addend problem, the children test out both solution methods. In one year only, did her students spend days explicitly devising and testing a class of problems until they were convinced of the general claim that any missing addend problem can be solved by adding and subtracting. Were the children in this class actually coming to a different, more general, conclusion than Lester's students in other years?

Thus, the episodes in this section lead us to the following questions: When and how do students become aware that the operations, themselves, are objects with predictable properties? What is the connection between reasoning within the sphere of arithmetic and the formation of such algebraic concepts as commutativity and the inverse relationship between addition and subtraction? What are indications that children are moving from a consideration of particular number relationships to exploration of the general patterns they instantiate?

HOW ARE GENERALIZATIONS EXPRESSED?

Certainly, much of the power of algebra comes from its concise notational system. At the same time, a major problem with the learning and teaching of algebra in its extant mode is that many students merely follow rules for manipulating symbols without meaning. One of the goals of an algebraic strand in elementary school would be to help students make meaning for algebraic symbolization. To start working toward such a goal, we must examine the language students already bring to their algebraic observations and consider how and when to introduce algebraic expressions.

Episode 4: Fourth-Grade Students Expressing the Relationship Between Consecutive Square Numbers

Belinda Knox related the following story from her fourth-grade class:

> The homework assignment was "to represent as many square numbers as possible on one piece of graph paper. They were to add a key on a separate piece of paper which included: the color of the pencil they used to represent a square, the square number, its square root, and the dimensions of the square." When the class met to discuss their homework, Adam said that he had discovered "something amazing and it worked every time."

Knox explained that Adam had difficulty articulating his idea "even though he knew exactly how to do it." But with help, he was able to share his discovery, which the group then began to explore: If you take two consecutive numbers, add the lower number and its square to the higher number, you get the higher number's square. For example, consider 2 and 3. Adam's rule says add 2 plus the square of 2 plus 3 to get the square of 3, or $2 + 2^2 + 3 = 2 + 4 + 3 = 9$. Knox writes: "[The group] had some discussion about whether it would work each time and Fred insisted that it wouldn't work with a higher number than that, so we tried 7 and 8: $7 \times 7 = 49$; $8 \times 8 = 64$; and $7 + 49 + 8 = 64$."

Adam then raised an additional question: What would you subtract from a square to find the square below the one you have?

Episode 5: Second-Grade Students Expressing Observations About Square Numbers

Although Rosemary Rigoletti (1991) had planned a set of lessons that would go in another direction, her second graders became curious about which numbers of cubes could be arranged into square shapes:

> Ted: I found out that four is a square number and so is nine.
> Rigoletti: How do you know that?
> Ted: Well, you see, if you put two blocks across the top and two below it, you get a square with four. Also, if you put three across the top

and three below it and then three more, you get a square and that's a square with nine. ...

Scott: There are certain numbers of cubes that when you put them together they will make a square, like with four and nine. But like with two and three you can't make a square no matter how hard you try.

These comments, which took place near the end of a whole-group discussion, prompted Rigoletti to suggest an investigation of square numbers when the class convened for math the following day. Rigoletti writes:

> After we reviewed the previous day's discoveries, we talked about making some predictions for the square numbers after 9. Evan said 16 for the square of 4. There were some giggles and heads nodding in agreement. I asked them how they knew, and Doug said, "Because I did it yesterday when I finished my paper." Others giggled and said, "Me, too!" Without any prompting, Nick said, "And I will predict 20 for the next square number." Other responses of 21, 24, and 25 were written on the board. The class was alive. Nick asked if we could make square numbers up to 100. When the answer was yes, there were hoots and howls. They couldn't wait to begin. They scurried off to get their unit cubes and multilinks.

Rigoletti organized her students into small groups to build and explore square numbers. Each group had its own way of accomplishing this: Some used addition to determine the number of cubes; some used multiplication; some added on cubes to an existing square to make the next larger square; some built new squares for each example. As they built the squares, the students noticed patterns in their work. Rigoletti, herself, was intrigued with their discoveries. She wanted the children to share their ideas, so she placed a chart on the board for them to record their findings. Throughout the remainder of the math time, children came up and wrote their conclusions:

- 1, 4, and 9 are square numbers.
- 16, 25, 36, 49, 81, and 100 are square numbers.
- Square numbers go odd, even, odd, even.
- If you times a square number by a square number, you get a square number ($4 \times 4 = 16$).
- Take any square number, add two zeros to it, and you will get another square number (4, 400).
- When you add a row at the bottom and a row to the side and make a corner, you get another square number.
- When you make a prediction for a bigger square, you always have to add a higher number to the square you just made.

Commentary: How Are Generalizations Expressed?

In these two episodes, children use English, a natural language, to describe relationships that are more frequently expressed with algebraic formalisms. Thus, many adults (including Knox and her colleagues when she shared this episode with them, as well as the authors of this chapter) who read Adam's "amazing" discovery—"If you take two consecutive numbers, add the lower number and its square to the higher number, you get the higher number's square"—feel the need to translate it into terms more familiar to them: $n + n^2 + (n + 1) = (n + 1)^2$ (which we would then be inclined to rewrite as $n^2 + 2n + 1 = [n + 1]^2$). In Rigoletti's class, children expressed a similar relationship in terms more closely associated with the representation they were using—"When you add [to a square] a row at the bottom and a row to the side and make a corner, you get another square"—which can be interpreted as $n^2 + n + n + 1 = m^2$.

The properties explored in the earlier examples could also be described in conventional algebraic terms, for example, $a \times b = b \times a$, and if $a + b = c$ then $c - a = b$. And, returning to Rigoletti's case, "if you times a square number by a square number you get a square number" can be written as $n^2 \times m^2 = k^2$ (Although, of course, $k = n \times m$, the children in the case are not so explicit. They simply state that the multiplication results in a square number.)

Although one may appreciate children's language to describe mathematical relationships, the ambiguities of natural language may, at times, cause concern (Ferrini-Mundy, 1996). For example, Rigoletti's students wrote, "Take any square number, add two zeros to it, and you will get another square number." In rigorous mathematical terms, this statement is false; the result of adding zero (or two zeros) to any number is that number, not another one. However, especially given their example—4,400—it seems clear that the children did not mean *add* in the mathematical sense, but in a colloquial sense, intending something like *concatenate*. In this sense, their statement is true. (In fact, readers might consider it a corollary to the children's previous statement. After all, we bring the understanding that the original number has been multiplied 100; that is, now $m = 10$.)

When is it appropriate to care about rigor in children's use of mathematical terms and conventional notation? Might the issue be analogous to invented spelling in language arts? One does not want concern about rigor to shut down students' expression of mathematical ideas; nor should issues of rigor be ignored. How should the balance be struck?

Further questions arise about when and how to introduce algebraic notation. For example, consider Brown's students' discussion of commutativity:

Might this have been an opportunity to introduce the notation "$a \times b = b \times a$," or is it more reasonable to wait until children are older? How do children connect their natural language descriptions of mathematical relationships with conventional symbolic forms? What tools for expressing general conjectures should they develop and when should these tools be introduced?

THE ROLE OF CONTEXT

Previous episodes featured math problems that were situated in story contexts. For example, introducing multiplication through a problem about pencil cases, as in Episode 1, or using a trip to Vermont for work on subtraction, as in Episode 3. The next episode offers an additional opportunity to consider the role of context. What is the purpose of setting a mathematical problem in a context that is familiar to the students and in what ways does the context limit their thinking?

Episode 6: Contextualizing Even and Odd Numbers

Consider Anne Hendry's (1988) story about her first graders. In preparation for the lesson, Hendry painted pictures of snowmen on dried lima beans. In class, she distributed these snowmen and told her students that the snowmen had received an invitation to attend the "Snow Ball," but they could not come unless they had a partner:

> This problem led to several days of thinking about odd and even numbers. They began to make observations about what numbers of snowmen could and couldn't go to the ball. [Using their lima-bean snowmen,] soon a few children began exploring why six [snowmen] could [go to the ball], but not seven; four but not five, and came up with a rule: "each time you add one number to a group that can go, you get a group that can't."
>
> We made journal recordings of this activity, recording rules named for the children who developed them. For instance, Mike noticed that if you add two to a number that could go (even), you got another even number. ... Zack showed that if you added together two groups that couldn't go, you would get a group that could.
>
> Probing further with questioning, I was amazed to see that by using these rules, the children soon became adept at applying them to larger numbers. There was a sense of self-satisfaction and empowerment when they were able to look at a number such as 69 and know not only that it was odd but how to make it even using their own rules. [*sic*] (p. 3)

Commentary: The Role of Context

Hendry's context, pairs of snowmen going to a ball, embody the mathematical issue her students are to explore—numbers that can be broken into pairs and those that can't. Listening closely to the first graders' words, it is not easy to determine whether their reasoning is limited to the given context or if they are thinking in more general terms about odd and even numbers. For instance, the statement "if you added together two groups that couldn't go, you would get a group that could" appears to be about the snowmen. Might the child be talking about snowmen groups as a model of number and actually have in mind that the sum of two odds is even? The statement "if you add two to a number that could go (even), you get another even number" begins as a claim about groups of snowmen, but ends as a claim about number. Does that offer more evidence that children are using the snowmen model to think about numbers more generally?

This episode raises broader questions concerning the role of context in mathematical explorations. In what ways do problem contexts provide a means for students to reason at a general level and when do problem contexts limit their reasoning to case specifics? What is the role of context-specific reasoning in the process of developing more general thinking patterns? Further, what do teachers need to understand about the process of representing numbers and operations in order to be able to analyze the limitations and strengths of various contexts?

EXTENSIONS BEYOND NATURAL NUMBERS

We began with the question of how general is general. That is, we observed that, when we listen to children's claims of generality, we do not always know over what domain they expect their generalization to hold. However, in the previous examples, it appears that the domain under consideration does not extend beyond natural numbers.

In our observations of elementary school classrooms, we find that questions arise about the behavior of zero or the meanings of operations with rational numbers. And it is here that we find powerful *applications* of generalizations previously made. This also reminds us that some of the more important consequences of generalizing may occur farther along in the students' mathematical development.

Episode 7: Generalizing to New Numbers—Is Zero a Square Number?

Jenny Richards, a fifth-grade teacher, wrote a message on the board to begin her math lesson:

> Dear Class,
> Yesterday we started a great debate about square numbers. We are thinking about 3 × 3 and 9 × 9 and 4 × 4 and 179 × 179. They create perfect squares

on graph paper. Then we wondered about 0 × 0. Hmm … what does that mean?
Your friend,
Ms. Richards

The discussion began:

Lindsay: Zero times zero is zero. You can't make anything with it. You have to imagine it in your mind and when I do that I imagine a square.
Richards: What does this mean? 4 × 4. I don't want the answer. I want to know what it means. … What would it look like?
Carolyn: Four rows with four in each row.
Richards: And 9 × 9?
Chris: 9 rows with 9 in every row.
Richards: And if you drew it on graph paper, you would get a perfect square? What does this mean, "0 × 0"?
Lesley: Zero rows with zero in each row.
Danny: When we were explaining 6 × 6, it was 6 rows with 6 in each. Zero times zero is zero rows with nothing in it. So there is nothing. It can't be a square.
Richards: Because there is nothing there?
Jake: If you count by two, you go 0, 2, 4, 6, 8. There is 0 there. If it is nothing, why would they put it there?
Melissa: I did 5 times 5 and it made a box. 0 times 0 makes a box. It's nothing, but the box is still there.
Katherine: To do 4 times 4 on graph paper you color four times four. For 0 times 0 you would color nothing. But it is a fact. It is still there.
Gail: It's simple. 6 times 6 equals a perfect square. 4 times 4 equals a perfect square. How come 0 times 0 doesn't?

The discussion ended with some students supporting zero as a square number, some disagreeing, and some not sure.

Episode 8: More Generalizing—Is Zero Even or Odd?

Beginning a lesson on odd and even numbers, Sally Gordon distributed both hundreds charts and multiplication charts to her third graders and told them, "Yesterday we noted that an odd and an odd make an even. Today I want you to find more rules like that and examples for each. You can think of adding, subtracting, multiplying, and even dividing."

After the children had some time to work, a TBI visitor approached a group of three girls and found that they had created a list of conjectures using addition, subtraction, and multiplication. For example, they had written out the statements, "Odd + Odd = Even, Even + Even = Even, Odd + Even = Odd, Even + Odd = Odd," and under each statement they had listed numerical examples to illustrate it:

Visitor: So those rules will work for all the numbers on the charts or just the numbers you have listed?
Sandra: It's for all of them.
Visitor: How do you know?
Lesley: Wait a minute. What about zero? We don't know if it is odd or even.
Sandra: How can it be anything? It's nothing.
Becky: But it has a line on the multiplication chart. It must be something.
Visitor: How can you decide? Can you use what you have already done?
Sandra: Let's see if it fits our rules.

The girls began to check out examples in which one of the numbers is zero. After testing a few cases, Lesley looked up:

Lesley: We think it is even, because it doesn't mess up anything that way. If you make it odd, it messes up the rules.

Episode 9: Rethinking the Meaning of Multiplication to Apply to Rational Numbers

When Joanne Moynahan (1996) began a unit on fractions with her sixth graders, she knew they would need to reconsider generalizations they had made from their work with whole numbers. To start, she gave them time to work on the following problems, using whatever strategies they chose:

1. The Davis family attended a picnic. Their family made up 1/3 of the 15 people at the picnic. How many Davises were at the picnic?
2. John ate 1/8 of the 16 hot dogs. How many hot dogs did John eat?
3. One fourth of the hot dogs were served without relish. How many were served without relish?

After working in pairs for some time, the class came together to share their solutions. She describes what happened next:

> As we discussed each problem I recorded a shortened version on the dry erase board. ... At the end of sharing the board looked like this:
>
> 1/3 of 15 = 5
> 1/8 of 16 = 2
> 1/4 of 16 = 4
>
> We didn't have much time left before the recess bell, but I thought I would ... give them something to think about and posed the following question:
>
> Does anyone know what they were doing with these numbers? (Long pause.) What operation did you use? Did you add, subtract, multiply, or divide? (Another long pause.) What symbol could we put in here instead of "of"? ...

Mary: I think we should put division in there.
Teacher: Why?
Mary: Well, the problem said 1/3 of the people were Davises. I drew a circle and divided the circle into three parts—then I put the people in.
Jeff: I agree. We divided our cubes into three groups. …

It really did seem like division. They took 15 people and divided them into three smaller groups, all the same size. However, that's dividing by 3, not 15. The number sentence $1/3 \div 15 = 5$ does not represent what they did when they divided the 15 cubes (representing 15 people) into three equal groups. They did not divide by 15.

Teacher: Does everyone agree? Should I erase the "of" and put in "÷"? Think about what you know about division. …
Rebecca: I don't think divide is right.
Teacher: Why do you think divide won't work?

Rebecca came to the board and wrote $1/3 \div 15 = 5$.

Rebecca: This (pointing to the 15) means how many 15s are in 1/3. That (pointing to the 5) means five 15s are in 1/3. I know that's not right. There aren't any! (pp. 28–29)

Rebecca's objection convinced her classmates, but they weren't yet ready to give up on their sense that their actions were represented by division. They rearranged the dividend and divisor to check $15 \div 1/3 = 5$, but then agreed that wouldn't work, either:

Rebecca: I think it's times.

I invited Rebecca to come to the board. She began writing a line of 1/3s.

Rebecca: That's 1/3 fifteen times. Now add them up.

I could see that Rebecca had moved to the abstract. She was considering the number sentence without connecting it to the Davises. Where in her diagram were the Davises? Rebecca could see that she was not convincing her classmates. She offered this final defense:

Rebecca: I didn't multiply. I'm just trying to prove that you can. I divided the 15 people. She (pointing to Mary) says divide and I'm trying to show that multiply works.

R-I-N-G! That marked the end of class. (pp. 30–31)

When the class reconvened for math the next day, Moynahan asked if anyone had further thoughts about yesterday's question. In fact, many of the students had done some thinking. Mark commented that, whatever operation they used, they needed to think of 5 as the result of that operation on 1/3 and 15. Jacob suggested that they try out addition, but everyone could quickly see that the result would be (expressed by the mixed number) 15 1/3. They also tried subtraction and, although it took longer to figure out the calculation, they were quite clear that this was not the appropriate operation, either. Then Mary had another idea:

Mary: Division is the opposite of multiplication. Take $12 \times 2 = 24$. Then $24 \div 2 = 12$. So. . . . If $1/3 \times 15 = 5$, then $5 \div 15 = 1/3$. Does that work?

I liked Mary's "if, then" strategy. She and many others were not sure what 5 divided by 15 was, but Mary was definitely trying to resolve this problem. She was using previously grounded concepts to make sense of a new situation.

Teacher: How many 15s are in 5?
Mark: There aren't any. You can't make any 15s if you only have 5. Wait. You could make a part of a 15.
Jeff: I got it! You would have 1/3 of 15! It does work. Rebecca was right—it is multiply! (Moynahan, 1996, pp. 32–33)

Commentary: Extensions Beyond Natural Numbers

In all of the examples in previous sections, children are exploring operations on the set of natural numbers. The fact that many of the tools and models they employ (e.g., cubes, arrays, paired objects, squares drawn on graph paper) apply exclusively to natural numbers is, however, not obvious to them. When children extend their explorations to include zero or fractions, as in the examples of this section, they confront the need to develop new tools, models, and criteria of justification.

For example, Richards' students are trying to decide whether zero is a square number. Because they rely on the criterion they used for natural numbers—can that number of objects be arranged in a square array?—they are unable to resolve the issue. Some students argue that they can imagine zero objects as a square; others say they cannot. As long as the class' justifications remain in the realm of natural numbers, the argument results in a stalemate.

Gordon's students, however, use other means to decide if zero is an odd or even number (after they decide that zero is, after all, a number). The issue arises as they are devising rules about operating with odd and even numbers and checking their rules against the hundreds and multiplication charts that had been handed out. When challenged to think about zero, they initially believe the category does not apply—"How can it be anything? It's nothing." On second thought, because zero appears in their charts, they decide it must be a number, "it must be something." Eventually they conclude, "We think it is even, because it doesn't mess up anything that way. If you make it odd, it messes up the rules." Thus, Gordon's students make the decision to consider zero even because that choice will maintain consistency of the patterns they had discovered while working with natural numbers. Their belief that mathematics should be consistent drives their decision.

Similarly, when Moynahan's students are challenged to decide what operation sign could replace *of* in "1/3 of 15 = 5," the class is finally convinced by Rebecca's argument relying on maintaining consistency of the system. The key for her is that multiplication and division are inversely related—one operation undoes the other—and that "5 ÷ 15 = 1/3" is a sentence that makes sense in the context of the problem they had been given. This convinces Rebecca and her classmates that "1/3 × 15 = 5" also models the problem they had been given.

So, what tools for reasoning need to be introduced as students move beyond the natural numbers? How are these tools extensions of previously developed approaches? In what ways can mathematical structure itself become a tool for reasoning? What experiences allow students to reason from the structures they have built? How do they determine which conclusions about the set of natural numbers remain valid as they expand their sense of what constitutes number?

PREREQUISITES TO ALGEBRAIC THINKING IN ELEMENTARY SCHOOL

This chapter has presented classroom episodes illustrating children making a shift from thinking in terms of specific arithmetic statements to exploring more general assertions. For us, these are examples of children engaging in early algebraic thinking. They express their generalizations using language, diagrams, and story contexts that capture the actions of the operations rather than in formal symbolic notation. The compact notation of symbolic algebra is in their future and should be built on these early algebraic experiences.

Based on these episodes, we have posed a set of questions that we feel must be considered by those who intend to support the development of algebraic thinking in the elementary school. However, although the children in the classrooms depicted here are typical, the classrooms, themselves, are not. That is, these classrooms in which children are invited to articulate their mathematical ideas are not representative of extant practice in the United States, although they do represent the kinds of practice that are envisioned in such reform documents as the *NCTM Standards* (1989, 1991, 1995a, 1995b, 2000).

To wit, the mathematics in these classrooms is conceived as much more than a sequence of facts and procedures to be memorized. Rather, mathematics is a realm of exploration, and doing mathematics is a social process: Children learn to actively and purposefully conjecture, revise ideas, offer proof, and argue mathematically. Children in these classrooms are seen as rational beings that have mathematical ideas worth the attention of their classmates and their teacher. The process of learning mathematics is largely a matter of engaging with ideas. The teachers in these classrooms believe that children create meanings for mathematical concepts when they work within contexts that already have meaning for them. For example, their understandings of addition, subtraction, multiplication, and division develop as they come to see them as models of situations. Thus, the teacher presents mathematical tasks grounded in familiar situations.

Whereas the teachers in these episodes had no specific intention of introducing algebra into their instruction (in contrast to the teachers depicted in chap. 16), they do believe their instruction should be organized to elicit students' mathematical ideas and those ideas should become the explicit focus of instruction. This sometimes means the teacher's planned agenda for the day is altered when an individual student brings an idea that is especially intriguing or when the class expresses curiosity about or enthusiasm for a particular question or investigation. In fact, as seen in Moynahan's case, students will continue to think about their mathematical ideas beyond the formal classroom setting.

Once classroom cultures are established in this manner, we see that children will express the regularities they note in their work as mathematical generalizations; these generalizations later will become codified and expressed in formal mathematical language, such as the commutative and associative laws. We see that as teachers establish a classroom practice in which children's arithmetic thinking is fostered, their thinking naturally takes the form of generalization.

The questions we have posed in this chapter 6 are intended to raise issues for all engaged in this work: researchers, curriculum developers, teachers, and staff developers alike. How might teachers actively support the development of algebraic thinking? We believe the answer to this question will be found by researchers and teachers working in partnership and must be built on what students already bring to this effort. In chapter 16, we look further into how this generalizing and justifying process can be supported in the materials that teachers can use and how the surrounding cultural views of mathematics interact with this process.

REFERENCES

Brown, V. (1996). Third graders explore multiplication. In D. Schifter (Ed.), *What's happening in math class? Vol. 1. Envisioning new practices through teacher narratives* (pp. 18–23). New York: Teachers College Press.

Ferrini-Mundy, J. (1996). Mathematical thought-in-action: Rich rewards and challenging dilemmas. In D. Schifter (Ed.), *What's happening in math class? Vol. 1. Envisioning new practices through teacher narratives* (pp. 77–86). New York: Teachers College Press.

Hendry, A. (1988, June). Snowman math. *The Constructivist, 3*(2), 3–4.

Lester, J. B. (1996). Is the algorithm all there is? In C. T. Fosnot (Ed.), *Constructivism: Foundations, perspectives, and practice* (pp. 145–152). New York: Teachers College Press.

Moynahan, J. (1996). Of-ing fractions. In D. Schifter (Ed.), *What's happening in math class? Vol. 1. Envisioning new practices through teacher narratives* (pp. 24–36). New York: Teachers College Press.

National Council of Teachers of Mathematics. (1989). *Curriculum and evaluation standards for school mathematics*. Reston, VA: Author.

National Council of Teachers of Mathematics. (1991). *Professional standards for teaching mathematics*. Reston, VA: Author.

National Council of Teachers of Mathematics. (1995a). *Algebra in the K–12 curriculum: Dilemmas and possibilities*. Reston, VA: Author.

National Council of Teachers of Mathematics. (1995b). *Assessment standards for school mathematics*. Reston, VA: Author.

National Council of Teachers of Mathematics. (2000). *Principles and standards for school mathematics*. Reston, VA: Author.

Resnick, L. B. (1992). From protoquantities to operators: Building mathematical competence on a foundation of everyday knowledge. In G. Leinhardt, R. Putnam, & R. A. Hattrup (Eds.), *Analysis of arithmetic for mathematics teaching* (pp. 373–429). Hillsdale, NJ: Lawrence Erlbaum Associates.

Rigoletti, R. (1991). *Second graders explore square numbers*. Unpublished manuscript.

Schifter, D. (1994). Voicing the new pedagogy: Teachers write about learning and teaching mathematics. *Center for the development of teaching. Paper series*. Newton, MA: Education Development Center.

Schifter, D. (Ed.). (1996a). *What's happening in math class? Vol. 1. Envisioning new practices through teacher narratives*. New York: Teachers College Press.

Schifter, D. (Ed.). (1996b). *What's happening in math class? Vol. 2. Reconstructing professional identities*. New York: Teachers College Press.

Schifter, D., Russell, S. J., & Bastable, V. (1999). Teaching to the big ideas. In M. Z. Solomon (Ed.), *The diagnostic teacher: Constructing new approaches to professional development* (pp. 22–47). New York: Teachers College Press.

7

Children's Reasoning About Change Over Time

Cornelia Tierney
TERC

Stephen Monk
University of Washington

Until recently, the accepted view of algebra among many mathematics educators has been that it is too abstract and too difficult for children to learn before secondary school. This chapter argues against this viewpoint. We believe this view arose from the narrow way the subject has come to be construed in U.S. schools and the way children are viewed in relation to it. In the past two decades, much effort has been devoted by mathematics educators in the United States toward determining what algebra and algebraic thinking are and when children are capable of doing it (Carpenter, Franke, & Levi, 2003; Chazan, 1996; National Council of Teachers of Mathematics, 1998; Schifter, 1999). A separate, but related, strand of mathematics education research and curriculum development has investigated children's understanding of the mathematics of change here and abroad (Barnes, 1992; DiSessa, Hammer, Sherin, & Kolpakowski, 1991; Krabbendam, 1982; Swan & Shell Centre Team, 1989).

Our work reported here is rooted in the latter strand. In exploring children's understanding of the mathematics of change, we have come to see that children in elementary school are capable of forms of thinking that underlie algebra. Among the themes common to the mathematics of change and algebra is the use of the notion of variable (Monk, 2003).

There is probably no richer area of children's life experience connected to variation, than change over time. The notion of varying events over time underlies stories that children tell from an early age. In our research on change over time, we engage children in telling stories about one variable changing over time, and in representing these stories in tables, graphs, and groups of additive changes. Students come to use these representations and reason about them as symbol systems with conventional meanings, arranged according to accepted rules to provide a description of phenomena that change over time.

We describe episodes from three classroom conversations and one individual interview in which children from age 8 to 10 interpret and create representations to tell a story about change. In our discussion of each episode, we elaborate on the varied means children have of carrying out the given problems and the particular thinking processes we believe are indicated by their solutions. By providing these descriptions and discussions, we hope to make real and convincing our view that algebra can be construed in ways that are not only possible for children to learn, but can be experienced as exciting and challenging during the learning process.

The data for the classroom vignettes (Episodes 1, 3, and 4) is taken from research done in developing a strand of curriculum materials about change over time for the *Investigations in N umber, Data, and Space* (Russell, Tierney, Mokros, & Economopoulos, 2004) mathematics program for grades K–5. In a progression of activities, children observe and represent varying speeds, heights, population, events in their lives, and changing number of objects. The interview (Episode 2) is part of an earlier research study in which we interviewed children about the relationship between the observation and measurement of physical phenomena and its mathematical representations (graphs, number sequences, etc.) in the context of learning the mathematics of change.[1] In further work with elementary children,[2] we continued looking at how children use their own activity with motion detectors, tables, and graphs to make sense of and analyze change over time.

EPISODE 1: FINDING THE MISSING BEGINNING NUMBER

This episode shows children sharing multiple approaches to the problem of finding a missing value, a type of problem at the very core of algebra. Solving such a problem almost always calls for making generalizations in one of several ways.

---

[1]Measuring and Modeling, Ricardo Nemirovsky, PI (NSF No. MDR-8855644).

[2]Student's Conceptions of the Mathematics of Change, Ricardo Nemirovsky, PI (NSF No. MDR-9155746).

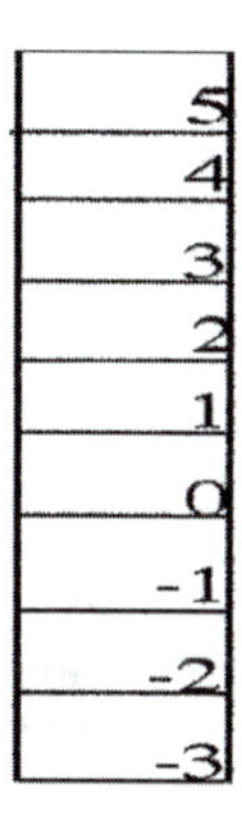

FIGURE 7.1.

Table 7.1

Starting Floor	*Changes*	*Ending Floor*
??	+2, –3	1

In the unit called *Up and Down the Number Line* (Tierney, Weinberg, & Nemirovsky, 1994), written as part of the third-grade *Investigations in Number Data, and Space* curriculum, students are asked to imagine a skyscraper that has floors "forever underground" and "forever above ground." In the elevator, illustrated by a vertical number line as in Figure 7.1, the button that is pushed (which can be any integer between –3 and +3) determines how many floors up or down the elevator will move rather than the floor it moves to; for example, pushing +2 moves the elevator up two floors, pushing –1 moves it down one floor. Students have been working on problems in which the starting floor and the changes are given and they must find the ending floor. Now the teacher challenges them with a backward problem:

Teacher: This time I don't know the starting floor. I go on the elevator. I push the plus two and the minus three and I end on floor one. [She begins to make a record on the board, Table 7.1]

Kadisha: We're supposed to end on floor 1. If I were to go +2 and –3, I would end up one below where I started, so I must be one above the ending floor, so that's plus two.

Sylvia: I just switched them around. I made the plus two a minus two and the minus three a plus three and then I did it.

Table 7.2

Starting Floor	*Changes*	*Ending Floor*
??	−1, +3	2

Teacher: How did you think of that?
Sylvia: If I'm starting on floor one and I don't know my ending place, I'd have to switch the whole way.
Mosi: If I started on the fourth floor and I did minus three and plus two, I'd wind up on three so I tried three. Then I tried plus two and it worked.
Teacher: That's called trial and error.
Luke: I did the same thing.

The teacher poses another problem and waits a few minutes for the students to figure it out (see Table 7.2)

At the end of this time, all have an answer except two students, who haven't found a way to get started. The teacher directs one of these students, Nora, in doing the problem by what she calls trial and error, trying a number and then trying a higher or lower number, depending on whether the ending number is above or below the target ending floor:

Teacher: Let me do this with Nora. Give me a floor. We'll do trial and error.
Nora: +1
Teacher: Okay. Let's do it. Minus one, plus three. Did it work?
Nora: Plus three. No.
Teacher: So what could we do?
Nora: Start at 0.
Teacher: Does that work?
Nora: Yes.

Next, Kadisha and Sylvia describe the methods they used to solve this problem, and then Holly says she used the teacher's trial and error method:

Kadisha: I started on plus two and then I said. . . . No, I actually didn't start anywhere. Let's see. Minus one and plus three makes the minus one cancel out one of the three so I have plus two and I want to end up at plus two so I must have started at zero.

Teacher: Sylvia, did you do it a different way?
Sylvia: Yes. I made minus one a plus one, and the three a minus three. I started at plus two.
Christina: And I did that to check it over.
Sylvia: And I start at zero and do it the regular way to check it.
Holly: I did trial and error for the starting point and kept adjusting downward.

Discussion of Episode 1

Mosi, Luke, and Holly used the method most common among third- and fourth- grade students we have worked with, as did Nora with the teacher's help. This is the method the teacher calls *trial and error*. However, this, in our judgment, is not just trial and error when the children recognize that, after starting with a number and ending up too high (or too low), they should adjust their starting number to one that is lower (or higher). As Holly puts it, they "kept adjusting downward." Sometimes this requires three or four guesses in all.

Most students we have observed eventually manage to fine-tune this method to get the correct starting number in two guesses. Instead of adjusting their guesses by one each trial, they jump directly to the exact starting number after seeing the result of their first trial. This is not evident in this episode. Had Mosi done this, after finding that starting with +4 left her at 3 (two higher than her goal of 1), she would have immediately tried +2. Children who perfect this method continue to start to solve these backward problems with a guess, rather than with a view of the whole problem. Although their initial guess may be quite sensible, they do not articulate reasons for that guess.

Kadisha and Sylvia have a broader view. They articulate ways of handling numbers that are not particular to the given numbers. As Kadisha says herself, she "doesn't start anywhere." She figures out the relationship between the starting and ending numbers before she considers the particular value of either. She determines the net effect of the changes and then looks at the ending number in order to determine her starting number: "I would end up one below where I started, so I must be (start) one above the ending floor." Sylvia works backward from the ending number, changing the signs of each change to its opposite. Her reasoning is similar to that of a student, Joseph in a different class[3] when solving the problem: $? + 1 - 3 + 1 - 1 + 2 - 1 = 6$ in the context of putting blocks in and out of a bag. Joseph sums the positives, on the one hand, and the negatives, on the other hand, and then works backward. He says: "That's putting in 4 and taking out 5. Eleven minus 5 would equal 6 so it's what number plus 4 would equal 11." Sylvia and Kadisha, as well as Joseph, generalize from knowledge about combinations of arithmetic operations.

The variety of approaches the children bring to this problem suggests that the question of what is or isn't algebra does not lie in the problem, but in the way the children think as they do the problem. Some of the children used an almost random process of trial and error, but still others

---

[3]A third grade not described in this chapter.

(Mosi and Holly) began to use a heuristic on their trial and error approach that because the elevator ended too high, they had better start lower. Is Mosi's and Holly's thinking algebraic? Perhaps it is not. Perhaps it is only a utilitarian rule based on repeated experience such as that often heard in algebra class: "When you move it [a term] to the other side of the equation, you change the sign." Kadisha did the problem by compressing the two operations of subtracting 1 and adding 3 into a single operation of adding 2, whereas Sylvia kept the operations separate. Both girls reversed the operation in order to find out what number one must start out at to end up at a given place. These are all, including Mosi's and Holly's short cuts, ways of thinking that involve making generalizations.

EPISODE 2: IDENTIFYING MAXIMUM AND MINIMUM ACCUMULATION

In the next episode, we see a student reasoning about the results of combining sets of numbers without finding specific numerical answers. This is a form of making generalizations, a common theme in early algebra.

This episode comes from an individual interview with Rose in the summer between her third- and fourth-grade years, on the day before her ninth birthday. It is from an interview that had a single session lasting about 1 hour, during which the interviewer put blocks in and out of a paper bag to illustrate problems similar to those the students in episode 1 did using the elevator model. In the previous task, Rose looked at the series 3 + 6 + 1 – 5 – 3 + 2 to decide, without computing, when the bag would have the most and the least blocks in it. She said of this example that the most is after the + 1 and "the least is between the minus three and plus two ... because you've taken away five and three ... and then you spoil it by putting in two."

Now the interviewer gives her a new task, of telling when the most and least number of blocks will be in the bag, based on whether blocks have been added or taken away, but without telling her how many blocks he has transferred:

> Interviewer: I want to write one out for you. I'm going to show you some changes but I'm not even going to tell you what they were exactly. [Okay.] Let's just say (writes 4) you start with four, okay? Here's a bunch of changes. I'm just going to show you the sign as you called it, of what each one was (writes series of signs without numbers) equals some number at the end.

$$4 + + + - - + +$$

Rose: What do I get to do, put in the numbers?

Interviewer: Well, not first off. First off I'd like you to tell me, if you can, where I had the most in the problem.

Rose: Right there (after the third plus) because you've just had a lot of plusses and then you have to go through minuses to get back to plusses so I think you would have the most right there.

Interviewer: Because "you've gone through." And what does "going through" plusses do?

Rose: It adds more and more and more. Because there's three plusses. More and more and more (touching each of the first three plusses). And then if you go through the minuses there's less and less and then you go back to the plus there's more but you have already taken out two of these guys. Say they're all one. I'm just trying to give an example. There's one all the way across.

Interviewer: By one, you mean each change is a one?

Rose: I mean every number is one. Anyway. You take away two of the ones and then you add one more. That's only, that's only six but here (after the initial 4 and + + +) you've got seven.

Interviewer: So if it's all ones, it works out like you said. That's where you've got the most?

Rose: Yes.

Interviewer: And is there any situation where it wouldn't work out that way, or it might not?

Rose, thinking aloud, seems to change her mind, but then sticks with her idea that the end of a series of pluses is when there are the most, even if you start with a minus:

Rose: I don't think so because if you. I think the first as far as you can go without going through minuses is where you are going to have the most. Unless, unless like the first sign is a minus and then there's a whole line of plusses all stuck together like this (draws + signs touching). And there's another minus and you'll probably have the most right there (before the minus).

Interviewer: Is there anything about this plus (at the end) that would make you change your mind, imagining all the numbers that it could be?

Rose: Mm. Only one thing—if that one (the last +) was a lot.

Interviewer: If the last one was a lot?

Rose: Yeah. Like 10 or so. Then it would probably be okay. Then probably that would be the most provided these (the first three plusses) are not too much and these (the minuses) are not too much either.

Interviewer: In other words, you have to have a real big one after the taking out part to end up with the most at the end. Okay.

Discussion of Episode 2

Although Rose describes what will happen when blocks are added or taken away in unspecified quantities, she moves freely back and forth from the general to the specific, from relative quantity to exact quantity. For instance, she suggests the specific case of considering every number to be 1 to illustrate the general problem, but does not insist that this must be the case. When she chooses "10 or so" as a lot, this is relative to the small numbers she has been working with. Rose seems to be working at first with a rule that says, "If you add positives and then subtract, you will have a maximum before the subtraction." But her thinking is quite flexible. She is able to imagine obtaining a maximum at the end of a series in quite a different way—it does not come as the result of a series of positives followed by a negative, but by one positive that is relatively much larger than the other changes. Her ability to imagine these possibilities is built on her concrete experience of actual numbers of blocks, perhaps on a mental image of the blocks, but it has gone beyond that so that she is able to imagine situations that she has not actually experienced. (These episodes are more fully described in Nemirovsky, Tierney, & Ogonowski, 1993.)

Whereas some of the children in Episode 1 were generalizing from patterns and regularities they had observed in many examples, the kind of generalization we see Rose making is closer to one in which a person sees a logical necessity of a general statement based on a sense that "it could be no other way." Thus, we might know that a multiple of 4 must be an even number, because a multiple of 4 is a multiple of 2 and not because we have studied many numbers of this kind. This type of generalization is the one associated with traditional algebra, which is based on the implications of formal laws, rather than on generalizations from many cases. However, once this distinction is made between generalization from patterns and regularities and generalization from a sense of logical necessity, we see, even in Episode 1, evidence of a generalization that combines elements of both kinds. Kadisha and Sylvia work with finding net change with particular numbers, perhaps generalizing several cases. Yet, when they work on the backward problem to infer what floor they would have to begin at in order to arrive at a given floor, there is sense of the logical necessity of their answer. The difference between their approach and Rose's is that they consistently work with specific numbers, while she works with unspecified symbols standing for numbers. Such movement on the child's part is widely considered an important step in the transition from arithmetic to algebra. The literature on children's early development establishes that young children are quite adept at generalizations from examples. In fact, they are often described as generalizing too freely from a few examples (Smith, DiSessa, & Rochelle, 1994). We claim that children have rich resources for these other kinds of generalizations as well.

EPISODE 3: SLOPE AS INDICATING VARIATIONS OF SPEED

In this episode, we see students analyzing graphs of plant height over time in order to compare changes in the plants' heights, as well as rates of change. This illustrates an early and productive use of a symbol system as a preliminary to the use of literal variables, formulas, and equations (Swan & Shell Centre Team, 1989).

This episode comes from a conversation in a classroom in which fourth-grade students are using a unit called *Changes Over Time* (Tierney, Nemirovsky, & Weinberg, 1994) from the curriculum series *Investigations in Numbers, Data, and Space*. The children have grown plants from seed and have recorded and graphed their plants' heights each day for two weeks. Now they are asked to interpret qualitative graphs of plant height over time in which no quantities are shown on the axes; only the shapes of graphs are provided and the labels *height* on the vertical axis and *time* on the horizontal axis (see Fig. 7.2). All the students interpreted steeper graphs as meaning the plant was growing faster and higher graphs as showing a taller plant. In the episode described here, they are working on a problem that has two graphs, one that is higher, but not steep, and the other that is lower, but steeper. The problem provokes disagreement in which students deal with the issues of change of height versus height and rate-of-change versus change.

When the teacher asks which plant is growing faster, Michelle, Sean, and James describe rate of growth. Michelle compares the growth of the two plants by comparing the changes in height in a fixed amount of time:

Michelle: The light line [is faster]. It started really small and got bigger and bigger and took the same amount of time to get to the same height.

At first, Sean and James respond directly to the shape by interpreting it in terms of comparative change:

Sean: The light one [grows faster], because it always going up. The dark one is kind of steady and kind of going across.
James: The dark one is slightly going up and it's not going fast.

When Darius disagrees, James adds an argument like Michelle's of considering growth in time:

Darius: It [the dark line] is going fast
James: It didn't grow high in a short time.

When the teacher questions him, James bases his answer on the shape of the line, describing it in a language appropriate for the plant it depicts:

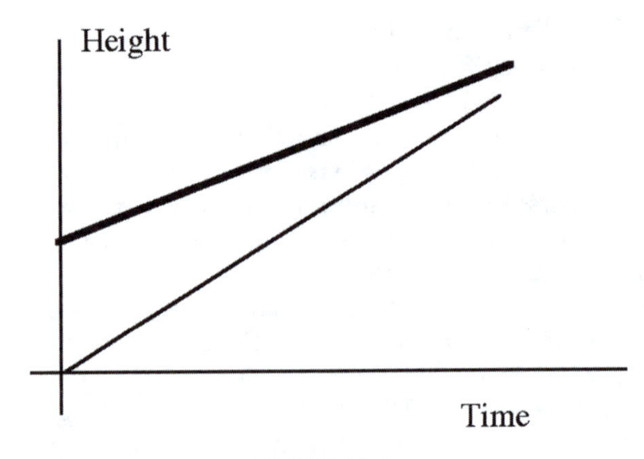

FIGURE 7.2.

Teacher: Tell me about the changes.
James: The dark line is only growing a little bit over a long time. The light line, the changes are bigger over the same amount of time.

Instead of focusing on change-in-height illustrated on the graph, other students focus on current height and how the plant reached that height before the part of the story shown on the graph. Bobby and Sarah speak of the plant before and after the time depicted on the graph:

Bobby: The dark line grew faster at the beginning, before the graph.
Sarah: I chose the dark line. The light line takes time to grow up. It's going to take it a long time to catch up with the black line.

The teacher asks Bobby to come up to the board and draw the dark line as he thinks it might have been before the graph began. He starts at left end of the dark line and extends it leftward, making a line that curves down to the horizontal axis (see Fig. 7.3):

Bobby: [Moving his finger from left to right along the line he drew] It grew fast, then still fast, then started to get steady.

Discussion of Episode 3

These children are involved in a lively discussion in which genuine issues about comparing changes and various meanings in a graphical symbol system[4] arise. Students agree about the interpretation of the graph, but disagree about whether to take into account data previous to the graphed

[4]By symbol system, we mean an arrangement of markings of various kinds that has conventionally accepted rules for interpretation within a given community.

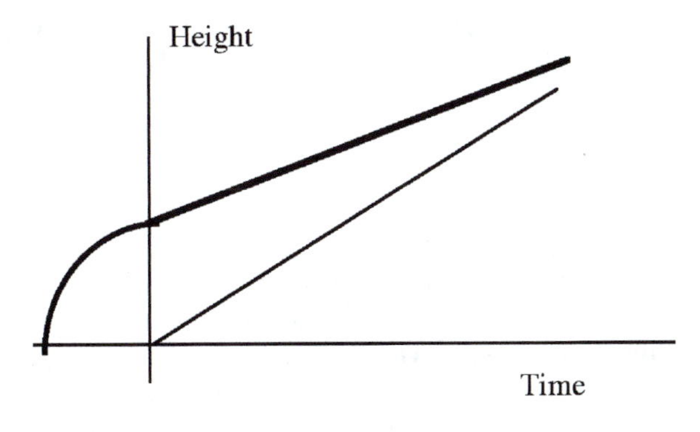

FIGURE 7.3.

data to answer the question of which plant grows fastest. They are using their understanding of how plants grow to make sense of the graphs, and they are interpreting the complex meanings in the graphs in order to elaborate richer possibilities for relations between sizes of plants, changes in size, and rates of change. For instance, to discuss the issue of which plant is growing faster, some children focus on its visual aspects, such as steepness, whereas others focus on quantitative information carried in it. (For a full discussion of students' use of the visual aspects of graphs, see Monk & Nemirovsky, 1994.) Bobby's view that there is a part of the graph that was missing from the original one raises an important issue in the use of symbols. What is the status of the graph in relation to the problem situation? Is it a complete record, the only source of available information about the event, or is it like an illustration that tells part of the story to be supplemented by other things we know and believe?

EPISODE 4: FROM CONTINUOUS TO DISCRETE—MAKING TABLES TO FIT STORIES

This episode shows a class deciding when two tables that were made by students to depict the same event are in fact the same or different.[5] In using any symbol system, even one as apparently straightforward as tables, questions often arise as to whether or not two different arrangements of symbols really have the same meaning and whether a given arrangement could possibly have two very distinct interpretations. This is the underlying problem in making abstractions: to decide which of the many aspects of a symbolic array or situation are to be paid attention to and which are to be ignored. Seeing anything is a matter of highlighting, organizing, and structuring (Arnheim, 1969).

In one of the activities in the *Investigations* unit for fifth grade, *Patterns of Change: Graphs and Tables* (Tierney, Nemirovsky, Noble, & Clements, 2004), students make tables to fit with trips they plan and act out along a 10- to 15-meter line. This activity takes advantage of children's strong inclination to use narrative to describe an event such as walking along a line at varying speeds. Each student chooses one of three stories provided of trips that could be made along a line and makes a table to go with the story. The students exchange their tables and then each is supposed to post this new table on a bulletin board near the story they believe the table belongs with. They discuss the tables posted with each of the stories to decide if they are posted with the right story and to compare tables that fit the same story.

Nadine has made a table (Table 7.3) that has been placed with the story: "Walk very slowly about a quarter of the distance, stop for about 6 seconds, and then walk fast to the end":

David: I don't think that Nadine's goes with the story. It says walk very slowly for about a quarter of the distance. But then it says stop for 6 seconds. She keeps going.

Teacher: [pointing the entries in Nadine's table: 5, 8, and 10], Then what would you have put here, here, and here instead?

David: Threes.

Nadine agrees with David that it does not belong there. The teacher asks April, who placed it improperly in the first place, to figure out which other story Nadine's table belongs to. She puts it with the story, "Run about halfway and then go slower and slower until the end." She indicates that this partly fits Nadine's table, although there is still a discrepancy because Nadine's table goes almost the same speed at the end instead of continuing to slow down. Thus, April considers ways in which this table and the three stories differ, and decides this story is similar enough that it might have been the one Nadine was thinking of.

The teacher then draws attention to the other tables (Table 7.4) that go with the first story ("Walk very slowly about a quarter of the distance, stop for about 6 seconds, and then walk fast to the end."), and asks if they are the same or different:

Elena: Me and Judith's are the same. We were in different parts of the room. Somehow it got the same.

Teacher: How does Anita's table differ? What does she point out?

Judith: There's more time. She figured that to get to 14 she needed 23 seconds.

Table 7.3
Nadine's Table

Time in Seconds	*Distance in Meters*
2	3
4	5
6	8
8	10
10	11
12	12
14	13

Table 7.4

Judith's Table

Time	*Distance*
2	2
4	3.5
6	3.5
8	3.5
10	3,5
12	9
14	14

Elena's Table

Time	*Distance*
2	2
4	3
6	3
8	3
10	3
12	9
14	14

Anita's Table

Time	*Distance*
2	1
4	2
6	3
8	4
10	4
12	4
14	4
16	6
18	9
20	11
22	13
23	14

Teacher: What part of the script did Anita take very seriously? Hers is a bit different. But it's got similarity. She has the person waiting four chunks of time (as Judith and Elena do).
David: Six seconds.
Nick: In it she goes only 1 meter in 2 seconds. Theirs is 2 seconds 2 meters.
Teacher: So they went a little further in walking slowly.
Judith: Her trip took a longer time. Her going slow was really slow.
Anita: It (the story) says "very slow."

Discussion of Episode 4

Elena expresses amazement that she and Judith have made the same table, although we see that the corresponding entries in their tables are not identical. However, although the three tables fit more or less the same story, the students recognize that Anita's table is different from the other two. This is not only because it is longer, but because Anita's tells a different story. "Her going slow was really slow." In making such distinctions about how things might be different while being the same, these children show an early capacity for abstraction. Elena sees a pattern in her and Judith's table. The students recognize a different pattern in Anita's. Earlier, April picked out a pattern in Nadine's table that was different from the slowing down pattern she expected for the story it was to fit. These overall patterns are features abstracted from the table, which the students connect to the relationship between distance and time.

Many of the tasks children do in our interviews and in the *Investigations* curriculum involve thinking qualitatively and constructing representations: How do two things compare? Where is the maximum? Which plant represented in this graph is growing faster? How are two stories of change alike or different?

The children's grasp of these situations suggests that it is possible to develop a curriculum that engages children, starting in the early grades, without getting enmeshed in the details of procedures of calculation, setting up of scales, or other issues associated with teaching conventional graphing. Such a curriculum might move children from the arithmetic of specific quantities to thinking about relationships among varying quantities through qualitative representation in stories, graphs, and literal variables.

This qualitative comparison appears in the kind of problem described in chapter 7 by Carraher, Schliemann, and Earnest (this vol.) as an invitation to think about the relationships among variables. This comparison and examination of relationships can be about quantities or just about generalized numbers. This work can include analysis of purely numerical relationships as children move from the specifics of a particular instance or a group of instances to a conjecture in which they establish a generalization of necessity being true. Thus, students can generate conjectures about and discuss open sentences that are always true ($a + b = b + a$), sometimes true ($a - b = b - a$), or never true ($a + 3 = a - 3$), an activity suggested by Davis (1964) for middle school many years ago. When students work at algebra of literal variables, the students build on their experiences of reasoning about these quantities in context.

Behind the argument that algebra cannot be taught to students before eighth or ninth grade is a view of school mathematics sharply divided into two worlds: a world of operations on specific, concrete numbers, and a world of operations and reasoning on unspecified, abstract variables.

The children we work with are exploring the vast middle ground in between. Students reason about the outcome of specific arithmetic operations on possible numbers and making generalizations on these (Episode 1); finding patterns and making generalizations of a logical kind in number sequences (Episode 2); making qualitative interpretations of symbolic but situated representations (Episode 3); and judging the possibilities that two symbolic representations might belong to different or same events (Episode 4). Through living in this middle ground and becoming confident and familiar with these mental processes, students can grow to be masters of the world traditionally called algebra.

REFERENCES

Arnheim, R. (1969). *Visual thinking*. Berkeley, CA: University of California Press.

Barnes, M. (1992). *Investigating change*. Melbourne, Australia: Curriculum Corporation.

Carpenter, T., Franke, M. L., & Levi, L. (2003). *Thinking mathematically: Integrating arithmetic and algebra in elementary school*. Portsmouth, NH: Heinemann.

Chazan, D. (1996). Algebra for all students? *Journal of Mathematical Behavior, 15*(4), 455–477.

Davis, R. B. (1964). *Discovery in mathematics: A text for teachers*. Palo Alto, CA: Addison-Wesley.

DiSessa, A., Hammer, D., Sherin, B., & Kolpakowski, T. (1991). Inventing graphing: Meta-representational expertise in children. *Journal of Mathematical Behavior, 10*, 117–160.

Krabbendam, H. (1982). Non-quantitative way of describing relations and the role of graphs. In G. van Barneveld & H. Krabbendam (Eds.), *Proceedings of the conference on functions* (Report 1, pp. 1245–1246). Enschede, the Netherlands: National Institute for Curriculum Development.

Monk, S. (2003). Representations in school mathematics: Learning to graph and graphing to learn. In J. Kilpatrick, W. G. Martin, & D. Schifter (Eds.), *A research companion to the NCTM standards* (pp. 250–262). Reston, VA: National Council of Teachers of Mathematics.

Monk, S., & Nemirovsky, R. (1994). The case of Dan: Student construction of a functional situation through visual attributes. *CBMS Issues in Mathematics Education, 4*, 139–168.

National Council of Teachers of Mathematics. (1998). *The nature and role of algebra in the K–14 curriculum*. Washington, DC: National Academy Press.

Nemirovsky, R., Tierney, C., & Ogonowski, M. (1993). *Children, additive change, and calculus* (Working Paper No. 2-93). Cambridge, MA: TERC.

Russell, S. J., Tierney, C., Mokros, J., & Economopoulos, K. (2004). *Investigations in number, data, and space*. Glenview, IL: Pearson Scott, Foresman.

Schifter, D. (1999). Reasoning about operations: Early algebraic thinking in grades K–6. In L. V. Stiff & F. R. Curcio (Eds.), *Developing mathematical reasoning in grades K–12* (pp. 62–81). Reston, VA: National Council of Teachers of Mathematics.

Smith, J., DiSessa, A., & Roschelle, J. (1994). Misconceptions reconceived. *Journal of the Learning Sciences, 3*, 115–163.

Swan, M., & Shell Centre Team. (1989). *The language of functions and graphs: An examination module for secondary schools.* Nottingham, UK: The Shell Centre for Mathematical Education & Joint Matriculation Board.

Tierney, C., & Nemirovsky, R. (1991). Young children's spontaneous representations of changes in population and speed. In R. G. Underhill (Ed.), *Proceedings of the 13th annual meeting of the North American Chapter of the International Group for the Psychology of Mathematics Education* (Vol. 2, pp. 182–188). Blacksburg, VA: Conference Committee.

Tierney, C., Nemirovsky, R., Noble, T., & Clements, D. (2004). Patterns of change: Tables and graphs (A unit for grade 5 of *Investigations in number, data, and space*). Glenview, IL: Pearson Scott, Foresman.

Tierney, C., Nemirovsky, R., & Weinberg, A. (1994). Up and down the number line: Changes over time (A unit for grade 3 of *Investigations in number, data, and space*). Glenview, IL: Pearson Scott, Foresman.

Tierney, C., Weinberg, A., & Nemirovsky, R. (1994). Changes over time: Graphs (A unit for grade 4 of *Investigations in number, data, and space*). Glenview, IL: Pearson Scott, Foresman.

8

What Is a Legitimate Arithmetic Number Sentence? The Case of Kindergarten and First-Grade Children

Nitza Mark-Zigdon
Dina Tirosh
Tel-Aviv University

There is a growing awareness nowadays in various countries, including Israel, that basic algebraic ideas should be enhanced as early as possible (Ministry of Education, 2005). The ability to represent symbols is essential to the learning of algebra. This chapter describes a study of the informal knowledge of addition and subtraction number sentences that preschool children in Israel bring with them to school. The study continued throughout the first year of schooling in order to follow the development of this knowledge.

The development of algebraic reasoning is taken by many, including several authors in this volume, as having a sound starting point in the use of number sentences to build generalizations of arithmetic operations and their properties. Others use the semiotic features of number sentences as bases for building the kind of symbol sense that serves algebraic reasoning. Hence, we feel that it is important to understand what students bring to this enterprise in terms of what they think number sentences are.

This chapter covers three main topics: the theoretical framework that we used to construct the research tools and to analyze the data, the study, and the main findings, and raises issues in need of further exploration.

THE DEVELOPMENT OF CHILDREN'S KNOWLEDGE OF THE SYMBOLIC ARITHMETIC SYSTEM: WHAT DO WE KNOW FROM RESEARCH?

The main function of the symbolic arithmetic system is to represent the arithmetic concepts, operations, and relations, to transmit information and to enable students to arrive at the mathematical meaning behind the symbol (Bialystok, 2000). There are two major components of the symbolic arithmetic system. First, there are symbols: namely, the numerals, which are the symbols of the basic operations; the equals sign; and the other two order-relation signs (">" and "<"). The second component is constituted by the rules for operating on the symbols. These rules are based on the positional principles of the decimal system and on the conventional ways of writing the basic operations. In our study, we attempted to assess kindergarten and first-grade children's knowledge about these two components of the symbolic arithmetic system, with particular focus on the differences between recognizing and producing addition and subtraction number sentences.

WHAT FACTORS INDICATE A CHILD'S ABILITY TO DERIVE THE MEANING OF A SYMBOL?

Certain symbols, such as the numerals standing for numbers, the operation symbols and the equals sign, stand in an arbitrary but conventional relationship to their referents, so students need to supply the connections between the symbols and their referents (Bialystok, 1992; Dorfler, 2000; Mandler, 1992). Children's ability to grasp the meanings of a symbol is determined by three major factors: the representation space of the user of the symbol, the development of symbolic thinking, and the conceptual mathematical knowledge base.

Representation Space

The representation space includes the database of symbols, metaphors, and the representation structures that are embedded in the personal experience of the interpreters of the meaning, and their knowledge of the principles for operating on the representation systems (Nemirovsky & Monk, 2000; von Glaserfeld, 1991). Included in this factor is the user's ability to differentiate between the components of the representation system that are relevant to the mathematical concepts and those components such as size, color, and font that might be visually prominent but are irrelevant from a mathematical point of view (Janvier, 1987; Kaput, 1987, 1991; Lesh, Post, & Behr, 1987).

Symbolic Thinking

To engage in a developing ability to perceive the symbol as representing the meaning of an object or idea, without the latter being literally expressed within the symbol itself, is at the heart of symbolic thinking. Here we focus on the changes from fusion of signs and objects to differentiation (Nemirovsky & Monk, 2000; Werner & Kaplan, 1963). In a situation of fusion, the symbols are perceived as the object of representation. Consequently, at the early stages of the development of symbolic thinking, children perceive the objects as possessing traits and, in various situations, they identify the features of the symbols with those of the object that it represents. Thus, for example, children will write names of large objects in large letters (Thomas, Jolley, Robinson, & Champion, 1999). When symbolic thinking is at the stage of differentiation of sign from referent, the children exhibit their ability to relate separately to the object being represented and to its symbol (Nemirovsky & Monk, 2000; Werner & Kaplan, 1963).

Conceptual Knowledge Base

In the absence of a proper knowledge base regarding the mathematical entities of arithmetic, the child will not be able to access the appropriate meaning of the representation (e.g., children who are unfamiliar with the structure of the decimal system will not be able to deduce the meaning of combinations or strings of numerals such as 23; Hart, 1981).

WHAT IS KNOWN ABOUT THE DEVELOPMENT OF THE SYMBOLIC ARITHMETIC SYSTEM?

Studies reveal that newborns are cognitively equipped from the very outset to recognize quantities and operations with quantities (Butterworth, 2000) and that the ability to symbolize begins to develop in children from very early life stages (DeLoache, Miller, & Rosengren, 1997; Mandler, 1992, Piaget, 1962). In many cultures, children are exposed, from a very early age, to conventional symbolic systems, including the symbolic arithmetic system and, consequently, they acquire various types of knowledge about this system (Bialystok, 1992). Tolshinsky-Landsmann and Karmiloff-Smith (1992) reported that children from about age 4 distinguish between symbols that belong to the number system and those that do not belong to it. They found that children at age 4 differentiate between letters and numerical symbols. They reported, for instance, that children at that age perceive a repetition of the same number as a number, but a repetition of the same letter is not considered a word. Still, several researchers noted that the ability to attribute meaning to numerals develops gradually (e.g., Bialystok, 1992; Hughes, 1986).

This study employed the detailed, four-stage hierarchical model of the development of the symbolic representation of numbers that is offered by Hughes (1986), who describes how, at the first stage, children represent numbers by means of idiosyncratic representations. These symbols are not linked to the shape and quantity of the objects that they represent. At the second stage, children represent the numbers by means of pictographic representations. At this stage, they employ the graphic expression appropriate to the quantity, shape, situation, color, or direction of the objects (e.g., the child draws five children to describe a given situation relating to five children). At the third stage, children employ iconic representations. Here they represent the numbers by means of a symbol system based on one-to-one correspondence between the number of shapes drawn and the given number of objects, such as lines or circles. At the fourth stage, children use the conventional symbols to represent the numbers. Here they employ number symbols (numerals) on the basis of their understanding of the meaning that these symbols represent, and their awareness that the very use of the symbols activates their meaning. The development of the symbolic knowledge related to the basic arithmetic operations, to the equals sign, and to the order relation signs occur later—and in that order. In this respect, A. Sinclair and H. Sinclair (1984) noted that when children reach school they are familiar with the numerals but not with the symbols of the basic arithmetic operations and of the order relations.

KINDERGARTNERS' AND FIRST GRADERS' KNOWLEDGE ABOUT NUMBER SENTENCES: THE STUDY

Participants

One hundred and fifty-four children from upper middle-class families (48 kindergarten children and 106 first graders) participated in the study. Half of the participants in each of these two age groups were male and the other half female. The kindergarten children (ages 5–6) attended two nursery schools where no formal instruction related to the symbolic arithmetic system was evident. The first graders (ages 6–7.5) studied in five classes at two schools. All the students came from upper-middle-class families.

Tools and Procedure

A structured, individual interview was developed for this study. The interviews consisted of two main sections, in accordance with the two facets of symbolic knowledge: production and recognition. The interviews were conducted by the first author, in a quiet room. Each child was interviewed twice: first on the production section and then on the recognition. Each interview lasted at least 30 minutes.

Production. Participants were asked to write four numbers (5, 8, 13, 20), two addition number sentences (i.e., $4 + 2 = 6$ and $13 + 4 = 17$), and two number sentence involving subtraction ($5 - 2 = 3$ and $12 - 3 = 9$). Each number and each number sentence was written on a separate, empty card.

Recognition. This section consisted of 12 cards, 6 with addition and subtraction number sentences: 2 in the canonical form ($3 + 2 = 5$ and $9 - 5 = 4$), 2 with the operation sentence on the right side ($8 = 12 - 4$ and $7 = 3 + 4$), 1 multioperational number sentence ($4 + 2 + 1 + 3 + 2 = 12$), and 1 addition number sentence written in vertical format. Six cards contained inappropriate writing of number sentences: two with missing symbols (i.e., 2 + 9 _ 11 and 3 _ 8 = 11) and four with letters or pictorial symbols instead of some of the numerals (e.g., _ + 3 = 8). The interviewee presented the child with one card at a time. The child was asked to determine whether what appeared on the card is a correct way to write a number sentence. The interviewee explained the task to each child in the following way: "I asked you, at the first meeting, to write on cards, like $4+2=6$ ... Do you remember? I asked other children to do the same. I will show you cards that the others wrote. Look at the card and tell me if what is written on it is a correct way to write addition/subtraction. If it is, put it in the red box, if it is not, put it in the blue box. Please, while you do this, explain why." At the end of the classification, the child was encouraged to look at the two piles of cards and to make changes, if they felt that such changes were needed.

Main Findings

Production. All the first graders who participated in this study correctly wrote the two numbers below 10, 90% correctly wrote the two numbers above 10, and 86% correctly wrote all four addition and subtraction number sentences. Those who wrote inadequate number sentences wrote the expressions from right to left. In the case of writing addition number sentences, such writing resulted in correctly written expressions. This, however, is not the case for the subtraction number sentences.

Almost all the kindergartners (93%) correctly wrote the number 8 and about half adequately wrote the numbers 5 and 13. The number 20 was the most difficult to write (21% kindergartners wrote it correctly). Only about 10% of the kindergartners correctly wrote the four addition and subtraction number sentences. The others did not write any of the four number sentences in a correct manner. About 50% of the children wrote the numbers and the number sentences from right to left. In the number sentences, they wrote only the numbers (the symbols of the operations and of the equals sign were not written). The children's comments during the interviews suggest that they assumed that numbers, words, and sentences are written in the same direction: from right to left. (In Hebrew, the direction of writing is from right to left.)

A substantial number of kindergartners (about 40%) avoided writing the numbers and the number sentences. The explanations given by these children were of the type: "I know that I don't know how to write this," "I know it's with two numbers but I don't know how to write it." This might imply an existence of an intermediate, awareness–avoidance stage between Stage 3 and Stage 4 (Hughes, 1986). This awareness–avoidance stage is characterized by "knowing that—but not how," that is, the awareness of the existence of the rules for writing numbers and number sentences is coupled with a lack of knowledge of these rules. Obviously, there is room to expand the research in this direction in order to test this hypothesis.

Recognition

Conventional Writing. Of the six cards that are included in this category, the two number sentences that were written in the canonical format were identified as correctly written addition and subtraction number sentences by almost all the children (98% and 89% of the kindergarten and in first grade, respectively). However, other representations of addition and subtraction number sentences were not accepted as such by the vast majority of the children. The representations of the two number sentences in which the operation sentences were written on the left side of the number sentence were identified as correct number sentences by about 40% of the children in each group. The explanations of the first graders who argued that these were incorrect representations of addition and subtraction number sentences revealed that they tended to interpret the equals sign as expressing a result of an operation (e.g., 2 + 3 results in 5) and not as an indication of equivalence between two expressions (see e.g., Kieran, 1981). The kindergartners typically argued that: "I don't know if this is how we write it." This could be regarded as another instance of the awareness–avoidance stage.

The vertical representation of the addition number sentence was accepted as an adequate representation of an addition number sentence by about 30% of the children in each group. Most children who did not accept this representation explained, "The numbers should be in a row" or "The numbers and the + are not arranged correctly, they are not in a row." Some first graders further commented, "The equals sign is missing." The addition number sentence $4 + 1 + 1 + 3 + 3 = 12$ was regarded as an adequate number sentence by about 60% of the children in both grades. The first graders who did not accept it as an adequate representation argued that "there are too many numbers and too many +" or that "there must be only three numbers." Most of the kindergartners that did not choose this as an adequate expression explained that they did not know if it is "OK to write in this way."

Inappropriate Writing

Missing Symbols. Two thirds of the first graders correctly argued that the expressions that did not include an operational symbol or an equals sign "are not number sentences." They clearly stated the signs that were missing in each expression. About half of the kindergartners were aware that "something was missing" in the expression without an operation sign, and about 20% regarded the expression that lacked the equals sign as an inappropriate number sentence. These findings are in line with previous research findings indicating that the development of the recognition of the necessity to include operation symbols in number sentences precedes that of the inevitability of including the equals sign.

Mixture of Arithmetic Symbols and Other Symbols. Almost all the first graders (about 90%) argued that expressions, including pictorial symbols or Hebrew letters, are not number sentences. Typical explanations were "When you write a drill you do not draw" or "When you write a drill you write numbers, you don't write letters." About one third of the kindergartners accepted these sentences as addition and subtraction number sentences. They explained that these expressions contain numbers and therefore they "Should be put in the red [number] box." Again, more than half of them explained that they do not know if this writing is "OK."

The overall picture regarding the kindergartners' and the first graders' ability to differentiate between expressions that are conventionally regarded as addition and subtraction number sentences and those that are not show that a substantial number of first graders accepted only canonical representations as adequate number sentences. Probably, these are the (only) number sentences that they encountered in class (see also, Franke, Carpenter, & Battey, chap. 13, this volume). About one third of the kindergartners accepted expressions that included numerals and other, nonmathematical symbols as addition and subtraction number sentences. It seems that first graders made a decision that a certain expression presents a number sentence on the basis of three criteria: The expression contains arithmetic symbols, the expression contains only arithmetic symbols, and the numerals and the other symbols are written in the canonical format. A substantial number of the kindergartners argue that they lack the knowledge needed to reply to these tasks. Those who did tended to relate only to the first criterion and, consequently, they considered expressions that included arithmetic signs together with other, nonarithmetical signs as numbers.

CONCLUSIONS

This research has expanded the knowledge base on kindergarten and first-grade children's informal and formal knowledge of the symbolic arithmetic system. The results suggest that for kindergartners, the recognition as such of canonically written number sentences was profoundly easier than the production of such number sentences. First graders, however, highly succeeded in both these tasks, although their performance on the recognition tasks was better than on the production tasks.

Had we stopped our analysis at this point, we might have concluded that, for both kindergartners and first graders, recognition of addition and subtraction number sentences is a less demanding task than their production. However, the data reveal a rather complex situation. The kindergarten children were indeed less successful in each of the production tasks than in each of the recognition tasks. This suggests that before formal instruction, production is a highly demanding task. However, for the first graders, all the recognition tasks included in the interview, apart from those related to number sentences written in the canonical format, were profoundly more demanding than those that involve the production of number sentences. One possible recommendation for instruction, accordingly, is to devote more attention to discussing, in class, the nature of addition and subtraction number sentences and of each of the arithmetic symbols, and to describing critical and noncritical properties of these sentences.

A phenomenon that was identified among kindergartners in this study is that they tend to avoid both producing and recognizing addition and subtraction number sentences. It seems that such avoidance reflects their awareness both of the existence of rules for writing number sentences and of their own lack of knowledge of these rules. Issues related to this phenomenon, such as whether this is a general, awareness–avoidance stage, should be explored further.

All in all, this study provides some indications that many kindergartners and first graders are beyond the third, iconic stage in the development of the symbolic representation of numbers described by Hughes (1986). The fourth stage, in which children symbolically represent numbers and number sentences should, however, be further explored. This exploration should take account of both the production and the recognition facets. This chapter could be viewed as a first step in this direction.

REFERENCES

Bialystok, E. (1992). Symbolic representation of letters and numbers. *Cognitive Development, 7*, 301–316.

Bialystok, E. (2000). Symbolic representation across domain in preschool children. *Journal of Experimental Child Psychology, 76*, 173–189.

Butterworth, B. (2000). *What counts: How every brain is hardwired for math*. New York: The Free Press.

DeLoache, J. S., Miller, K. F., & Rosengren, K. S. (1997). The credible shrinking room: Very young children's performance with symbolic and non-symbolic relations. *Psychological Science, 8*, 308–313.

Dorfler, W. (2000). Means and meaning. In P. Coob, E. Yackel, & K. McClain (Eds.), *Symbolizing and communicating in mathematics classrooms: Perspectives on discourse, tools, and instructional design* (pp. 100–132). Mahwah, NJ: Lawrence Erlbaum Associates.

Hart, K. (Ed.). (1981). *Children's understanding of mathematics*. London: Murray.

Hughes, M. (1986). *Children and number: Difficulties in learning mathematics*. Oxford, England: Blackwell.

Janvier, C. (1987). Translation processes in mathematics education. In C. Janvier (Ed.), *Problems of representations in teaching and learning of mathematics* (pp. 27–32). Hillsdale, NJ: Lawrence Erlbaum Associates.

Kaput, J. (1987). Representation systems and mathematics. In C. Janvier (Ed.), *Problems of representations in teaching and learning of mathematics* (pp. 19–26). Hillsdale, NJ: Lawrence Erlbaum Associates.

Kaput, J. (1991). Notations and representations as mediators of constructive processes. In E. von Glaserfeld (Ed.), *Radical constructivisim in mathematics education* (pp. 53–74). Dordrecht, the Netherlands: Kluwer Academic.

Kieran, C. (1981). Concept associated with the equality symbol. *Educational Studies in Mathematics, 12*, 317–326.

Lesh, R., Post, T., & Behr, M. (1987). Representations and translation among representations in mathematics learning and problem solving. In C. Janvier (Ed.), *Problems of representations in teaching and learning of mathematics* (pp. 33–40). Hillsdale, NJ: Lawrence Erlbaum Associates.

Mandler, J. M. (1992). The foundation of conceptual thought in infancy. *Cognitive Development, 7*, 273–285.

Ministry of Education. (2005). *The new version of the National Israeli curriculum for elementary schools*. Jerusalem, Israel: Author.

Nemirovsky, R., & Monk, S. (2000). "If you look at it the other way ... ": An exploration into the nature of symbolizing. In P. Coob, E. Yackel, & K. McClain (Eds.), *Symbolizing and communicating in mathematics classrooms: Perspectives on discourse, tools, and instructional design* (pp. 177–224). Mahwah, NJ: Lawrence Erlbaum Associates.

Piaget, J. (1962). *Play, dream and imitation in childhood*. New York: Norton.

Sinclair, A., & Sinclair, H. (1984). Preschool children's interpretation of written numbers. *Human Learning, 3*, 173–184.

Thomas, G. V., Jolley, R. P., Robinson, E. J., & Champion, H. (1999). Realist errors in children's responses to picture and words as representations. *Journal of Experimental Child Psychology, 74*, 1–20.

Tolchinsky-Landsman, L., & Karmiloff-Smith, A. (1992). Children's understanding of notation as domain of knowledge versus referential-communicative tools. *Cognitive Development, 7*, 287–300.

von Glaserfeld, E. (1991). Radical constructivism. In E. von Glaserfeld (Ed.), *Radical constructivism in mathematics education* (pp. 53–74). Dordrecht, the Netherlands: Kluwer Academic.

Werner, H., & Kaplan, B. (1963). *Symbol formation*. New York: Wiley.

9

Visualizing Algebraic Reasoning

Timothy Boester
University of Wisconsin–Madison

Richard Lehrer
Vanderbilt University

In this volume, algebra and algebraic reasoning are proposed as a core constituent of a general mathematics education that extends throughout schooling. We share this conviction with the contributors to this volume, but we aim to extend the franchise to space and geometry as a complementary strand in a general mathematics education (Lehrer & Chazan, 1998). Rather than place these two strands in competition for curricular space and time, we propose synergy: Visualization bootstraps algebraic reasoning and algebraic generalization promotes seeing new spatial structure (Goldenberg, Cuoco, & Mar, 1998).

We explored prospective relations between geometry and algebra by conducting a 2-year sequential design study (Brown, 1992; Collins, 1992) with two cohorts of a sixth-grade classroom. Design studies are conducted to explore prospective trajectories of student learning along with the means to support learning (Cobb, Confrey, DiSessa, Lehrer, & Schauble, 2003). This study investigated students' reasoning about Cartesian graphs, linear functions, and tables when these forms of representation were deployed as tools for describing visual patterns. In this instance, students characterized similar two-dimensional figures, namely, rectangles. The design was informed by previous studies in which younger students (third and fifth graders) were introduced to algebraic reasoning via study of geometric similarity (Lehrer, Strom, & Confrey, 2002). Foreshadowing our conclusions here, the results of these earlier endeavors with younger children were much like those we later describe in this chapter, suggesting that synergies between algebraic and spatial reasoning, and opportunities to learn about them, may be far more important than distinctions based on age.

META-REPRESENTATIONAL COMPETENCE

The emphasis in both the previous and current work was on supporting student learning by fostering the development of *meta-representational competence*, that is, competence to represent similarity in multiple ways (e.g., ratios, graphs, equations) and to develop conceptual relationships among these different representational forms (DiSessa, 2002, 2004; Lehrer et al., 2002). Although traditional accounts of learning mathematics tend to view representational forms as mere adjuncts to learning, we accord them a more central role (Kaput, 1991; Lehrer & Lesh, 2003). In our view, each representational form embodies a different conceptual sense, or niche (Hall, 1990), and mathematical reasoning evolves as a coordination or resonance among these different senses and associated representations (Lehrer et al., 2002). For example, one might view a ratio as a quotient, or as occupying a slot in the equation of a line (e.g., Schoenfeld, Smith, & Arcavi, 1993). Conventionally, these are equivalent shadows of the same referent. That is, there is an underlying construct of ratio, and one merely inscribes (following Latour, 1990) the referent differently. In contrast, our stance (Peirce, 1960) interjects the interpretant between the sign and the object (the signifier and the signified), thus providing space for the process of signification, that is, the telling of in what respect the sign stands for the object. In Peirce's (1960) words:

> A sign or *representamen* is something which stands to somebody for something in some respect or capacity. It addresses somebody, that is, creates in the mind of that person an equivalent sign, or perhaps a more developed sign. That sign which it creates I call the *interpretant* of the first sign. The sign stands for something, its *object*. It stands for that object not in all respects, but in reference to a sort of idea, which I have sometimes called the *ground*. (p. 135)

This triadic model of semiosis suggests a niche view of symbolization, where different notations convey different senses of the same mathematical object. Later, we describe how a ratio participating in an equation (one notational system) or in a line (yet another notational system) has very different meanings for students, although from a conventional perspective, the ratio signifies the same relation (the ratio of sides of rectangles).

In short, symbolic media have conceptual consequences because they afford different patterns of reasoning: For example, a figure constructed with traditional tools and media is reasoned about differently than one constructed with electronic tools and media (e.g., Lehrer, Randle, & Sancilio, 1989), even though the products are identical from a disciplinary point of view (e.g., both are rectangles).

Designing to Promote Representational Competence

The instructional design in this study was intended to create opportunities for students to develop competence with (at least) four representational forms typically introduced to students during their first course in algebra, each of which provided entry to a way of thinking—a mathematical sense—of similarity. The four different representational forms explored were rules in the form of verbal descriptions (e.g., long side is twice short side), re-expressed and re-interpreted as symbolic equations (e.g., $LS = 2 \times SS$), Cartesian graphs, tables, and quotients representing ratios.

We introduced similarity in a context of modeling, employing an overhead projector and changing its distance from a screen as it projected an image of a rectangle. We asked students to characterize what stayed the same and what changed as the distance varied. Our intention was to place additive (the sides were growing or shrinking additively) versus multiplicative (the sides were growing or shrinking multiplicatively) relations in competition as explanations in order to promote multiplicative thinking for the rest of the curriculum.

We employed classification as a second context for considering similarity. Students sorted cutout models of rectangles into groups using whatever criteria they liked. Our intention was to support the notion that similarity was one kind of invariance among many potential ways of classifying the same objects, but that classifications based on similarity were in accord with the behavior of the projector. We employed variations of classification three times during the instructional sequence, beginning with whole number ratios and eventually including rectangles that were not similar, yet that were described by the same linear function (one group of nonsimilar rectangles followed the equation $LS = SS + 5$). The intention for the latter was to generalize the concept of line, so that students could come to see a line described conventionally as a two-part schema.

Rules were introduced as a way of expressing students' conjectures about the nature of the visual pattern observed (e.g., by growing and shrinking images with the projector, rectangles could be superimposed to form a visual trace of similarity in the classification task), such as "one side is always twice as long as the other side." These rules were encapsulated by more formal, symbolic expressions, such as "$LS = 2 \times SS$," which expressed a generalization about the nature of the structure observed in a few instances. The symbolic expressions could be rearranged and otherwise manipulated in ways that were comparatively more difficult to consider informally, and these syntactic manipulations turned out to be important for establishing the sensibility of considering a rule as expressing a quotient (e.g., $LS/SS = 2$).

Yet another sense of the general structure of similarity was supported by Cartesian graphs (e.g., an invariant ratio as a property of a locus of points on a line), and a step emerged during instruction as a bridge between the systems of equations, on the one hand, and the visual appearance of steepness of lines, on the other. Tables too served as an alternative means of describing similar ratios, especially consideration of first-order differences between successive entries. These were employed to explore the behavior of graphs and of equations, and as bridges between them. Because these forms of representation are employed more generally to support algebraic reasoning, we included opportunities for students to use them in broader contexts (e.g., modeling relations between body measures).

Promoting Meta-Representational Competence

Many classroom activities were designed to support developing meta-representational competence, to examine the connections between different senses. Most of these activities focused on moving between grouping the physical rectangle cutouts, forming rules or equations, and graphing, although tables and steps were also explored in concert with other senses of similarity.

After grouping the physical rectangle cutouts in each collection, students were encouraged to make tables of each group of rectangle dimensions. They were also asked, using the table, to write down the dimensions of a few new rectangles that would also belong in the same group. Some students used the tables to help form rules for the groups of rectangles, and these rules also helped in finding the new rectangles that belonged to the group. Occasionally, after initially grouping the rectangles and creating tables, a few students regrouped the rectangles once the graph had been created, finding that the graph did not support their initial conjectures about the ratio groups. Thus, the rectangle cutouts (affording practical visualization of similarity) were expected to typically influence the formation of the groups and their respective tables, which in turn would help to create the rules and graphs, although the reverse process also occurred.

When forming rules and equations from rectangle groups, students were asked to state their rules in words and/or symbols using the terms

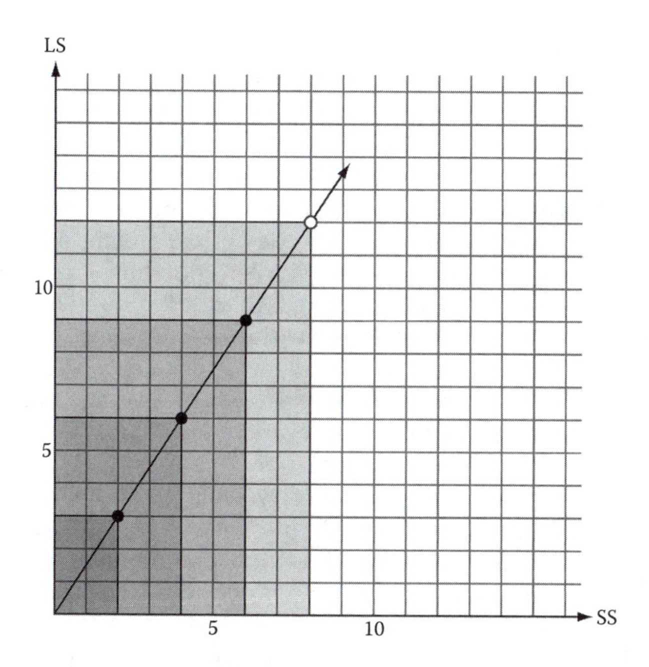

FIGURE 9.1. Coordinate graph representation of similar rectangles.

long side and *short side* to designate the different sides, respectively. Using these rules, students were asked to create new rectangles that would fit their rules, just as they were asked with the tables. Students also had to determine, with the third collection of rectangles, how to adjust their rules and equations to fit nonsimilar groups, those with a nonzero intercept. For example, given the rectangles 1×6 and 2×7, students would form the group $LS = SS + 5$ and might create the new rectangle 3×8 to fit the group. We chose collections that emphasized the intercept as a translation, so that students would encounter a dilemma: What accounted for parallelism with a corresponding similar group? Continuing the previous example, students would have been simultaneously given a rectangle set which followed the rule $LS = SS$ in order to compare the two sets.

When forming graphs from rectangle groups, students were asked to put the physical rectangle cutouts onto the grid paper, with the bottom, left-hand corner of each at the origin, and mark off the top, right-hand corner as the coordinate. The coordinates of each group were connected by a line to form the graph (see Fig. 9.1). Students created new rectangles that would fit on their graphs, an activity that paralleled creating new rectangles from the tables and equations.

Different conceptions of infinity were examined when students indicated where and how many different new rectangles could be placed on the graph. Students also considered which types of rectangle groups were represented by vertical and horizontal lines, and compared the graphs of nonzero intercept groups with those of similar groups.

A large portion of the activities in the curriculum focused on providing opportunities to strengthen meta-representational competence by comparing the same action over multiple senses of similarity. Students were asked, given a graph, what would be the corresponding equation and, given an equation, what would be the corresponding graph. Students also considered how the adjustments for intercept groups for equations and graphs are similar. Comparing the three senses of physical rectangle cutouts, equations, and graphs, students formed new rectangles from each group using each sense, and then established correspondences among each system. Students considered how procedures for creating new rectangles within one representational system related to creating new rectangles in other representational systems. As another way to generate new rectangles, given one length measure of a rectangle and a rule, equation, or graph, students found the corresponding length. They then were asked to consider how these methods of finding the other dimension with one particular form related to means employed for another. For example, given an equation, some students might generate new rectangles by substituting different values for a short side and then use the function to find the corresponding long side. Others might use the graph to locate a point and read off the corresponding coordinates. How were these very different actions related? In the graphical case, the method could be perceptually, albeit informally, tested: Did the result look as if it belonged (i.e., Was it similar?)? Of course, we did not rely on informal visualization only: Given a few rectangles, students used any of their representational systems to determine whether or not the rectangles comprised a similar group. Our intention was to foster ratio as a concept unifying these alternative descriptions of membership in the same group.

Finally, students created steps from their graphs, by moving from one coordinate to the next, making note of the up and over of the step. Some students actually drew in the steps on their graphs, making them look like a staircase, whereas others simply traced out the steps with their fingers and noted the dimensions. A few students used the rectangle cutouts to form their steps; others used the tables to find the differences between the coordinates. Students also examined the different ways that steps could be created, from the physical rectangle cutouts, graphs, or tables. For example, with graphs, a step was often interpreted as a physical movement (going up, then going over), but in a table, a step was a coordination of differences. Students looked for commonalities and differences between these representations. Commonalities were not always obvious (e.g., Why would physical movement on the graph have anything to do with tabular, first-order, differences?) and were compounded by the fact that the values generated via various representations for the steps were not always the same (students tended to retain the smallest step generated by the graph, while table calculations had no such interpretation).

ASSESSING STUDENT CONCEPTIONS

We developed a flexible interview (Ginsburg, 1997) to probe students' conceptions of each form of representation deployed during the course of their investigations.

Baseline

To get a sense of basic representational competence, students sorted cylinders in two different whole number ratios of height to circumference into similar collections. Tools provided to students included paper, pencil, and graph paper. Students had no experience with three-dimensional objects during the course of study, but we expected that the simple whole number ratios employed to generate each collection would be readily recognized. Students also commented on whether or not cubes would be considered similar, and why. We recorded students' reasoning in each task and noted which, if any, tools they employed to reach their conclusions.

Complicated Sorting

Mirroring classroom activity, but increasing its complexity, students sorted a collection of 15 rectangles (four ratio groups and one nonzero intercept group: 1:1, 1:2, 1:5, 2:3, and $LS = SS + 7$ arranged randomly, into groups, however they chose to define them. Figure 9.2 displays a Cartesian representation of the groups (1:1 group with dimensions 2×2, 3×3, 5×5; 1:2 with dimensions 1×2, 2×4, 3×6; 1:5 with dimensions 1×5, 1.5×7.5, 2×10; 2:3 with dimensions 3×4.5, 4×6, 6×9; nonsimilar but linear with dimensions 1×8, 2×9, 3×10). Students again were provided with paper and pencil, and graph paper. We noted how students accomplished this classification, especially their spontaneous use of representational forms. We followed up with probes of students' understanding of relations between different representational forms, and also of their understanding of two senses of infinity expressed on a graph: infinitely many (the most basic understanding expressed as moving away from the origin in typically whole number multiples of a seed rectangle) and infinitely dense (moving toward the origin). We employed these senses of infinity as indicators of how students understood the generalization inherent in graphs. Would

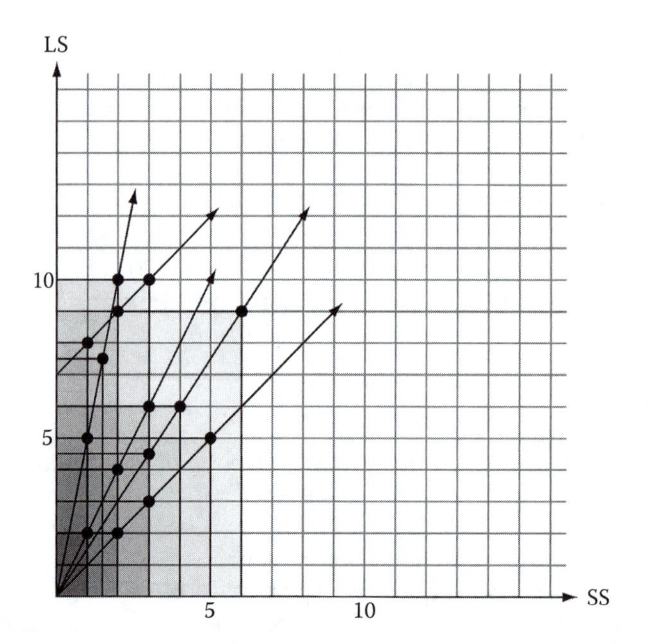

FIGURE 9.2. Cartesian representation of rectangles employed in sorting task.

students think of graphing primarily as connect the dots or would they conceive of a line as representing a generalization?

Representational Competencies

Four additional tasks probed students' representational competencies. The first assessed their strategies for using a table as a tool to identify if three rectangles were similar. What strategies did students employ? The second assessed their ability to use scaling (up and down) as a strategy for identifying additional similar rectangles. We included this task to probe students' understanding of similarity as growth from a seed, a view supported by the projector experiments, but one that received less attention during the classification activities. The remaining two tasks probed students' understanding of the equation of a line. In one task, students saw two equations with the same slope. The interviewer pointed out that the slopes of two lines were identical, so why weren't the lines the same? In the final task, students saw an inscription of a line without a frame of reference. This last task involved presenting a line drawn on a sheet of 8.5×11 paper without any other markings (but with the convention of the paper arranged vertically with a top and bottom). We looked to see how students might elaborate the inscription to render it sensible, asking students how they might write an equation to represent the line (and what they would have to do in order to accomplish this goal).

DESIGN STUDIES

We conducted the design studies over two successive years. Our aims included documentation of the growth of student reasoning during the course of instruction, documentation of relations between teaching practices and student activity, and assessment of student learning following instruction. Following analysis of design failures in the first year, we redesigned instruction during the second year with an eye toward improving instructional efficiency.

In the first year, 20 sixth-grade students and their teacher participated. The students consisted of 11 females and 9 males. In the second year, 14 sixth-grade students, 2 females and 12 males, and their teacher (the same as the first year of study) participated. During each year of the design study, occupations of parents were diverse, ranging from unemployed and homeless to high status professionals.

Instructional Practices

The pedagogical structure of the classroom was based on using student group work and collective thinking to generate and assess mathematical ideas. On most days, the teacher would start the class by summarizing previous work and previewing the day's activities. Students would break up into two to four student table groups, and the teacher would walk around and assist the groups when they needed help, direction, or additional instructions. The teacher would call for whole classroom discussions at the end of class, when enough students had made progress on the activity, or when an important discovery had been made. During the first year, sessions varied in length from 1 to 3 hours for 30 days across 10 weeks beginning in March. During the second year, sessions varied in length from 1 to 2 hours for 14 days during 3 weeks beginning in May. Students kept track of all their work, notes from whole classroom discussions, and summaries of thinking in a math notebook that was reviewed each week by teachers and student peers.

FIRST ITERATION OF DESIGN

We first summarize some of the obstacles and opportunities we observed in the classroom during the first iteration of the design (our first attempt). We go on to summarize students' competencies and forms of reasoning as revealed by their responses to the flexible interview.

Design Failures as Opportunities

We selectively present some of the observations we made during the course of instruction. Our observations are selected to emphasize failure, because it is failure that motivates re-design.

The initial context of modeling with the overhead projector provoked much conversation and also much contest between additive and multiplicative accounts, just as we anticipated. However, because the instrumentation had error and because students had no prior experience with modeling (teacher report), the error in measure made either account indistinguishable from the other. For some students, the image of the projector was one of motion, so they understood the task as designing an animated picture of the progress of the image, not developing a mathematical account. This made for a poor bridge to the classification context, and, indeed, there was little evidence of transfer (in word or deed) from the projector to classification contexts. The apparent inability to come to any kind of consensus also proved frustrating for the classroom teacher, who appeared ready to abandon the study at its inception.

A second obstacle was our failure to account for students' lack of knowledge about the mathematics of measure. In many classroom activities, we relied on a sense of relation as measured: the relation of one side as measured in units of the other, slope as a measure of steepness, and a sense of fraction as measured quantity. Instead, we found that students had little experience with measurement and their predominant sense of fraction was a part–whole relation (Thompson & Saldanha, 2003). This was especially apparent when the dimensions of the rectangles were not whole numbers. For example, during the course of one the classroom activities, we asked students whether or not 2/3 could ever be the same as 3/4 and, if so, in what sense? Students took this question as nonsensical. They then proceeded to explore a situation involving two different lengths, splitting the lengths to develop respective measures of 2/3 and 3/4, and eventually considered the importance of unit when considering the question. Nevertheless, students' conceptions of measure and of fractions as anything other than part–whole relations proved a significant obstacle throughout the course of the study because the design relied on coming to see one side of a rectangle as measured by the other.

A third obstacle consisted in our overreliance on classroom norms that we had observed in the classroom prior to this study. The teacher always elicited students' thinking, nearly always insisted on justifications for that thinking, and generally conducted a classroom emphasizing mathematical conversations. She also promoted mathematics as a form of literacy, and students kept mathematics journals that she employed to keep track of transitions in student reasoning (and she encouraged students to do the same).

In many ways, the teacher's norms were ideal from a mathematics reform perspective (e.g., the communication and reasoning strands of the NCTM *Principles and Standards*). Yet the teacher had little experience with these forms of mathematics and, in fact, was learning them along with her students. Practically, this meant that she often had difficulty orchestrating mathematical conversation in the classroom, although conjectures and justifications were abundant. One of the major transitions we noted during the course of instruction was the development of the teacher's repertoire for weaving students' often-disparate conversations into mathematical wholes. This process is discussed extensively elsewhere (Seymour & Lehrer, 2006), so here we simply note that the teacher's pedagogical content knowledge appeared to be transformed during the course of the first year's design study.

Opportunities

Despite these design failures, we also noted several recurrent opportunities presented by the sequence of tasks and tools developed for this study. First, visualization often supported the development of reasoning. For example, during the first classification task, some students noticed that the upper right vertices of superimposed rectangles formed a line, and this imagined line had a counterpart in the Cartesian coordinates. Furthermore, the perceptual difference between groups of similar and nonsimilar rectangles described by the same slope supported sense making: It was (perceptually) clear that the figures did not conform to the same system of description, but nonetheless something was common to both sets. This led to generation of some alternative perspectives. Some students suggested cutting off the extra part of the long side coordinate to get back to a ratio line with no intercept ($LS - b = m \times SS$), and others suggested translating from the ratio line, adding the intercept point to get to the nonzero intercept line ($LS = m \times SS + b$). Uses of tables clarified what might be the same (ratio) yet different about the two systems of description.

Second, the use of natural language bridged easily to symbolic expression, so that, for example, "the long side is twice the short side" was readily re-expressed symbolically as $LS = 2 \times SS$. This made sense because the symbolic reference was easily associated with natural language, which was in turn supported by perception of a figure (stacks of paper cutouts of rectangles). Many alternatives were spontaneously proposed by students, such as $1/2 \times LS = SS$ or $SS/LS = 1/2$.

Third, the activity structure of sharing solutions across small groups put features of representation "in play." For example, when comparing the graphs generated by different groups to describe the results of the first classification task, some groups extended their lines down to the origin, whereas some stopped at the smallest rectangle coordinate of the group.

Similarly, some lines stopped at the last data point, whereas others continued past the largest rectangle coordinate of the group. This prompted a discussion about infinity, using the extension past the final coordinate to talk about infinitely many, and using the extension down to the origin to talk about infinitely dense. Many students quickly recognized that they could keep drawing the line forever (provided they had enough paper), and that more and more rectangles would fall on the line, and thus would be in the same group. Many students relied on scaling or multiplying a seed rectangle by whole numbers to generate larger and larger rectangles, whereas others appealed to symbolic expression as a generator by using the ratio represented by an equation and generating a long side from a short side (or vice versa).

Fourth, the orientation toward coordinating representations and re-describing the same action in terms of different systems of representation often led to bootstrapping, where one system of representation served as a bridge to another. For example, when students first attempted to determine a symbolic expression for rectangles in the ratio of 1 to 4, they were stymied because it was one of the first groups to contain a rectangle with non-whole number dimensions. Whereas there was a 1 × 4 rectangle in the group, the next largest rectangle was 1 1/2 × 6. Students found that a step on the graph between these two rectangles (2 up by 1/2 over) did not translate as easily into a ratio as all whole number steps had for past rectangle groups. The teacher helped one student overcome this by first suggesting making a table, then measuring how many short sides fit into a long side of each rectangle to find the 1:4 ratio. Instead of going directly from the graphical to the symbolic representation, this student used a table and measurement to bootstrap her understanding. Over time, the teacher came to recognize these opportunities as especially fruitful and became skilled in orchestrating conversations around them.

ASSESSMENT

As we described previously, we interviewed all students at the end of instruction. Our aim was to document individual conceptions in contexts with less assistance than was typical of the classroom. We present the results of the assessment in three parts. First, we focus on students' degree of representational competence in a baseline condition involving finding groups of similar cylinders by attending to ratios of circumference to height. This was a near-transfer task because students had not worked with three-dimensional forms. We supplement this baseline with results obtained from a portion of the more complicated sort of rectangles that employed whole number ratios (1:1, 1:2). We then present results from the portions of the second classification task where students were confronted with non-whole number ratios and also nonsimilar figures. We go on to describe students' conceptions of infinity (infinitely many and infinitely dense), their ability to employ tables to generate alternative representations, and conclude with probes aimed more specifically at students' conceptions of the equation of the line.

Baseline Performances

All students generated similar groups of cylinders in ratios of 1:1 and 1:2. Students in both instances primarily employed an internal ratio strategy, thinking of the relation between the circumference and height for a particular ratio within an instance, and then testing for that ratio in other instances. All students moved with relative ease between representations, and each readily generated equations and graphs to check on the results of their initial sorts. The same tendencies were observed for rectangles with simple whole number ratios of sides, although these were embedded in a set that included other ratios (1:5, 2:3, and $LS = SS + 7$). All students discovered all members of the 1:1 and 1:2 groups. The predominant strategy for the 1:1 group was again anchoring to an internal ratio, but students used either a graph or an internal ratio as their initial strategy for the 1:2 ratio group. Again, students appeared to move freely among representations, so that we considered students' to have a fused or highly overlapping sense of the systems of representation employed (with the exception of two students who were not able to generate an equation for the 1:2 group of rectangles when embedded in the more complicated sort, described next).

Complicated Sorting

Although students were largely successful in finding the remaining groups of similar rectangles (85% for 1:5 and 90% for 2:3), we noticed that representational performance (here, being able to classify rectangles into groups through various representations) was more often characterized by bootstrapping (60%) than by fluid translation. For example, students had difficulty classifying the $1\ 1/2 \times 7\ 1/2$ rectangle, and would leave that rectangle aside while trying to initially group by symbolic expression. Most commonly, students had difficulty with fractional lengths (e.g., $3 \times 4\ 1/2$ as a member of the 2:3 group), and often located membership for some of these instances by employing the Cartesian system. Thus, through graphing, students would clean up these residual rectangles left over from their initial symbolic sort. They then usually checked on the membership with an equation, determining whether or not the dimensions given fit an equation already developed to describe the ratio groups. The remaining participants were split equally between those who again translated fluidly between representations and those who seemed to experience each system as distinct, although when information conflicted, the graphical sense tended to dominate.

Students experienced the most difficulty with the nonzero intercept group. Only 45% of the students spontaneously assembled the three members of this group and of these, only one third (15% of all participants) could write an equation without assistance. When assistance was rendered in the form of a partial equation ($LS = SS + \_\_$), students who had grouped these rectangles rapidly completed the intercept. When we pointed out the members of the group to the remaining students, we found that half had assigned at least one of the members to another group because in fact the dimensions of at least one rectangle were consistent with members of another group.

Infinity

Because all students could construct a graph of the line describing the 1:2 rectangle group, we asked students how many rectangles would fall on the line. Most students (85%) recognized that there were an infinite number of larger and larger rectangles as one moved away from the origin (typically expressed as whole-number multiples of a seed rectangle). Most (75%) also recognized that as one moved toward the origin, the number of potential rectangles was infinitely dense. Perhaps most interesting were the 5 (25%) students who suggested that lines consisted of a locus of (infinite) points, meaning that the whole line had infinitely many points (which are not restricted to whole number multiples of a seed rectangle) and they were infinitely dense.

Patterns in Tables

When asked if three nonsimilar rectangles, whose dimensions (1×3, 3×6, 4×9) were listed in a table, belonged to the same group, all but two (90%) answered no, and either used an internal ratio strategy or a graphing strategy. Those who tried to use a between ratio strategy found that there was no consistent pattern. Four students considered nonzero intercept groups. These students recognized the possibility that, whereas the rectangles did not form a similar group, they might form a group with a line that did not pass through the origin.

Comparing the Slopes of Lines

Students were asked to explain that, while two lines ($y = x$ and $y = x + 4$) have the same slope, when you divide the *LS* by the *SS* for two different rectangles on the nonsimilar line, the answers are not equal to each other or to the slope. This task assessed students' comprehension of the relation between slope and intercept. Most students (60%) recognized that moving the line creates a group of rectangles that is no longer similar. The students commonly expressed this idea by saying that the rectangles were no longer equivalent fractions, by comparing their internal ratios. Although the intercept group does not follow the same litmus test for classification (checking the internal ratio), students had shown in class discussions that they were comfortable with the combination of slope and intercept. Five students (25%) gave no reasoning beyond recalling that the $y = x + 4$ line is moved up four units. A small number of students (15%) could not generate any explanation.

Removing the Cartesian Grid

To determine just how far students could stretch their graphical competence, students were asked to determine the equation for a line without a grid. Three students (15%) superimposed a sheet of graph paper onto the line (due to interviewer error), thus changing the task, and these were eliminated from this analysis. Most students (65%) created a set of axes by drawing them on the sheet of paper along side the line. Different subsets of this group thought about how the placement of origin and intervals would affect the equation of the line. Thus, these students rendered the line by imposing their own version of the Cartesian grid and then proceeded to create linear equations, either with or without intercepts. The second group (29%) drew in a step to find the slope of the line, measuring each length and constituting the ratio. They noted the slope and remarked that the intercept was indeterminate. The remaining students either weren't sure how to go about the task or simply estimated a value for the slope.

Summary

As illustrated with classroom examples from the first year (and will be further shown with dialogue in the second year), opportunities to juxtapose visual and algebraic patterns, and to re-describe actions taken in one representational system (e.g., equations) in another (e.g., graphs) were important stepping stones for the development of understanding. At the end of the unit, the majority of students appeared to be able to fluidly coordinate relations among different systems of representation for describing geometric similarity for familiar, whole number dimensions. Nearly all students appeared to understand the senses of infinity inherent in the Cartesian line (infinitely many and infinitely dense), and many could develop equations of a line even without a Cartesian grid. This was especially impressive in light of the history of students' difficulties with the equation of the line (e.g., Schoenfeld et al., 1993). However, as is typical of design studies, we encountered many unanticipated obstacles along the way, and we were less satisfied with students' ability to coordinate different systems of representation fluidly when the dimensions and ratios of the dimensions of the figures involved were not whole numbers. Moreover, although many students apparently understood more about the slope–intercept form of the equation of a line than might be typical, nonetheless, a significant minority appeared to have but a tentative grasp. This set the stage for our redesign of instruction the following year.

SECOND ITERATION OF DESIGN

In the second year, we re-designed instruction to ameliorate some of the obstacles encountered in the first iteration. We began by introducing a series of activities in the beginning of the year based on linear measure. The sequence had been repeatedly tested with younger children (e.g., Lehrer, Jacobson, Kemeny, & Strom, 1999; Lehrer, Jaslow, & Curtis, 2003) and, in addition, had been the subject of repeated iterations of professional development (e.g., Koehler, 2002). The linear measure sequence emphasized student invention of units of length measure, and the organization of these units (and composites of these units), in tools constructed by students. We took care to ensure that not all measures were in whole numbers, which motivated development of operator conceptions of fractions. For example, multiplication of fractions was conceived of as repeated splitting of a unit of measure (e.g., 1/2 of 1/2 of 1/2). Equivalence was addressed as equal measure, and some activities stressed translation of one student's unit into those of another student's unit.

We also eliminated the initial emphasis on modeling, reverting to a previous instructional design (Lehrer et al., 2002) featuring similarity ratio as one way of classifying planar figures. Hence, students explored different ways of classifying figures first, and projectors and magnifying lenses were used as models of one of them.

Perhaps the most important re-design was not of our making. The teacher's efforts during the first design cycle included many attempts to orchestrate classroom conversations, and when these efforts failed, she tended to reflect on the sources of failure and try new pedagogical moves. Over time, she became adept at orchestrating classroom conversations around the fulcrum of the design: developing meta-representational competence. This progress over the course of the successive iterations of the design cycle is documented in Seymour and Lehrer (2006), but we provide an illustration because we believe the teacher's mediation of learning accounts for some of the improvements in student learning (described in the next section) that we found in the second year.

TEACHER MEDIATION OF META-REPRESENTATION

During the course of the first year, the teacher, Ms. Gold, developed a repertoire of conversational moves aimed at supporting translations between systems of representation. During the second year of the study, she frequently deployed these tactics. In this instance, she is questioning two students about relations between an equation and its counterpart on the graph. In the turn immediately preceding this one, Lucas states that multiplication is found on the graph as a means of getting from one coordinate to another (the rectangles are 1×4 and 2×8, expressed as $LS = 4 \times SS$):

T: Ok, show me on the graph where the times [multiplication] is.
Ryan: Umm, between here [points at the coordinate (1, 4) on the graph] and here [points at the coordinate (2, 8) on the graph].
T: Ok, but show me how the graph shows that. How does the graph show multiplication, from one point to the next?
Lucas: Does it?
T: You just told me it did.
Ryan: Lucas ... [The conversation stumbles for a moment, so the teacher encourages the two students that this is exactly where she wants the conversation to go.]
Lucas: I don't know how the graph goes ...
Ryan: Well ...
Lucas: Well, I think you can fit, like this here [he uses his fingers to duplicate the space between the origin and the first coordinate to the first coordinate and a second] ...
T: You can do what?
Lucas: You can fit 4 by 1, the corners, you have another 4 by 1 right here [he repeats his gesture] ...
T: Ahhh ...
Lucas: So you're going to get times two, you're putting two in.
Ryan: Oh yeah, times two.

We believe that Lucas (see Fig. 9.3) noticed that two of the 1×4 rectangles fit on the graph, one between the origin and the coordinate (1, 4), the second between the coordinates (1, 4) and (2, 8). This pacing out of the distance needed to fit two of the same rectangle used the concept of measure to explain why the rectangle represented by the coordinate (2, 8) was twice that of the rectangle represented by the coordinate (1, 4).

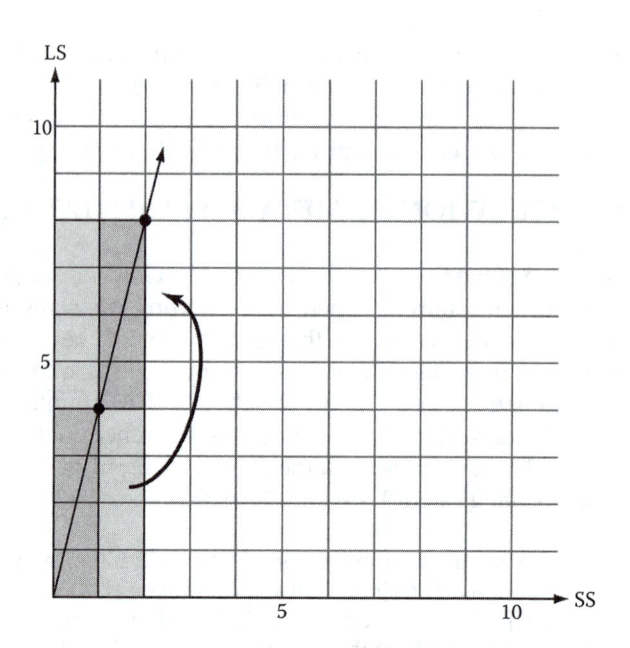

FIGURE 9.3. Lucas sees growth of rectangles as movement of a ratio rectangle.

The teacher began to draw on their graph the rectangles that Lucas had created with his fingers:

T: You said, right here, ok. Hold on, so, you said there's another rectangle right here. [the teacher lightly draws a 1 × 4 rectangle between the coordinates (1, 4) and (2, 8)] Ok, and if we drew it out, yeah, ok, where's your first rectangle?

Ryan: Right here. [the teacher lightly draws a rectangle between the origin and the coordinate (1, 4)]

T: Tell me what's happening.

Ryan: It's stairs?

Lucas: Stairs [traces the staircase pattern with his finger]

T: Ok, and, what are, what are your stairs?

Both students immediately noticed that the teacher's retracing of the gesture made by Lucas looked like the stairs or steps that had emerged in previous group activity (see Fig. 9.4). When the dialogue resumed moments later, Lucas didn't answer the teacher's question about "what are your stairs," but instead discovered that there were more stairs to be found, developing a new notion of infinitely dense:

Lucas: You can, kind of, break it into smaller stairs, right here, and then it goes right here. And then, within this stair, you have one, like, right here, and they're all on the line.

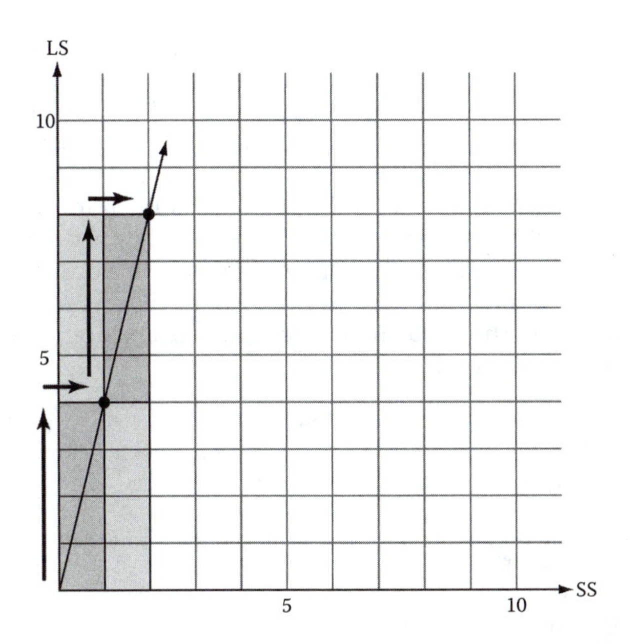

FIGURE 9.4. Students coordinate steps (slope) with internal ratio of the sides of a rectangle.

T: They are?
Lucas: And then, if you keep getting smaller and smaller, the stairs are so tiny, it looks like a line.

Lucas subdivided the steps into smaller and smaller ones, first drawing in examples with his pen, and then simply pointing them out with his finger. Not only did he recognize that each step could be subdivided into smaller ones, but he also saw that, as more and more steps were packed into the same space, the steps would make a better and better approximation to the line itself. He was simultaneously using senses of physical rectangle cutouts, graph, step, and the idea of infinity. Ms. Gold followed up with a probe that we believe was intended to draw attention again to the components of slope (and thus the sense of ratio-based multiplication that she was hoping to help the students develop):

T: But what do they have to have regardless of how small they are? What do they have to do?
Ryan: Ohh, oh, uhh, the, uhh, this part right here [he measures out the step over, or what you would actually step on if these were real stairs, on the graph with two fingers], the actual step, has to be able to equal one fourth of the [gestures toward the step up part] ...

Ryan noticed that the important part of the diagram that has been drawn is the relationship not between two particular rectangles or coordinates, but between the two parts of the step, the *up* and the *over* (the over part being measured by the up part).The students go on to conclude that this is another way to describe the steepness of the line: a measure. However, this conversation is not enough. Ms. Gold asks them to summarize the conversation in their notebooks (math journals). They had some difficulty summarizing the interaction in their notebooks for the class because there are two senses of multiplication in the conversation, as scaling and as a ratio within sides. They had another, shorter conversation about this with Ms. Gold a few minutes later:

T: So, then we said, we wanted to get from this [(1, 4)] to another point on the line. Ryan said, well, as long you keep this, um, that one is 1/4 of the other one. Right?
Ryan: Yeah.
T: Alright? So, and then we looked at, and that comes right back to our, oh, 4 to 1. Right?
Lucas: Mm-hmm.
T: Up 4 over 1.
Lucas: Mm-hmm.
T: And so I asked you, so where is the multiplication? 'Cause that was the next part you told me, you have to, we could multiply.
Ryan: Well, Lucas said ...
T: Ok, and I said, where's the multiplication? And you said, right here [the teacher traces the height and then the length of the step]. Up four over one, each time. Ok, so, this side, how much bigger is this side than this side?
Ryan: Two times.
Lucas: Four times.
T: Four times. Four *times*.
Ryan: Ohhh, that's, ohhh ...
T: Four times.
Ryan: Ohhhhh!
Lucas: You don't always have to do it by four.
T: No you don't.
Lucas: So that's why it goes ...
T: Which goes back to ...
Lucas: It's the relationship between this and this [points to the two parts of the step], not this and this [points to two of the rectangles].
T: Correct.

At the end of this conversation, Lucas seems to map between the equation and the graph. We suspect that Ryan's "ohhhh!" also signifies this relation.

Second-Year Assessments

We halved the time allotted for instruction, and at the end of the instructional unit, we again interviewed students. We focus here on differences that signaled significant improvements in student learning. First, the non-whole rectangle dimensions were not difficult for most students; only four students (28%) had to resort to bootstrapping representations to locate these instances of rectangles during the sorting task. The overwhelming majority of students moved fluidly between representations, relying primarily on calculating ratios within instances by writing symbolic expressions, and using the other systems of ratio as checks. Second, most students (78%) spontaneously generated nonzero intercepts for the nonsimilar group and wrote the corresponding equation. Only three students continued to employ an internal ratio strategy for this group as well, with the consequence that two of the three proposed a separate group for each nonsimilar ($LS = SS + 7$) rectangle. Third, looking across the tasks posed to students, no student interpreted the different systems as unrelated. The predominant interpretation of the different systems of representation was as redundant, even if during the course of learning they were not originally so perceived. Fourth, the percentage of students invoking the idea of a line as a locus of points to justify different senses of infinity increased (from 25% to 73%). Fifth, when comparing the LS/SS of two parallel lines (one similar, one nonsimilar), all but one student in the second year pointed to the intercept as causing the difference between the various ratios, whereas half of the first year class did not.

CONCLUSIONS

The conduct of this sequential design study suggests that spatial structure serves as a potentially important springboard to algebraic reasoning. The spatial structure of similarity provided a perceptual apparatus for algebraic descriptions and served as bedrock for the development of representational competence. Students came to see spatial relations as susceptible to algebraic description; algebra was a tool for describing equivalence classes of figures compactly. Conversely, algebraic description prompted rethinking the nature of space. Students originally perceived but several instances of similar figures. Algebraic operations cued the development of new qualities of this class: Figures were imagined as an infinite class by reasoning about the nature and implications of a line in the Cartesian system.

Developing representational competence is a habit of mind that sustains mathematical reasoning in any field of endeavor. We were especially impressed with students' capacities for disciplining (Stevens & Hall,

1998) their perceptions of the Cartesian system. When presented with a line without the support of a grid or axes, or for that matter, any supporting context, many proceeded to provide the necessary elaboration. This suggested an emerging sense of the line as an object in its own right.

The conduct of this study also served as an important reminder of the leitmotiv of designing a learning environment defined jointly by the tasks posed to students, the mediational means available to them, and the forms of argument privileged (here, generalization). Yet, all of these elements of design are contingent on the activity of teaching, and there were significant transitions in the tactics employed by the teacher to support students' developing meta-representational competence. Although a curriculum can juxtapose representational systems and even suggest that students translate between them, the teacher's activity created a dialogic space that made this feasible and even fruitful. Spatial structure again played its role, because the teacher and the student could index objects visually and mutually consider their properties. As shown in the segments of classroom dialogue, symbols were hooked to visual perceptions and to gestures that traced imagined spatial structures, generating common ground for teacher and students. The teacher improvised during the course of the conversation to draw relations between slope and similar figures, and students invented the notion of subdividing steps while preserving ratio to arrive at a sense of infinite density of similar figures. The teacher also showed thoroughly practiced pedagogical moves as she helped students render a history of discovery in a way that revisited coordinations among representations via ratio, and acutely aware of the fragility of oral histories, she had students develop a textual retelling.

Finally, we conclude with a brief comment about the nature of explanation in design studies. Design studies are employed to test the feasibility of new approaches to teaching and learning. Clearly, such a commitment generates contingencies among teaching and learning that cannot always be reliably explained. But, by conducting comparative case analysis, a method employed in other sciences (e.g., field biology), we can begin to trace patterns of stability. For the approach documented here, we now have accumulated several cases at different ages and grades (e.g., Kaput, 1999; Lehrer et al., 2002). Each case supports the utility of co-originating algebra and geometry and sustaining the interaction across multiple years and grades, beginning in the primary grades and ideally sustaining the effort across elementary school. Most suggest the important role that measurement can play in developing knowledge of rational numbers. All suggest the importance of coupling sustained professional development with prolonged looks at the development of mathematical reasoning.

REFERENCES

Brown, A. L. (1992). Design experiments: Theoretical and methodological challenges in creating complex interventions. *Journal of the Learning Sciences*, *2*(2), 141–178.

Cobb, P., Confrey, J., DiSessa, A., Lehrer, R., & Schauble, L. (2003). *Design experiments in education research.* Educational Researcher, *32(1), 9–13.*

Collins, A. (1992). Toward a design science of education. In E. Scanon & T. O'Shey (Eds.), *New directions in educational technology* (pp. 15–22). New York: Springer Verlag.

DiSessa, A. (2002). Students' criteria for representational adequacy. In K. Gravemeijer, R. Lehrer, B. van Oers, & L. Verschaffel (Eds.), *Symbolizing, modeling and tool use in mathematics education* (pp. 105–129). Dortrecht, the Netherlands: Kluwer Academic.

DiSessa, A. (2004). Metarepresentation: Native competence and targets for instruction. *Cognition and Instruction*, 22(3), 293–331.

Ginsburg, H. (1997). *Entering the child's mind. The clinical interview in psychological research and practice.* Cambridge, England: Cambridge University Press.

Goldenberg, E. P., Cuoco, A. A., & Mark, J. (1998). A role for geometry in general education. In R. Lehrer & D. Chazan (Eds.), *Designing learning environments for developing understanding of geometry and space* (pp. 3–44). Mahwah, NJ: Lawrence Erlbaum Associates.

Hall, R. (1990). *Making mathematics on paper: Constructing representations of stories about related linear functions.* Unpublished doctoral dissertation, University of California, Irvine.

Kaput, J. (1991). Notations and representations as mediators of constructive processes. In E. von Glasersfeld (Ed.), *Radical constructivism in mathematics education* (pp. 53–74). Dordrecht, the Netherlands: Kluwer Academic.

Kaput, J. (1999). Teaching and learning a new algebra. In E. Fennema & T. A. Romberg (Eds.), *Mathematics classrooms that promote understanding* (pp. 133–155). Mahwah, NJ: Lawrence Erlbaum Associates.

Koehler, M. J. (2002). Designing case-based hypermedia for developing understanding of children's mathematical reasoning. *Cognition and Instruction*, *20*(2), 151–195.

Latour, B. (1990). Drawing things together. In M. Lynch & S. Woolgar (Eds.), *Representation in scientific practice* (pp. 19–68). Cambridge, MA: MIT Press

Lehrer, R., & Chazan, D. (1998). *Designing learning environments for developing understanding of geometry and space.* Mahwah, NJ: Lawrence Erlbaum Associates.

Lehrer, R., Jacobson, C., Kemeny, V., & Strom, D. (1999). Building on children's intuitions to develop mathematical understanding of space. In E. Fennema & T. A. Romberg (Eds.), *Mathematics classrooms that promote understanding* (pp. 63–87). Mahwah, NJ: Lawrence Erlbaum Associates.

Lehrer, R., Jaslow, L., & Curtis, C. (2003). Developing understanding of measurement in the elementary grades. In D. H. Clements & G. Bright (Eds.), *Learning and teaching measurement. 2003 yearbook* (pp. 100–121). Reston, VA: National Council of Teachers of Mathematics.

Lehrer, R., & Lesh, R. (2003). Mathematical learning. In W. Reynolds & G. Miller (Eds.), *Comprehensive handbook of psychology* (Vol. 7, pp. 357–391). New York: Wiley.

Lehrer, R., Randle, L., & Sancilio, L. (1989). Learning pre-proof geometry with Logo. *Cognition and Instruction, 6*, 159–184.

Lehrer, R., Strom, D., & Confrey, J. (2002). Grounding metaphors and inscriptional resonance: Children's emerging understanding of mathematical similarity. *Cognition and Instruction, 20*, 359–398.

Peirce, C. S. (1960). *Collected papers of Charles Sanders Peirce. II.* Cambridge, MA: Harvard University Press.

Schoenfeld, A. H., Smith, J. P., & Arcavi, A. (1993). Learning: The microgenetic analysis of one student's evolving understanding of a complex subject matter domain. In R. Glaser (Ed.), *Advances in instructional psychology* (Vol. 4, pp. 55–175). Hillsdale, NJ: Lawrence Erlbaum Associates.

Seymour, J., & Lehrer, R. (2006). Tracing the evolution of pedagogical content knowledge as the development of interanimated discourses. *The Journal of the Learning Sciences. 15*(4), 549–582.

Stevens, R., & Hall, R. (1998). Disciplined perception: Learning to see in technoscience. In M. Lampert & M. L. Blunk (Eds.), *Talking mathematics* (pp. 107–149). Cambridge, England: Cambridge University Press.

Thompson, P. W., & Saldanha, L. (2003). Fractions and multiplicative reasoning. In J. Kilpatrick, G. Martin, & D. Schifter (Eds.), *Research companion to the principles and standards for school mathematics* (pp. 95–114). Reston, VA: National Council of Teachers of Mathematics

10

Early Algebra Is Not the Same as Algebra Early

David W. Carraher
TERC

Analúcia D. Schliemann
Judah L. Schwartz
Tufts University

Many mathematics educators recognize that algebra has a place in the early grades. But they can also identify with Russell's (1967) remarks:

> The beginnings of Algebra I found far more difficult [than Euclid's geometry], perhaps as a result of bad teaching. I was made to learn by heart: "The square of the sum of two numbers is equal to the sum of their squares increased by twice their product." I had not the vaguest idea what this meant, and when I could not remember the words, my tutor threw the book at my head, which did not stimulate my intellect in any way. (p. 34)

To move algebra-as-most-of-us-were-taught-it to elementary school is a recipe for disaster. If algebra is meaningless at adolescence, then why should it be meaningful several years earlier? Why are increasing numbers of today's mathematics educators embracing early algebra? What guarantees that early algebra will not turn into lumps in the gravy, hostile bacteria in inflamed tissue, excess luggage for our already overburdened syllabi? What is early algebra, if it is not the algebra most of us were taught?

Early algebra differs from algebra as commonly encountered in high school and beyond. It builds heavily on background contexts of problems. It only gradually introduces formal notation. And, it is tightly interwoven with the following topics from the early mathematics curriculum:

1. *Early algebra builds on background contexts of problems.* The idea that rich problem contexts can support the introduction of algebra may appear to undermine the goal of getting students to use formal notation without having to "translate" the meaning to mundane contexts. Why immerse students in nuanced discussions about problem contexts if we want them to think ever more abstractly? The justification for building on rich problem contexts rests on how most young students (and many adults) learn. They do not draw conclusions solely through logic and syntactical rules. Instead, they use a mix of intuition, beliefs, and presumed facts coupled with principled reasoning and argument. We discuss the problem of contexts at length elsewhere (Carraher & Schliemann, 2002a; Schliemann & Carraher, 2002; Schwartz, 1996). We would also like to draw attention to the insightful analyses of colleagues (Smith & Thompson, chap. 4, this volume; Verschaffel, Greer, & De Corte, 2002). In treating situations as one of the three defining characteristics of mathematical and scientific concepts, Vergnaud has made seminal contributions to the role of contexts in additive and multiplicative reasoning (Vergnaud, 1982, 1994, 1996).[1] Starting from rich problem contexts and situations, one hopes that at some point students will be able to derive conclusions directly from a written system of equations or an *x–y graph* drawn in a plane. But what assures us that they will ever arrive at this point? This is where the role of the teacher can be decisive.
2. *In early algebra formal notation is introduced only gradually.* Young students will not reinvent algebra on their own, and without a certain degree of guidance they are unlikely to express a need for a written notation for variables. Algebraic expressions need to be introduced, but introduced judiciously, so as to avoid "premature

[1]The somewhat vague expressions, additive structures, and multiplicative structures are widely used among mathematics educators to encourage thinking about arithmetical operations as subsuming far more than the computational routines. They would emphasize, for example, that a young student may learn to multiply and divide long before showing a deep understanding of ratio, proportion, rational number, and related concepts that comprise the multiplicative conceptual field (Vergnaud, 1994).

formalization" (Piaget, 1964). Teachers need to introduce unfamiliar terms, representations, and techniques, despite the irony that in the beginning students will not understand such things as they were intended.[2] The initial awkwardness vis-à-vis new representations should gradually dissipate, especially if teachers listen to students' interpretations and provide students with opportunities to expand and adjust their understandings.[3]

3. *Early algebra tightly interweaves existing topics of early mathematics.* It makes little sense to append early algebra to existing syllabi. Algebra resides quietly within the early mathematics curriculum—in word problems, in topics (addition, subtraction, multiplication, division, ratio and proportion, rational numbers, measurement), and in representational systems (number lines and graphs, tables, written arithmetical notation, and explanatory structures). The teachers help it emerge; that is, they help bring the algebraic character of elementary mathematics into public view.

This chapter discusses these three distinguishing characteristics of early algebra,[4] drawing on examples from our longitudinal investigations of four classrooms in an ethnically diverse school in the Greater Boston area. From the second half of Grade 2 to the end of Grade 4, we designed and implemented weekly early algebra activities in the classrooms. Each semester students participated in six to eight activities, each activity lasting for 90 minutes. The activities related to addition, subtraction, multi-

---

[2]Some might argue that symbols should only be introduced when students know what they mean. Were this reasoning to be applied to the case of first language learning, adults would never speak to newborns on the grounds that infants do not already know what the words mean!

[3]We use *representation* in a generic sense here to include any expression of mathematical ideas, but especially those that are observable to others and not merely private and mental. Students' own *representational forms* include natural language ("their own words"), diagrams, and written mathematical stories (although even in these cases the student relies on linguistic and graphic conventions already acquired through cultural transmission). Conventional representational forms in mathematics are those sanctioned by modern mathematicians: graphs, tables, various types of written notation, and so on. Over time students will increasingly work conventional representations into their expressive repertoires—they will "make them their own." A *representational system* (or symbol system) refers to not only the forms themselves but also to associated underlying structure and processes.

[4]Early algebra does not touch on certain advanced topics of algebra. But the qualifier, early, alerts the reader to this, so there is no need to mention this as a fourth characteristic of early algebra.

plication, division, fractions, ratio, proportion, and negative numbers. The project documented how the students worked with variables, functions, positive and negative numbers, algebraic notation, function tables, graphs, and equations in the classroom and in interviews (Brizuela & Schliemann, 2004; Carraher, Brizuela, & Earnest, 2001; Carraher & Earnest, 2003; Carraher, Schliemann, & Brizuela, 2001; Schliemann & Carraher, 2002; Schliemann et al., 2003). To highlight the nature of the progress students can make in early algebra, we will compare the same students' reasoning and problem solving at the beginning of Grade 3 and in the middle of Grade 4. We will show how mathematics educators can exploit topics and discussions so as to bring out the algebraic character of elementary mathematics.

FROM PARTICULAR TO GENERAL: THE CANDY BOXES PROBLEM

To exemplify how young students make initial sense of variables and variation in mathematics, we begin with our first lesson in one of the classes from Grade 3. The students are 8 years old. In one of the classes, David (the first author of this chapter and instructor) holds a box of candies in each hand. He tells the students that:

- The box in his left hand is John's, and all of John's candies are in that box.
- The box in his right hand is Mary's, and Mary's candies include those in the box as well as three additional candies resting atop the box.
- Each box has exactly the same number of candies inside.

He then invites the students to say what they know about the number of candies John and Maria have. At a certain point, he passes around one of the boxes so that students can examine it; rubber bands secure it shut and students are asked not to open it.

What Students Focused On

Students invariably shake the box (Fig. 10.1), seeking to appraise its contents. After having held and shaken the box, most make a specific prediction, even though the instructor has not requested that they do so. One student holds a box in each hand and concludes that the boxes hold differing amounts of candy. Several others appraise the weight of the two boxes. Another conjectures that there are no candies at all in the boxes. David explains that he placed tissue in the boxes to muffle the sound and

FIGURE 10.1. Student shakes one of the candy boxes listening closely to estimate the number of candies.

make guessing difficult; the students find this preventive measure amusing. Eric ventures, with a gleam in his eye, that there is a doughnut in one of the boxes—a comment his classmates find delightfully funny, perhaps because it violates given facts David is trying to establish. David insists that he put the exactly same number of candies in each box; he appeals to the students to accept his word. The students appear to accept his claim and to take pleasure in having raised reasonable doubts.

After 15 minutes of discussion, David asked the students to express in writing what they knew about the amounts John and Mary had. If students balked or stated that they did not know how many candies the two had, David encouraged the students to show what they did know and to show how they were thinking about the story. Fifty-six out of 63 children produced drawings. Two distinct foci emerged.

A Single Instance. The first focus consisted in ascribing a particular value to the amounts. Forty of the 63 children (63.4%) focused on a single case. That is, they used drawings, labels, or prose to assign particular numbers to the amounts John and Mary had.

Figure 10.2 illustrates such a focus on a single case. In Erica's drawing, the candies are shown in a top row of a table;[5] the respective owners, John and Mary, are identified through drawings and labels in the next row. Erica's representation considers only one case or instance: the case of six candies in each bag, giving John and Mary 6 and 9 candies, respectively.

---

[5]It is curious that Erica organized her data in a table; normally a table is used to capture the results of many cases.

FIGURE 10.2. Erica's representation of John and Mary's candies.

Erica also expresses the single case through the number sentences, "Six + three = nine" and "●●●●●● + ●●● = nine."

Under questioning, Erica recognizes that there are other possible answers:

[1] Darrell [interviewing Erica in class immediately after she produced her drawing]: Do you think there are any other guesses you can make?
[2] Erica: Yeah.
[3] Darrell: Yeah? What are some other guesses?
[4] Erica: You can put seven, and you can have eight. You can have more.

At the beginning of Grade 3, many students conceptualize the situation as Erica does: Given indeterminate amounts, they assign particular values, even though they realize they are hazarding a guess.

An Indeterminate Amount "Keeps Options Open." Twenty-three children (36.5%) refrained from assigning values to the amounts of candies in the box. Vilda's annotated drawing (Fig. 10.3) illustrates this approach: "I thick Mary's have more den John because Mary's have 3 more candy den John. If you taek 3 She vill have the Same of John (*sic*)."

Although Vilda does not speculate on the amounts John and Mary have, she states that the children would have the same amounts if three

FIGURE 10.3. Vilda leaves indeterminate the number of candies in the boxes.

were taken away from Mary. Several children simply stated in writing that Mary had three more candies than John or that John had three less than Mary, much as Vilda did. We considered these responses to be similar insofar as they leave options open.

Occasionally, a student will attempt to explicitly represent the indeterminate amounts: In another classroom, Felipe (Fig. 10.4) chose to highlight with question marks the fact that the amounts in the boxes are unknown. When the teacher asked him to provide more detail, he ventured a guess: "My guess for the candies is that their (*sic*) are 8." Felipe's thinking nonetheless emphasizes the indeterminate nature of the amounts. This is not quite the same as conceptualizing the amounts as variable quantities, but a representation of an indeterminate amount is a placeholder for the eventual introduction of a variable.

It might appear to be of no significance that a student expresses the indeterminate nature of amounts. After all, the amounts were presented by the instructor as indeterminate: There was "some amount" in each box.[6] However, students show restraint in deciding to leave the possibilities open. And, as we shall see, the students who insist on leaving the values indeterminate provide an important opportunity for the teacher to introduce new notational forms that will prove useful not only in this setting but also in many other future discussions.

---

[6]The *relationship* between the amounts was determined, but not the amounts themselves.

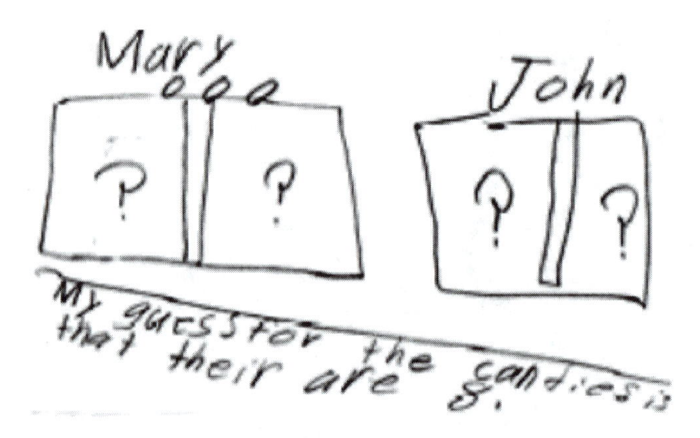

FIGURE 10.4. Felipe's question marks explicitly represent the unknown amounts in the candy boxes. The vertical lines drawn in the middle of each box are the rubber bands that hold the boxes shut.

Intervening to Shift the Focus and Broaden the Discussion

So far the class appears to be discussing a particular story about two children and their candies. Yet, it is possible to conceive of the candy boxes as a collection of many possible stories. The teacher hopes to gradually move the focus of discussion toward algebra by capitalizing on this shift in thinking.

Tables Draw Attention to Multiple Possibilities. Back in the original class, David proceeds to summarize certain features of students' representations in a data table with columns for students' names and the amounts they suggested for John and Mary. Later, he adds an additional column for keeping track of the differences between John and Mary's amounts (see Fig. 10.5).

Students who already made predictions in their representations merely need to restate the amounts; some changed their responses as a result of having a new hunch. Students who had not specified particular amounts in their drawings are now asked to suggest possible values.

It might seem that a prediction table would merely reinforce the students' natural tendencies to focus on a single case. Surely this is not the aim of a lesson designed to elicit algebraic thinking among students, where the emphasis should presumably be moving toward generalization. However, listing the individual cases (students' predictions) serves to highlight multiple possibilities. Furthermore, issues of logical consistency come to the fore under such circumstances. Both of these characteristics are desirable in bringing out the algebraic character of the story.

FIGURE 10.5. A table of possible outcomes. Students' names are on the left.

By listening to other students' conjectures, all the students have the opportunity to think more deeply about the problem. For example, Dylan writes "5" in John's column and "5" in Mary's column as well. This leads David to ask whether John and Mary could have the same total amounts. Dylan has been thinking only of the amounts inside the boxes, without taking into account the three extra candies Mary had on top of the box. When Dylan realizes that column three should list the "total amount Mary has," he amends his answer to 8.

Students occasionally give answers that violate the given premises. For example, Chris suggests that John has 7 and Mary has 13 candies (see Fig. 10.5). Several students notice the inconsistency and eagerly explain why this cannot be the case. But, because Chris himself is not yet convinced, David leaves Chris's predictions in the table for the time being. When the table is almost completed, David shifts attention to the differences in amounts of John and Mary. Several students insist that the differences have to be three. David tries to assume the role of a devil's advocate (Can't John have 7? [yes] Can't Mary have 13? [yes]. So, what's the problem?). Students argue that even though they do not know what Mary and John have, some answers (ordered pairs) are not right. Soon all students appear to agree that, although John and Mary could in principle have any amount, once an amount is assigned to one of them, the other amount can no longer be anything; that is, it is no longer free to vary because the variable has been constrained to a single solution.

So it is not the case that anything goes when values are indeterminate. We don't know how many candies John has. We don't know how many candies Mary has. Yet, it cannot be true, for example, that John has 6 candies and Mary has 7 candies. By drawing attention to this, issues of a more general nature begin to emerge.

LETTERS CAN NAME INDETERMINATE AMOUNTS, SETTING THE STAGE FOR VARIABLES

Students who draw attention to the indeterminacy of the amounts offer an excellent opportunity for instructional intervention. In another classroom taught by David, Kevin writes that John has three candies fewer than Mary without assigning values to either amount. Likewise, he becomes silent when David asks him to state possible values for the amounts of candy. When it is Matthew's turn to predict a possible outcome, he also balks. David recalls that Matthew, like Kevin, had preferred not to make a prediction in his drawing. So he turns to Matthew, hoping to introduce the algebraic convention that a letter can represent an indeterminate or a variable amount:

[5] Matthew: Actually well I ... I well ... I think without the three maybe it's ... [having a change of heart] Yea, I pretty much don't wanna make a prediction.

[6] David [seizing the opportunity to introduce a new idea]: Okay, but let me offer you an alternative and see if you're willing to do this.... What if I tell you, Matthew, that John has *N* ... *N* pieces of candy. And *N* can mean any amount. It could mean nothing. It could mean 90. It could mean 7. Does that sound okay?

[7] Matthew [cautiously, without a lot of conviction]: Yea.

[8] David [writing *N* on the blackboard]: All right, so why don't you write down *N*. [Addressing the remaining students:] He's willing to accept that suggestion.

[9] David [wondering to himself, "What will the students call Mary's amount?"]: Well, now here's the problem, and this is a difficult problem. Matthew, how many should we say that *Mary has* if John has *N* candies, and *N* can stand for anything?

Several students suggest that Mary's amount also be called *N*. This is not unreasonable: after all, David had just told them that "*N* could stand for anything." However, this is going to lead to trouble.[7] David prolongs the discussion a bit more:

---

[7]Clearly, $N = N$. But if the first N is 5, whereas the second N is 8, we produce the bothersome expression, $5 = 8$.

[10] Cristian [raising his hand with great energy, suggesting he has just discovered something]: Oooh. Oooh!
[11] David [suspecting that Cristian is going to say "*N* plus three," but hoping to first give the other students more time to think about the problem]: Hold on Cristian.
[12] David: What do you think we should do? How would we call ... how would we call ... if *N* stands for any amount that he happens to have ... okay ... that John happens to have, then how much would Mary have? You think she'd have ...?
[13] Joseph: *N*.
[14] David: *N*?
[15] Joseph: Yes.
[16] Student: Yes.
[17] David: Well, if we write *N* here [on the blackboard, next to the *N* assigned to John] doesn't that suggest that they [John and Mary] have the same amount?
[18] Several students: No.
[19] Briana [defending the use of *N* to describe Mary's amount]: It could mean *anything*.
[20] Another student: She could have *any* amount like John.
[21] David: Yea, it could be anything. I know just what you mean. But some people would look at it and say it's the *same anything* if you're calling them both *N* ... Is Mary supposed to have more than John or less or the same?
[22] Students: More.
[23] David: How many more?
[24] Student: Three more.
[25] David: Three more, so could ... how could we write down "three more than *N*" if *N* is what John has? How could we do that?
[26] Joey: Three ... cause *N* could stand for nothing.
[27] David: It [*N*] *could* stand for nothing, but we're telling you that we're gonna use it to stand for *any* possibility.
[28] Another student: Nothing.
[29] David [clarifying to that student]: Okay, it *could* stand for nothing.
[30] Joseph: *N* plus three.
[31] David: *N* plus three?
[32] Joseph: *N* plus three.
[33] David [amazed]: Wow! Explain that to us.
[34] Joseph [dazed, as if he had been speaking to himself]: Huh?
[35] David: Go ahead.
[36] Joseph: I thought, 'cuz she could have three more than John. Write *N* plus three 'cuz she could have any amount plus three.
[37] David: So any amount plus three. So why don't you write that down, *N* plus three.
[38] Anne [a member of the research team]: Cristian had his hand up for a long time too. I was wondering how he was thinking about it.
[39] David: Cristian, [do] you want to explain?

[40 Cristian: I was ... I was thinking the same thing.
[41] David: Go ahead, explain. You think the same thing as Joseph, *N* plus three? [Cristian nods.] Why don't you explain to us your reasoning and let's see if it's just like Joseph's.
[42] Cristian [exemplifying the general relation through a particular case]: 'cuz if it's any number, like if it's 90.
[43] David [encouraging Cristian to continue explaining]: Yea ... [44] Cristian: You could just, like add three and it'd be 93.
[44] David: Yes. Yea, this is really, really neat. You guys are ... are ... What ... what should we call that ... should we call that the "Joseph and Cristian Rule?"
[45] Students: [laughter]

After mulling over the possible confusion engendered by the use of *N* to designate both John's and Mary's amounts, and prodded by the teacher to find a better way [9], Joseph and Cristian reached the conclusion that Mary's amount should be called "*N* plus three" [32–44]. This is a significant step in the direction of using algebraic notation. Joseph and Cristian are using *N* not simply as a label.[8] In appending the expression "+3" to the "N," they are effectively operating on *N*.

Summary of Findings From the Candy Boxes Problem

By asking the students to make predictions about numbers of candies, we may have encouraged some of them to construe their task as having to guess accurately. However, this same activity served as an opportunity to discuss impossible answers, such as when a student suggested that child had 8 candies and the other 10 candies. As the prediction table was completed, students could try to describe what features were invariant among the (valid) answers. In a sense, the data table encouraged students to generalize.

The Candy Boxes task is ambiguous, that is, subject to alternative interpretations. It calls to mind a particular empirical state of affairs as well as a set of logical possibilities. The former empirical viewpoint gains prominence when one wonders how many candies are actually in the box. The logical view emerges as children attempt to find multiple solutions and express this in some general way.

Each viewpoint has its own version of truth or correctness. Empirically, there is only one answer to the issue about the number of candies John and Mary have. By this standard, only students who ascertain the precise numbers of candies in the boxes can be right. However, the logical standard to which algebra aspires treats as valid all answers consistent with the

[8]Their expression evokes conventional notation for a function, as in $f(N) := N + 3$.

information given, regardless of whether they correspond to the actual case at hand. Jennifer expressed this point of view clearly at the end of the lesson when asked to say who had given correct answers: "Everybody had the right answer ... Because everybody[9] ... has three more. Always."

Some readers may regard such ambiguity as merely a source of confusion that should have been minimized. Yet it turns out to have advantages. The story's ambiguity allows teacher and students to hold a meaningful conversation even though they may have markedly different initial interpretations.

By engaging in such conversations, students can begin to appreciate the tension between realistic considerations and theoretical possibilities (Carraher & Schliemann, 2002a; Schliemann & Carraher, 2002). This tension arises whenever one uses mathematics to model worldly situations (Carraher & Schliemann, 2002b). For example, one can ask whether a host of a party will ever run out of refreshments if, starting with a full liter, she distributes half to the first guest, half of what remains to the second guest, and so on (Stern & Mevarech, 1996). In the physical world, the drink eventually runs out when a guest receives the last drop (or molecule). In the world of mathematics, the host can serve refreshments without ever running out because the remaining amount, $(1/2)^n$ liters, never reaches zero liters no matter how great n becomes. Rather than regard this as a shortcoming of the problem, one can treat it as a useful illustration of how models serve as simplified approximations that break under certain conditions.

The results from the Candy Boxes task suggest that young students may be able to shift their focus from individual instances to sets and their interrelations. In this new conceptual framework, the mathematical object is no longer the single case or value but rather the relation, that is, the functional relationship between two variables.

But we should be careful not to overinterpret promising first steps. The Candy Boxes lesson represents the beginning of a "long conversation about *N*" that will extend over several months and years and in a wide variety of contexts. Let us revisit the students 1 1/2 years later to see how their thinking changed in the ensuing period.

COMPARING FUNCTIONS: THE WALLET PROBLEM

The following episodes come from a unit we implemented at the beginning of the second semester of fourth grade in the same four classrooms. This unit was an extension of the children's work on functions. Once again, we

[9]By the time Jennifer said this, all the predicted outcomes in the table were consistent with the information given.

asked them to provide us with a look at how they understood a situation that might be construed in a variety of ways. Here the students, now 10 years old, and their instructor considered the following situation:

> Mike has $8 in his hand and the rest of his money is in his wallet;
> Robin has exactly 3 times as much money as Mike has in his wallet.
> What can you say about the amounts of money Mike and Robin have?

In each classroom, we projected the problem with an overhead and asked students to read the problem out loud. Sometimes, while the projector was off, we asked students to recount the story in their own words.

At the outset, two opinions typically arose. Some students took the view that Robin has three times as much money as Mike. Others insisted that Robin had only three times the amount in Mike's wallet. After discussing the various interpretations (and re-reading the story out loud several times), the students reach a general consensus around the second interpretation. We then asked the students to provide us with drawings and explanations showing their understanding of the problem, much as we did for the Candy Boxes task. By this time, the students were accustomed to such an open-ended request and they quickly went to work making representations.

From Candy Boxes to Wallets: The Evolution of Children's Representations

In the intervening 18 months, the students' thinking has undergone remarkable transformations. Consider, for example, the case of Lisandra. When asked to represent the Candy Boxes at the beginning of grade 3, she essentially drew pictures showing specific amounts of candies (see Fig. 10.6); she also made a statement about the relation between the amounts.[10]

But when asked to represent the Wallets problem, her drawings take on a very different role. She draws three wallets for Robin to convey the notion that Robin has three times as much as Mike has in his wallet. Even more striking, she has written the letter *N* on each wallet. Finally, she has expressed Mike's total as "$N + 8 = \square$" and Robin's amount as "$N \times 3 = 3N$."

---

[10]Lisandra's Candy Boxes drawing (Fig. 10.6) is actually difficult to classify. Is she leaving the amounts indeterminate? Or is she trying to depict specific amounts? She stated that Mary had three more than John, but she has drawn 17 in John's box and 13 in Mary's (not counting the 3 candies on top of her box). Regardless, there is little doubt that her depiction of the amounts in the wallet problem (Fig.10.7) is considerably more advanced.

What can you show about John and Mary's candies? Draw or write something below.

FIGURE 10.6. Lisandra's representation of the Candy Boxes Problem at the beginning of Grade 3.

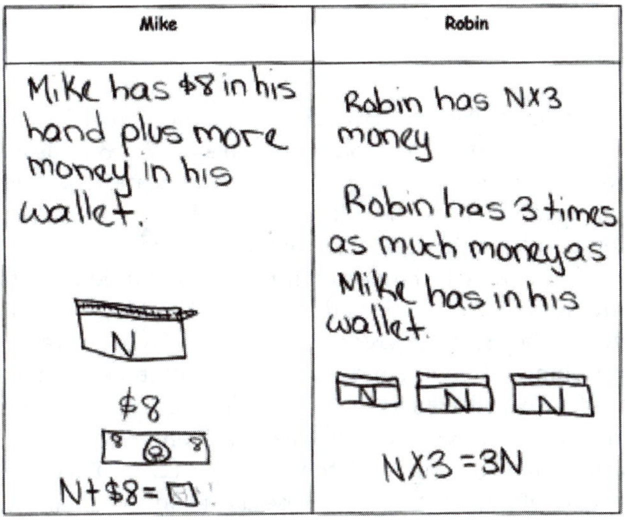

Mike	Robin
Mike has \$8 in his hand plus more money in his wallet. N \$8 N+\$8= ☐	Robin has NX3 money Robin has 3 times as much moneyas Mike has in his wallet. N N N NX3 =3N

FIGURE 10.7. In the middle of Grade 4, Lisandra represents the amounts of money Mike and Robin have. Note the symbolic use of the wallets with N dollars.

Lisandra's progress is not exceptional. In fact, most students (74.6%, or 47 of 63) made substantial progress between grades 3 and 4. Like Lisandra, 39 of the 63 students (61.9%) provided general, algebraic representations of the Wallet problem. Here is a breakdown of the algebraic answers:

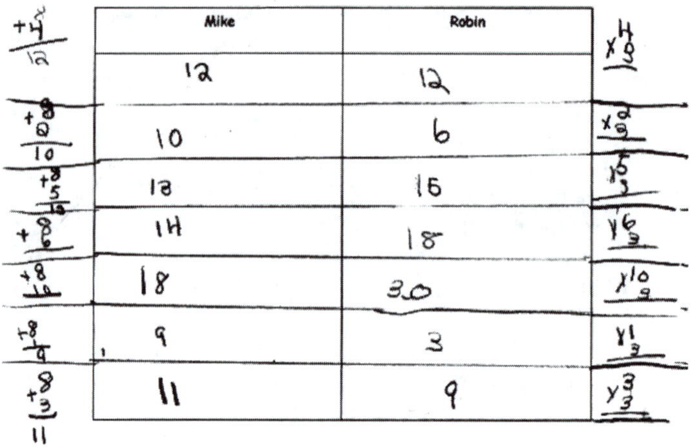

Mike	Robin
12	12
10	6
13	16
14	18
18	30
9	3
11	9

FIGURE 10.8. Erica decided to draw a table showing multiple possibilities for the amounts Mike and Robin had. N.B. The lines were provided by her, not given as part of the problem.

1. *Conventional notation*: Sixteen of the children (25.4%) represented Mike's amount as $N + 8$ and Robin's as $N * 3$ or as an equivalent expression such as $N + N + N$ or $3N$ (in some cases they used w or r, instead of N). We included Lisandra in this group; her representation also exhibited characteristics associated with iconic variables.
2. *Implicit operations*: Fifteen children (23.8%) used algebraic notation but omitted the + sign in their account of Mike's amount; in other words, they simply wrote "$N\ 8$" (or "N \$8"). However, in the case of Robin, only two children left the operation implicit writing "$N\ N\ N$." The others chose to write $N \times 3$ (nine children), $3N$ (two children); and $N + N + N$ (two children).
3. *Iconic variables*: Eight of the children (22.7%) used wallet icons instead of a letter. Some of these used conventional signs for addition and multiplication; others used the implicit operations described previously under implicit operations.

Approximately one student in eight (12.7% or 8 of 63), produced drawings or tables with multiple possibilities for the amounts (see Fig. 10.8). This type of representation highlights variation and covariation. Sometimes considerably more is conveyed. For example, Erica's computations on the left and margins highlight what varies (the amount in the wallet) and what remains invariant (in Mike's case, the $8; in Robin's case the "× 3").

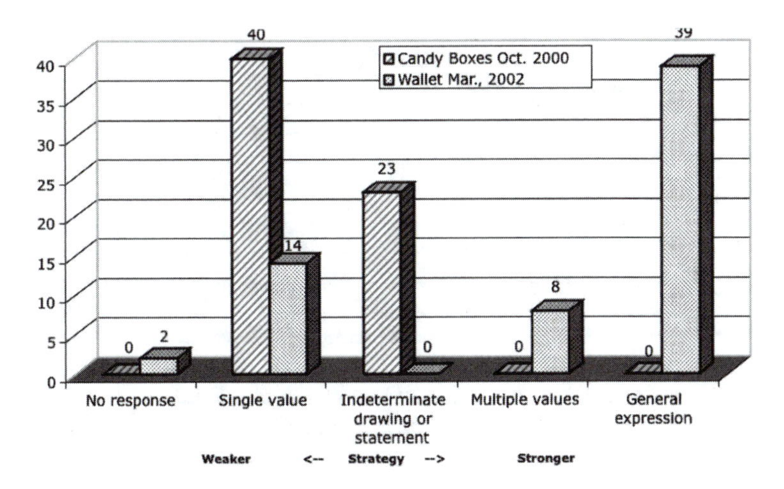

FIGURE 10.9. The growth of students' thinking over 1 1/2 years.

Fewer than one in four students (22.2%) represented the amounts through a single possibility or instance. This compares to nearly two thirds of the students (63.5%) in the lesson given at the beginning of Grade 3.

Figure 10.9 shows how students' thinking changed over the 18-month period. There was a dramatic shift in focus. At the beginning of Grade 3, students thought of the Candy Boxes word problem as a story about two children who had either specific or indeterminate amounts of candies. By the middle of Grade 4, most of the children conceptualized the problem as a story involving multiple possibilities. Many of those (39 children, or 62.2%) used algebraic notation to capture the functional relationships among the variables. We know from many other studies, including our own, that fourth-grade students in the United States do not show this sort of shift in thinking without having learned about algebra; they do not invent such things on their own.

This shift in conceptualization allowed the students to further deepen their understanding and technical mastery of mathematics.

Intervening to Enrich the Discussion

Algebraic Table Feaders Identify Functions, Streamlining Thought. A week later, David reviewed the wallet problem by having the students help fill in a three-column table projected onto an overhead screen at the front of the class. The original column headings were: "In Mike's Wallet," "Mike (in wallet and hand)," and "Robin." Because the students had discussed the

Complete the table:

W	W + 8	3W
In Mike's Wallet	Mike (in wallet and hand)	Robin
0	8	0
1	9	3
2	10	6
3	11	9
4	12	12
	13	
		18
		21
8		
	17	
		30
11		
	20	

FIGURE 10.10. The table discussed by the whole class (via overhead projector). Note the use of algebraic notation for column headings.

problem and provided their personal representations in the prior class, David expected this to be a routine task intended merely to refresh their memories before he would turn toward the graphing of the functions. However, he noticed that the students repeatedly asked to be reminded about the details of the word problem (What was Michael holding in his hand? What did Robin have?). Once an amount was suggested as the value in the wallet, $0 for instance, the students appeared to need to reconstruct in their minds the situation involving the story's two protagonists. To expedite the process, David added algebraic headers above the original headers: "*W*," "*W* + 8," and "3*W*," corresponding respectively to the independent variable, Mike's function, and Robin's function (Fig. 10.10).

These short inscriptions had a noticeable effect on the collective activity of filling in the table. Once the algebraic column headers were inserted into the table, students were able to quickly supply the values for Mike's and Robin's amounts, given a value for the independent variable, *W*. In explaining their reasoning, it became clear that they no longer needed to think through the problem by imagining Mike's holding $8 in his hand, with the rest of the money remaining in his wallet; similarly, they didn't have to reconstruct Robin's amount by parsing the story once again. To obtain Mike's total, they simply added 8 to the value of *W* in column 1. To obtain Robin's total, they simply multiplied the value of *W* by 3. Thus, the algebraic expressions served as more than column labels. Students used them as cognitive mediators for producing output values for the functions without

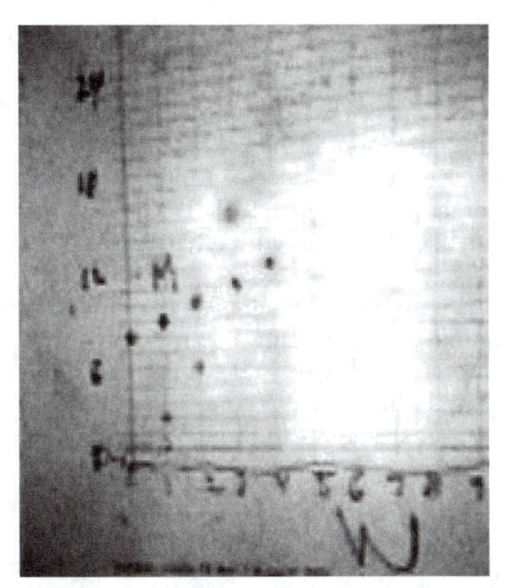

FIGURE 10.11. Mike's and Robin's amounts plotted as a function of W, the amount in Mike's wallet.

having to concern themselves with the situation-specific meaning underlying the computations. This procedure is considerably more efficient. It is also very different: Using it, students can temporarily disregard the story problem, instead focusing on, and operating on, the written symbols.

This shift, away from semantically driven and toward syntactically driven problem solving, does not signal the end of semantics. Those who use mathematics to model worldly situations (i.e., engineers, students, applied statisticians, scientists, and just plain folks, as opposed to pure mathematicians and statisticians) cannot consign semantics and background contexts to the trash bin, because they continue to have important roles in mathematics. Nonetheless, the word *shift* is appropriate here because young students are gaining familiarity with a domain of mathematical thinking where there can be considerable (and meaningful) inference making that does not ask for immediate translation back to mundane reality (Resnick, 1986).

Graphs Highlight Covariation. The students completed the table of values on individual worksheets. David guided them in plotting Mike's total for the cases when the wallet holds $0, $1, $2, $3, and $4; they also plot the total values for Robin when Mike's wallet contains $0, $1, and $2. Figure 10.11 shows a (poor quality) picture of the projected image at the moment the following dialogue starts. The x-axis was used to represent the amount in Mike's wallet. The y-axis was used to represent the total amount.

A student has just plotted the point (2, 6), corresponding to the case where Mike's wallet contains $2 and Robin's total is $6:

[47] David (drawing attention to the colinear points that are beginning to map out two broken lines, one for Mike's and the other for Robin's amounts of money as a function of *W*): What's happening to these lines? Does anybody notice anything happening? They don't look parallel to me. Yeah, William?
[48] William [referring to the increments as one proceeds rightward]: That Mike is going one by one and uh, Robin is going three by three.
[49] David: Yeah. Robin is going three by three. Can you show us where the "one by one" and "three by three are," William? Cause people might not understand what you mean by that. Where is the one by one that you see?
[50] William [pointing to the line representing Mike's amounts]: Like uh, Mike's not, see, he's going one more up.
[51] David: He goes one up. And then next time he goes one more up, like he goes from six, I'm sorry, from eight to nine to ten to eleven and then to twelve. And what's happening with, uhm, Robin?
[52] William: She [Robin] starts at zero. She goes three and then up to six.
[53] David: Ok. She's only going up by threes.

William appears to be describing something like the slopes of the two lines according to the size of the increments by which they grow [48–52]. When it is time to plot the point (4, 12), for Robin's function, David asks the class:

[54] David [ingenuously]: Wait a minute, but I thought we already used up that point [The point (4, 1) was contained on Mike's graph]. Can I put another one on there? Can I give the same point to Robin that we give to, to Michael?
[55] Student: Yeah.
[56] Erica: Yeah, 'cuz on number four they were even.
[57] David: Oh, they're even. So how do you know that they're even by looking at the graph? How do you see that they're even? They all look different to me. But how do you know that they have the same amount of money?
[58] Erica: Cause on, on number four they're like, in the same place.
[59] David: The same place? Yes, they are in the same place. Ok.

The realization that the two lines cross when "they are even," expressed by Erica [56] and by other students, is an important step toward equations. It is also a clear example of how the students can interpret the graph in terms of the word problem, that is, to attribute semantics of quantity (Schwartz, 1996) based on the syntax (Resnick, 1982) of the graph.

The children then move on to complete their tables of possible values and the corresponding graphs. This was easily achieved and, at the end of the lesson, in the four classrooms we worked with, 62 of 63 students (98.4%) completed the table successfully. Thereafter, the students completed the graphs on their own worksheets, referring to the values they had entered in their function tables: Fifty-seven (90.5%) students correctly plotted Mike's and Robin's values.

Graphs Can Clarify Tables and Vice Versa. As they finished their work, Anne, one of the researchers present in the class, asked Jessie to explain his graph:

[60] Anne: Ok. What do you notice about that graph?
[61] Jessie [Focusing on the intersection of the two graphs]: That it crosses over here.
[62] Anne: Can you explain why?
[63] Jessie: Because over here, in the table, it's four and it's twelve, twelve, so they're equal. And then over here it's four and then over here, that's why they cross.
[64] Anne: That's why they cross because what?
[65] Jessie: Cause they are equal in the table.

This is a clear example of what some authors (Brizuela & Earnest, chap. 11, this volume) refer to as navigating between diverse representational forms or coordinating diverse representations:

[66] Anne: Ok. What happens down here in the graph? Who has more money on this part of the graph?
[67] Jessie: Uhm, Mike.
[68] Anne: How do you know? How does the graph show that?
[69] Jessie: Because, uhm, Robin's down here and then Mike's all the way up there.
[70] Anne: And then what happens after they meet?
[71] Jessie: Robin goes higher.
[72] Anne: So what does that mean?
[73] Jessie: That Robin gets more money.

Prompted by Anne's questions, Jessie also makes use of the convention that, in a graph, "higher means more" and that in the particular context they are working with it means "more money" [69–73]. Other children provided similar explanations as they were interviewed in class and during the whole class discussion that followed.

Next is another case where a student used the table as a mean for verifying her analysis derived from the graph:

[74] Anne: How much do they have when they each have the same amount?
[75] Lisandra: Twelve.
[76] Anne: Twelve. Then what happens after that?
[77] Lisandra: Uh, they get different.
[78] Anne: What?
[79] Lisandra: They get different, I mean—
[80] Anne: They get different. Who's gonna have more money after?
[81] Lisandra: Uhm, I think ...

Here Lisandra flips back to the page showing her table, carefully inspects it, then turns back to the graph and further inspects the lines:

[82] Lisandra: Uhm, Robin [will have more money after they each have the same amount].
[83] Anne: Robin. How does ... How do you know?
[84] Lisandra: Cause Robin's all they way up here [showing the highest point drawn on Robin's line].

SOLVING EQUATIONS: THE WALLET PROBLEM REVISITED

Because the Wallet Problem involves the comparison of intersecting functions, it is suitable for delving into equations. Nonetheless, it is important to realize that the problem was not originally put forth as an equation. Doing so would have subdued the functional relations we wished to emphasize. This is easily understood by considering two distinct interpretations of equation $w + 8 = 3 \times w$.

A Numerical Interpretation of the Equation

One might construe the equation, $w + 8 = 3 \times w$, as an equality of the left and right terms, each of which stands for a single number (or measure). If it turns out that the number on the left is the same as the number on the right, then the equation is true. If the number on the left is different from the number on the right, then the equation is false. We can refer to an unsolvable equation as indeterminate.

A Functional Interpretation of the Equation

There is a strikingly different way of thinking about an equation, namely, as the setting equal of two functions. This is the interpretation of equations we build toward throughout the early algebra instruction. In this framework, $w + 8$ is a function that is free to vary (take on diverse values) within a specified domain,[11] say, the non-negative integers; $3 \times w$ is a different function, presumably in the same domain.

Setting the two functions equal can be expressed by the following equation written in standard symbolic notation: $w + 8 = 3 \times w$. What does this mean? What consequences do setting the functions equal have?

The equation $w + 8 = 3 \times w$ is true only for the case that $w = 4$. This is the case when the functions "are even" [56], when "they are equal in the table" [65], when they (the graphs) "cross" [63] or are "in the same place" [58], when they "have the same amount" [74], and so on. Otherwise, the equation is not true, the graphs separate, "they get different" [77], and so on. In these latter cases, an inequality such as $3 \times w > w + 8$ holds [see 80–84].

It is consistent with the present view that the letter a in the equation, $5 + a = 7$, represents a variable, not a single value. The equation is true only when $a = 2$. But a is still a variable.

Likewise, the equation $b = b + 1$ is a perfectly sensible equation, even though there is no value of the variable, b, for which the equation is true.

This all may appear to be unnecessary mental gymnastics, but holds a number of important implications for mathematics education. For one thing, it implies that there is no need to treat unknowns and variables as fundamentally different. We prefer to think of an unknown as a variable that for some reason or other happens to be constrained to a single value. This is precisely what happens to w when $w + 8$ is set equal to $3 \times w$. The equation holds only for certain values of w—actually, only one value. It does not transform w from a variable into a single number or instance.

In the example that follows, we invited the children to consider the case where Mike and Robin have the same total amounts of money. The children already knew that this corresponded to the case where Mike's wallet had $4 in it. Accordingly, they already knew the solution to the equation before they were asked to solve it. So, at best, working with the equation would appear to offer them no more information. However, the students still had much to learn about how to draw inferences in a new representational system, and it is in this spirit that we introduce the next section.

Setting 8 + W Equal to 3W

After discussing the tables and the graphs, David writes the equation "$8 + w = 3W$" on the blackboard and asks one student to represent Mike by holding

[11]For most young children, the domain will be the non-negative integers, which is the set of natural numbers including zero. Many adults will by default consider the rational numbers (those that can be expressed as a/b where a, b are integers and b is not zero), or at least the non-negative rationals, to be the default domain. Those with advanced training in mathematics tend to treat real numbers (irrationals and rationals) as the default domain.

a 3 × 5 card on which $8 was written and another on which "*W*" was written to represent the variable amount in Mike's wallet. Another child, playing Robin, is given three cards to hold; each one has "*W*" written on it. In the ensuing discussion, David sometimes mistakenly addressed the two actors by their actual names. This did not appear to be a source of confusion for the students. So in the transcription that follows we have replaced the children's real names with Mike's and Robin's names to facilitate reading:

[85] David: If these are equal, if the money here in his hands is equal to, altogether, is equal to all the money that she has, do you know how much money is in the wallet? [Hoping to find a volunteer] Do we have any Sherlock Holmes here?
[86] Students: No.
[87] David: Jacky?
[88] Jacky: Four.
[89] David: And how do you know?
[90] Jacky [thinking of Robin's case]: Because uh, four times three is 12.
[91] David: Four times three is 12? And also ... So three times four is 12 and?
[92] Jacky [realizing that he it needs to work for Mike also]: Eight plus four is 12.
[93] David: Eight plus four is 12.
[94] David: So that's the only way that, that they can have the same amount?
[95] Student: Mh-hm.

They continue:

[96] David [aiming to simplify the equation by eliminating like amounts for each actor]: ... Can I have them spend some money?
[97] Students: Yes.
[98] David: Ok. I want Mike here to spend everything that's in his wallet. What should I do with the amount that he has?
[99] Student: Take away w.
[100] David [taking the "W" card from "Mike's" hand]: Take away the w. You spent it. Thank you.
[101] David: I picked your pocket, Ok? Just for fun. Are they equal now? Do they have the same amount of money?
[102] Students: No.
[103] David: Well, I wanna keep them equal. How can I keep them equal?
[104] Students: Take away Robin's.

It is not immediately clear to the students how much should be taken from Robin, so that her amount will be equal to Mike's diminished amount:

[105] Students: Take away all.
[106] David: Take away all three from Robin? [David takes away her three "wallets"] You think they're equal now? [Unclear what the students responded.]
[107] Students [laughing]: No ... No ...

[108] David: He's got $8. She's left with nothing.
[109] Students: Take, take—
[110] Student: Just take two away.
[111] David: Just take two away? I took one [David means one W, but this is ambiguous] away from him when they were equal, so what should I do to [Robin]?
[112] David: I'll go back, remember what I did. I took away his wallet, but I wanna do the same thing to her so that they stay equal.
[113] David [after returning all the cards to "Mike" and "Robin," who have 8 & *W* and *W* & *W* & *W*, respectively]: They're equal now, right? I told you that they're equal. So I took away this [showing the *W* card in Mike's hand]. What should I do to Robin's money?
[114] Student: You take away two.
[115] David: How much did I take away?
[116] Student: Two.
[117] David: But I took away one *W* from him, why should I take double from her?
[118] Student: Because she—
[119] Student [Still trying to make sense, forgetting that, although Robin has more cards, she and Mike are said to have the same amount of money]: She has more to loose.

As the discussion proceeds they finally agree on what to do to keep Robin's amount equal to Mike's:

[120] David [capitalizing on the fact that the students already know the solution to the equation]: Hold on. How much is in [Mike's] w?
[121] David: They're equal, remember? They're equal. So how much do we know is in the wallet?
[122] Nathan: Four.
[123] David: That's right, Nathan. Does everybody agree there's $4 here?
[124] Students: Yes.
[125] David: Ok, cause that's the only way they're equal. So how many dollars am I taking away from him?
[126] Student: Two.
[127] Students: Four!
[128] David: $4, Ok? I just took $4 from him. So how many dollars do I have to take from her?
[129] Student [It is possible there is a momentary confusion of dollars and cards; but this doesn't explain the answer, "two"]: Two. One. Four
[130] David: I have to take the same amount!
[131] Students: One! One!
[132] David: One dollar?
[133] Students [finally getting their referent straight]: One wallet! One w!
[134] David: Ok, one wallet. Did I take the same amount away from each of them?
[135] Students: Yes.

[136] David: Ok.
[137] Anne: How do you know that?
[138] David: Did I? I took how many dollars away from Mike? How many dollars did I take away?
[139] Students: Four.
[140] David: And how many dollars did I take away from her?
[141] Students: Four.
[142] Student: She has the same amount cause I can see it in her hands.
[143] David: . . . so [Mike] has $8 and [Robin] has . . .
[144] Student: $8.
[145] David [There would be nothing to "solve" if w is removed from the conversation, so he insists on referring to what is actually written on Robin's remaining two cards.]: $2w$.
[146] David: Can you tell me what w has to be equal to?
[147] Student: Four, cause four plus four is eight.

A New Equation: 100 + W = 3W

David repeats the same process now with a different function. His aim is to put the students in a situation where they do not already know the answer. He writes the equation $100 + w = 3w$ on the board. He explains that the situation is now completely different. Then he hands William a "100 card" and a "w card." He hands Nancy three "w cards." Now the students are dealing with an equation for which they don't know the solution. After some discussion, the teacher recommends that they take away one "w card" from each student:

[148] David: Now, they've still got the same amounts cause we took away the same amounts from each of them.
[149] Students: Oh! Oh!
[150] David: Oh. Oh. William, go ahead.
[151] William [realizing he can now infer the amount in each "w card"]: Uh, Nancy has uhm, 50 each in w.
[152] David: Really? So how much is Nancy holding altogether?
[153] William: A hundred.
[154] David: And how much is he [William] holding?
[155] William: A hundred.

Jessie's Representations

When David asks, "Did anybody else realize that it was fifty?," Jessie shows (Fig. 10.12) and explains how he solved the problem in writing:

[156] David: Jessie, how did you do it?
[157] Jessie: Three ws stand for 100.

FIGURE 10.12. Jessie's representation of the $3w = 100 + w$ and subsequently $3w - w = 100 + w - w$.

[158] David: I'm sorry, three?
[159] Jessie (showing his drawing): Three *w*s and I crossed out a *w*.
[160] David (explaining to the whole class): He crossed out a *w*. That was like taking away the *w*. This is a really, really nice way of doing it. He wrote *w*, *w*, *w*, and that was to stand for what Robin has, right? And then you wrote 100 and *w*, to stand for what Michael has. You took away a *w* from each of them, and you were left with two *w*s and 100. And if two *w*s is $100, each *w* has to be equal to ... $50.

Jessie's written work shows that he understands how to solve an equation that has sprung from the setting equal of two functions. But this is also the case for students who have managed to solve the equations in the form of index cards and statements, a point we shall soon revisit.

ALGEBRA IN EARLY MATHEMATICS

Reprise: Early Algebra Is Not Algebra Early

We noted at the outset that early algebra is not the same as algebra early. Early algebra builds on the background contexts of problems, only gradually introduces formal notation, and tightly interweaves existing topics of early mathematics.

Both the Candy Boxes and the Wallet lessons immersed students in particular *background contexts* for which they attempted to describe the relationships between physical quantities and ultimately to make mathematical generalizations. As the conversations progressed, we gradually introduced *formal representations* (tables, graphs, and algebraic symbolic notation)—

where possible, as extensions to students' own representations. Algebra served as a thread that weaves through and helps establish tight bonds across diverse topics (arithmetical operations, variables, sets, additive differences, composition of quantities) and representations (tables, diagrams, number lines and graphs, verbal statements, that written symbolic notation).

Now let us shift our attention to the kinds of reasoning that early algebra calls for. As we shall see, it engages students in a special kind of generalization.

Deduction Cannot Be the Whole Story

> The very idea of a science of mathematics seems to raise an insoluble contradiction. If this science is deductive only in appearance, from where does its perfect rigor come—a rigor that no one would deny? If, on the other hand, all the propositions mathematics puts forth can be derived from each other through formal logic rules, will mathematics not be reduced to an immense tautology? (Poincaré, 1916/1968, p. 31, translated by the authors)

Mathematics is not entirely deductive. Sometimes it involves thinking about unspoken premises. Sometimes it involves conjectures.

UNSPOKEN PREMISES

Consider the statement that arose in the context of the Candy Boxes problem: "It cannot be the case that John has a total of 6 candies whereas Mary has a total of 13 candies." At first glance, this statement would appear to be necessarily true, given the information that Mary has three more candies than John. But this ignores the fact that students need to think about unspoken premises. The students had to disregard, for example, the possibility that the instructor had misled them or made a mistake in loading the candy into the boxes. They further had to assume that the number of candies put into the box remained invariant (e.g., none fell out or melted).

As students were passing around the candy boxes for inspection in one of the classes, Joey, a student, accidentally dropped one of Mary's loose candies onto the floor where it shattered. David only noticed the broken candy several minutes later; the shattered pieces could be easily seen through the candy's transparent protective wrapping. Joey admitted apologetically that he had accidentally dropped the candy. He seemed concerned that David, the teacher, might be upset.

After assuring Joey that this caused no harm, David asked Joey whether this made a difference for the discussion they were having about the amounts of candies. Keep in mind that the boxes still had not been opened. Joey reflected for a moment, and then said that Mary had more candies than before. This prompted immediate denials by several students. Others sided with Joey. After giving the matter more thought, Joey concluded that the amount had not changed because if he were to put the shards back together again, they would yield the original amount.

It may appear that conservation of matter cleared up Joey's confusion. But the class discussion had been focusing on the number of pieces without regard to differences in size. By this criterion, Mary arguably had more candies after the candy shattered. Or was the real issue the weights of the candies? If so, what does it mean for two weights to be equal if one only measures weight within a certain margin of error?

Whenever mathematics is used to make sense out of data,[12] decisions need to be made about the premises that will be honored or disregarded. Even when these matters are settled, decisions need to be made about the mathematical tools useful for making sense of the data. Deductive logic cannot settle all of these issues of modeling because matters of usefulness, cost, and fit depend on human judgment. Furthermore, context-specific considerations may constrain a problem's domain and co-domain. Mathematics education cannot avoid these issues. On the contrary, it needs to raise their profile so that students can assess their germaneness to problems at hand.

CONJECTURAL GENERALIZATION

> Mathematicians ... always strive to generalize the propositions they have obtained, and ... the equation we have been using,
>
> $$a + 1 = 1 + a$$
>
> serves to establish the following equation,
>
> $$a + b = b + a$$
>
> WHICH IS DEMonstrably more general. Mathematics thus proceeds just like the other sciences, namely, from the particular to the general. (Poincaré, 1916/1968, p. 42, translated by the authors, equation captions added)

Poincaré certainly knew that no amount of deduction would justify the leap from Equation A to Equation B. In fact, this is precisely his point:

---

[12]We mean here both data collected from the world as well as invented tables of numbers

Mathematicians look for opportunities to generalize even when not entitled by the laws of logic.

Now let's imagine that Poincaré had begun with a case involving no variables:

$$7 + 1 = 1 + 7$$

But why stop here? By focusing on each side of the equation, we notice that 7 + 1 can be expressed more generally as $a + 1$, which can in turn be expressed more generally as $a + b$. Thus, a numerical expression can be regarded as a particular instance of a function. More boldly, any arithmetical statement can be regarded as a particular instance of a more general, algebraic statement and expressed as such through the notation of functions. Any situation involving arithmetic affords an opportunity for thinking about algebraic relations.

The generalization of interest here consists in treating an instance (e.g., 7 + 1) as a case of something more general (e.g., $a + b$). We refer to this as *conjectural generalization* to highlight its nondeductive nature and its relevance to the formulation of mathematical conjectures. The scope widens considerably as variables replace particular values. Attention shifts from number operations to functional relationships. The new, yet familiar, notation belies the profound shift that has taken place. This is precisely the sort of shift we attempted to promote among our students throughout the *Early Algebra, Early Arithmetic Project* (e.g., Carraher, Brizuela, & Earnest, 2001; Carraher, Schliemann, & Brizuela, 2001; Schliemann, Carraher, & Brizuela, 2001, 2007).

Because we worked with indeterminate amounts, the tasks can be interpreted at various levels of generality. This is curious. Mathematics is widely acclaimed for its precision, rigor, and clarity. But ambiguity can be an important resource in teaching and learning. Working algebra successfully into the early mathematics curriculum often hinges precisely on the deft exploitation of ambiguity in problems.

Functions Enable the Shift to Algebra

We have noted how functions such as $a + 1$ and $a + b$ bring to light the general, algebraic character of elementary mathematics. We are not the first to note the critical role of functions. A. Seldon and J. Seldon (1992) drew attention to the integrative role functions played in the history of modern mathematics in the introduction to an important work about the suitability of functions as an organizing concept in mathematics education (Dubinsky & Harel, 1992). And Schwartz and Yerushalmy (1992, 1995) developed a broad middle and secondary mathematics curriculum centered around functions.

Although there is general agreement that algebra should become part of the elementary school curriculum (National Council of Teachers of

Mathematics, 2000; Schoenfeld, 1995), there are varying views regarding the most promising approach for integrating algebra into the early mathematics curriculum. Some have proposed generalized arithmetic (Mason, 1996); others focus on the representation of quantities and the solution of equations (Bodanskii, 1991). Still others have defended pluralism on the grounds that no single approach can do justice to the range and complexity of algebra (Kaput, Blanton, & Moreno, chap. 2, this volume).

We would claim that functions are special and deserve a careful look. All issues of generalized arithmetic can easily be subsumed under functions, but the converse is not true. (The Candy Boxes and Wallet problem are two cases in point. So are most issues from geometry.) Functions are at home in pure mathematical endeavors such as number theory, but they are equally at home in applied mathematics, science, engineering, and cases where modeling and quantitative reasoning are critical. Functions even provide the tools for data analysis and statistics. Pluralism has a certain appeal, and we would be the last to argue for a "one-size-fits-all" approach to mathematics education. Nonetheless, the topic of functions merits a top spot as a general organizing theme for early mathematics.

It is nothing short of remarkable that the topic of functions is absent from early mathematics curricula. Although the concept of function arrived late in the history of mathematics, we are finding that students can work with and understand functions at surprisingly early ages. We suspect that the concept of the function can unite a wide range of otherwise isolated topics—number operations, fractions, ratio and proportion, formulas, and so forth—just as it served a unifying role in the history of modern mathematics. It seems to us that curriculum developers, teachers, and teacher educators have much to gain by becoming acquainted with functions for mathematics education in the early grades. It will likely take many years for this to happen. And it will require a program of research that puts to the test new ideas for early mathematics education.

Functions Need to Be Distinguished From Their Representations

Functions are normally introduced in such a limited fashion that a few words are in order about what they are and are not. As a warm-up exercise, consider the concept of number. In daily life, it makes perfect sense to say that the following are numbers: 8, 7, 0, –43, 3/4, 3.14159, ... and so on. However, in mathematics education, equating numbers with their written forms can lead to serious problems such as the mistaken view that 3/4 and 0.75 are different numbers. But the issue goes deeper. In Figure 10.13, there are several representations of the same number; not one of them is the number.

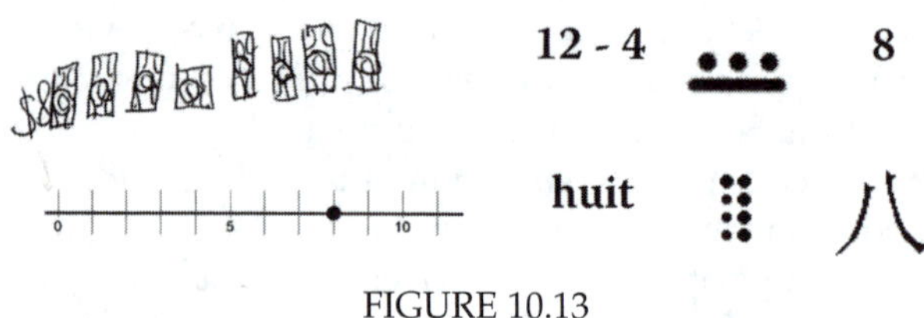

FIGURE 10.13

A similar predicament arises in the case of functions. Figure 10.14 shows four distinct means of representing the same function. Admittedly, these representations are not fully interchangeable. Each representation is likely to highlight certain characteristics of the function. The table tends to be a poor representation for conveying the continuity of a function. The graph conveys continuity, but it can be ill-suited for displaying precise values of the function.

Going one step further, Figure 10.15 shows various representations of the same equation. This may surprise those readers who think an equation is a written symbolic expression.[13] But, if we confuse the written form with the equation itself, that is, a setting equal of two functions, we will fail to recognize when students are working with equations in other formats, for example when our students were making drawings (see Fig. 10.7) of the Wallet problem or trying to describe the equation in their own words. This is not a matter of condescending to accept students' "personal yet inferior answers." Research mathematicians acknowledge that functions are validly expressed in language, written notation, graphs, and tables and they rely on these symbolic systems for representing functions in their professional work.

The Growth of Algebraic Understanding

Our approach highlights the shift from thinking about relations among particular numbers and measures toward thinking about relations among sets of numbers and measures, from computing numerical answers to describing and representing relations among variables. Whereas our main interest continues to lie in student reasoning, we have found ourselves thrust into the additional role of curriculum developers and teacher educators by virtue of the fact that many young students only show a proclivity to algebra when offered conditions that encourage them to

[13]This perspective leads to the unfortunate situation in which many students come to regard as equations only those equations that are analytically and symbolically soluble. Although the symbolic representation of the equation $x = \cos x$ gives no hint as to whether there are *any* solutions—and if there are, how many there are—the graphical representation of the function makes clear that there is exactly one solution and even gives a rough estimate of its magnitude.

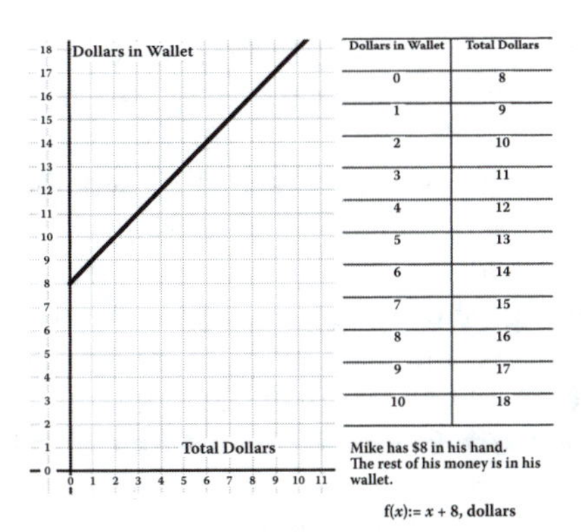

Dollars in Wallet	Total Dollars
0	8
1	9
2	10
3	11
4	12
5	13
6	14
7	15
8	16
9	17
10	18

FIGURE 10.14

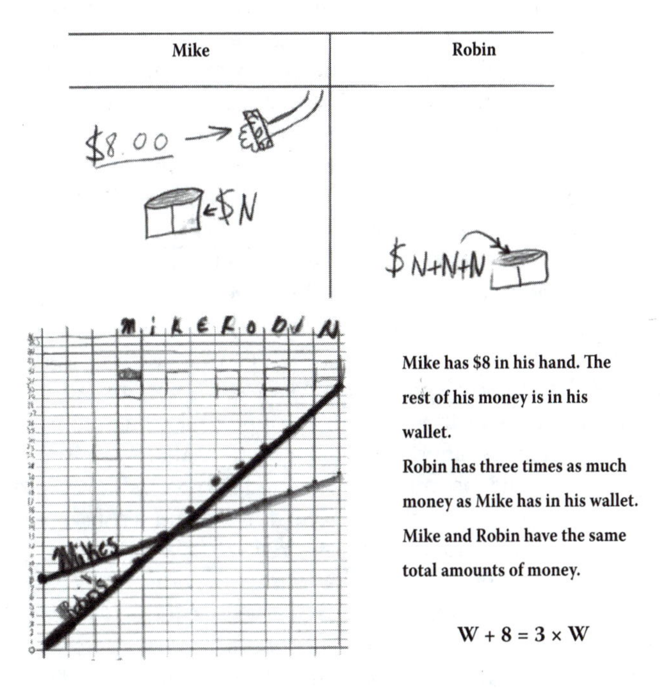

FIGURE 10.15

make mathematical generalizations and use particular representations (e.g., graphs and algebraic notation) normally introduced much later.

We witnessed a dramatic shift in students' thinking over 18 months. At the beginning of Grade 3, students interpreted a story with indeterminate quantities as a single tale about particular people and amounts. By the middle of Grade 4, most students construed this sort of situation as entailing many possible stories involving variable quantities in an invariant functional relationship. Many of the students made use of algebraic notation to convey the variations and invariance across the stories.[14]

The decisive changes in their thinking surprised us. Several years ago, when invited to assess the evidence regarding whether young students could "do algebra," we downplayed the discontinuity between arithmetic (as generally taught in K–8) and algebra (Carraher, & Brizuela, 2001, 2007). Our initial findings showed that young students could make mathematical generalizations and express them in algebraic notation (Carraher, & Brizuela, 2000; Carraher, Schliemann, & Brizuela, 2001; Carraher, & Brizuela, 2001). Our view at that time was that there did not exist the enormous conceptual leap from arithmetic to algebra that other researchers had proposed; otherwise, young students would not have been able to make such progress in so short a period.

We were opposed to the notion of a cognitive gap (Collis, 1975; Filloy & Rojano, 1989; Herscovics & Linchevski, 1994) because we were skeptical about the underlying claim that the transition from arithmetical to algebraic thinking was inherently developmental. We had repeatedly seen authors appeal to the concept of "developmental readiness" to argue that it was unreasonable to expect students to learn *topic x* at a given moment. We had witnessed this in Brazil a quarter of a century earlier, when many people, appealing to developmental readiness (or the lack of it), found nothing particularly surprising in the fact that, for every 100 children who entered the first grade, only 50 moved ahead to Grade 2 one year later.[15] What we would call the "developmental readiness syndrome" was well captured by Duckworth (1979): "Either we're too early and they can't learn it or we're too late and they know it already." This syndrome afflicts many adults, including quite a few educational theorists, developmental psychologists, and teachers. Piaget (Inhelder & Piaget, 1958) once thought that in order for students to master proportional reasoning they needed to have achieved the stage of formal operations, generally thought to arrive around adolescence and, even then, for a minority of students. He later revised his view to apply only to inverse proportion (Piaget, 1968; Schliemann & Carraher, 1992).

Our present findings have convinced us that there is indeed a large leap from thinking in terms of particular numbers and instances to thinking about functional relations. But, the fact that most students throughout the United States do not make this transition easily, nor early, may well say more about our failure to offer suitable conditions for them

[14]Our repeated use of the letter N in instruction turns out to have some benefits. Like a researcher who follows the distribution, across the continents, of genetic markers on the y-chromosome in order to infer about the paths of human migration (Wells & Read, 2002), we can trace the fourth-grade students' preferential use of the letter N to particular discussions about the Candy Boxes task in Grade 3.

[15]Approximately half of the students who stayed behind repeated first grade; the rest of these dropped out of school altogether.

to learn algebra as an integral part of elementary mathematics than it does about the limitations of their mental structures.

What Are Suitable Conditions?

We are still beginning to understand the conditions that promote early algebra learning. It has already become apparent, however, that certain representational forms play a major role. The first of these are the representations students themselves bring to bear on problems. We gave various examples in the present chapter of children's drawings, tables, and verbal comments. We tried to show how they are important as points of departure for the introduction of conventional mathematical representational forms.

Tables play an important role in urging students to register multiple instances of a function; hence the expression, "function table." Even when students learn about functions while working with highly engaging instruments such as a pulley, little may be learned unless students take care to transcribe the data to a table (Meira, 1998). But filling in a table is of little use in itself. Students need to scan the table locally and globally in search of generalizations that can be used to predict outputs from inputs and extend the table to cases for which data do not yet exist. In this regard, it can be very helpful to have students register the data as uncompleted calculations; such expressions facilitate the detection of variation and invariance throughout the table. The ultimate test of whether a table is being employed as a function table is whether or not a student can express in general fashion an underlying rule for an arbitrary entry. This amounts to the recognition that the underlying rule is a recipe (i.e., a function) for computing an output for any allowable input in the domain of the recipe (see e.g., Schwartz & Yerushalmy, 1992, 1995).

Once students are comfortable with symbolic notation for functions, symbolic table headers can be employed, allowing students to reason about relationships with a diminished need to verify the meaning of the data in terms of the semantics of the situation. Eventually, the syntactical moves will acquire a logic of their own, and the student can temporarily disregard the meaning of the symbols, deriving conclusions from the structure of the written forms and the current rules of inference.

Over the course of time, problems can be introduced in the form of written symbolic notation, graphs and tables, for which the students are asked to generate appropriate meaningful situations. Furthermore, students can be asked to envision how transformations within one representational system manifest themselves in another (e.g., if the graph of a function is displaced upwards by three units, how does this change the associate real-life story underlying the graph?).

As significant as our students' progress may be, algebra is a vast domain that can allow for continued learning and intellectual growth over many years. There will be new functions and structures with which to become familiar. The shift from thinking about instances to functional relations may well resurrect itself at other moments along students' intellectual trajectories. Rasmussen (2004) found, for example, that in learning about differential equations, university students first focus on local differences in values (deltas) along x and y, and slopes at a single point. Only later do they conceive of slope as a derivative, that is, as a function that comprises all the particular instances of slopes all along the graph. As educators increasingly implement programs of early algebra at the elementary level, there will be many opportunities for helping how early mathematical learning evolves over many years and helps set the stage for later learning.

ACKNOWLEDGMENTS

The authors thank the National Science Foundation for support for this research through REPP grant no. 9909591, "Bringing Out the Algebraic Character of Arithmetic," and ROLE grant no. 9909591, "Algebra in the Early Grades," both awarded to the TERC–Tufts Early Algebra Project. The data are from the first project; the conceptual analysis was developed in both projects.

REFERENCES

Bodanskii, F. (1991). The formation of an algebra method of problem solving in primary school. In V. V. Davydov (Ed.), *Psychological abilities of primary school children in learning mathematics* (Vol. 6, pp. 275–338). Reston, VA: National Council of Teachers of Mathematics.

Brizuela, B. M., & Schliemann, A. D. (2004). Ten-year-old students solving equations. *For the Learning of Mathematics, 24*(2), 33–40.

Carraher, D. W., Brizuela, B., & Earnest, D. (2001). The reification of additive differences in early algebra: Viva La Différence! In H. Chick, K. Stacey, J. Vincent, & J. Vincent (Eds.), *The future of the teaching and learning of algebra* (Proceedings of the 12th ICMI study conference, pp. 163–170). Melbourne, Australia: University of Melbourne.

Carraher, D. W., & Earnest, D. (2003). Guess my rule revisited. In M. Pateman, B. Dougherty, & J. Zilliox (Eds.), *Proceedings of the 27th Conference of the International Group for the Psychology of Mathematics Education* (Vol. 2, pp. 173–180). Honolulu: University of Hawaii.

Carraher, D. W., & Schliemann, A. D. (2002a). Is everyday mathematics truly relevant to mathematics education? In J. Moshkovich & M. Brenner (Eds.), *Monographs of the Journal for Research in Mathematics Education* (Vol. 11, pp. 131–153). Washington, DC: National Council of Teachers of Mathematics.

Carraher, D. W., & Schliemann, A. D. (2002b). Modeling reasoning. In K. Gravemeijer, R. Lehrer, B. van Oers, & L. Verschaffel (Eds.), *Symbolizing, modeling and tool use in mathematics education* (pp. 295–304). Dordrecht, the Netherlands: Kluwer Academic.

Carraher, D. W., Schliemann, A. D., & Brizuela, B. (2000). *Children's early algebraic concepts.* Plenary address presented at the 22nd annual meeting of the North

American Chapter of the International Group for the Psychology of Mathematics Education, Tucson, AZ.
Carraher, D. W., Schliemann, A. D., & Brizuela, B. M. (2001). Can young students operate on unknowns? In M. van der H.-P. (Ed.), *Proceedings of the 25th conference of the International Group for the Psychology of Mathematics Education* (Vol. 1, pp. 130–140). Utrecht, the Netherlands: Freudenthal Institute.
Collis, K. F. (1975). *The development of formal reasoning*. Newcastle, Australia: University of Newcastle.
Dubinsky, E., & Harel, G. (1992). *The concept of function: Aspects of epistemology and pedagogy*. Washington, DC: Mathematical Association of America.
Duckworth, E. R. (1979). Either we're too early and they can't learn it or we're too late and they know it already: The dilemma of "applying Piaget." *Harvard Educational Review, 49*(3), 297–312.
Filloy, E., & Rojano, T. (1989). Solving equations: The transition from arithmetic to algebra. *For the Learning of Mathematics, 9*(2), 19–25.
Herscovics, N., & Linchevski, L. (1994). A cognitive gap between arithmetic and algebra. *Educational Studies in Mathematics, 27*, 59–78.
Inhelder, B., & Piaget, J. (1958). *The growth of logical thinking from childhood to adolescence: An essay on the construction of formal operational structures*. New York: Basic Books.
Mason, J. (1996). Expressing generality and roots of algebra. In N. Bednarz, C. Kieran, & L. Lee (Eds.), *Approaches to algebra: Perspectives for research and teaching* (pp. 65–86). Dordrecht, the Netherlands: Kluwer Academic.
Meira, L. (1998). Making sense of instructional devices: The emergence of transparency in mathematical activity. *Journal of Research in Mathematics Education, 29*(2), 121–142.
National Council for Teachers of Mathematics. (2000). *Principles and standards for school mathematics*. Reston, VA: Author.
Piaget, J. (1964). Development and learning. In R. E. Ripple & V. N. Rockcastle (Eds.), *Piaget rediscovered: A report on the conference on Cognitive Studies and Curriculum Development* (pp. 7–20). Ithaca, NY: Cornell University.
Piaget, J. (1968). *Episémologie et psychologie de la fonction* [Epistemology and psychology of function];. Paris: Presses Unversitaires de France.
Poincaré, H. (1968). Sur la nature du raisonnement mathématique. *La science et l'hypothese* [On the nature of mathematical reasoning: Science and hypothesis] (pp. 31–45). Paris: Flammarion. (Original work published 1916)
Rasmussen, C. (2004). *Students' evolving understanding of differential equations*. Report presented at the Workshop on Design Research in Mathematics Education. Fairfax, VA: George Mason University.
Resnick, L. B. (1982). Syntax and semantics in learning to subtract. In T. P. Carpenter, J. M. Moser, & T. A. Romberg (Eds.), *Addition and subtraction: A cognitive perspective* (pp. 136–155). Hillsdale, NJ: Lawrence Erlbaum Associates.
Resnick, L. B. (1986). The development of mathematical intuition. In M. Perlmutter (Ed.), *Perspectives on intellectual development: The Minnesota symposium on child psychology* (Vol. 19, pp. 159–194). Hillsdale, NJ: Lawrence Erlbaum Associates.
Russell, B. (1967). *The autobiography of Bertrand Russell*. (Vol. 1, pp. 1872–1914). London: Allen & Unwin.
Schliemann, A. D., & Carraher, D. W. (1992). Proportional reasoning in and out of school. In P. Light & G. Butterworth (Eds.), *Context and cognition* (pp. 47–73). Hemel-Hempstead, England: Harverster-Wheatsheaf.

Schliemann, A. D., & Carraher, D. W. (2002). The evolution of mathematical reasoning: Everyday versus idealized reasoning. *Developmental Review, 22*(2), 242–266.

Schliemann, A. D., Carraher, D. W., & Brizuela, B. M. (2001). When tables become function tables. In *Proceedings of the 25th conference of the International Group for the Psychology of Mathematics Education* (Vol. 4, pp. 145–152). Utrecht, the Netherlands: University of Utrecht.

Schliemann, A. D., Carraher, D. W., & Brizuela, B. (2007). *Bringing out the algebraic character of arithmetic; From children's ideas to classroom practice.* Mahwah, NJ: Lawrence Erlbaum Associates.

Schliemann, A. D., Carraher, D. W., Brizuela, B. M., Earnest, D., Goodrow, A., Lara-Roth, S., & Peled, I. (2003). Algebra in elementary school. In N. A. Pateman, B. J. Dougherty, & J. Zilliox (Eds.), *Proceedings of the 27th conference of the International Group for the Psychology of Mathematics Education held jointly with the 25th conference of PME-NA* (Vol. 4, pp. 127–134). Honolulu, HI: CRDG, College of Education, University of Hawaii.

Schoenfeld, A. (1995). Report of Working Group 1. In C. B. Lacampagne, W. Blair, & J. Kaput (Eds.), *The algebra initiative colloquium* (Vol. 2, pp. 11–18). Washington, DC: U.S. Department of Education, Office of Educational Research and Improvement.

Schwartz, J. L. (1996). *Semantic aspects of quantity.* Cambridge, MA: Harvard University Press.

Schwartz, J. L., & Yerushalmy, M. (1992). Getting students to function in and with algebra. In G. Harel & E. Dubinsky (Eds.), *The concept of function: Aspects of epistemology and pedagogy* (Vol. 25, pp. 261–289). Washington, DC: Mathematical Association of America.

Schwartz, J. L., & Yerushalmy, M. (1995). On the need for a bridging language for mathematical modeling. *For the Learning of Mathematics, 15*(2), 29–35.

Seldon, A., & Seldon, J. (1992). Research perspectives on conceptions of function: Summary and overview. In E. Dubinsky & G. Harel (Eds.), *Concept of function: Aspects of epistemology and pedagogy* (pp. 1–21). Washington, DC: Mathematical Association of America.

Stern, E., & Mevarech, Z. (1996). Children's understanding of successive divisions in different contexts. *Journal of Experimental Child Psychology, 1,* 153–172.

Vergnaud, G. (1982). A classification of cognitive tasks and operations of thought involved in addition and subtraction problems. In T. P. Carpenter, J. M. Moser, & T. A. Romberg (Eds.), *Addition and subtraction: A cognitive view* (pp. 39–59). Hillsdale, NJ: Lawrence Erlbaum Associates.

Vergnaud, G. (1994). Multiplicative conceptual field: What and why. In G. Harel & J. Confrey (Eds.), *Multiplicative reasoning in the learning of mathematics* (pp. 41–59). New York: SUNY Press.

Vergnaud, G. (1996). The theory of conceptual fields. In L. P. Steffe, P. Nesher, P. Cobb, G. Goldin, & B. Greer (Eds.), *Theories of mathematical learning* (pp. 219–239). Hillsdale, NJ: Lawrence Erlbaum Associates.

Verschaffel, L. G., Greer, B., & De Corte, E. (2002). Everyday knowledge and mathematical modeling of school word problems. In K. Gravemeijer, R. Lehrer, B. van Oers, & L. Verschaffel (Eds.), *Symbolizing, modeling, and tool use in mathematics education* (pp. 249–268). Utrecht, the Netherlands: Kluwer Academic.

Wells, S., & Read, M. (2002). *The journey of man: A genetic odyssey.* Princeton, NJ: Princeton University Press.

11

Multiple Notational Systems and Algebraic Understandings: The Case of the "Best Deal" Problem

Bárbara M. Brizuela
Tufts University

Darrell Earnest
University of California, Berkeley

Children's introduction to oral and written language happens in natural and spontaneous ways. When speaking to young children—even infants—we rarely speak to them non-stop in baby talk, or babble back at them. We do not use their own language, but instead speak to them using conventional vocabulary and grammar, expecting them to gradually pick up on the complexities of conventional oral language. The same is true of written language. Infant and children's books are written using conventional language, vocabulary, sentence structure, spelling, and grammar, never trying to mirror children's language, but instead providing a scaffold on which children can develop and learn. Even esteemed authors such as bell hooks (1999) and Toni Morrison (1999) have written books for children—introducing them to the complexities and beauties of written language in a natural way. Although children may use unconventional sentence structure and invented spelling, the books we read to them do not.

In a way, we might say that we introduce children to language—both in its oral and written forms—in all of its complexities. We do not shy away from introducing language as such. In fact, we would probably reject proposals of children's books written in baby talk, or using invented spelling throughout. Whereas theories of language development vary on exactly how this process takes place, there is a consensus among leading linguists that language is a generative system through which children learn to represent spoken, written, and mental forms based on a common set of rules.

However, when we move from written and oral language to the field of mathematics, these lessons learned are sometimes forgotten. We explicitly avoid complex mathematical terminology when speaking to students (some high school students keep talking about input and output, and never encounter x and y, domain and range), finding more simple, transitional language that we deem will be easier for them. Closer to the topic concerning us in this chapter, we also avoid presenting students with notations that we deem would be too difficult for them to comprehend or adopt (e.g., Cartesian coordinate graphs, function tables, and algebraic notation). However, drawing a parallel from the case just presented in the area of language, can we say that we are being true to the discipline and to the nature of the mathematical content if we are not presenting the conventional and appropriate terminology, or the notations that are an integral part of the mathematical content? Would algebra be algebra without graphs, tables, and notation? Can we truly say that we are teaching children mathematics or algebra if we do not also teach them the notations that are part of this content?

We are convinced that notations need to be a part of the content being taught. In our work with young elementary school children in early algebra, graphs, tables, and notation are an integral part of the curriculum, and at times take on protagonist roles in our teaching. In fact, we are also convinced that introducing children to these notations enhances their conceptual understandings in important ways. We would even dare to say that their understandings of the content are incomplete if students do not have access, interact, and understand the notations as well. Moreover, as discussed in this chapter, students' moving across different notations enhances and complexifies their understandings even further.

Moving across modes of representation heightens mathematical understanding and provides children with opportunities to infer, confront, and refine ideas. The dynamic relationship among multiple representational systems pushes mathematical thinking to enhance one's overall understanding. By withholding accepted mathematical notations from children until a traditionally accepted age, we in effect deprive them of an enriched understanding of the number system, operations and computations, and functional nuances provided by the comparison of multiple notations. Any isolated representational form tells only a part of the mathematical story.

With any given system, certain mathematical facets are brought to the fore while others fall to the background. This is also true with language. For example, the word *pigs* involves a noun with an added morpheme, -s, to make it plural, and requires a connection between written symbols and their culturally accepted meanings; the word also conjures a mental image that is a unique internal representation, while also being a group of farm animals to whom we attach the word. Whereas one of these representation accurately portrays *pigs*, these three systems together present a more robust meaning created by the comparison and consideration of multiple systems of representations. This is also true in mathematics. A lone representation brings clarity to some part of the mathematics, but this clarity hides an indistinct treatment of other mathematical attributes. The inherent ambiguity of any one representation necessitates the embodiment and support of additional representations and their underlying mathematical constructs or concepts to fully appreciate the nuances of a mathematical situation, and thereby resolve some ambiguity in any one system. Just as children learn to connect written words with a mental image with a real-life situation, they can learn to connect mathematical symbols to a tabular, graphical, or verbal representation to create a deeper understanding and connection with the mathematics.

Children's work with and understanding of notations plays a pivotal role in their emerging algebraic knowledge. Various representational systems serve to provide an entryway into the bigger and generalizable ideas of the mathematics. Children can use mathematical notations not only to register what they understand, but also to structure their thinking. That is, notations can help further children's thinking (Brizuela, 2004). A variety of representational systems (e.g., tabular, graphical, verbal, and iconic) augment mathematical understanding, allowing children's algebraic thinking to continue to emerge through the consideration and comparison of multiple structures. This facility of connecting ideas among systems allows children to make inferences about mathematical attributes and their various manifestations that they might otherwise not have made.

Much of our early algebra research has focused on the issue of introducing various mathematical symbols in meaningful ways. Our approach relies on introducing new notations as variations on students' spontaneous notations (Brizuela & Lara-Roth, 2002; Carraher, Schliemann, & Brizuela, 2001). Although symbolic reasoning is traditionally associated with the syntactical manipulation of written expressions, other systems of representation play a role, highlighting otherwise hidden mathematical attributes. When meaningfully structured, these additional notations bring new understandings and corroborate previously learned mathematical information.

Researchers and policy guidelines have focused on the importance of multiple representational systems, a focus that has carried over to early algebra research (National Council of Teachers of Mathematics, 2000). However, a dynamic relationship among representational systems goes beyond the recommendations made by guidelines. Different representations not only "illuminate different aspects of a complex concept or relationship," as explained in the National Council of Teachers of Mathematics (NCTM) policy guidelines (p. 68), but also shed light on new understandings through the consideration and comparison across representations. Conventional notations help extend thinking (Cobb, 2000; Lerner & Sadovsky, 1994; Vygotsky, 1978), but if they are introduced without understanding, students may display premature formalization (Piaget, 1964). For these reasons, students need to be introduced to mathematical notations in ways that make sense to them.

RESEARCH ON MULTIPLE REPRESENTATIONS

In the past few years, increased attention has been given in the field of mathematics education to the power of establishing multiple connections and relationships among different notations. Perhaps it is no coincidence that the increased attention on multiple notational systems has roughly paralleled the increased focus on bringing algebra into the elementary school curriculum. As research and policy continue efforts to improve children's mathematical understanding, we have come to accept that algebraic thinking involves the consideration and understanding of multiple notational systems; conversely, the consideration and use of multiple notational systems sets the groundwork for algebraic understanding.

Research shows that moving across notations for a single concept or problem implies understanding the nuanced ways each iteration embodies that concept or problem (e.g., Behr, Lesh, Post, & Silver, 1983; Brenner et al., 1997; Dreyfus & Eisenberg, 1996; Goldin & Shteingold, 2001). Regarding this point, Behr and his colleagues (1983) point out that "it is the ability to make translations among and within ... several modes of representation that makes ideas meaningful to learners" (p. 102). Further, each notation highlights different aspects of a mathematical concept by stressing some aspects of information while hiding others (Dreyfus & Eisenberg, 1996). Moving among notations allows access to these different aspects of a mathematical concept, becoming more powerful through the links established among representations.

Regarding this generalizing across various notations, Goldin and Shteingold (2001) point out that in order for students to be able to move across different representations, they must be able to develop "adequate internal representations for interacting with various systems" (p. 9). For concepts to be fully developed, children will need to represent them in various different ways. Goldin (1998) points out that ambiguity in one system is resolved by means of *unambiguous* features of another system.

Brenner and her colleagues (1997) have explored students' abilities to translate among representations in the area of algebra (see also Yerushalmy, 1997). They carry out training experiments with students and explore the impact on their algebraic understanding of being able to move across representations. To the kinds of representations identified by Williams (1993; i.e., algebraic, graphical, and tabular), Brenner adds verbal representations as another primary way in which students should be expected to understand functions. In their training, they emphasize a guided-discovery approach in which students are encouraged to explore different representations and to develop their own understanding of each one. Students from a mathematical community that encouraged multiple notational systems were more likely to use appropriate tables, diagrams, or equations to represent functions. The authors conclude that supporting students with instruction on different kinds of external representation enhances problem-solving skills.

Brenner and her colleagues also explore the ideas about flexibility with representations. They make a distinction between *flexibility within a representation* and *flexibility across formal symbolisms* (but still within a single representation system). The ability to translate among different kinds of written representations is believed to contribute to greater conceptual knowledge and enhanced problem solving.

The role of multiple representations extends beyond mathematical education and language acquisition. We can further appreciate the power provided by moving across representations by examining its role in music. Bamberger's (1990) work in the area of musical notations provides an argument for the use of multiple representations. Similar to mathematics, music can be represented using a type of written syntax, through verbal description, in terms of speed and rhythm, and even based on an emotional response to overall musical composition. Bamberger emphasizes these various modes of representation because of the confrontation of "differences and similarities that [emerge] as [the students] [move] across materials, sensory modalities, and kinds of descriptions" (p. 39). Her reflections about the use of multiple representations are valuable lessons for the field of mathematics, even though developed in the area of music specifically.

In mathematics, as in music and language acquisition, multiple representations provide a *generative value*, as Bamberger describes it with respect to music. A body of musical representations contains immanent potential in a child's ability to make sense of musical expressions (Bamberger, 1990). Similarly, multiple notational systems in mathematics provide a generative value resulting in critical transformations in children's sense making of symbolic expressions and an enhanced understanding of the overall mathematics. Bamberger described both the relationships one person might establish between different symbolic expressions, as well as the active confrontation of representations made by different people. In both music and mathematics, this latter confrontation process can help us become increasingly aware not just of what we have noticed, but of what we have not noticed that others have found meaningful (Bamberger & Ziporyn, 1992).

This idea of a generative value echoes the mathematical research of Dreyfus and Eisenberg (1996). The movement and links between different notations provide what they call flexibility of thought: "acquaintance with various representations related to a set of concepts, the establishment of strong and detailed links between these representations, and the ability to translate and switch between them is equivalent to a deep understanding of these concepts and enables their flexible use in problem-solving situations" (p. 282).

POLICY GUIDELINES

Recently, organizations such as the NCTM (2000) have also encouraged teachers to foster the development of relationships among different mathematical representations. The NCTM has acknowledged the importance of translating or establishing relationships among multiple modes of representation. Representations have been recognized as one of the process standards, meant to highlight ways of acquiring and using content knowledge in the area of mathematics. NCTM states that the "different representations often illuminate different aspects of a complex concept or relationship.... Thus, to become deeply knowledgeable about [a specific mathematical concept]—and many other concepts in school mathematics—students will need a variety of representations that support their understanding" (p. 68).

The present chapter illustrates these premises, using the specific example of the representation of piece-wise defined linear functions. A piece-wise defined function is one that is defined differently for different x-values (values of the domain of the function).

DEFINING A TERMINOLOGY

This chapter focuses on what Martí and Pozo (2000) have called *external systems of representation*, to differentiate from mental representations. Thus, in this chapter, whenever we refer to *representations*, we mean external representations, made with pencil and paper and having a physical existence. Goldin (1998) also refers to external representations (Goldin, 1998; Goldin & Shteingold, 2001) as distinguished from internal representations. These external representations are "the shared, somewhat standardized representational systems developed through human social processes" (Goldin, 1998, p. 146). Our definition of *mathematical representations* borrows from Goldin, Hughes, Kaput, Lehrer, and Martí. They relate to what Lehrer and Schauble (2000) call *representational models*: material inscriptions that sometimes form part of representation systems, but can also be nonconventional and nonsystematic. Using Kaput's (1991) words, representation systems are the "materially realizable cultural or linguistic artifacts shared by a cultural or language community" (p. 55). Hughes (1986) also refers to these kinds of representations as *symbolic* representations: those representations that correspond to widely adopted conventions.

With these caveats in mind, however, a further distinction between notations and representations is important. In keeping with an understanding of representations as internal, or mental (see Freeman, 1993), Lee and Karmiloff-Smith (1996) distinguished between notations and representations in the following way: "We reserve the term 'representation' to refer to what is internal to the mind and the term 'notation' to what is external to the mind. ... While representation reflects how knowledge is constructed in the mind, notation establishes a 'stand for' relationship between a referent and a sign" (p. 127).

Lee and Karmiloff-Smith (1996) argue that external representations include writing, numerical notations, drawings, maps, and any form of graphic marks created intentionally. These kinds of external representations are characterized by having an existence independent of their creator, having a material existence that guarantees their permanence, and constituting organized systems. According to Martí and Pozo (2000), to be considered a system, there must at least be a relationship between a graphic mark and what it represents. Following this definition, almost any notation can be considered as part of a system. Nemirovsky's (1994) definition of what counts as a symbol system is helpful to clarify what is meant by system: "With 'symbol system' I refer to the analysis of mathematical representations in terms of rules. For example, Cartesian graphs can be considered as a symbol system; that is, a rule-governed set of elements, such as points being determined by coordinate values in specific ways on scales demarking units regularly" (p. 390).

Given the variations in the definitions of representation, we have chosen to use the term *notation* throughout this chapter. Thus, notations fall under what some researchers have called external representations. Furthermore, the inevitable relationships or rules established by creators of notations between their graphic marks and what they intend to represent, lead these notations, be they idiosyncratic or conventional, to form part of larger notation systems.

RESEARCH AND METHODOLOGY

In a 3-year longitudinal study, we followed 70 students in four classrooms from the second through fourth grades. Our goal over the course of the study was to examine how, as they participated in early algebra activities, the students would work with variables, functions, positive and negative numbers, algebraic notation, function tables, graphs, and equations. Students were from a multiethnic community (75% Latino) in Greater Boston, where over 83% of the students were eligible for free or reduced lunch. From the beginning of their second semester in second grade to the end of their fourth grade, we implemented and documented six to eight early algebra activities each semester in their classrooms, each activity lasting about 90 minutes.

The data we analyze here is taken from interviews administered at the end of third grade. Students' previous work up to the time of these interviews focused on: more and less; additive comparisons; addition and subtraction as functions; generalized numbers and variables; multiplication and division; graphs and number lines; tables for organizing information and looking at functions; and conventional notations, including algebraic inscriptions. We chose to focus on group interviews as a way to document and assess their progress, difficulties, and the impact of our work in the classroom. The interviewed students represented a range of mathematical ability as assessed by their classroom teacher and the research team. With each group, we were striving for diversity in ways of thinking as well as performance levels in mathematics.

The interviewer was a member of our research team; thus, there was both a research and a teaching component to these interviews. The interviewer provided guidance and suggestions to the students, making pedagogical decisions about when to introduce or prompt for a different notation based on students' ideas and real-time assessment.

These end-of-year interviews involved eight groups (two from each of four classrooms) of three children—24 interview participants. The eight groups broke down as outlined in Table 11.1, with the interviewer from our research team listed in parentheses.

The interviews, which lasted between 60 and 90 minutes each, revolved around a single problem. Students explored the four types of notations identified by Williams (1993) and Brenner et al. (1997): verbal, algebraic/written, tabular, and graphical. Although students had worked with functions before this interview, this was the first problem that required them to consider more than a single function at one time. The problem they worked on is shown in Table 11.2.

We can algebraically define four functions that appear in this problem: the two overt linear functions of each deal and two piece-wise defined functions for the best and worst deals (see Table 11.3).

Table 11.1
Eight Groups Interviewed, With Interviewer Listed in Parentheses

Group 1 (David)	*Group 2 (Bárbara)*	*Group 3 (Jerry)*	*Group 4 (Bárbara)*
Albert	Briana	Carolina	Jaime
Eric	Cristian	Emily	Jeimy
Erica	Nancy R.	Jimmy	Katherin
Group 5 (Bárbara)	*Group 6 (Susanna)*	*Group 7 (Bárbara)*	*Group 8 (Bárbara)*
Jeofrey	Jesie	Joey	Jeffrey
Nicole	Paul	Joseph	Jennifer
William	Vilda	Nancy A.	Nathan

Table 11.2
Let's Make a Deal!

Raymond has some money. His grandmother offers him two deals:
Deal 1: She will double his money.
Deal 2: She will triple his money and then take away 7.
Raymond wants to choose the best deal. What should he do?
How would you figure out and *show him* what is the best to do?
Is one deal *always* better? Show this on a piece of paper.

The goal of this chapter is to explore ways in which children's understanding about a certain problem and the functions associated with it can be enhanced by moving across different notations for the problem. At the same time, their understanding of each individual notation is also enhanced when compared and contrasted with other notations. As we explore the various responses from these eight interviews, we conceptually divide the interview into four stages that were common across all interviews. Each of these stages brought with it a unique notational manifestation of the mathematics of the problem, as defined by Brenner (Brenner et al., 1997) and Williams (1993), allowing students to make meaningful translations among and within these different notations (Behr et al., 1983). The four stages and their featured notations, in sequential order, are: verbal representations and instantiations with manipulatives, initial written/algebraic notations using words or pictures, tabular notations, and graphical notations. Each stage and its notation illuminated a different aspect of the mathematics involved in the "Best Deal" problem for the students, resulting in a more meaningful overall mathematical understanding. The analysis cites examples from the eight interviews that illustrate how students confront ambiguities of one system using another system (e.g., Goldin, 1998), the generative value of multiple notational systems for the mathematical problem (e.g., Bamberger, 1990; Bamberger & Ziporyn, 1992), as well as the flexibility of thought fostered by the

Table 11.3

Four Functions Related to the Best Deal Problem

Deal 1	*Deal 2*	*Best Deal*	*Worst Deal*
$2x$	$3x-7$	$2x$ $x<7$ $3x-7$ $x>7$	$3x-7$ $x>7$ $2x$ $x>7$

movement across different notations for one same problem or concept (e.g., Dreyfus & Eisenberg, 1996).

THE FOUR STAGES

In the eight interviews, the four stages marked both a shift in approach to the problem, and a shift in the notational manifestation. These stages are not mutually exclusive; once introduced, the nuances of the notation used or focused on confronted and resolved the nuances in another notation, throughout the rest of the interview. We identify the stages as follows: Interviewed students would (a) react to the problem, giving their guesses as to which of the two deals Raymond should choose. During this stage, the interviewer would introduce colored chips to instantiate what would happen to Raymond's starting amount for each of the two deals. After (b) representing on paper their understanding of the problem, which in many cases involved writing a message to Raymond with their recommendation or drawing a picture of the situation, the interviewer would ask them (c) to make a table showing what was happening in the problem. Then, they moved on (d) to discuss the graphical representation of the problem.

Throughout the interview, the interviewer made pedagogical decisions based on assessments of the students about when and how to move on to the next stage. In some cases, individual groups spent more (or less) time on a given stage. Our following analysis will show some of the results of moving across notations, and how the interaction among students provided a confrontation of ideas that allowed them to learn about what others had found meaningful in the problem and their own notations (Bamberger, 1990; Bamberger & Ziporyn, 1992). As their work and responses progressed, the explicit and implicit qualities of the notations continually refined and enhanced their understandings of the problem.

Stage 1: Verbal Reactions and Instantiations

After reading the problem two times, each group stated verbal reactions and used the chips to make instantiations based on Raymond's starting amount. This first stage of the interview encouraged students to predict and instantiate for the better deal. Students answered the question, "Which deal yields the most money?" Their work in this stage thereby highlights the piece-wise function that expresses the best deal. Accordingly, one deal is labeled as the better deal, whereas the other is discarded. Unlike the other three stages in children's reaction to the problem, the verbal reactions bring in real-life nuances through dialogue, which remain hidden or do not exist in written, tabular, and graphical notations. These verbal representations highlight the best deal function. In addition, children also brought up the contextually undesirable thought of having money taken away from Raymond whatever the end result, in the second deal.

In all of the groups, some or all students stated that Deal 1 would always be the better deal. In fact, three groups—Groups 1, 3, and 5—were in full agreement that Deal 1 would be the better deal. Through instantiations, either mentally or with manipulatives, students began to consider that sometimes Deal 2 would be the better deal, although in some cases students still wanted to choose one deal over the other. At the end of this stage, students concluded something about the piece-wise function, or the best deal function, in which Deal 1 is better until 7, and Deal 2 is better after 7. They were also able to conclude that at 7, the deals are equal.

In Group 6, Jesie stated that Deal 1 is better because the grandmother doesn't take away $7. Although he did not seem convinced that Deal 1 would always be the better deal, his concern brought out a nuance of the verbal representation—that Raymond wouldn't want money taken away from him. Paul felt uncomfortable with Jesie's statement about Deal 1 being better after mentally instantiating for $15 as the starting amount. The group remained focused on the identity of the best deal, and confronted their idea of a single function versus pieces of two functions:

Jesie: He could pick Deal 1.
Susanna (interviewer): Do you think Deal 1 will be better? Why do you think it will be better?
Jesie: She [the grandmother] doesn't take away $7.
Susanna: Do you think it will always be better?
Jesie: No.
Paul: 'Cause he could have $15, and triple it, and she only took away 7.
Susanna: So what would that mean?
Paul: That means he'd have more money.
Susanna: With which deal would he have more money?
Paul: Deal number 2.

Group 5 went on to instantiate for other values starting with 4, each time identifying which would be the better deal. The verbal statements in this stage of the interview shed light on only the parts of the Deal 1 and Deal 2 functions for which the deals yield the most possible money for Raymond.

A few minutes into Group 8's interview, Jennifer, Nathan, and Jeffrey instantiated different starting amounts for Raymond using the manipulatives, and the idea was put forth that sometimes Deal 1 would be Raymond's better choice, and sometimes it would be Deal 2. After the group used the chips to determine that Deal 1 is better when Raymond starts with $4 and then $5, the interviewer, Bárbara, asked them if there was a case for which Deal 2 would be better. Nathan used the chips to instantiate when Raymond started with $8:

Nathan: I started with 8.
Bárbara (interviewer): So if you double it, how much do you end up with, Jeffrey?
Jeffrey: 16.
Bárbara: 16. Let's try it on this side. Put 8 [yellow chips for Deal 2], Jennifer.

Jennifer tripled the 8 yellow chips, and took away 7. She counted the total chips.

Jennifer: 17.
Bárbara: 17 [for Deal 2], and here [for Deal 1] you have?
Nathan: 16.
Bárbara: So which deal is better?
Jennifer: This one [Deal 2].
Bárbara: So when is [Deal 2] better?
Nathan: When you use 8.
Bárbara: Okay, so Deal number 1 is still better with 6. And how about 7? Can you do 7, the two of you?

They then used the chips to determine each deal when Raymond started with 7. As they state their findings, notice Nathan's puzzlement [But did you triple them?] that Deal 2 has the same result as Deal 1:

Nathan: 14 [for Deal 1].
Bárbara: And how many did you [Jeffrey] end up with?
Jeffrey: 14 [for Deal 2].
Nathan: But did you triple them??
Jeffrey: Yes.

Bárbara asked them to pretend they were starting over again, and re-read the original problem. Although they have determined the deals to be equal when starting at 7, they continued to face the unpleasant idea of having money taken away from them in Deal 2. As they again verbally represented the problem, the students gave conflicting answers regarding which was the better deal:

Bárbara:	Raymond wants to choose the best deal. What would you say?
Nathan:	Deal 1.
Jennifer:	Deal 1.
Jeffrey:	Deal 2, because it worked out with 8.
Bárbara:	What happens with 7?
Jeffrey:	With 7 it's equal.
Bárbara:	So again, I'm going to ask you the question. What would you do?
Jennifer:	Deal number 2.
Nathan:	Deal number 2.
Bárbara:	Always?
Jennifer:	No. Like, first, like, if his grandma asks you those questions, if you want Deal number 1 or Deal number 2, you should try it out first.
Nathan:	Depending on how much money he has.
Bárbara:	So what would you say to your grandma, if you were Raymond?
Nathan:	If I have like $7, I'll take number 1 because you'll double it to 14. Not Deal 2, because you'll just take all the $7 away.

With the interviewer questioning their selection (Bárbara: Always?), they moved toward thinking about Raymond's choice as depending on his starting amount. However, they still faced the problem of having money taken away from them. When verbally representing the information, Nathan stated that Deal 1 was still better than Deal 2 with the starting Output of 7, even though they result in the same value. Nathan kept his mind on the context, and preferred that no money be taken away from him:

Bárbara:	If you have 7 then it doesn't make a difference, does it?
Nathan:	She'll just take it all away then.
Jennifer:	If you have $7, you could pick either one.
Bárbara:	And what if you have less than $7?
Nathan:	Then you go with Deal number 1.
Bárbara:	And when would you pick Deal number 2?
Jeffrey:	8.
Jennifer:	So if you have seven, it's right in the middle. It's equal. And if you have higher than 8 you go with number 2, and lower than 7 Deal number 1. (While making these statements, Jennifer is making gestures with her hands, showing a midpoint [7 on a number line?], sweeping to the right of this midpoint [more than 7 on a number line?], and then to the left of the midpoint [less than 7 on a number line?].)

In all of the groups, students brought up the idea that Deal 1 was better than Deal 2 because in Deal 2 the grandmother takes away $7. As Group 8 worked through the instantiations, confirming for themselves that the two deals yielded the same result when Raymond started with 7, Nathan still believed that Deal 1 was better because the grandmother would "just take all the $7 away." This theme carried over in other interviews, as well. For example, in Group 2, Briana read the problem, then immediately said that Deal 1 was the better deal because "he won't have $7 taken away." Cristian stated that Deal 2 was better after instantiating with a starting value of 10:

Briana:	Deal 1.
Bárbara (interviewer):	Why?
Briana:	Because he won't have $7 taken away.
Bárbara:	What do you think, Cristian?
Cristian:	Deal 2. Because if he had $10, it would turn into 30 and then 27 [*sic*] then if he had $10 with Deal 1, it would only turn into 20.

This idea that at 7 Deal 1 is better than Deal 2 came up only in this first stage. Through verbal representation, students discussed this contextual nuance of the problem, and were not always in agreement about the deals being equivalent, although they agreed that they both yielded the same outcome.

Stage 2: Initial Written/Algebraic Notations (Jennifer's Vector and a Number Line)

After the groups had instantiated the problem and discussed when Raymond should pick Deal 1 or Deal 2, the interviewers asked students to represent on paper using any method they chose to show how they figured it out. At times, this was rephrased to the students as what advice they could provide for Raymond about the two deals. Some students chose to write a message to Raymond, whereas others used pictures, numbers, or number lines.

The methods students chose varied from group to group, although four groups (Groups 1, 3, 5, and 7) skipped this recording stage in the interest of time or at the decision of the interviewers. In Group 5, Paul chose to write a message to Raymond: "Anything above 7 you pick Deal number 2. Anything below 7 you pick Deal number 1. On 7 you can pick

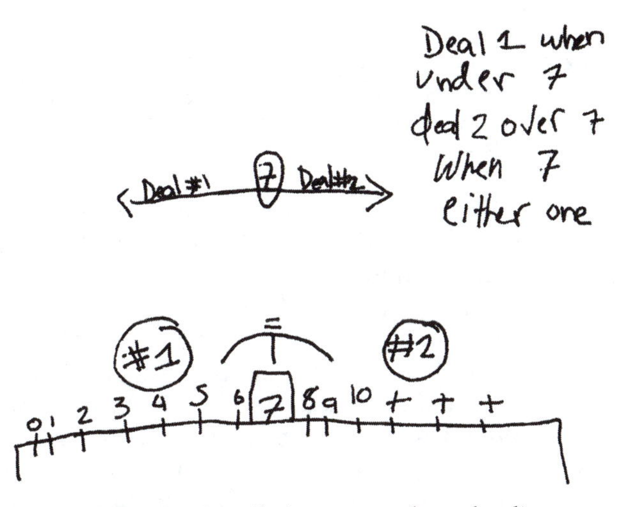

FIGURE 11.1. Jennifer's vector and number line.

either," while Jesie wrote out a dialog between Raymond and his grandmother. The contextual nuance that surfaced in the first stage about the deals' equality at 7 faded away in the second stage's notations. No student claimed that Deal 1 was better than Deal 2 at 7 during this stage.

In our analysis of the second stage, we focus mostly on ideas that surfaced in Group 8, in which Jennifer decided to use first a vector and then a number line to show which deal Raymond should choose. The number line notation brings with it a different way to view the problem. Like verbal representations, it highlights and hides various qualities of the four functions at play. Specifically, it focuses on the piece-wise best deal function based on Raymond's inputs. Unlike other types of notations, the number line hides information about the range of the best deal function, providing nominal information for each starting value.

Reflecting her group's discussion from the first stage, Jennifer first drew a vector that provided information about Deal 1, Deal 2, and the number 7. She also went on to draw a number line that showed that either Deal 1 or Deal 2 could be the better deal. The number line (see Figure 11.1) has a series of tick marks from 0 to 10, with plus signs written after the 10 at the three subsequent tick marks. Above the area for the numbers for 1 to 6, Jennifer wrote "#1" to reflect Deal 1, whereas above 8 to 10 and then over the "+" marks, Jennifer wrote "#2" to reflect Deal 2. She also marked the 7 with an equal sign:

Bárbara (interviewer): What are you going to do with the number line? Let's talk about it.

Jennifer: The center of the number line is 7. And if it's higher than 7, it's number 2.

In her notation, Jennifer has chosen to respond to the real-life question carried over from the first stage, "Which deal should Raymond choose based on his starting amount?" Her notation organized the information based on the input value for the functions. Above the 7 on the number line, Jennifer wrote an equal sign to show that when Raymond started with $7, each deal yielded the same amount of money. By her decision to show only Raymond's potential starting amount and not the output values, Jennifer provided a visual generalization of the problem. Her number line referenced Raymond's starting amount of money without indicating the output based on that starting amount. We may know at which points Deal 1 or Deal 2 yield more or the same amount of money, but we have no indication as to how much more or less. The number line shows the pertinent pieces of Deals 1 and 2 without comparing the two functions.

At the same time that Jennifer's number line explicitly illustrates one way to answer the problem's question, it also implicitly answers the opposing, complementary question about the "worst" deal (Brizuela & Lara-Roth, 2002). With the "#1" and "#2" that she wrote over the two portions of her number line, Jennifer addressed which deal would be better. There was also an implicit meaning to this about which deal would not be the better deal—or, which was the worst deal. If Deal 1 was the better deal from 0 to 6, then Deal 2 must be the worst deal. If Deal 2 was the better deal from 8 to 10 and beyond, then Deal 1 must be the worst deal. With this in mind, her notation explicitly addresses the input value for the best deal function, but could also implicitly address the input value for the worst deal function. However, we are assuming that this is something that she has decided to keep implicit. This information seems implicit to us in her notation, but most likely she was simply not focusing on it, as opposed to not wanting to make it explicit.

We could speculate about what kinds of notations could illustrate or highlight all the other deals involved in the problem—the worst deal, Deal 1, and Deal 2. Jennifer's choice to highlight the best deal is reflective of the discussion that occurred in the first stage of the interview. The written number line supports the preceding discussion. At this stage, the worst deal, Deal 1, and Deal 2 have not been represented as their own functions. As we move to the next two stages, we will get a sense of how tables and graphs not only represent the best deal function in a different manner, but also serve as methods to showcase the other three functions at play that have thus far remained hidden or implicit.

Stage 3: Tables

As the interviews progressed, the interviewer asked students to represent the information in a table. The students, who had worked with tables throughout their grade 2 and 3 intervention, constructed their own tables. They decided which pieces of information to include in their columns and rows and what information to exclude.

As a new kind of notation, tables bring a generative construct for looking at the problem. We call it generative because it builds on previously established information, showcasing the functions in a new light and extending them to include previously hidden information. As we will see, children answered in their tables the same question from Stages 1 and 2: "Which deal should Raymond choose based on his starting amount?" Some students, such as Paul of Group 5, represented only this information, whereas Group 2's Briana and Cristian created additional columns in their tables, highlighting different aspects of the functions. Some of the information students provided in their tables overlaps with that provided in Stages 1 and 2, whereas other information is handled explicitly for the first time. Students used the table in some cases to provide information about Deals 1 and 2. Thus far, the verbal representations and written notations provided in Stages 1 and 2 have served as communicators only of the best deal function. Various students used the table in Stage 3 to show three different functions—the best deal, Deal 1, and Deal 2.

Paul's table (see Fig. 11.2) in Group 5 highlights the best deal function. It has two columns, the first for the amount of money Raymond starts out with, and the second for which deal Raymond should choose for each corresponding starting amount of money.

Paul's table provides information about which deal Raymond should choose (found in Column 2) based on the possible starting amount. As in Jennifer's number line, this table gives a nominative range of Deal 1, Deal 2, or "either deal." We contrast Paul's table with Briana's. Similar to Paul's, Briana's table (see Table 11.4) shows only the best deal function. However, her table goes one step further, providing the output value in addition to naming the better deal.

Briana constructed her table with two columns; one for Raymond's starting amount, and the other to name the better deal for each starting amount. She added a third column labeled *Dollars*, giving the output value for the better deal. Utilizing the table to extend what is known about the problem, Briana states the dollar amounts that the better deal would yield for Raymond.

Table 11.4
Briana's Table

Table		*Dollars*
1	Deal 1	2
2	Deal 1	4
3	Deal 1	6
4	Deal 1	8
5	Deal 1	10
6	Deal 1	12
7	same	14
8	Deal 2	17
9	Deal 2	20
10	Deal 2	23
11	Deal 2	26
12	Deal 2	29
13	Deal 2	32

Briana further extends this with two written statements beside her Table: (a) "You can count by 2s when you have less money than 7"; and (b) "You can count by 3s when you have more than 7." She has extracted an attribute from each of the two functions that comprise the best deal function and described them, providing additional information about the piece-wise nature of the function she shows in her table. The Deal 1 function could be written as $2x$, x being the amount Raymond starts with. A quality of this function is that it "counts by 2s." The Deal 2 function could be written as $3x - 7$, with x again standing for Raymond's starting amount. Once again, a quality of this function is that it "counts by 3s," although the latter description ignores the subtrahend of 7 in the Deal 2 function.

Briana's table enhances the description of the functions' identities by including both the input and the outputs for the piece-wise function. Her table provides a value for the domain and range, illuminating the interplay among the parts of the Deal 1 and Deal 2 functions that comprise the best deal function. The table provides two types of information—which deal is better and the amount of dollars yielded by the better deal—each addressing a different aspect of the original problem. Similar to Stages 1 and 2, Briana addresses in Stage 3 the question of which is the best deal based on the starting amount by nominally stating in the second column, which is the better deal. She is able to go a step further with the use of the tabular notation, adding the column for the better deal's output. This gives the piece-wise function a range to go with its domain. Her added

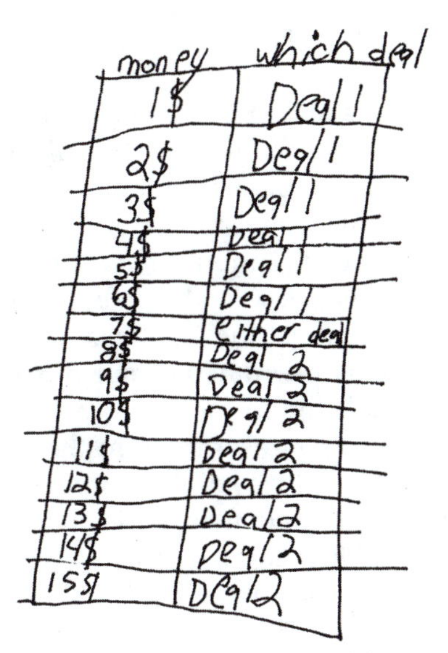

money	which deal
1$	Deal 1
2$	Deal 1
3$	Deal 1
4$	Deal 1
5$	Deal 1
6$	Deal 1
7$	either deal
8$	Deal 2
9$	Deal 2
10$	Deal 2
11$	Deal 2
12$	Deal 2
13$	Deal 2
14$	Deal 2
15$	Deal 2

FIGURE 11.2. Paul's table.

statements beside the table about the additive trend she is seeing in each of the two functions emphasize the dual component of the piece-wise function. The kind of information she added could not have been provided in a number line, highlighting how different notations illuminate different aspects of a problem or concept.

Briana's table has some qualities that overlap with the previous written notations and verbal representations. Similar to these previously used representations, Briana's table hides the properties of the worst deal function. Unlike her explicit treatment in Column 3 of the best deal function, her notation does not illustrate the full input–output quality of its piece-wise counterpart. Someone looking at her table might infer the existence of a complimentary function; if one deal is in fact the better in the second column, then another must exist that is worse. Moreover, the tabular information hides the continuous nature of Deal 1 and Deal 2 functions. Her table seems to state that Deal 1 exists only when the input is from 1 to 7, and that Deal 2 exists only when the input is from 7 to 13. As we later move forward to Stage 4, we will see this idea come out in Briana's interpretation of the functions on the graph.

The table generated by Cristian (see Table 11.5), also in Group 2 with Briana, provides more information in it, containing one more column than Briana's table. Unlike Paul and Briana, he gives Deal 1 and Deal 2 their own output columns, thus illuminating the continuous nature of the

Deal 1 and 2 functions. He adds a final column that states nominally the best deal.

In all stages thus far, including Stage 3, the students chose to nominally present the best deal function, highlighting the salience and importance of this function to the students. Building on the previously established information, Cristian's table contains the most extensive amount of information about the Deal 1 and 2 functions' inputs and outputs. The nuances of this notation allow Cristian to notice something about the differences between the two functions, an attribute of the mathematics that verbal representations and written notations thus far had not brought forth. By making explicit that there is a Deal 1 function and a Deal 2 function, Cristian, with the help of the interviewer, notices a trend in the comparison of the two functions: Until 7, the difference between outputs of Deal 1 and Deal 2 decreases by $1 as the starting amount increases by $1; after 7, the difference increases by $1 as the starting amount increases by $1. He writes the difference between the two deals' outputs (see Table 11.5):

Bárbara (interviewer): How about the difference this way [going across the row]? Remember we said that when you start with 5, we said that the difference between the two deals is 2?

Cristian: This one, these two [row 8] is 1, this one [row 7] is the same, these two [row 9] is 2, this is [going down the rows] 3, 4, 5, 6, and then if we do 14, the difference will be 7.

Bárbara: Let's try it. If we put 14 here [for the starting amount], what's double 14.

Cristian: 28.

Bárbara: And what's triple 14?

Cristian: 30 ... I mean, 42.

Bárbara: Minus 7?

Cristian: Equals 35.

Bárbara: So what's the difference there?

Cristian: 7.

Cristian then goes on with Bárbara's suggestion to write on the table the difference between the two deals, between the columns for Deal 1 and Deal 2:

> Bárbara: Why do you think that's going on? ... Why is the difference always changing in that way? ... > 0, 1, 2, 3, 4, why do you think that's going on? Do you have any ideas?

The group does not respond to this question. Although this notation has highlighted the difference between the two functions, this meaning and

Table 11.5
Re-Creation of Cristian's Table: Cristian Writes in the Difference Between the Two Deals as the Dialog Unfolds

		Deal 1	*Difference*	*Deal 2*	
	0	0	7	–7	Deal 1
I	1	2	6	–4	Deal 1
II	2	4	5	–1	Deal 1
III	3	6	4	2	Deal 1
IV	4	8	3	5	Deal 1
V	5	10	2	8	Deal 1
VI	6	12	1	11	Deal 1
VII	7	14	0	14	Both
VIII	8	16	1	17	Deal 2
IX	9	18	2	20	Deal 2
X	10	20	3	23	Deal 1
XI	11	22	4	26	Deal 1
XII	12	24	5	29	Deal 1
XIII	13	26	6	32	Deal 1

connection to the context still eludes the children. This issue reappears in Stage 4 during the discussion of the graph. Bárbara continues with the interview.

Cristian's final column once again goes back to addressing the piece-wise function; he answers nominally which deal is the better deal. Again, we see the continued presence of the best deal function. In Group 5, because time was running short, Bárbara, the interviewer, constructed the table (see Fig. 11.3) on chart paper herself with Jeofrey, William, and Nicole providing inputs and outputs at her prompts. She did not include a separate column to name the best deal, as students in other groups did. Although we can infer the best deal from the information the table provides about Deals 1 and 2, the information does not state which deal Raymond should choose. We see this tension with the information provided in the table come out in Jeofrey's dialogue.

After the group instantiates for 100, Nicole says that Raymond should choose Deal 2, but the group quickly decides that this wouldn't always be very good advice for Raymond. Jeofrey then verbalizes the idea of a piece-wise function, that there is a part of Deal 1 and a part of Deal 2 that we need to answer this question. Jeofrey interprets the information in the table, noticing that the outputs for Deal 1 are better up until 14, and then the outputs for Deal 2 are better:

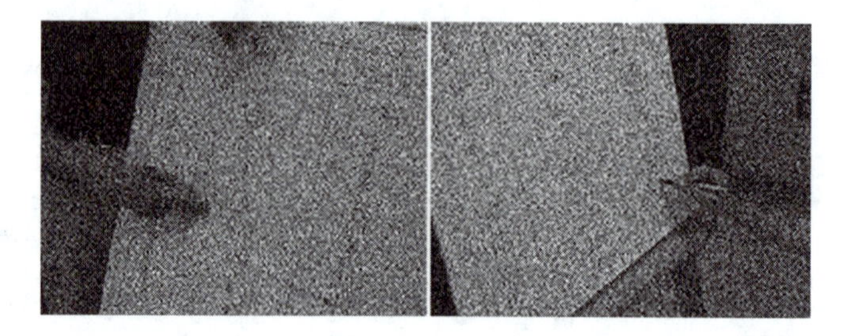

FIGURE 11.3. Group 5's table from two different angles.

Jeofrey: I think he'll get half of it [the table]. Half of Deal 1 and half of Deal 2.

Bárbara (interviewer): What do you mean?

Jeofrey: Like, get half. Like, if you *cut* 14 off [from the Deal 1 column], right? And put that from Deal 2, that part [below 14], put it next to Deal 1, maybe it will go like that.

We interpret Jeofrey's suggestion to "cut" the tables as being reflective of his desire to show succinctly the deal that Raymond should choose. Throughout the stages, this information recurs, underscoring the importance of the piece-wise function as being essential to answering the original problem. Compared to the kinds of notations and verbal representation used in previous stages, tables generate a more robust sense of the mathematics by providing more information about the functions at play; nevertheless, the piece-wise best deal function is used across the representations to ground the mathematics in the context of the problem and the question posed for Raymond regarding what function he should choose.

Stage 4: Graphs

The graphical notation of the best deal problem offers a new lens to view the four underlying functions and their nuances. When the functions are graphed, students must reinterpret the problem to make sense of this notation. For example, students have to reinterpret where the best deal function is, similar to Jeofrey in Stage 3, when his table provided the information but did not explicitly name the better deal. Unlike notations thus far, the Deals 1 and 2 functions are at the forefront of the graphical representation, because red and blue markers were used to draw the lines for the two deals. These two deals became more explicit in Stage 4 than they had been in earlier stages and notations.

Students bring their previous understandings of the problem into the discussion during Stage 4. When Group 2 moves on to plot the graph, the interviewer asks Cristian to read the values for Deal 1 from his table, because his table gives both the input and output values necessary to plot the line. Remember that Briana's table includes inputs for each deal, but the output for the best deal only. As we see through the ensuing dialog, Briana's thinking about the functions at play continues to evolve through the introduction of the graph and the confrontation brought on by Cristian's notation. As the group constructs the graph, the students' individual tables each factor into their interpretation of the graph's nuances and inform their thoughts on what remains ambiguous for them in the graph:

Bárbara asks, after the Deal 1 line is plotted and drawn:

Bárbara: So what line is this [line]?
Briana: The money line.
Cristian: Deal 1 line.
Briana: No! ... Some of it is Deal 2.

Coming from Stage 3, Briana seems to be thinking about this line as representing the piece-wise, best deal function. She calls it simply "the money line." We infer that she's thinking only about the best deal function:

Bárbara: What did you read off, Cristian?
Cristian: Deal 1.
Briana: ... I thought [Deal 1] *stopped* at 7.
Bárbara: ... But look at Cristian's table. [Deal 1] goes on and on and on.

Briana continues to use language that could have been used to describe her table, or Jennifer's vector, in particular the words "starting" and "stopping" ("I thought it stopped at 7"). The graphical notation brings forth the Deal 1 and Deal 2 functions more explicitly than notations used thus far. The graph does not name the best deal function as the other notations do. The piece-wise function, which came out naturally in the table, is represented as two line segments with different slopes. Deal 1 is shown with a red line and Deal 2 with a blue line. A trained eye may be able to visualize that Raymond should choose the function whose line falls higher than, or to the left of, the other.

Cristian, the only one in this group to represent the Deal 1, Deal 2, and piece-wise functions in his table using both inputs and outputs, draws a horizontal line at (7; 14) to mark where the two deals are equal. The interviewer asks the group what the areas above and below this horizontal line mean, attempting to get at the best deal function. Briana reuses her language of *starting* and *stopping*, as if she is thinking only about the best deal function:

Briana: This [below the 14] is where Deal 1 is, and the Deal 1 stops [at the 14 line]. And then Deal 2 starts right here [above the 14].
Bárbara: What do you mean, "it stops?"
Briana: Like, Deal 1 stops and then, like, Deal 2 starts.
Bárbara: Starts what? There's no more Deal 1? No more Deal 2?
Briana: [Shakes head] ... Deal 1 stops right here [at the 14 line] and Deal 2 stops [sic] right there, I mean starts.
Bárbara: Starts what? What do you think she means, Cristian?
Cristian: Like, there [at 14] it starts to be better, Deal 2 starts to get better and better than Deal 1.

The graph forces Briana and Cristian to interpret the best deal function in a different way than they did in the previous stage, in which the notation allowed them to name the best deal. Throughout the discussion, Briana is trying to transfer the information she has already learned through previous notations into the current one, and with some difficulty due to the graph's implicit treatment of the best deal function, relative to previous representations. As the dialogue continues, Briana works the nuances of this new notation into her overall understanding of the mathematics, ultimately incorporating not only the best deal function, but also the Deal 1 and Deal 2 functions, highlighting what Bamberger (1990) might call the generative value of the new graphical notation:

Bárbara continues to ask about the parts of the graph before and after the two lines meet:

Bárbara: Why do you think that right here [below 14] Deal 2 is below Deal 1, and here [above 14] it's above Deal 1?
Briana: Because both of them can work ... in each one of them.
Bárbara: Where does Deal 2 work better?
Briana: [pointing to the graph above the 14 line] Up here. But maybe, maybe it can work down here [below the 14 line] sometimes. ... Um, maybe because this [the Deal 1 line] is on top [of the Deal 2 line] it's better. And then this [the Deal 2 line] is on the bottom down here [below 14], but on this [above 14] Deal 2 is on top.

David Carraher, the project's principal investigator who videotaped the interviews, asks Briana to clarify what she means by saying a line is on top. Prompted by this question, Briana reuses the terms *starting* and *stopping* to explain the significance of one line being higher than another. We see in her last statement in the next excerpt that her language acknowledges the continuous nature of Deals 1 and 2:

David: Being on top, what does that mean?
Briana: Like, Deal 2 is starting. But over here [at the 14 line] Deal 1 was stopping right here and then the [Deal 2] line was getting to be first.

David: What does it mean that it's first?
Briana: That Deal 2 is now better in this one [above 14] and that Deal 1 is better in this one [below 14].

Through Briana's dialog, we can see that this particular notation has shed light on a different aspect of the problem. She knows when on the graph Deal 2 works better, which in this problem means a higher output value, and she continues to state when Deal 1 would be better.

When the discussion centered around the table representation in the third stage, Cristian wrote the differences between Deals 1 and 2 on his table, but did not answer Bárbara's question about explaining what the difference meant. In this fourth stage, Bárbara asks the group why the lines for Deals 1 and 2 get closer together until reaching point (7; 14), and then start getting further apart. Prompted by this question, Cristian recalls the earlier question about differences, and is able to address it. The ambiguity in the graph is confronted by Cristian's earlier thoughts about the changing difference:

Bárbara: Why do you think the lines are getting first closer together, right? They're far apart, right? And then they get closer, closer, closer, closer.
Cristian: Oh!
Bárbara: And then they start going further, further, further, further. What?
Cristian: It's like this [on the table], when we said ... This part [at 7 where the differences are], 0, 1, 2, 3, 4, 5, 6, 7.
Bárbara: So what's going on?
Cristian: That the difference is getting bigger.

The graphical notation provides a generative value for Cristian's developing understanding of difference with respect to Deals 1 and 2. On Bárbara's question about the visual feature of the two lines moving closer together and then further apart, Cristian gasps ("Oh!") and then retrieves his table, immediately pointing to the differences he had written between the Deals 1 and 2 outputs. Although he wasn't able to explain the difference in Stage 3 with the table, he looks at the graph in Stage 4 and, with the help of his table, states that "the difference is getting bigger." The graph helped illuminate for Cristian what this means and how it relates to a visual component of the graph.

The graph served as a generative tool for Jennifer and Jeffrey in Group 8, also building off of work with the table. In the dialogue about the graph, Bárbara asks the group why the two lines meet at one point on the graph. Thinking back to her table, Jennifer realizes that she also showed on her table that Deal 1 and Deal 2 were equal at a point:

Bárbara: Where do you think [Deals 1 and 2] cross?
Nathan: At 14 and ...
Jennifer: [gasps!]
Nathan: 14 and 7.
Bárbara: At 7, 14.
Jennifer: That's what I said on this paper over here!

Jennifer is referring to her table, which she picks up and compares to the graph:

Bárbara: Why do you think ... Let's see, why do you think they cross at that point and not at any other point? Why do you think the two lines cross at 7, and not at 8, or at 9, or at 10?

Jennifer walks over to the graph, and using a marker, writes an equal sign next to the point (7; 14):

Bárbara: Why?
Jennifer: They're equal. ... When you do the Deal 1 and 2 they're equal to each other.
Bárbara: They're equal to each other at that point? Is that why you think they cross each other there?
Jennifer: Yeah. At the same point because they're both right here [at 7, 14].

Jennifer confirms this nuance of the graph by comparing it with the table, calling her table the answer sheet for the graph. Bárbara then asks Jeffrey if he got the same answer or if he had arrived at the same or similar conclusion. Jeffrey refers to his own table, which he constructed similarly to Cristian, listing the inputs and outputs for Deals 1 and 2:

Bárbara: Did you get the same answer on your table?
Jeffrey: Yeah! 7 [as a starting value].
Bárbara: 14, 14 [that he wrote as outputs].

Not only is the graph becoming clear through previously built understandings, but phenomena from the graph are also clarifying previously constructed information.

There is another instance in Group 8 that highlights the graph's generative value. When Bárbara asks the group about the positioning of the Deal 1 line and the Deal 2 line, Jeffrey once again goes back to his table. Jeffrey shows his surprise ("Whoa!") on realizing how the two notations complement each other:

Bárbara: Look at what's going on before 7 and after 7. What happens?
Jeffrey: Oh!
Bárbara: Do you see what's going on, Jeffrey? What are you trying to figure out?

Jeffrey picks up his table, and compares it with the graph:

Jeffrey:	Whoa!
Bárbara:	What's going on?
Jeffrey:	After this half [that falls above (7;14)] ... This is ... The Deal number 2 is greater. And there [below (7; 14)], Deal number 2 ...
Bárbara:	Deal number 1, you mean.
Jeffrey:	I mean the Deal number 1.

Whereas these ideas about the piece-wise function have come up in previous notations, the students refine their understanding about the four functions overall through their interpretation of the newest notation. The graph presents the mathematical information in a different manner, forcing the students to confront their understandings about the functions at play.

CONCLUSIONS

This chapter provides a detailed description of the kinds of thinking and understanding highlighted by different notations, as children progress through different steps in solving a linear function best deal problem. It is not our intent to highlight particular children, but to underscore the different kinds of conceptualizations that can be explored by navigating across different representations for the same problem. This chapter illustrates both the research and the policy suggestions and recommendations for the use of multiple representations, showing the generative value (Bamberger, 1990) of each particular notation.

ACKNOWLEDGMENTS

This research has been supported by the National Science Foundation through Grant No. 9909591, "Bringing Out the Algebraic Character of Arithmetic," awarded to David Carraher (TERC) and Analúcia D. Schliemann (Tufts University). Special thanks to Gabrielle Cayton and Patty Chen for careful readings of this chapter; to Jim Kaput, David Carraher, and Maria Blanton for helpful editorial suggestions; and to Jim Kaput for inspiration.

REFERENCES

Bamberger, J. (1990). The laboratory for making things: Developing multiple representations of knowledge. In D. A. Schön (Ed.), *The reflective turn* (pp. 37–62). New York: Teachers College Press.

Bamberger, J., & Ziporyn, E. (1992). Getting it wrong. *The World of Music*, *34*(3), 22–56.

Behr, M. J., Lesh, R., Post, T. R., & Silver, E. A. (1983). Rational-number concepts. In R. Lesh & M. Landau (Eds.), *Acquisition of mathematics concepts and processes* (pp. 91–126). New York: Academic Press.

Brenner, M. E, Mayer, R. E., Moseley, B., Brar, T., Duran, R., Reed, B. S., & Webb, D. (1997). Learning by understanding: The role of multiple representations in learning algebra. *American Educational Research Journal, 34*(4), 663–689.

Brizuela, B. M. (2004). *Mathematical development in young children: Exploring notations*. New York: Teachers College Press.

Brizuela, B. M., & Lara-Roth, S. (2002). Additive relations and function tables. *Journal of Mathematical Behavior, 20*(3), 309–319.

Carraher, D. W., Schliemann, A. D., & Brizuela, B. M. (2001). Can young students operate on unknowns? In M. van der H.-P. (Ed.), *Proceedings of the 25th conference of the International Group for the Psychology of Mathematics Education* (Vol. 1, pp. 130–140). Utrecht, the Netherlands: Freudenthal Institute.

Cobb, P. (2000). From representations to symbolizing: Introductory comments on semiotics and mathematical learning. In P. Cobb, P. E. Yackel, & K. McClain (Eds.), *Symbolizing and communicating in mathematics classrooms. Perspectives on discourse, tools, and instructional design* (pp. 17–36). Mahwah, NJ: Lawrence Erlbaum Associates.

Dreyfus, T., & Eisenberg, T. (1996). On different facets of mathematical thinking. In R. J. Sternberg & T. Ben-Zeev (Eds.), *The nature of mathematical thinking* (pp. 253–284). Mahwah, NJ: Lawrence Erlbaum Associates.

Freeman, N. (1993). Drawing: Public instruments of representation. In C. Pratt & A. F. Garton (Eds.), *Systems of representation in children* (pp. 113–132). New York: Wiley.

Goldin, G. (1998). Representational systems, learning, and problem solving in mathematics. *Journal of Mathematical Behavior, 17*(2), 137–165.

Goldin, G., & Shteingold, N. (2001). Systems of representations and the development of mathematical concepts. In A. A. Cuoco & F. R. Curcio (Eds.), *The roles of representation in school mathematics. NCTM 2001 yearbook* (pp. 1–23). Reston, VA: National Council of Teachers of Mathematics.

hooks, b. (1999). *Happy to be nappy*. New York: Hyperion.

Hughes, M. (1986). *Children and number*. Cambridge, MA: Blackwell.

Kaput, J. (1991). Notations and representations as mediators of constructive processes. In E. von Glasersfeld (Ed.), *Radical constructivism in mathematics education* (pp. 53–74). Dordrecht, the Netherlands: Kluwer Academic.

Lee, K., & Karmiloff-Smith, A. (1996). The development of external symbol systems: The child as a notator. In R. Gelman & T. Kit-Fong Au (Eds.), *Perceptual and cognitive development: Handbook of perception and cognition* (2nd ed., pp. 185–211). San Diego, CA: Academic Press.

Lehrer, R., & Schauble, L. (2000). Model-based reasoning in mathematics and science. In R. Glaser (Ed.), *Advances in instructional psychology* (Vol. 5, pp. 101–159). Mahwah, NJ: Lawrence Erlbaum Associates.

Lerner, D., & Sadovsky P. (1994). El sistema de numeración: Un problema didáctico [The number system: A didactical problem]. In C. Parra & I. Saiz (Eds.), *Didáctica de matemáticas: Aportes y reflexiones* (pp. 93–184). Buenos Aires: Paidós.

Martí, E., & Pozo, J. I. (2000). Más allá de las representaciones mentales: La adquisición de los sistemas externos de representación [Beyond mental representations: The acquisition of external systems of representation]. *Infancia y aprendizaje, 90,* 11–30.
Morrison, T. (1999). *The big box.* New York: Hyperion.
National Council of Teachers of Mathematics. (2000). *Principles and standards for school mathematics.* Reston, VA: Author.
Nemirovsky, R. (1994). On ways of symbolizing: The case of Laura and the velocity sign. *Journal of Mathematical Behavior, 13,* 389–422.
Piaget, J. (1964). Development and learning. In *Piaget rediscovered: A report of the conference on cognitive studies and curriculum development* (pp. 7–20). Ithaca, NY: Cornell University.
Vygotsky, L. S. (1978). *Mind in society. The development of higher psychological processes.* Cambridge, MA: Harvard University Press.
Williams, S. R. (1993). Mathematics and being in the world: Toward an interpretive framework. *For the Learning of Mathematics, 13*(2), 2–7.
Yerushalmy, M. (1997). Designing representations: Reasoning about functions of two variables. *Journal for Research in Mathematics Education, 28,* 431–466.

12

Signed Numbers and Algebraic Thinking

Irit Peled
University of Haifa

David W. Carraher
TERC

We suggest that signed numbers[1] and their operations belong in early grades. If carefully introduced, signed numbers can make fundamental algebraic concepts such as equality and function accessible to young students. In turn, signed numbers can be learned more meaningfully when taught within an *algebrafied* curriculum.

We first identify some of the problems related to the learning of signed numbers. Then we show how algebraic contexts can facilitate the learning of this problematic topic. Finally, we look at how signed numbers provide a supportive context for learning algebra.

The present discussion of signed numbers and algebra builds on mounting evidence suggesting that young children can learn algebra (Blanton & Kaput, 2000; Brizuela & Earnest, chap. 11, this volume; Carpenter & Franke, 2001; Carraher, Schliemann, & Brizuela, 2001; Carraher, Schliemann, &

[1]Signed numbers are positive and negative numbers. They refer to integers, not merely the natural numbers. They refer to rational numbers, not merely the non-negative rational numbers. They refer to real numbers, not merely non-negative real numbers. Students may learn to accept negative numbers in the co-domain (output of computations) before accepting them in the domain (input of computations). When they are comfortable with both, we say they have learned not only that signed numbers exist, but they can serve as the input for functions.

Schwartz, chap. 10, this volume; Davydov, 1991; Dougherty, chap. 15, this volume; Schliemann et al., in press).

USING DIDACTICAL MODELS TO CONSTRUCT A MATHEMATICAL MODEL

In working with children to construct meaningful algebraic structures, keep in mind that we are not merely creating mathematical models to "play around with functions." One of our main goals is to make the emerging algebraic structures available to and actively used by children in analyzing and modeling situations. The algebraic concepts will serve as mathematical models, namely, tools with which different phenomena can be conceived and organized (Gravemeijer, 1997; Greer, 1997; Shternberg & Yerushalmy, 2003).

Teachers use *didactical models*—that is, manipulative materials used in a specifically defined language within a planned teaching trajectory. The teaching trajectory might employ a sterile model such as Cuisenaire Rods, a situation model such as the Realistic Mathematics models (Gravemeijer, 1997), or anything in between as a means of helping children construct their toolbox of *mathematical models*. These mathematical models are then applied in solving problems, and their conception is changed and expanded by the application process. Figure 12.1 represents the construction and application of the mathematical model. The double-headed arrow drawn between the application and the mathematical model stands to convey that even when one applies an already acquired (a somewhat misleading term) mathematical concept, the concept's image keeps changing and expanding following each application.

We discuss the learning of signed numbers within the general framework provided by Figure 12.1, focusing on situations involved with the construction and the application of the mathematical concepts of signed number operations. We give several examples of situations that, according to our analysis, can affect the senses and constructs of these concepts and their predisposition to become activated modeling tools.

There are different kinds of didactical models for teaching signed number operations (Janvier, 1985). One favorite among teachers in Israel is the Witch model, which involves adding or taking away warm cubes or cold cubes to and from the witch's potion bowl. Similar models can be found in U.S. textbooks. For example, Ball (1993) mentions a "Magic-Peanuts model," which she suspects might create the impression that mathematics involves some kind of hocus-pocus. Indeed, it is doubtful whether such a model would facilitate future use of signed numbers in modeling realistic situations.

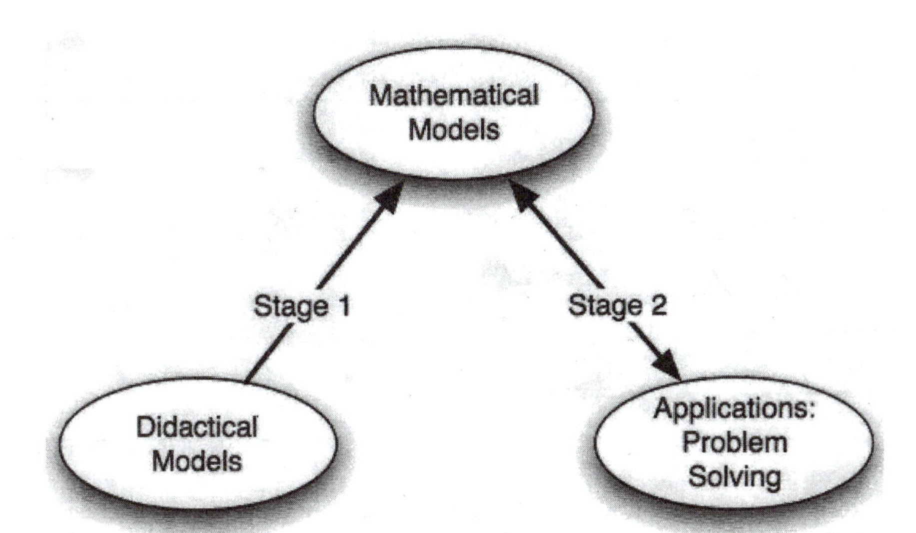

FIGURE 12.1. Mathematical models, didactical models, and situations.

Some teachers appear to believe that children will find it easier to remember a set of strange rules in such context better than a set of number rules. In a course for preservice secondary school teachers, students interviewed teachers about models they use for teaching operations with signed numbers. One of the interviewed teachers explained: "I use a model with piles and ditches. The minus sign stands for a ditch, the plus sign stands for a pile. Subtraction means: cancel. Addition means: put more. So, if I have –(–3) it means that I get a cancellation of a ditch that is 3 m deep, it means that I have a 3 m pile. . . ."

If performance is evaluated strictly in terms of computational fluency, then perhaps all these models have a similar effect and just teaching the rules would not be much different. Indeed, Arcavi (1980) showed that there is no difference in computational performance between four different instructional models for signed number multiplication.

However, signed numbers entail more than computational skills. They are supposed to become an addition to the set of mathematical lenses with which we model problem situations. Thus, students need to construct these mathematical models; teachers need to introduce these abstract models through didactical models. By and large, didactical models for signed numbers have not been very successful. Pitfalls associated with money models are discussed in one of the following sections. A promising

Table 12.1
Word Problems That Pre-Service Teachers Created for the Expression "2-7"

Problem / *Context*	*Consistent with "2-7"*	*Consistent with "7-2"*	*Incomplete answer (no question asked)*	*No answer*	*Total*
Money/debt	2	6	1	–	9
Temperature	1	–	–	–	1
Height	1	1	–	–	2
No context (no answer)	–	–	–	3	3
Total	4	7	1	3	15

direction emerges from Realistic Mathematics Education (RME; Gravemeijer, 1997). Based on the RME approach, Linchevski and Williams (1999) have constructed and tested several models, analyzing where the models work and where they fail. We would like to suggest that a combination of an RME approach with algebraic tasks (in the spirit of the examples in the following section) may offer an even more promising direction.

After being introduced to signed numbers and operations with signed numbers, children are expected to apply their new tools in solving a variety of arithmetic problems. Unfortunately, as we show, most of these problems do not facilitate conceptual growth. The following section illustrates that algebraic problems are more suited than arithmetic problems to promote meaningful learning of signed numbers.

The Challenges Posed by Signed Number Problems

Before we criticize the common signed number textbook problems, we should admit that it is not easy to compose good-signed number problems. With some exceptions (Rowell & Norwood, 1999), children and teachers do not have much trouble finding everyday situations that correspond to expressions such as $3 + 4 =$ __ or $4 \times 5 =$ __. However, they are often at a loss for finding contexts involving negative numbers and measures—for example, when trying to write a story for $2 - 7 =$ __ or $7 - (-5) =$ __. Temperature and money (credits and debts) are favorite examples, and yet even these contexts pose special challenges. For example, in composing a word problem to exemplify the expression, $3 - 5 = -2$, a student may suggest, "Johnny had three apples and he had to give five apples to his friend, so now he has minus two apples." The student is able to employ the negative numbers, but only in a contrived, artificial way. Table 12.1 summarizes the answers

15 preservice teachers gave when asked to compose a story problem for the expression, "2 – 7 = __."

The table shows that:

- Most preservice teachers (9 of 15; 9 of the 12 who gave answers) used a money/debt context. An example of a reasonably appropriate problem (one that can indeed be solved by 2 – 7) that a teacher wrote is the following: I have $2. I owe you $7. How much money do I [really] have? (A somewhat clearer, although awkward, version of the final question would be: How much do I really have, considering my assets and debts?)
- Only three preservice teachers used an alternative context. Of those, two teachers composed appropriate problems using contexts dealing with measurement of heights and temperature. For example, one teacher wrote: In the middle of the winter my thermometer read 2°. Overnight it dropped 7°. What temperature was it the next morning?
- Debts were used in a pseudonegative role. Consider the following example: If Jennifer owes Matthew $7 but she only has $2 to give him, how much money does she still owe Matthew? While being composed as a problem that can be solved using 2 – 7, this problem is more likely solved by using 7 – 2. This was confirmed when preservice teachers, given some of their own word problems, solved them by using 7 – 2 rather than 2 – 7.
- Some of the preservice teachers' word problems made use of the measures given, but included additional assumptions that went beyond the data given. Consider the following word problem created by a preservice teacher to reflect the expression, 2 – 7.
  Steve has $2, but owes his friend $7, if Steve pays back his friend, how much money will he have?
  We wonder: How can Steve pay back his friend $7 if he only has $2?

How People Avoid Signed Numbers

Part of the difficulty in creating signed number tasks stems from the fact that many problems that come to mind can be solved without negative numbers, relying instead on work-arounds.

Mukhopadhyay, Resnick, and Schauble (1990) compared children's performance on problems posed in the context of a story with their performance in number problems (calculations) that according to the authors' conception, correspond to the contextual problems. The authors found that children's performance "is far more complete and competent" in a narrative story about a person whose monetary standing goes up and down over time than in what they term "isomorphic problems presented as formal equations with mathematical notations." It may thus appear that the everyday context helps in dealing with signed numbers. However, for the children these were not isomorphic cases. Although the story situations could be matched by experts to signed number expressions, in dealing with them children did not use negative numbers to add up debts; rather, they performed simple addition and subtraction of non-negative quantities.

In interviews with strong and weak-performing sixth graders a year after they learned addition and subtraction of negative numbers, the first author found that students exhibited overall low computational performance (Peled, 1991). In another unreported part of the study, the children were given word problems that could be solved using signed number computations. Most children solved the problems correctly while circumventing signed number operations.[2]

One of the word problems in the study involved a context often used in textbooks to teach the concept of the difference between signed numbers: elevation with respect to sea level. The children were asked to find the difference in height between two cities—one located below sea level, at –200 meters and the other above sea level, at +300 meters. Rather than subtracting the numbers to find the distance by: $300 - (-200) = 500$, the children simply added the (absolute values of the) distances from sea level: $300 + 200 = 500$.

Students who correctly solve sea level problems show: (a) They understand the directed nature of the measures in the story context but (b) they have not mastered, or are not yet fully comfortable with, signed numbers. To appreciate the significance of the first point, it is important to recognize that the young child views numbers as counts: The natural numbers are used by them exclusively for representing the cardinality of sets (how many?). Extending the concept of number to include measures (how much?) is a major achievement. Students who solve the sea level problems not only treat numbers as measures. They also display a careful distinction between above and below measures, similar to the distinction between assets and debts, in the case of money problems. Expressed another way, they exhibit some understanding of two measure worlds separated by the zero point, similar to the divided number line model suggested by Peled, Mukhopadhyay, and Resnick (1989). But, eventually they will need to extend this conception further allowing for a number to

[2]There is nothing inherently wrong with solutions that do not use signed numbers. However, if our goal is to increase understanding of expressions such as $3 - (-5)$ and the conditions for their application, we need to look someplace else.

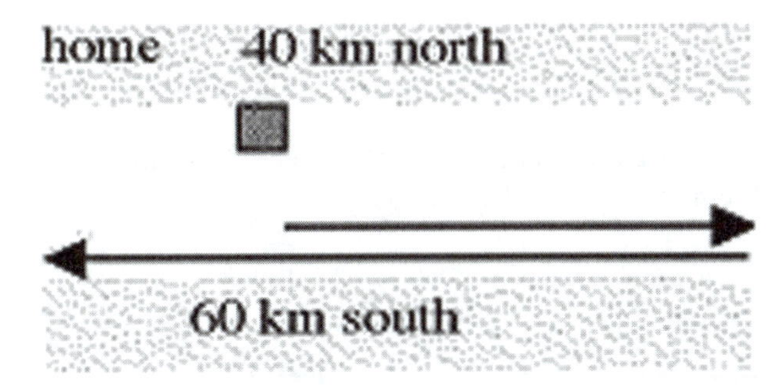

FIGURE 12.2. Trip A.

possess two characteristics: measure and direction. They also need to regard the number system as a single coherent system (rather than two separate worlds) with unified operations that hold regardless of the sign of the numbers.

LEARNING ABOUT NEGATIVE NUMBERS IN AN ALGEBRAIC CONTEXT

Whereas arithmetic problems present a relatively simple *specific case*, algebraic problems can introduce a challenge that calls for using mathematical tools to model *general algebraic structures*. Two examples involving trips serve to illustrate.

Algebraic Tasks for Learning Signed Numbers

Trips Along a Straight Line

Trip A: An Arithmetical Trip. Anne drove 40 kilometers north from her home to an out of town meeting. She then drove back going 60 kilometers south to another meeting. After both meetings were over, she called home asking her husband, Ben, to join her.

a. How far will Ben have to go and in what direction?
b. Write an expression for finding the length of Ben's trip.

A graphical diagram of the trip would look like the drawing in Figure 12.2.

From looking at Trip A (Fig. 12.2), one can see that Anne went south more than she went north, and Ben will have to travel that excess amount, namely, 20 kilometers south. The symbolic expression that best reflects the actual operation is:

Equation 1: An expression for Trip A

$$60 - 40 = 20$$

The symbolic expression teachers might have had in mind in composing such a problem is shown in Equation 2:

Equation 2: What teachers may have thought of for Trip A

$$40 - 60 = -20$$

As discussed earlier, one can avoid Equation 2, opting instead for Equation 1 while mentally keeping track of the direction. The following version, however, makes it harder to get away with this "partially explicit, partially implicit" approach.

Trip B: An Algebraic Trip. Anne drove a certain number of kilometers north from her home to an out of town meeting. She then headed 60 kilometers due south to another meeting. After both meetings were over, she called home and asked her husband, Ben, to join her. How far will Ben have to go and in what direction?

a. Write an expression for the length of Ben's trip.
b. Could Anne have driven less than 60 kilometers north on her first trip? If not, explain why. If she could have, give an example and explain its meaning.

Trip B is more general than Trip A: The initial part of Anne's journey corresponds to a mathematical variable. Using a number line representation with Anne's home marked as 0 and the direction to the right as distance in km due north, a typical figure might look like the drawing in Figure 12.3.

One legitimate expression for the husband's trip would be "$X - 60$ kilometers north." Notice that if x, the distance Anne first traveled northward is greater than or equal to 60, the answer is non-negative ("Go north! or if $x = 60$, stay put!"). If x is less than 60, the answer will be negative ("Go south!"). In principle, this should not be difficult. However, many solvers neglect to check the constraints on this expression. Question B was designed to serve as a hint, explicitly raising the possibility that $X - 60$.

The minus sign retains the same "old" definition (moving left on the number line, as depicted in Fig. 12.4), even when the end point is a negative number. If, for example, $X = 40$, then Ben will have to travel 20 km south, as shown by Equation 3.

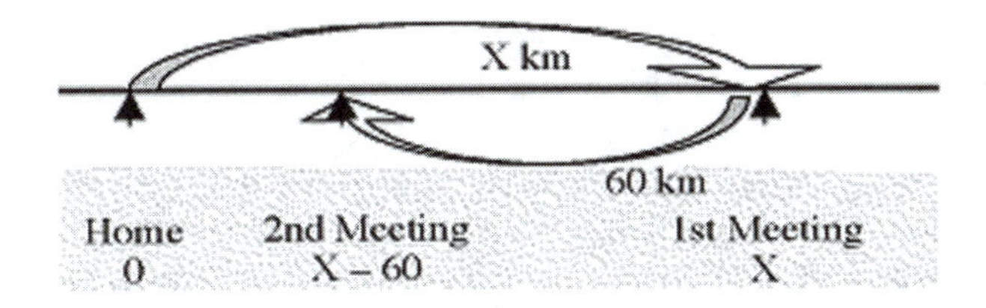

FIGURE 12.3. Version 2 of the "On the Road" problem.

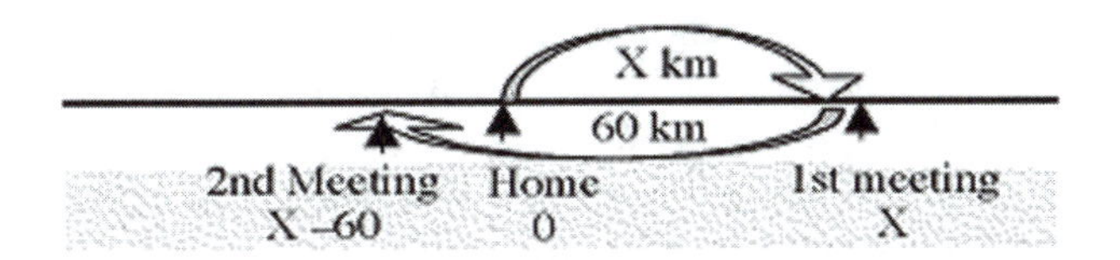

FIGURE 12.4. The "On the Road" problem for X < 60.

Equation 3:

$$X - 60 = 40 - 60 = -20$$

This simple description, $X - 60$, holds for a wide variety of cases. Furthermore, one can check its boundary conditions, investigate different cases using informal knowledge, and then discover generalizations and connections to formal knowledge. In this sense, the algebraic nature of the expression facilitates an understanding of operations with signed numbers.

Changes in Temperature

Temperature Story 1: Arithmetical. On the February 24th last year the morning temperature was –5° and on the same date this year the morning temperature was 3°. What was the change in temperature from last year to this year on February 24th?

The problem can be solved with the circumvention strategy used in the elevation above sea-level problem described earlier. In the sea-level problem, children added the absolute differences in height from the two cities to zero (one city was below sea level and the other above sea level). Here, one can add the absolute differences around zero temperature: 5 and 3, or consider going up to zero and then up from zero to 3°. The total change amounts to 8°. Accordingly, one would conclude: On February 24th this year, the temperature was by 8° higher compared to the same date last year.

The symbolic expression that best reflects the solution is: $5 + 3 = 8$. However, the expression the teacher was probably hoping students would use was: $(+3) - (-5) = 3 + 5 = 8$. Let us now see why a more general framing of the problem is likely to encourage students to use signed numbers in their expression.

Temperature Story 2: Algebraic. A computer with a special measuring device records the morning temperature in your office each day at a specific hour. As a meteorologist, you want it to calculate the difference in temperature between the same day last year and this year's record. The computer holds the data, but you have to tell it how to make the calculation. What would be your instructions? Here are the temperature data of the last 2 weeks and the corresponding data from last year.

The strategy one uses in solving this problem depends on one's experience with algebraic problems. An expert solver can use a top-down solution. He would choose some variables, such as T1 to denote the temperature on a given date last year, and T2 to denote the temperature on the same date this year. Then express the change in temperature from last year to this year (on a specific date) by the generalized expression: $T2 - T1$.

Depending on his experience, a problem solver might check if the expression is valid for negative as well as positive numbers. For example, if the temperature on a chosen date last year was -5 and the temperature on the corresponding day this year is 3 (as in Version 1), the change in temperature can be calculated informally as in version 1 and then compared to what one gets by substituting the variables in the general expression: $T2 - T1 = (+3) - (-5) = +8$, meaning there was an increase of 8°.

A novice has to search for patterns, work bottom up, organize the situation, and make generalizations. Specifically, a novice would look at a sequence of temperature changes, and come to recognize the advantages of using signed numbers to differentiate between temperature decreases and temperature increases. The task requires the use of variables, generalization of an expression, awareness of different possible cases, and use of available mathematical tools, in this case, the use of operations with signed numbers. These are exactly the skills we want a novice to develop.

Our analysis is theoretical and should, of course, be tested by refining and implementing the tasks. Yet, there is some evidence that an effective learning trajectory can be designed. Researchers from the Early Algebra Early Arithmetic Project (Carraher et al., 2001; Carraher et al., chap. 10, this volume) have shown that third-grade children in the Early Algebra

Project can express relations between heights and compare differences in cases where heights were variables and not specific values.

A Word of Caution About Money Contexts

As noted earlier, in-service and preservice teachers often use money and debts to construct signed number problems. The assumptions are that this context is meaningful to children and that children will use signed numbers in modeling the situations.

As mentioned earlier in discussing the study by Mukhopadhyay et al. (1990), children solve money problems by using their knowledge about the situation circumventing the need to use signed numbers. Still, this study showed that children have some understanding of debt situations and might lead us to conclude that it would be helpful to find problems that use this context and require the use of signed numbers (e.g., by using an algebraically structured context as discussed earlier).

However, even a very familiar context can be tricky. The following, previously unpublished, episode (TERC Tufts Early Algebra Early Arithmetic Project, 2000) exemplifies the difficulty of mapping situations involving money to a mathematical representation.

Two children, Filipe and Max, are enacting mathematical expressions as movements along a number line at the front of the classroom. The number line is represented as a clothesline with the integers, –10 to +20, written on labels hanging by clothes pins and spaced approximately one foot apart. Filipe and Max's positions are to indicate how much money they have; their displacements are to signify the spending or obtaining of money.

Here is a brief synopsis of the episode. At a certain point, Filipe and Max are told they have each spent $3, which they show by moving three units leftward: Max moves from 8 (i.e., $8) to 5 ($5), and Filipe from 3 ($3) to 0 ($0). The teacher then informs them they are each to buy another item (a hamburger) costing $2. Max moves correctly to 3, as depicted in Figure 12.5. Filipe appears to be puzzled by the fact that he has no money left to pay for the item. The teacher offers to lend him $2.

Before reading the transcript of the episode, the reader is asked to consider the following question: Where should Filipe stand on the number line after receiving the $2 loan from his teacher to purchase the hamburger?:

- At +2? (He holds the $2 loan in his hand.)
- At 0? (The money is not his own.)
- At –2? (He owes the teacher $2.)

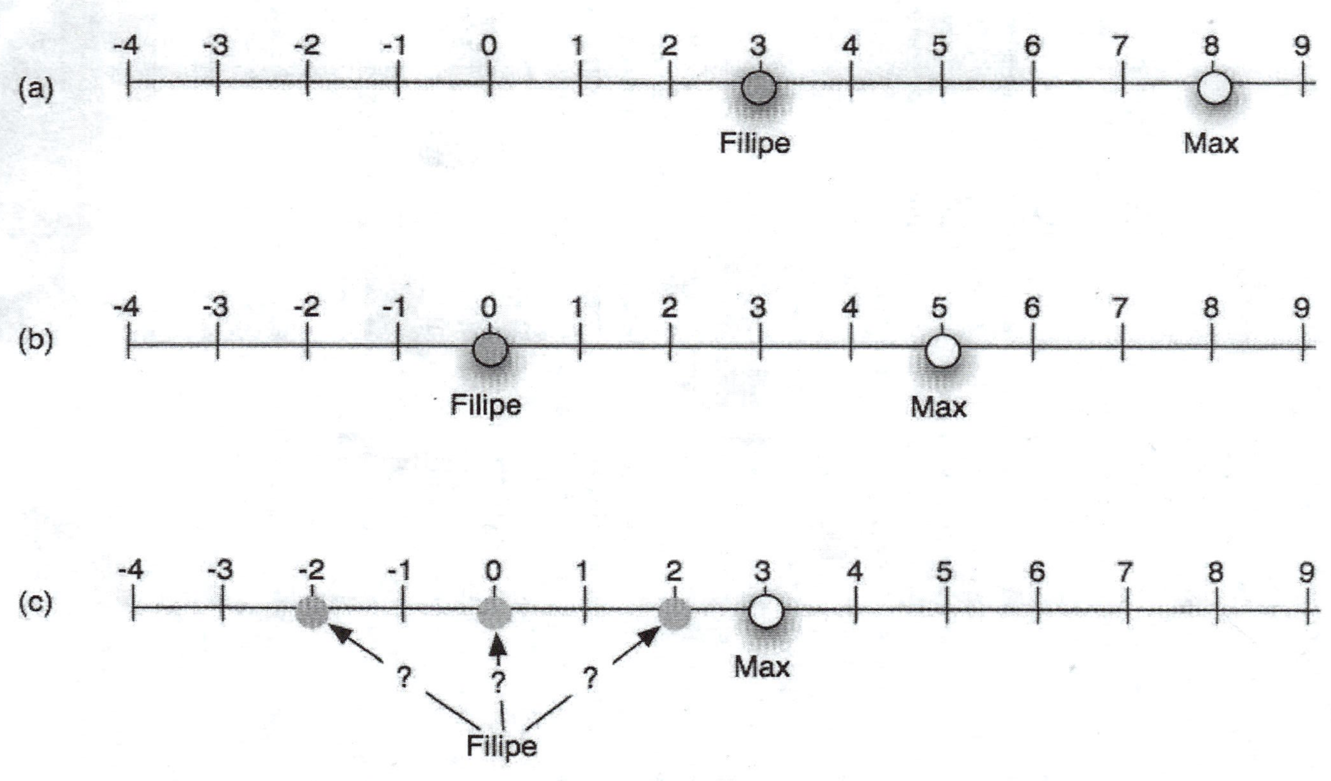

FIGURE 12.5. Position of Filipe and Max: (a) at beginning, (b) after spending $300, (c) after spending $2.00 more. In (c), Filipe has received a loan of $2.00 from his teacher.

All three possibilities arise in the following excerpt:

Barbara [the teacher] : ... and they spend another $2. Where would they end up? ... Max, where would you end up?
Max: [moving to three]
Barbara: And where would ... where would Filipe end up?
Filipe: Washing dishes if I didn't pay. [Class erupts in laughter]
Barbara [Repeating his words]: Washing dishes if you didn't pay.
Barbara: But where ... where would you be on the number line?
Filipe: I'm at it. [Filipe is still positioned at zero.]
Barbara: You're at it. So you would *stay at zero*?
Filipe: Yea.
Barbara: How come?
Filipe: Because I have no more money.
Barbara: You have no more money. Well, Anne [Barbara notices that Anne, a member of the research team, has something to say.] ... yea?
Anne: What if he borrowed $2 from somebody?
[Max pretends to offer $2 to Felipe.]
Filipe [accepting the offer]: Thanks Max.
Barbara [concerned that this offer by Max will require that he reposition himself]: Oh, but what if, if Max lends you $2, where will Max end up at?
Student: The zero. [Another student: One.]
Barbara: At one. So *I* am going to lend you $2. Okay? Max, you stay at three.
Barbara: I will lend you $2.
Filipe: [accepts the imaginary $2 and moves to +2]
Barbara [addressing the whole class]: Do you think he should go up to two? Does he actually have $2?
Students: No ... Yes.
Barbara [to Filipe]: How much money do you have of your own ... your own money?
Filipe: [moves back to zero, apparently interpreting Barbara's question suggesting that he should not have moved.]
Barbara: Okay. And he *owes* me $2.
Barbara: Okay. Are you all going to keep track of that money?
Students: Yes.
Barbara [realizing that the information about the loan is not apparent from Filipe's position]: You have to all be my witnesses. He owes me $2. Where would he go?

Filipe [looking for a practical solution]:	I'd get a job.
Barbara:	You'd get a job, but where would ... if I had to show with numbers ...
Ariana:	*Stay at the zero!*
Barbara:	Ariana, that's actually very good because that's how much money ... that's how much money he has.
Barbara:	He doesn't have any money, but he owes me $2 and we should have to show that, in some way, on the number line.
Filipe:	[moves to –2]

At first Felipe feels no need to move to the left of the origin because the least he can have is zero dollars [6–11]. Barbara, the teacher, had expected him to move to –2 under the assumption that the number line was being used to register his balance of credits and debits. Anne suggests [13] that someone lend Felipe $2, presumably to keep him solvent and make it easier for Felipe to conform to Barbara's expectations. When Max graciously offers $2 to Felipe [14–15], Barbara realizes that this would affect Max's amount [16] and require that he move; so she makes the loan [18–19] herself.

The matter is still not settled. How should Felipe respond to the fact that he has received the loan? Should he move to another position on the number line? Or stay at zero? Felipe and several classmates believe [20–22] he should move from zero to +2; after all, he now holds $2 in his hand. Ariana thinks [31] that Felipe needs to stay at zero. Barbara acknowledges the reasoning underlying Ariana's answer [32] "because that's ... the amount of money [Felipe] has" in terms of his balance.[3] (Felipe holds a different, more optimistic, view of "how much money he has."). Yet, Barbara's successive comments about Felipe's debt [33] together with her previous effort to enlist the class as her witnesses [28] show that she still believes he should move to –2, and eventually he gives in and takes the hint [34] moving to –2.

As we can see, the situation is quite complex and a teacher who thinks in terms of debits and credits may fail to recognize when a student takes into account only the tangible assets that are present on his person. As a matter of fact, not only young students choose to focus on the latter. The following example shows that the situation can be confusing and ill-defined for adults as well.

We asked 15 preservice teachers to think about the following situation:

> Let's imagine that you are an obsessively organized person and you keep track of the exact amount of money that you have in your checking account

---

[3]This is a reasonable conclusion if Felipe has not yet purchased the hamburger.

Table 12.2
Pre-Service Teachers' Answers for the Money Recording Problem

Answer	*Explanation*	*N*
–20	Because even though I have no money I cannot write "zero" since I owe them twenty dollars. Therefore by writing –20 I am accounting for what I have to pay back. It doesn't belong to me and I will need to give it back.	8
20	I will record $20 on the calendar because I have $20 in cash, even though I owe this to my parents who never ignore a debt. Twenty will be recorded that night on the calendar. You record the total of what's in your account and your wallet, so since you have $20 in your wallet that is what you record.	6
20 or –20	–20, because it was borrowed, I still owe that money, I'm in debt. Maybe I'm 20 because that's what I have.	1
0	During class discussion one of the students suggested 0 as another possible option.	

> and wallet altogether. Every night you record the amount of money you have by writing the relevant number in your calendar.
> One day you have no money left and you visit your parents and borrow $20 from them. What number will be recorded that night in your calendar? Explain.

The variety of answers in Table 12.2 speaks for itself. Even adults, including teachers, sometimes get confused in this context. We conclude by saying that the money context is not straightforward and does not automatically make signed numbers more accessible.

LEARNING ALGEBRA IN THE CONTEXT OF SIGNED NUMBERS

We have tried to show that an algebrafied curriculum offers good opportunities for meaningfully introducing signed numbers. This section argues that signed numbers can benefit the learning of algebra. Our claim rests on the idea that fundamental algebraic concepts can be enriched and more abstract concept images can be developed.

Table 12.3
Three Inequalities (Comparisons of Functions)

Expression 1	2xx
Expression 2	X + 3X
Expression 3	X + 33

Table 12.4
Truth Table for the Three Inequalities in Table 12.3

When x	*Expression1*	*Expression 2*	*Expression 3*
0	True	True	True
4.5	True	True	True

Functions and Graphs

Functions are an important topic in mathematics—arguably, the very cornerstone of an algebraic curriculum. Basic functions such as $f(x){:}= x$, $f(x){:} = x + 3$, and $f(x){:} = 4x$ play important roles in elementary mathematics even though they are likely to be implicit in discussions about matching expressions, adding, multiplying, and the like. The examples in this section demonstrate the importance for algebra of extending the number system to include signed numbers. The principal idea is that certain questions concerning algebra can be more suitably explored when negative numbers are taken into account.

Consider the following problem: What can you say about the following expressions? Are they valid? Under what conditions? (See Table 12.3.)

Table 12.4 and Figure 12.6 show some of the conditions under which the three inequalities hold. It might seem that the inequalities are generally true. Unfortunately, this picture is misleading. The graphs in Figure 12.7 demonstrate that the constant function $f(x){:} = 3$ is greater than $x + 3$ for $x < 0$.

Likewise, the function $2x$ is greater than x only for $x > 0$. The issue goes beyond extending the domain and co-domain of functions. In the case of $f(x){:} = 2x$, it means that "multiplying by two does not always result in a greater number." So it provides evidence that can potentially challenge the belief (Greer, 1992) that "multiplying always makes bigger and division makes smaller." The extended number system thus supports a more complete knowledge of basic functions such as x, $x + 3$, and $2x$.

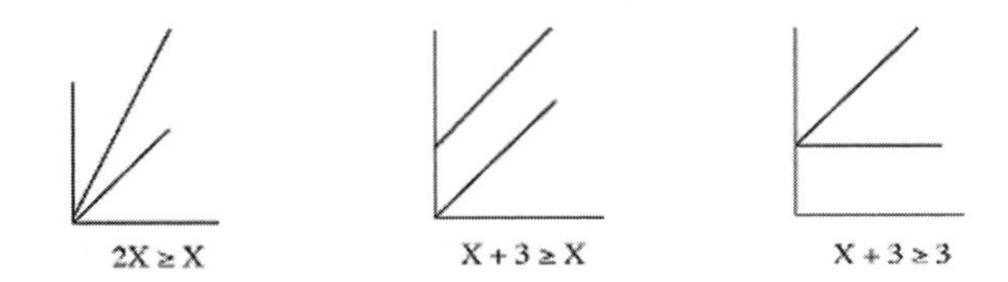

FIGURE 12.6. Showing order relations among various functions defined over positive real numbers. Note that $2x > x$.

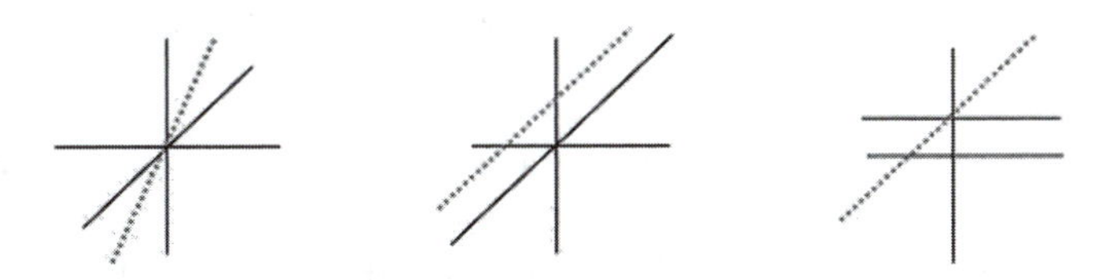

FIGURE 12.7. Showing the order relations among functions from Fig. 12.6 when defined over real numbers. Note that $2x > x$ for $x < 0$.

Comparing Functions

The following problem was given to preservice secondary school mathematics teachers, to graduate students in mathematics education, and to a few mathematics education experts.

Comparing Elevations With Respect to Sea Level.

Three friends are vacationing in a resort that has beautiful mountains and impressively deep canyons. One day they were conducting a conference call in order to decide about a meeting place for lunch. Before making a decision, they informed each other where they were by indicating the height of their location in sea level terms (don't ask us how they know it ...):

Anne tells her friends her current elevation.
John says that his elevation is 50 meters more than Anne's.
Sophia says that the height where she stands is 2 times that of Anne's.

a. Who is standing at the highest place? Explain.
b. Is it possible that John and Sophia are at the same height?
c. Is it possible that Anne stands at a higher place than Sophia?
d. Did you use a graph to answer the above question? Yes/No

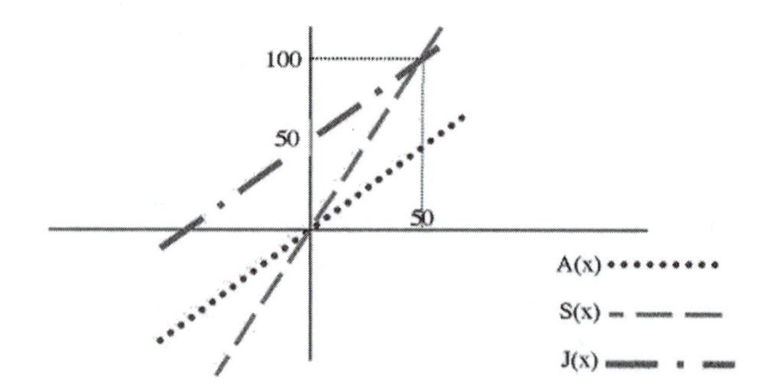

FIGURE 12.8. The sea level elevation problems.

If you did not, try to use a graph now, explain if it supports or changes your previous answers.

This problem can be solved by drawing the functions describing the height relations. If we represent Anne's height by X meters, then John's height is $J(x) = X + 50$, and Sophia's height is $S(x) = 2X$. In order to make height comparisons we need to represent Anne's height by using the identity function $A(x) = X$. Because elevation at sea level can also be negative, the functions should be drawn for any x E R, getting the drawing presented in Figure 12.8.

Some of the characteristics found in the solutions suggested by the different problem solvers (preservice teachers, graduate students, and mathematics education experts) were:

- Most of the problem solvers did not use a graph in the process of investigating the different possible cases in this situation. This population included secondary school teachers who were teaching linear graphs at that time.
- When asked to draw a graph many of the solvers had trouble realizing that in order to compare $2X$ with X one should draw the identity graph $f(x) = X$ and not use the X axis as a representation of X. . . .
- When asked to draw a graph many of the solvers ignored the interval $x < 0$ and made limited comparisons.

These findings support the present argument about the need to work with situations that offer opportunity to investigate many cases and enhance fluency in making comparisons. Another conclusion relates to the use of graphs. Many of the problem solvers showed that graphs had not become a natural tool for them in their investigation. Apparently, they use graphs only when asked to do so, only for certain familiar types of problems, and with specific instructions on the choice of axis.

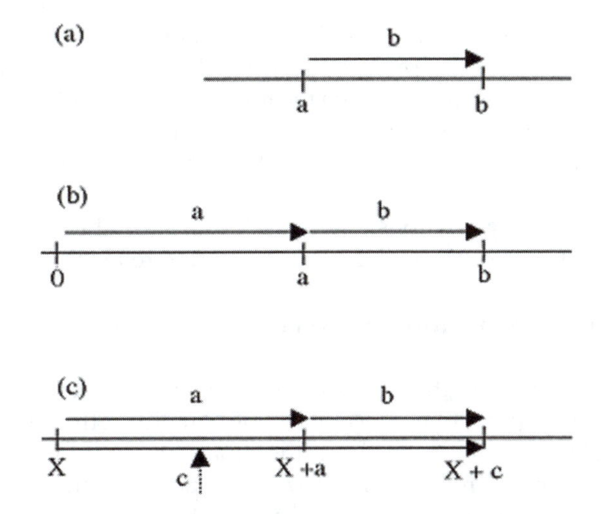

FIGURE 12.9. Defining a + b as one or two displacements.

Going back to our original argument, the extension of the number system opens up possibilities for constructing problems with rich investigations that can promote the analysis of functions through the use of graphs.

Composing Transformations

When numbers have a sign, the sign can assume the sense of direction, and the numbers can be represented as vectors rather than points on the number line. In the expression $a + b = c$, "a" can be perceived as a starting point, transformed by the function "$+ b$" (a unary operation with the operator $+ b$) to point "c," as depicted in Figure 12.9a.

The expression can also be perceived as a sequence of displacements transforming 0 to the point "a" (thus $+a$ is both an operator and a point) and then transforming "a" to the point "c," as shown in Figure 12.9b.

Several authors have argued that children need to think of the operations as operations on transformations (Janvier, 1985; Thompson & Dreyfus, 1988; Vergnaud, 1982). Symbolically: $X + a + b = X + (a + b) = X + c$ or $(+a) + (+b) = +c$, as represented in Figure 12.9c.

Vergnaud (1982) suggests that some cases can be perceived as a composition of two transformations that yields a third transformation. For example, he describes a situation where: Peter won 6 marbles in the morning. He lost 9 marbles in the afternoon. Altogether he lost 3 marbles. This situation is modeled by: $(+6) + (-9) = (-3)$ (Vergnaud draws the + sign differently to denote that it stands for addition of signed numbers). In situations like these, the problem solver will develop a sense of signed numbers as changes, similar to the meaning suggested by others (Davis & Maher, 1997). Although there is no information on the initial amount or the final amount, one can figure out the total change, which would be valid for any starting point.

In the following sections, the implications of these meanings are discussed in terms of their possible contribution to algebraic concepts.

Broadening the Meaning of Equals

Extensive practice with addition and subtraction leads children to develop certain primitive conceptions of equality. Children in primary grades tend to view the two sides of the equal sign nonsymmetrically: The left side is taken as a request to carry out an operation, the result of which is displayed on the right side. Accordingly, they view expressions such as $8 = 3 + 5$ as illegitimate (Carpenter & Levi, 2000; Filloy & Rojano, 1989; Kieran, 1981).

Equality expressions in early grades involve mainly addition and subtraction with amounts represented by natural numbers and quantity relations that obey the part–part–whole structure. As suggested by Freudenthal (1983), this structure can quite naturally (although not easily) be extended to include fractions.

The extension to negative numbers is a different story. Both the meaning of a number and the meaning of operations defined on the numbers have to undergo a drastic change. The image of a number as representing a physical quantity or a measurement has to change or at least get a new dimension that differentiates the quantity –2 from +2. The new constructs have to account for the order relation according to which –4 is smaller than –2, although there is more of that quantity in –4.

As to the extension of addition and subtraction (and later multiplication and division), a great part of previous knowledge has to be accommodated. The part–part–whole structure relations stating that addition makes bigger, subtraction makes smaller (regarding the first number as a starting point), and the whole is bigger than each part no longer hold. For instance, in $(+7) + (-2) = (+5)$, +7 is transformed to +5, which is smaller, and in $(+2) - (-3) = +5$, +2 is transformed to +5, which is bigger. In other words, addition can make smaller and subtraction can make bigger.

The gap between naive and advanced conceptions of the part–part–whole structure, and the gap between naive and advanced conceptions of equality might be bridged by allowing the operation extension to be defined within algebraic situations, as demonstrated in the following example.

Let us consider again the On the Road problem, modifying it to have Anne traveling X km and then Y km northward (X or Y can be < 0, in which case the respective segment of the trip will be southward). The general expression for Ben's trip (to meet his wife), marked as S, would be: $S = X + Y$. When $Y = -60$, as in the original story, we get $S = X + (-60)$, and in the specific case $X = 40$ we get: $40 + (-60) = -20$.

Ben's trip, S, is mathematically equivalent to Anne's complete trip consisting of $X + Y$ because the net effect, in each case, is to displace a person from home to the meeting point (see Figs. 12.10–12.12). However, Anne and Ben's trips are not conceptually equivalent when Anne's two segments represent different directions (see Figs. 12.10 and 12.11): Ben's trip is a shortcut. Because Ben proceeds directly to Anne at the end point of her trip, Ben's trip requires less driving, but, paradoxically, the net displacement is the same for Anne and Ben, thus we can now suggest a new extended meaning to the addition of signed numbers at end of sentence and subtraction.

Figures 12.10, 12.11, and 12.12 show some possible combinations of directions and sizes of the two journey parts. In all three examples, X is positive and rightward is northward, we can get symmetrical examples for $X < 0$ by reflecting the given examples (or regarding right as south). In the first two examples, Y is negative but has a different absolute value relative to X, and in the third case Y has the same direction as X.

In all the examples that can be generated for this situation, Ben's trip is expressed as the sum of the two parts of Anne's trip regardless of the signs of the parts, that is, $S = X + Y$. Similarly, we can express the second part of Anne's trip by looking at the difference between the whole (Ben's trip) and the part (Anne's) subtracting the part from the whole regardless of the number signs, that is, $Y = S - X$.

By offering this extension, we stretch the existing senses that children have about the operations and about equality. This bridge to an equivalence meaning of equation is different from the approach discussed by Filloy and Rojano (1989), who identify a didactical gap between arithmetic and algebra and view the transition to algebraic thinking as requiring the ability to operate on unknowns on both sides of an equation.

"The equivalence of routes" is a big idea because it requires suppressing otherwise important psychological features of situations in order to highlight a certain mathematical invariance. An analogous sort of reasoning is required when one deals with a string of transactions, each of which may correspond to a deposit or a withdrawal of funds in a bank account. One might sum for example 37 transactions over one week to determine a net effect, say, –$50 on one's bank balance. When a friend asks, "Has your bank account changed much in the past week?", one might answer, "Yes, I spent $50." This "short answer," as opposed to communicating the details of 37 transactions, corresponds to Ben's shortcuts in the first two trips.

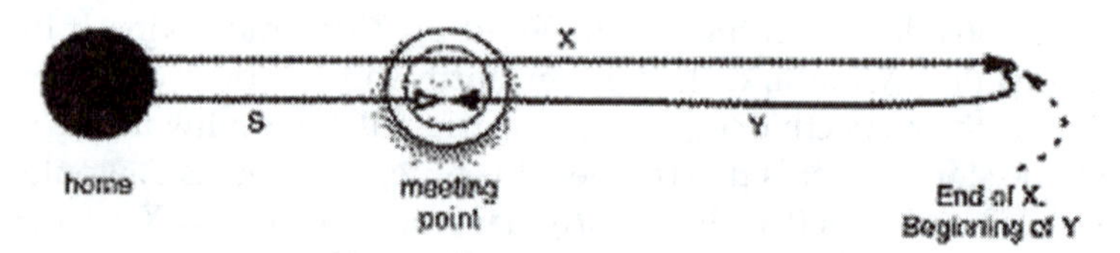

FIGURE 12.10. Equivalent routes I.

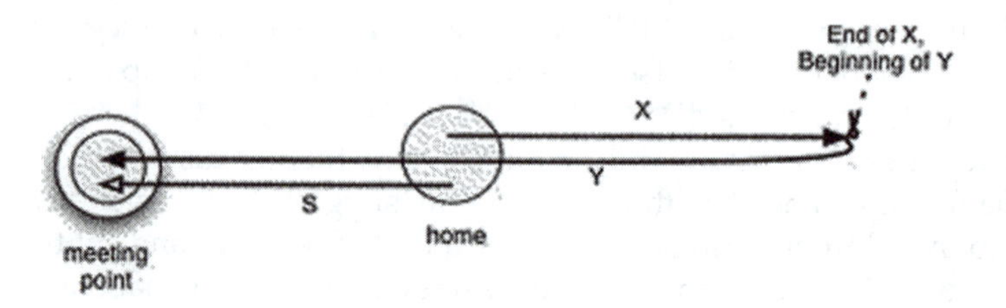

FIGURE 12.11. Equivalent routes II.

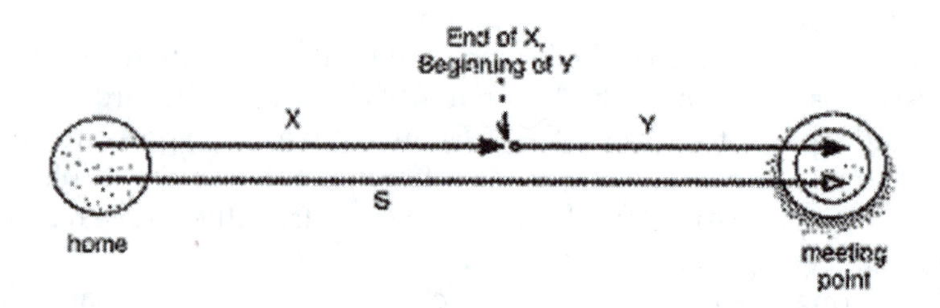

FIGURE 12.12. Equivalent routes III.

Can young children understand the idea of equivalent transformations? The following section details an example of children's investigation of alternative routes and discusses its implications.

EQUIVALENT TRANSFORMATIONS: THE LONG WAY AND SHORTCUTS

The following episode from the Early Algebra Project (2000) demonstrated that third graders could discover shortcuts on the number line (i.e., a displacement that does the work of two consecutive displacements). At the same time, this episode exposes some difficulties in this process.

The Episode

Secret *start numbers* were disclosed to two children, each of whom performed two (known) displacements on the secret number and showed where they ended up on the number line. When Carolina ended at 3 and James at 5 after performing a +2 and then a –1 transformation, several children claimed (correctly) that they started at 2 and 4 correspondingly. At this point, Filipe suggested an interesting explanation:

> Filipe: I was just thinking it because you had a plus two and then you minused the one from the two, and then that was only plus one.
> Barbara: Ohh. Did you hear what Filipe said? He just gave us a shortcut. Did you hear Filipe's shortcut?

With the class-shared experience of "Filipe shortcut," the teacher assumes that the children understood the idea of finding one transformation as a shortcut alternative to two given transformations. The children, however, were more impressed with Filipe's role as the fairy godfather (in enacting one of the problems) than with his mathematical ideas. Thus, despite being exposed to the shortcut idea, several investigations were needed before more children understood it.

In the following task, the children were given a list of starting points. For each point they were expected to find where one would end up following two transformations: –1 and +4 and were also asked to think about any kinds [of] shortcuts you can find to do all of these problems. At first, children perceived the task as requiring the actual performance of the two moves (i.e., the ritual of moving by –1 and +4 was perceived by them as an integral part of finding $N - 1 + 4$, just as a young child sees the counting ritual as an integral part of answering the question, "How many are there?"). Performing the transformation in a different way still needed to be made legitimate. When the input number was 5 and Nathan wanted to give it a try, Barbara (the teacher) was expecting him to suggest a shortcut but, instead, he said:

Nathan:	Eight
Barbara:	How did you figure that out?
Nathan:	I did it on my ruler, cause five minus one equals four. Plus four. One ... two ... three ... four ... is eight.
Barbara:	Okay, ... you did find a use for your ruler. You're using it like a number line, right? You're using all the numbers.

The next task involved pattern identification. Different input numbers were transformed by –5 and +4 and the children were asked to tell what they discover by looking at the input numbers, the corresponding output numbers, and the displacements. Most children could identify the –1 shortcut and the relevant connections. It should be noted that while doing the transformations, children had no trouble moving below zero. It was also interesting to note that some confusion was caused by zero as an input. Zero turned out to be a strange number to operate on, and involved –1 both as an output and a shortcut (i.e., as a point on the number line and as a transformation).

During several investigations of this nature, the children discovered the shortcut concept at different points along the sequence of tasks. Some discovered the repeating pattern but did not see the connection between the emerging shortcut and the two original transformations, that is, between the computation $(-5) + (+4) = -1$ and getting –1 as a shortcut in each specific case. Keeping in mind these were third graders, we can look at the glass half full and conclude that the equivalence concept is attainable, but the teaching trajectory should provide appropriate tasks.

CONCLUSIONS

In this chapter, we suggested that there is an interdependent relationship between algebra and signed number operations. In the first part, we argued that algebra provides a helpful context for introducing signed numbers. In the algebraic modeling of certain situations, one makes full use of operations defined on signed numbers and of the numbers as a combination of measure and direction. Arithmetical counterparts of such problems, on the other hand, do not require the full use of signed numbers. Thus, algebraic problems have the potential to facilitate the construction of a richer mathematical signed number operation model. Modified versions of problems such as the road problem and temperature problem can be used to generate teaching trajectories for introducing signed numbers and for introducing signed number operations. Similar problems can be later used to further enrich the mathematical model by applying it in a variety of situations where signed number operations genuinely contribute to the organization and analysis of these situations.

The second half of this chapter argued that signed number tasks can contribute to the understanding of algebraic concepts. The study of functions is more complete and meaningful with the extension of the number line to include negative numbers promoting the habit of checking different cases. Signed number tasks can help students move beyond the conception that equations display "an action on one side and its result on the other" to an "equivalence relation of transformations." Similarly, signed number tasks can support the transition from arithmetical additive equations that have a part–part–whole structure to algebraic equations by providing an extended equivalence meaning for this structure.

REFERENCES

Arcavi, A. (1980). *Comparison of the effectiveness of four different approaches for teaching the multiplication of negative numbers.* Unpublished master's thesis, Weizmann Institute of Science, Rehovot, Israel.

Ball, D. L. (1993). With an eye on the mathematical horizon: Dilemmas of teaching elementary school mathematics. *Elementary School Journal, 93*(4), 373–397.

Blanton, M., & Kaput, J. (2000). Generalizing and progressively formalizing in a third grade mathematics classroom: Conversations about even and odd numbers. In M. Fernández (Ed.), *Proceedings of the 22nd annual meeting of the North American Chapter of the International Group for the Psychology of Mathematics Education* (p. 115). Columbus, OH: ERIC Clearinghouse.

Carpenter, T. P., & Franke, M. (2001). Developing algebraic reasoning in the elementary school: Generalization and proof. In H. Chick, K. Stacey, J. Vincent, & J. Vincent (Eds.), *The future of the teaching and learning of algebra* (Proceedings of the 12th ICMI study conference, pp. 155–162). Melbourne, Australia: University of Melbourne.

Carpenter, T. P., & Levi, L. (2001). *Developing conceptions of algebraic reasoning in the primary grades* (Research Report). Madison, WI: National Center for Improving Student Learning and Achievement in Mathematics and Science. Retrieved November 2003 from http://www.wcer.wisc.edu/ncislapublications/index.html

Carraher, D. W., Schielmann, A. D., & Brizuela, B. (2001). Can young students operate on unknowns. In M. V. D. Heuvel-Panhuizen (Ed.), *Proceedings of the 25th conference of the International Group for the Psychology of Mathematics Education* (Vol. 1, pp. 130–140). Utrecht, the Netherlands: Freudenthal Institute, Utrecht University.

Davis, R. B., & Maher, C. A. (1997). How students think: The role of representations. In L. D. English (Ed.), *Mathematical reasoning, analogies, metaphors, and images* (pp. 93–115). Hillsdale, NJ: Lawrence Erlbaum Associates.

Davydov, V. (1991). *Psychological abilities of primary school children in learning mathematics.* Reston, VA: National Council of Teachers of Mathematics.

Filloy, E., & Rojano, T. (1989). Solving equations: The transition from arithmetic to algebra. *For the Learning of Mathematics, 9*(2), 19–25.

Freudenthal, H. (1983). Negative numbers and directed magnitudes. In H. Freudenthal (Ed.), *Didactical phenomenology of mathematical structures* (pp. 432–460). Dordrecht, the Netherlands: Reidel.

Gravemeijer, K. (1997). Solving word problems: A case of modelling? *Learning & Instruction, 7*(4), 389–397.

Greer, B. (1992). Multiplication and division as models of situations. In D. A. Grouws (Ed.), *Handbook of research on mathematics teaching and learning* (pp. 276–295). New York: Macmillan.

Greer, B. (1997). Modelling reality in the mathematics classroom: The case of word problems. *Learning & Instruction, 7*(4), 293–307.

Janvier, C. (1985). Comparison of models aimed at teaching signed integers. In L. Streefland (Ed.), *Proceedings of the ninth international conference for the Psychology in Mathematics Education* (pp. 135–140). Utrecht, the Netherlands: Program Committee.

Kieran, C. (1981). Concepts associated with the equality symbol. *Educational Studies in Mathematics, 12*, 317–326.

Linchevski, L., & Williams, J. D. (1999). Using intuition from everyday life in "filling" the gap in children's extension of their number concept to include the negative numbers. *Educational Studies in Mathematics, 39*, 131–147.

Mukhopadhyay, S., Resnick, L. B., & Schauble, L. (1990). Social sense-making in mathematics; Children's ideas of negative numbers. In G. Booker, P. Cobb, & T. N. de Mendicuti (Eds.), *Proceedings of the 14th international conference for the Psychology in Mathematics Education* (Vol. 3, pp. 281–288). Oaxtepec, Mexico: Conference Committee.

Peled, I. (1991). Levels of knowledge about negative numbers: Effects of age and ability. In F. Furinghetti (Ed.), *Proceedings of the 15th international conference for the Psychology in Mathematics Education* (Vol. 3, pp. 145–152). Assisi, Italy: Conference Committee.

Peled, I., Mukhopadhyay, S., &, Resnick, L. B. (1989). Formal and informal sources of mental models for negative numbers, In G. Vergnaud, J. Rogalski, & M. Artigue (Eds.), *Proceedings of the 13th international conference for the Psychology of Mathematics Education* (Vol. 3, pp. 106–110). Paris, France: Conference Committee.

Rowell, D. W., & Norwood, K. S. (1999). Student-generated multiplication word problems. In O. Zaslavsky (Ed.), *Proceedings of the 23rd international conference for the Psychology in Mathematics Education* (Vol. 4, pp. 121–128). Haifa, Israel: Conference Committee.

Schliemann, A. D., Carraher, D. W., & Brizuela, B. M. (2007). *Bringing out the algebraic character of arithmetic: From children's ideas to classroom practice*. Mahwah, NJ: Lawrence Erlbaum Associates.

Shternberg, B., & Yerushalmy, M. (2003). Models of functions and models of situations: On design of a modelling based learning environment. In H. M. Doerr & R. Lesh (Eds.), *Beyond constructivism: A model and modelling perspective on teaching, learning, and problem solving in mathematics education* (pp. 479–500). Mahwah, NJ: Lawrence Erlbaum Associates.

TERC Tufts Early Algebra, Early Arithmetic Project. (2000, September). *Number line conversations about money among students in grade 3* (Classroom transcript). Cambridge, MA: TERC.

Thompson, P. W., & Dreyfus, T. (1988). Integers as transformations. *Journal for Research in Mathematics Education, 19*(2), 115–133.

Vergnaud, G. (1982). A classification of cognitive tasks and operations of thought involved in addition and subtraction problems. In T. P. Carpenter, J. M. Moser, & T. A. Romberg (Eds.), *Addition and subtraction: A cognitive perspective* (pp. 39–59). Hillsdale, NJ: Lawrence Erlbaum Associates.

III

ISSUES OF IMPLEMENTATION: TAKING EARLY ALGEBRA INTO THE CLASSROOM

The previous sections developed the philosophical and epistemological groundwork for algebra in the early grades and instantiated these ideas in more practical terms through empirical classroom studies of children thinking algebraically. But, the early algebra story entails more than an epistemology of content, or even evidence of children's algebraic skills. It involves understanding the complexities of teacher professional development and how curriculum, whether as a prescribed agenda or as an enacted phenomenon, can build children's algebraic thinking. *What* and *how* teachers teach is critical to early algebra reform. In large part, the success of early algebra rests on elementary teachers and their ability to work flexibly with students' thinking and curricular resources to build classrooms where early algebra is a routine part of children's mathematical experience. This section looks at how curriculum and professional development can help teachers bring algebra into the early grades.

Chapters 13 and 14, written from the perspective that one cannot separate learning from either the mathematical or the sociological context in which it occurs, underscore how content and context, respectively, matter in teacher professional development. In chapter 13, Franke, Carpenter, and Battey contrast their professional development in cognitively guided instruction with their more recent work in early algebra to make explicit the particular ways early algebra content affects professional development and how its intrinsic cultural mathematical practices are brought to bear on teacher learning. The authors explore how key principles of their professional development, such as using student thinking as a tool that engages teachers, make explicit their own and their students' participation with the mathematics and provide a trace of the communities' ideas.

Blanton and Kaput (chap. 14), drawing on a district-based research and professional development project, widen the lens to examine the institutional context in which teachers learn and how this setting, including its leadership practice and ability to grow professional communities, can uniquely support teacher development so that algebraic thinking is increasingly supported within the larger district enterprise. They elaborate how affordances of the institutional setting, such as the development of a professional community network, a distributed approach to district leadership practice, or the integration of district professional development initiatives, can support teacher learning.

Chapters 15, 16, and 17 draw on the authors' extensive work in elementary classrooms to look at how curricula can more deeply scaffold algebraic thinking. The authors frame their ideas around a range of perspectives on learning, from developmental psychology to sociocultural theories, in building the case that early algebra is more a way of thinking and acting mathematically (and for the teacher, pedagogically as well) than just a particular set of tasks.

The *Measure Up* Project (Dougherty, chap. 15) builds on Davydoviian and Vygotskiian traditions by using mathematical generality as a starting point for developing students' understanding of structure. In this approach to algebraic thinking, students begin by comparing abstract quantities of physical measures as a way to generalize mathematical relationships about them. Dougherty uses insights from observations of students' mathematical understandings of number, operation, and measurement, to examine curricular and implementation issues associated with *Measure Up*.

In chapter 16, Schifter, Monk, Russell, and Bastable make the case that curriculum can and should deliberately call attention to the generality inherent in arithmetic, build on students' interest and curiosity toward generalizing, and account for how children's thinking develops. The authors explore how children engage with ideas related to commutative, associative, and distributive properties and use this to think about implications for curriculum design.

Goldenberg and Shteingold (chap. 17) examine both generalized arithmetic and patterns and relations in a project rooted conceptually in developmental psychology. They use their work to portray children's natural tendency toward algebraic thinking and to articulate principles of the algebraic trajectory through which the *Math Workshop* curriculum takes students.

The chapters in this section raise our awareness of the complexities of implementing algebra in the early grades. But they also provide concrete details on how to navigate these complexities. What they offer is not only evidence that children can think algebraically, but also how profoundly they can develop this habit of mind when the curriculum draws out (as Mason describes it) the general rather than the particular. They offer ideas about how to design professional development so that elementary teachers, schools, and districts can transform the mathematics of rote skill and procedure to the mathematics of generalization. The reward is students with a deeper, more connected understanding of mathematics who are better prepared for the mathematical years after the elementary grades.

13

Content Matters: Algebraic Reasoning in Teacher Professional Development

Megan Loef Franke
UCLA

Thomas P. Carpenter
University of Wisconsin, Madison

Dan Battey
Arizona State University

Throughout this book, you have been reading about new conceptions of algebraic thinking for elementary school students. These new conceptions of algebraic thinking have implications for how we think about teacher professional development. This chapter examines how our conceptualization of algebraic thinking influences opportunities for teacher learning and what that means for the design of professional development. We draw on our work in cognitively guided instruction (CGI) focused on the development of children's mathematical thinking primarily around operating with whole numbers to make sense of what it would mean for teachers to learn about developing algebraic thinking. We situate our views of professional development focused on the development of students' algebraic thinking in relation to those of our whole number operation work and we accomplish this by drawing on our years of research on teacher learning and professional development, the work of our colleagues, our own practices as we attempted to engage students in the algebraic ideas, and our time in teachers' classrooms as they figured out how to develop algebraic reasoning. Particularly, the change in terrain has allowed us to see where differences in the mathematical content of the professional development influences opportunities for teacher learning and impacts how we design professional development.

This chapter argues that general principles outlining how to design professional development, although potentially helpful, are not sufficient for meeting the needs of teachers.[1] Our work over the last 20 years consistently highlights the complexity of professional development and the necessity of understanding the details around engaging teachers in learning opportunities that lead to generative growth. Our algebraic thinking work provided the opportunity to think carefully about the role of the mathematical content in the design and implementation of professional development. We wondered about how the algebraic thinking professional development work would differ from our earlier whole number professional development. As we have engaged in professional development, and studied hundreds of teachers and their students, we have found there are differences in what we do. This chapter details some of the critical issues and how we have addressed them. We begin the chapter by characterizing our whole number professional development and contrast it with our algebraic thinking work. We use the contrast to highlight the differences and similarities we see in conceptions of student thinking, teacher content knowledge and stance toward the content, the existing cultural practices around teaching the content, and the way the content plays out in the sociopolitical context.

EARLY CGI CONTENT: BASIC NUMBER CONCEPTS

The initial CGI work focused on the development of basic number concepts. At the time we developed CGI in the early to mid-1980s, there existed a substantial body of research that provided a consistent and coherent picture of the development of basic number concepts (Carpenter, 1985; Carpenter, Fennema, Franke, Levi, & Empson, 1999; Fuson, 1992). This research documented that most children enter school with a rich store of informal knowledge and problem-solving strategies that can serve as the basis for developing much of the mathematics of the primary school curriculum. Building on these intuitive problem-solving strategies both enhances students' problem-solving abilities and provides a basis for constructing meaning for addition, subtraction, multiplication, and division concepts and procedures.

---

[1]See Loucks-Horsley, Hewson, Love, and Stiles (1998) for an example of these types of general principles for professional development.

Table 13.1
Subtraction Problem Situations That Generate Different Strategies

1. Twelve children were playing on the playground. Five children went home. How many children were left on the playground?
2. Viviana has $5. How many more dollars does she need to save to have $12 to buy a basketball?
3. Raymond earned $5 babysitting. When he put it with the money he had already saved, he had $12. How much money did Raymond have before he earned the money babysitting?
4. Marsha scored 12 points in the class basketball game. Anisha scored 5 points. How many more points did Marsha score than Anisha?

Detailing Problem Action and Relations

One principle underlying our model of students' mathematical thinking is that children naturally solve problems posed in real or imaginary contexts by representing the action and relations described in the problems. Thus, in order to understand how children think about and solve a specific problem, we needed to be explicit about the types of action and relations that distinguish different problems and corresponding student solution strategies (see Carpenter, Fennema, & Franke, 1996; Carpenter et al., 1999). The four problems in Table 13.1 illustrate some of the critical distinctions among problems that result in different solution strategies. All of the problems, however, can be solved using a range of strategies from direct modeling through recall.

Although all four problems could be solved by subtracting 5 from 12, young children consistently use quite different strategies in solving them. A first-grade student might solve the first problem using counters by making a collection of 12 counters and taking 5 from it. The same child might solve the second problem by first making a collection of 5 counters, adding counters until the total reached 12, and then counting the number of counters added to figure out the answer. The strategies are quite different, but in each case the strategy directly models the action described in the problem.

Over time, children become more flexible and begin to abstract these strategies to make them more efficient. For example, a child might solve the second problem using a counting strategy. The child would start counting at 5 and count up to 12, keeping track of the number of counts on her fingers. Children move beyond modeling the problem and using related counting strategies to rely directly on number facts. Children often learn doubles (e.g., 6 + 6, 9 + 9) and sums to 10 (e.g., 4 + 6, 8 + 2) earlier than other facts, so they may use number facts they already know to generate solutions to the problems in Table 13.1 as follows: "Five and 5 is 10 and 2 more is 12, so 5 + 7 is 12. So the answer is 7." The same progression characterizes the development of basic multiplication and division concepts and addition and subtraction with two and three digit numbers. Children extend their strategies to larger numbers by using units of 10 to model addition and subtraction involving two- and three-digit numbers.

Critical Features of the Content for Teacher Learning and Classroom Practice

Teachers engaged in the CGI early number work readily adapted to the idea that children come to school with a range of intuitive correct ways of solving problems that can be built on in developing skill and understanding. Because strategies children bring to school are reflected in different word problems, teachers need to understand the differences across word problems, how the differences are reflected in children's solutions, and then how children abbreviate these strategies to construct increasingly efficient, abstract, and sophisticated strategies. The evolution of strategies occurs without too much specific intervention by teachers. Certainly, teachers can scaffold the progression, and as teachers become more skilled they do, but teachers can accomplish a great deal in supporting student learning by posing a variety of problems and asking students to share a range of strategies. Teachers can support students by asking questions that focus on the details of the strategy students used and thus provide scaffolding for students to construct well-conceived sequences of strategies without worrying too much about the sequencing of problems.

The number concepts content we address directly relates to the early grades curricula already in place in most schools. However, the approach we took that focused on solving problems in context where the actions and relations were made more explicit stood in contrast to traditional curricula and many teachers' typical practices. Most often, in existing curricula, the word problems were relegated to the end of the chapter as an application task or an assessment measure. Word problems were not often used as a way to introduce a concept. So teachers would not have had much opportunity to engage in the practice of using word problems as a way to introduce concepts and so it required the development of new practices and the adaptation of existing ones. The CGI whole number content asked teachers to think differently about the relationship between skills and understanding and about student development.

Teachers felt confident they knew enough about the whole number content to engage with their students, but were often worried that they could not catch student strategies in the rapid real-time interchanges of their classrooms, especially initially when the strategies seemed so foreign to the teachers.

CONSIDERING ALGEBRAIC REASONING

Our focus on algebraic thinking in the elementary grades again draws on the research around the development of student thinking.[2] Here we are particularly interested in the ways students come to reason relationally, making use of the fundamental properties of arithmetic. Our goal in focusing on algebraic thinking is not to move high school algebra into the elementary grades, nor is it to cover all of algebra. Rather, we attempt to reconceptualize algebraic thinking for teachers and create a way to support teachers to engage in algebraic reasoning in elementary school (as do many authors in this book). We focus on extending arithmetic in ways that build algebraic reasoning (Carpenter, Franke, & Levi, 2003; Faulkner et al., in press; Kaput, 1998). We want to help teachers do more with arithmetic, to extend these ideas for themselves and for their students in ways that engage them all in algebraic thinking. Our focus on the development of elementary school children's algebraic reasoning is on the ability to generate, use, represent, and justify generalizations about fundamental properties of arithmetic. Generalizing and representing generalizations involves the articulation and representation of unifying ideas that make explicit important mathematical relationships. We focus on four primary themes: equality as a relation, relational thinking, articulating fundamental properties of number and operations, and justification.

Equality

Kieran (1992) characterizes the distinction between arithmetic thinking and algebraic reasoning as a shift from a procedural perspective of operations and relations to a structural perspective. One of the hallmarks of this transition is a shift from a procedural view to a relational view of equality, and developing a relational understanding of the meaning of the equal sign underlies the ability to make and represent generalizations. Behr, Erlwanger, and Nichols (1980), Erlwanger and Berlanger (1983), Kieran (1981, 1989), and Saenz-Ludlow and Walgamuth (1998) have documented, however, that children in the elementary grades generally consider that the equal sign means to carry out the calculation that precedes it, and this is one of the major stumbling blocks when moving

---

[2]See, for example, the work of Bastable and Schifter (chap. 6, this volume); Carraher, Schliemann, and Schwartz (chap. 10, this volume); Chailkin and Lesgold (1984); Clement (1982); Collis (1975); Driscoll (1999); Kieran (1992); Schifter (1999); and Tierney and Monk (chap. 7, this volume).

from arithmetic to algebra (Kieran, 1981; Matz, 1982). We have found similar results. When asked to solve a problem like 8 + 4 = _ + 5, students want to put 12 in the box. Some want to include the 5 in their total, so they put 17 in the box. Others create a running total by putting a 12 in the box and an " = 17" following the 5. In our work, we help teachers to understand the range of solutions students provide for problems like these and address what their solutions tell us about their understanding of the equal sign. We draw on Davis' (1964) work and introduce true–false number sentences (true or false: 7 = 7) as a tool to help challenge students' notions and create an understanding of the equal sign as a relation meaning "the same as."

Relational Thinking

A fundamental goal of all of our algebraic thinking work focuses on helping teachers understand and engage their students in relational thinking. By relational thinking we mean examining expressions and equations in their entirety rather than as a process to be carried out step by step. Doing so requires using fundamental properties of number and operation to transform the mathematical expression. For instance, in solving 78 + 34 – 34 = _, knowing that 34 – 34 = 0 and starting there rather than going in a linear fashion and starting with 78 + 34 involves relational thinking. Using relational thinking requires an awareness of properties of number but not the formal definition.

We begin building ideas of relational thinking with the equal sign and continue building them as we work on conjectures, generalization, and justification. If we return to the problem 8 + 4 = _ + 5, our goal is not only to have children calculate correctly 8 + 4 = 12 and 12 – 5 = 7 so a 7 goes in the box, but also to solve the problem by looking across the equal sign at the relationship between the 4 and the 5. Here, students see that 5 is one more than 4, so a number one less than 8 must go in the box to make the equation true. The student implicitly used the associative property of addition to transform the equation.

Ideas of relational thinking fit well into the arithmetic curriculum because they often make computation easier and provide an opportunity to make explicit much of what students are doing when they operate on numbers using standard algorithms. We see this when students add multidigit numbers and use an invented algorithm, for instance, on the problem 25 + 37 a child may respond, "I know that 20 and 30 is 50. Then I added the 5 and the 7. That's like 5 and 5 is 10, and 2 more. So that's one more 10, so 60 and 2 is 62." Students also can be seen engaging in this type of thinking when solving a multidigit multiplication problem. Take the problem 28 × 5. The child can multiply 20 times 5, figure out that is 100, then multiply 8×5, and know that is 40 getting 140. In the first example, the child is using repeated applications of the associative and commutative properties. In the second, the child used the distributive property. We are not suggesting that the children explicitly recognize that they are using these properties at this point. The next section discusses making these properties explicit.

Relational thinking is not learning a set of computational tricks or memorizing a set of mathematical properties. Relational thinking is about reasoning; understanding why particular transformations are possible entails understanding important ideas about the relations between operations and the fundamental properties. Many students do not have the opportunity to engage in this type of thinking in elementary school. They do not understand how basic properties of number operations are applied in their computations and, as a consequence, they do not recognize that arithmetic and algebra are based on the same fundamental ideas (teachers also do not see this connection). By failing to take advantage of the structure of the number system, the learning of arithmetic has been made much harder.

Articulating Fundamental Properties of Number and Operations

When students make conjectures about properties of numbers or operations, they make explicit their mathematical thinking. Conjectures provide a class with fundamental mathematical propositions for examination and open up students' thinking for analysis and discussion. Students have a great deal of implicit knowledge about properties of arithmetic operations, but they generally have not made their thinking explicit and the object of discussion in ways that provide opportunity for systematic examination and detailing of their ideas. We have found that many opportunities arise in the context of classroom practice that can provide entry into the conversation. However, this requires being able to notice and take up opportunities in ways that support the development of the conversation. We have also helped teachers to think about how to seed conversations where students make conjectures about properties of number and operations. One way to seed the conversation is to pose a true–false number sentence such as $16 + 9 = 9 + 16$ and follow up with questions about whether this would be true no matter what numbers you used. This provides students an opportunity to make explicit their ideas about commutativity and reason about how they know it is true and when it is true. We find that students need the opportunity to return to these conversations over time as they continue to challenge and deepen their understanding.

Justification

The most challenging algebraic thinking work for teachers and students involves justification of conjectures—answering the "How do you know that would always be true?" question. Here we focus on what counts as an adequate response and a more sophisticated response to statements such as $a + b = b + a$. We help teachers come to see the range of types of responses and the evolution of student thinking in response to these kinds of questions.

Balacheff (1988) specifies justification as discourse that aims to establish for another individual the validity of a statement and proof as an explanation that is accepted by a community at a given time. In negotiating norms for what counts as a justification in an elementary class, it is not, however, the case that anything goes, and classes exploring the use of justification have to negotiate the norms for what counts as a legitimate justification.

Consistent with the analyses of Sowder and Harel (1998) and others, we identify three general schemes for justifying mathematical propositions: appeal to authority, justification by example, generalization. At the lowest level, appeal to authority, students use or accept as justification sources other than their own reasoning. The outside source may be an authority such as a teacher, textbook, a ritual form of argument, or a meaningless manipulation of symbols. Students who make their justifications by offering examples are using empirical examples-based proof schemes. Examples illuminate an idea and can lead to higher levels of justification. However, justification based on examples can be misleading. When students rely on empirical examples they do not consider all cases necessary for more principled generalization. Generalizable arguments attempt to show that a claim is true for all relevant cases.

Naive generalizable arguments may essentially be based on assumptions that a given mathematical idea is self-evident: "When you add zero to a number you always get the same number." But students can learn to appreciate that they need to provide generalizable arguments for their ideas that are based on a set of accepted assumptions. For example, a student may argue that multiplication is commutative by showing that a rectangle can be rotated and so 3×4 and 4×3 represent the same number. This example only applies to specific numbers, but the argument is generalizable to any array. Thus, although elementary students lack notation to represent generalizations about numbers, they can engage in argument that represents general forms of justification.

Teachers may employ all the levels of justification described previously, and a goal of our professional development is to help teachers to examine their own proof schemes in the process of considering the types of arguments that students may employ.

Critical Features of the Content for Teacher Learning and Classroom Practice

The content of algebraic thinking we engage teachers with is quite different from the early number concept work of CGI. Equality focuses on a prevalent misconception. Teachers must consider how to challenge existing notions about equality, not build on what students know about it (although they can build on what they know about number relations). Using relational thinking and articulating conjectures build on and provide opportunity for students to use valid intuitive knowledge. However, these ideas of relational thinking and conjecture are not explicit aspects of the typical elementary mathematics curricula and as such are not areas where teachers have had much opportunity to hear students work with the ideas. It is often difficult, then, for teachers to initially see how to incorporate the algebraic ideas in the required mathematics curriculum. Moreover, justification involves getting students to engage in a practice that is often unfamiliar. Here students are asked to challenge what they see as a perfectly good form of justification (by example) to engage with other more general forms of argument that apply to all numbers.

The way we have chosen to focus our algebraic thinking work also has implications for the types of conversations that support the development of student thinking. In our early number work, the classroom conversations generally focused on sharing and detailing strategies. The interactions revolved around establishing norms for deciding whether two strategies were different and how strategies were alike. In our algebraic thinking work, conversations are about reasoning together around whether a mathematical idea is always true, or under what parameters it is true. These types of conversation require students to put forth arguments and support them, challenge each other's arguments, and continue the conversations over time. Students need to find ways both to agree and disagree with each other, to take on the truthfulness of mathematics and look beyond a set of procedures or strategies. Thus, algebraic thinking requires different work by the teacher and different kinds of knowledge and skills to support that work.

CONTENT AND PROFESSIONAL DEVELOPMENT

Although it may seem clear after reading the different descriptions of the mathematical content that varying opportunities for professional development would be necessary, we did not recognize when we began this work the extent to which this would be true. We purposely set out to keep some elements of our professional development the same and to adapt other aspects of our work with teachers. It is only after years of work with teachers and examining carefully the impact on both teachers and students across schools and communities that we have a better understanding of the professional development that can support teachers—particularly teachers in urban schools—to make sense of algebraic thinking within both the contexts of professional development and their classrooms. So, here we articulate our view of professional development generally (which itself continues to evolve and has evolved since our earliest CGI work) and then contrast our professional development work across the respective content domains of number concepts and algebraic thinking.

General Conceptions of Professional Development

For us, offering professional development has always involved offering ongoing opportunities for learning connected to the practices of teaching. As we plan professional development, we design space for teachers to come together with us and explore ideas of students' mathematical thinking, content, and pedagogy, where we can all make our practice public and develop ways of learning from each other (Little, 1999). In conceptualizing how to create these opportunities, we draw heavily on notions of apprenticeship within communities of practice, the development of artifacts to support learning, the role of language, the respective identities of those coming to the work, and the political and social context in which the work occurs (Greeno & Middle School Mathematics Through Applications Project, 1998; Lave, 1996; Rogoff, 1997; Wenger, 1998; Wertsch, 1998). We work to develop relationships with teachers and support the development of relationships of teachers with each other (particularly relationships around the work of teaching and learning mathematics). We appropriate and develop artifacts that open opportunities for participation by as many participants as possible (Wenger, 1998; Wetsch, 1998).

Our professional development focuses on the details of student thinking, mathematics, and classroom practices. However, details without structure that allow one to make sense of them and the relations among the ideas they represent, leaves little opportunity for continued learning. So we look for ways to provide and enable teachers themselves to create structures for making sense of the details. We pay close attention to who the participants are, what their experiences have been and why and how they see themselves both learning and contributing to the learning of the groups. We help teachers make explicit their views and their histories. We take seriously what teachers' share and use their ideas and experiences as opportunities for ourselves to reflect on the learning of the group and to make sense of the extent we are providing opportunities for learning that can be taken up in ways that are productive for the participants.

Contrasting Approach to Professional Development

An initial comparison between the structure and substance of the early number professional development and the algebraic thinking professional development highlights many similarities. In both cases, we structure the professional development to work in ongoing ways with teachers and to create overlap between the professional development sessions and classroom practice. In both cases, we attend to the details of students' mathematical thinking as a focal point of the professional development work, and we find ways to help teachers organize the relations between different aspects of student thinking, as well as different mathematical ideas. We create opportunities to challenge teachers' existing notions about students' mathematical thinking, the content itself, and their classroom practices. However, because the content differs, we find some very specific differences in how we have conceptualized teacher learning and professional development. We have noted ideas particular to our focus on algebraic thinking that led to adaptations to professional development. There are other ways in which one must more generally examine any content domain in designing professional development (considering the existing cultural practices for instance). We start first with examples of the specific adaptations we noted based on our conception of the mathematical content. We provide two specific examples of how the mathematical content matters in professional development. The first example relates to adapting existing practice and focuses specifically on posing problems. The second addresses issues of teacher content knowledge and specifically teachers' stances toward the content.[3]

Posing Problems

In our number concept professional development, teachers learn about the range of strategies elicited by particular problem types. They learn

---

[3]These differences apply specifically to our focus on algebraic reasoning. We are not suggesting that the same analysis would apply to other perspectives. In fact, our arguments that content matters specifically imply that different conceptions of algebraic reasoning would entail different analyses of professional development.

about how problem types build on each other and what progressions of strategies look like. And while we may focus on the progression, we find that teachers engage in much productive work with students by posing a single problem to students and getting students to detail their thinking. Where teachers put much of their effort here is in supporting students in detailing their thinking and explaining it to others. Whereas some teachers attend to construction of a careful sequence of problems, many teachers do not. We have observed teachers use a range of approaches for sequencing problems, all with positive outcomes for students. Sequencing of problems turns out to be less important in this case. Students are able to make connections within and across problems without careful sequencing. So, to get started in the number concept work, teachers could pose a single problem, then work on improving how they support students' thinking. They could draw on their many experiences with posing single problems to students. So productive problem posing begins quite readily in number concept professional development.

Our algebraic thinking work demands a different type of problem posing that does not build on existing practices of teachers and demands careful attention to the problems themselves. Because the focus of our algebraic thinking work is not on detailing a specific strategy (and its relation to other strategies), but rather reasoning about a key mathematical idea, multiple problems are often necessary to engage in the discussion and the sequence of the problems can be used to highlight elements of the mathematical idea and/or the students conceptions of the mathematical ideas. The sequence of tasks and the interaction around those tasks as a community become very important to developing algebraic thinking. Take for example $8 + 4 = \_ + 5$. If the students put a 12 in the box because they believe the answer comes next, one might pose a true–false number sentence like $7 = 3 + 4$. If the students respond that it is backward and you cannot write it that way, one might pose $7 = 7$ or $7 + 0 = 7 + 0$. The idea is that the number sentence posed is intended to challenge students' ideas and, although many different number sentences and sequences of number sentences could do this for students, working on the algebraic ideas does require thinking about sequence, not about posing a single problem and letting students work through it. Focusing on the sequence of problems requires the teacher to think ahead while keeping the mathematical idea central and at the same time listening to the students. This requires quick, real-time decision making while also orchestrating conversation among the students (which also may be a new practice).

Creating a sequence of problems to engage in algebraic thinking occurs not just in working through ideas about the equal sign, but also in working on ideas of relational thinking and creating opportunities for students to think relationally requires opportunities for conversation and that conversation can be productively scaffolded by choosing problems that particularly draw out children's current thinking and challenge them to think relationally. We saw one such sequence in a situation where a student had developed good computational strategies for solving problems like 43 + 28 = _ + 42 (for a full account of this interaction, see Carpenter et al., 2003). The student needed to work with problems that would encourage her to not use those computational strategies and think across the equal sign. The teacher posed a several similar open number sentences, which the student solved by computation. In order to encourage the student to consider an alternative strategy, the teacher posed the following problem in which the relations are more transparent 15 + 16 = 15 + _. The student saw the relation and recognized that she did not need to calculate. This was followed by a problem similar to the ones the student had solved by calculating (28 + 32 = 27 + _). With a little support, the student soon realized that she could also use relational thinking for this problem and for all the problems she had solved earlier. Here the teacher used a sequence that included a critical problem that pointed to the relational aspects of the number sentences. The student was not showing any progress in moving to use relational thinking until this key problem was posed.

Teacher Content Knowledge and Stance Toward That Content

The fact that algebra in the form of generalizing, representing, and justifying arithmetic ideas proves to be a more challenging content domain than number concepts for elementary teachers would not be surprising. However, we find that whereas content knowledge is an issue within both domains, there are particular differences across the domains in the opportunities that teachers have had to deepen their knowledge and in the stances teachers take to that knowledge.

In our professional development work, we have watched while teachers worked through the content. Teachers not only want to work on solving the problems themselves, but they also want to think about how to elicit that range of thinking in their classrooms, how to support students in making progress in their thinking, how to think about sequencing problems in productive ways, and how the content of algebraic thinking is connected to other mathematical ideas that we work with in our classrooms. Here is where we see the need for engagement around the content of algebraic thinking. Instead, we want to support teachers to deepen their understanding in ways that allow them to design the sequences of problems they want to pose, to tweak the follow-up questions that they ask students, and to notice when opportunities for algebraic reasoning occur within their ongoing mathematics work. So, whereas the professional development needs to address the content knowledge of teachers in algebra and early number content, there are differences in the support teachers need to teach them. In particular, there are differences in what teachers need to know to organize their understandings and make explicit the relationships across students' thinking and problem sequences. Engaging in this type of mathematical work requires deep understanding of the content and because elementary school teachers have had less opportunity to work themselves or with their students around algebraic reasoning, often teachers have developed different stances toward the content and require more support.

Teachers' stance toward the content makes a difference in the practices they appropriate, the ways they go about learning the content, and then how they teach the content. Most teachers we have worked with do not see themselves as knowing algebra or as ever having needed to know algebra. And although we ran into issues of content knowledge in our earlier CGI work, particularly around the structure of the base 10 representation system, the teachers seemed to feel confident that they could master the content issues that might arise as they taught that content. Teachers engaged in algebra lacked that confidence and worried about being able to productively engage students in algebraic thinking. This became exceedingly clear in the very first algebraic thinking professional development session we carried out in Los Angeles. We began by working with a group of grade K–5 teachers experienced in CGI as part of ongoing professional development. These teachers were all from the same urban elementary school (a low performing, predominately ELL school with over 1,400 students) where we had worked together for more than 3 years on CGI. The teachers as a group were skilled in talking about students' mathematical thinking. They worked in their classrooms to develop mathematical understanding by building on their students' existing mathematical ideas; they could talk with and learn from one another, and they were developing confidence in themselves as mathematics teachers and advocates for their students.

We were struck immediately by how the teachers responded to the algebra professional development. The teachers were both anxious about the algebra they anticipated engaging in and excited about doing something new. We started with the equal sign. We asked them to predict how their students would solve some problems of the sort discussed earlier (e.g., $8 + 5 = \_ + 9$) and then shared with them data on one of the problems from first through sixth graders. The data that revealed students' limited procedure-based understanding of the equal sign made them stop and think, and they raised a number of options for addressing the equal sign in their classrooms. But the tone of the interactions changed when we seeded a conjecture about commutativity. The teachers worked to create a written conjecture, edit it, and then justify it. As we circulated among the small groups working together, teachers whispered to us things like, "I'm not sure we know what you want" or "You know, I am not too good at math," and so on. But the teachers worked diligently on the justification with those sitting near them. The looks on their faces when they found they could justify it, were of pride and amazement. They surprised themselves that day with their math knowledge.

We have continued to see teachers surprise themselves with how much algebraic thinking they are capable of. However, many teachers remain hesitant. They do not see themselves as strong mathematically. Some see being strong mathematically as being able to give answers quickly, immediately knowing what to do, and knowing what the rules and procedures are—and, according to this criterion, some teachers are not strong mathematically. We have teachers who come to the algebra work with views of mathematics as a predetermined set of procedures to be memorized with confidence that they know they can share with their students. We have other teachers who fear algebra. They see algebra as something they have never been and never will be good at. They see it as an abstract manipulation of symbols that does not make sense. The algebra work seems to bring out a level of anxiety that does not exist in our earlier work. It also brings out more of a rule orientation to mathematics.

Considering the particular content knowledge that teachers need, along with a detailed understanding of how to use that content knowledge in teaching, as well as the stance toward that content knowledge that teachers bring to the professional development, we have made particular adaptations to our professional development. In particular, we address how the teachers see themselves in terms of the content, we create norms to help them feel comfortable with engaging in content conversations and to see what they do know about the content, and we support the creation of tasks that help teachers develop their knowledge in practice.

Issues of Teacher Learning

Beyond these two specific examples of the relationships between professional development and content, we have also considered some more general issues related to teacher learning that we take into account when we design professional development that play out a bit differently depending on the content addressed: existing cultural practices, the appropriation of practices, and the social and political context of the work.

Cultural Practices

Various communities have their own long-established ways of doing the mathematics. These cultural practices form the basis for how students and teachers engage with the content and highlight the values, beliefs, and knowledge that teachers, students, and families bring to the mathematical work. Understanding these cultural practices and making them explicit then enables us to make them sites for teacher conversation. We need to understand how particular cultural practices may interact differently with different mathematical domains to create a more subtle adaptation to the design of our professional development.

Solving multidigit addition and subtraction problems provides a good example of a cultural mathematical practice related to our number concept work. Teachers and students engage in solving multidigit subtraction problems by using the standard regrouping algorithm; this is the way to solve multidigit problems. The procedure defines the mathematical norm, what's right, explicitly and completely. The standard algorithm is passed on from older brother to younger sister, from parent to child, from teacher to student, and so on. It exists both inside and outside the school (although different algorithms are standard in different countries). We have heard students describe how their parents made them erase all the mathematical work from their homework page and do the standard algorithm, so they can turn it in right. Teachers ask us again and again to show them how to teach the algorithm right, as if we have a simple and elegant way to teach it so that every child, if they listen, will get it the first time.

The standard algorithm for regrouping exists across settings and is viewed as the method to use—it is what we all know to do. The standard algorithm for regrouping serves as the basis for engaging in practice around multidigit addition and subtraction, and engaging teachers in thinking about alternative invented algorithms—as we do in our work—must take these long-established practices into account by making them explicit sites for conversation and learning.

Algebra carries its own set of standard cultural practices. Algebra is often seen as bounded by rules; algebra involves doing a series of steps and making sure you do not forget any: "Do the same thing to both sides." So many of the teachers we work with in solving problems themselves repeat this mantra and cannot solve problems in any other way. The cultural practices around algebra are also evident in how we notate and how we use language. Multiplication is notated as $3a$ instead of $3 \times a$. The word *equal* means the answer comes next, whereas in our work we use the words *the same as* for the equal sign to highlight the relational aspect of the symbol. The notation and language associated with the content plays out in how teachers talk about algebra, how our professional development conceptualization of algebraic thinking may be different from the norms of teacher communities, and how those coming to professional development may be used to different sets of norms and language.

But, maybe even more important than these particular culture practices, algebra carries with it a sense of who can participate. Smart kids do algebra and smarter kids do algebra earlier in their academic career.

Teachers report that their students do not yet have the skills necessary to do algebra. Girls tell us it is okay if they are not in the group doing algebra because they are not ready yet and need to be in a group that goes more slowly, because algebra is hard. Parents who hear that their child is learning algebra are proud and hopeful.

The culture around doing mathematics helps determine the stance that teachers and students take toward the work of algebra, and so our understanding of this culture of algebra should influence how we design learning opportunities.[4] We must consider the practices that exist in and outside of the classroom and school, as well as the notation and language used to support the practices, and use them to shift the substance of the conversations in professional development.

Teacher Appropriation of Practices

As teachers begin engaging with new ideas in the professional development, they immediately begin to wonder how to make sense of the ideas in terms of their classroom practice. Teachers begin to understand the development of students' mathematical thinking and want to figure out how to make use of it in their classroom practice. As teachers do this work, they immediately draw on those practices available to them at that time, practices the teachers have used for many years with many different students, practices that have been refined to accomplish particular goals under particular conditions. Figuring out which teaching practices can support the development of student thinking or algebraic reasoning constitutes significant mathematical work. Teachers must draw on their existing practices, tweak them, and adopt new ones. And the practices that support the development of student thinking in place value may not be the same as those that support students' algebraic thinking. We have learned in our studies of teachers across the content domains that there are some significant differences in the practices that support the development of student thinking, and we have learned that the kinds of adjustments teachers need to make to their existing practices are also different. We have seen this particularly in the kinds of questions teachers ask, and the ways that existing practice influence how teachers implement the ideas of the professional development in their classrooms.

---

[4]We have chosen here to highlight the aspects of the cultural practice that are challenging for us as professional developers. There are also ways in which existing cultural practices can be quite productive.

Teacher Questioning

In our number concept professional development, we spent time providing opportunities for teachers to develop skill and understanding around how to elicit student thinking. As teachers developed a sense of the potential range of student strategies and what to listen for, they also developed ways of questioning to elicit the strategies. Teachers saw eliciting student thinking as challenging work until they developed questions that would support students and learned how and when to ask them. Most of the questions the teachers asked students required students to explain a strategy, to in some way explain to the teacher what they did.

In our algebraic thinking work, again teachers elicit student thinking, but here teachers are not trying to elicit what strategy students' used, but rather they need the student to explain more about why they did what they did. Take, for instance, the problem we have discussed throughout the chapter, $8 + 4 = \_ + 5$. Often teachers begin by asking, "How did you solve the problem, tell me what you did." This will productively get the student to explain how they added the 8 and the 4. If they put a 12 in the box, then more questions are necessary to find out why they put 12 in the box. If they put 7 in the box, asking students to explain how they got to the 7 will not help students who compute both sides to think relationally. The same would be true for a problem like, $75 + 28 - 28 = \_$. Teachers may begin by asking, "How did you do that?" This is a question they have asked before and they appropriate here. And, whereas this question begins to get students to talk about their thinking (I took 28 – 28 and got zero and then 75 – 0 is 75), it does not as readily lead students to thinking about a conjecture they might be able to make about the properties of zero. Over time, the teachers come to ask more often, "How did you know that? Will that work for all numbers?" These questions get students to reason why particular approaches work and lead to potential for generalization. However, figuring out how and when and what questions to ask is not simple and takes significant work on the teacher's part. It is not about learning a predetermined set of questions to ask. It is figuring out how these questions fit with and do not fit with the questions they already ask, which of these are best asked when, of which students, under which conditions, and so on. Having questions to draw on helps the teachers get started in engaging students in algebraic thinking, but many of the questions they ask need tweaking to get at the algebraic ideas we address.

How Existing Practices Influence Teachers' Implementation

When working with teachers who have little experience in eliciting student thinking, we found that their existing practices lend themselves to not eliciting or engaging in conversations around the ideas of algebraic thinking. Often the existing practices of teachers do not include ways of getting at student thinking, but do include many ways of getting at students' answers and making sure students get the correct answer. One of our biggest professional development surprises came when we worked with a group of teachers on the equal sign and true–false number sentences and they turned our jointly constructed lists of potential true–false number sentences designed to challenge students' notions about the equal sign into a worksheet. The teachers were very concerned about the data around students' understanding of the equal sign and they had collected their own data only to find their students also had many incorrect notions about it. The teachers wanted to "fix" this situation immediately. We discussed true–false number sentences and created together sequences we could use to challenge these notions; we watched conversations other students had as they engaged with the true–false number sentences, and we talked about the kinds of discussions that may ensue. Yet, after leaving the professional development, teachers relied on existing practices to help them. They created worksheets of true–false and open number sentences so they could practice getting the correct answers to problems like $8 + 4 = \_ + 5$. The teachers used the worksheets for continued practice. They were collected, graded, and returned to the students. The worksheets were not used to promote discussion, to challenge existing ideas, or to figure out why 12 would not go in the box. The worksheet was an artifact of existing practice. The teachers used the sequences we discussed in class in the context of an existing practice and changed the purpose of the number sentences themselves. Here the teachers appropriated a practice that was not helpful in relation to developing algebraic thinking, particularly relational thinking. In retrospect, we should not have been surprised by this. The work around the equal sign did lend itself to this existing practice, and it was a case where teachers saw themselves as having to get to it immediately because they were appalled that their students were getting these problems wrong. We need to recognize the teacher's need to quickly address the issue as well as the practices they have to draw on, discuss them, and create tools to support different practices teachers can use.

The Social and Political Context of the Classroom and School Community

Certainly the way that the broader community defines what is important in mathematics makes a difference in how teachers come to the professional development and make sense of it in their classroom. We worked in one district where the district administrators were thrilled to have us because they saw the importance of engaging elementary school students in algebra.

They had a large high school drop out rate and high failure rate in algebra. They saw our work as a way to solve their problems. However, the district as a whole had a procedural view of mathematics, and their goal was to raise test scores. They had done little work in developing mathematical understanding. We began our partnership by working with all teachers in one elementary school within this district. Because the school had a history of teachers alienated and not working together, our work there was a struggle. We had designed professional development where teachers needed to work together and make their practice public. We learned a great deal in this setting about how teachers appropriate existing practices that are not productive for engaging in algebraic reasoning. We also learned how algebra opened administrative doors for our work, but at the school level, teachers' struggles to see how algebraic thinking fit; the skills-based testing and accountability agenda of the school and district made our work more challenging.

We also work in districts where the curricula is mandated and the teachers are held to a daily pacing plan with benchmark accountability measures at the end of each academic quarter. In this environment, seeing the "fit" of the professional development work with the pacing plan and assessments proves challenging. This has not been the case in our number concept work, where all teachers see the operations as central, and the strategies are so robust that almost regardless of what teachers do, some of the strategies arise as teachers work through these problems with students. Teachers may not see how central the role of word problems is, but they are already in the textbooks quite explicitly so our work requires supporting teachers to see how to make use of them as resources for the development of children's number and operations concepts.

In our algebraic thinking work, fit becomes an issue that requires attention and a different kind of work. Teachers need support in noting where these ideas emerge. We have found with the algebra work that teachers have to strive to make space for algebraic thinking in their classrooms. We are asking teachers to notice when opportunities arise in doing arithmetic to extend these ideas to algebraic thinking. This happens when writing number sentences and taking the time to talk about the equal sign; it happens when asking how many legs 6 cows have, and the students all solve the problem by making 4 groups of 6, providing an opportunity to make a conjecture about commutativity. Noticing were algebraic ideas can be addressed is challenging work. Openings for algebraic reasoning come up quite often, but they are not readily apparent nor is it easy to know how to take them up.

In each of the professional development settings, the political and social culture played out in our work. In some of the cases, like that described earlier, algebraic reasoning may not have been the best starting place because of the fit issue. In other cases, the political face validity of algebra and the severe difficulties in later algebra courses motivated the district to ensure teachers had time to participate, which provided us entry and time to build relationships with schools and teachers in a way that could lead to ongoing learning.

ADAPTING PROFESSIONAL DEVELOPMENT

As we reflect on the changes to our professional development and our shift from addressing number concepts to algebraic thinking, we recognize that we have maintained our general approach to professional development. But we have also adapted the artifacts that we use, and the ways in which we interact around the artifacts. Our final section provides specific examples of the adaptations we have learned to make in our algebraic professional development. These examples include the development of new artifacts and the adaptation of others to support teacher learning through practice and the creation of opportunities for teachers to enrich their knowledge in practice.

Development of New Artifacts

Index Cards. As we mentioned earlier, the teachers often appropriated practices that focused on whether students got the correct answer and not how they were thinking about the mathematical idea (as they did when creating worksheets to help students learn about the equal sign). Teachers also took the true–false number sentences and used them with students without thinking about how to productively sequence the number sentences in relation to student responses. So, we created a tool to support a different type of classroom practice. As we worked in professional development in creating true–false number sentences to challenge students' notions about the equal sign, we had teachers write the number sentences on index cards. As we discussed sequencing, we talked about how one might move through the set of index cards and how different responses from students could justify different card selections by the teacher. The focus became creating sequences of number sentences that could vary. We created the image of carrying around a stack of cards, rather than creating pre-made worksheets. Our goal was not necessarily to get each teacher to carry a stack of cards, but rather to use the index cards as ways to communicate a nonlinear, flexible approach to working with the true and false sentences that required teachers to focus on sequencing and to emphasize student conversation over student answers.

We used index cards throughout the professional development when we created true–false number sentences to foster the learning of number facts, or different number facts to seed a conjecture, that teachers added to their set. As we learned, we added additional index cards to our sets, and whereas teachers collected their own set, the cards were co-constructed within the group. We often used these cards for review, to reflect on how we were thinking about and using the ideas in the classroom. The index card artifact was designed to be used both within the professional development and the classroom and to help carry new ideas and practices between the settings. The artifact then lent itself to the work of algebraic thinking both in the professional development and in the classroom.

Creating Video Excerpts. The index cards were not the only new artifacts were supplemented by predesigned video segments that depicted a sequence of a student's responses as their mathematical ideas developed. These sequences were not about a single answer to a single problem, but focused on how a student responded to a series of true–false number sentences and how the responses changed throughout the engagement. We also included video episodes of classroom interactions to provide opportunities for conversation about teachers and students engaged in discussion around the mathematical ideas. The content of the algebra work develops often as students engage with and challenge one another. Disagreement and argumentation, as well as agreement, become part of developing algebraic ideas. These ways of engaging are often different from how elementary teachers think about teaching mathematics and different from how they think about the mathematics. We wanted to support talk about the pedagogical moves teachers could appropriate to support this learning.

Supporting Knowledge in Practice

Drawing Examples from Teachers' Classrooms. As we attempted to build teachers' classroom practice into the professional development and to help them develop knowledge about identifying opportunities for algebraic thinking, we bring examples of interactions we observed in teachers' classrooms to the group for discussion. For example, in a third-grade classroom, we watched the teacher pose the following problem to her class as a warm-up: If 6 cows have 4 legs each, how many legs are there altogether? She asked the students to share their strategies at the board after having solved the problem at their desks. The teacher elicited four different strategies. All of the students represented the problem as 4 groups with six in each group. Two drew pictures (four circles with six tallied in each), one counted by 6, four times and one added 6, four times.

The teacher acknowledged that the children had done a good job and asked a follow-up question about the number sentence that could be written. But she did not pursue commutativity. We brought the story of this example to this teacher's next professional development session, as well as to other professional development groups, to discuss what might have been the array of options at that point and why you may or may not pursue the different options. The goal here was to help teachers notice when opportunities for algebraic thinking come up and how they might pursue them (as do Blanton & Kaput, 2002). We see numerous examples as we watch teachers engaged in teaching mathematics that create great opportunities for discussion, sometimes because of the exciting things a student may have said or a question the teacher asked that opened up discussion, or a missed opportunity for engagement in algebraic thinking. We did not find that we did as much of this in our CGI work. Often the teachers would bring these examples themselves, and when they did it was often more about what a child had done and less about how the teacher moves in relation to what the student said. And whereas we can prompt teachers to bring them up, the prompting has been less successful in algebra. Having conversation around teachers' own practices (real-life situations that come up) engages the teachers in sense making around their own and their colleague's practice. It helps teachers learn to make their algebraic thinking practice public and over time we do see some teachers bring up these examples on their own. We do think having to work more carefully on noticing opportunities for algebraic reasoning relates to the difficulty with creating a space for algebra in contrast to the whole number work, which is everywhere.

Build in Ways to Reflect on Student Understanding. In the second year of our work with over 100 teachers in a large urban district, we have found ourselves creating structured opportunities for teachers to reflect on where their own students are in their understanding of the various ideas of algebraic thinking. This group of teachers uses the same mathematics textbook and follows a district pacing plan that is connected to quarterly benchmark assessments. However, although opportunities exist to use algebraic thinking in textbooks and on some of the benchmark assessments, the teachers often do not have a sense of where their students are in their ability to engage in algebraic thinking. We created a midyear student assessment that covers key ideas we had been working with using a limited number of problems.

In one professional development session, we asked the teachers to give the assessment to their students and determine if they could notate, as they walked around and asked questions, what their students were doing in attempting to solve the problems. Then, in the subsequent professional development, we looked them over and discussed where we might want to go with the different ideas. Our goal is to support teachers to create ways of knowing more about their students' algebraic thinking and to generate possible next steps. We hope to provide opportunities for the teachers to create structures for making sense of student understanding with the algebraic ideas. This is not as clear in the algebra work as in the early number work. Moreover, the existing curricula does not support teachers to build these ideas because the district assessments focus so much on whether students can do mathematical operations and solve narrowly defined types of problems.

CONCLUSIONS

We have always focused on the substance of what occurs during professional development. We agree with our colleagues that we must make explicit the content of teachers' inquiry as we characterize professional development (Grossman, Wineburg, & Woolworth, 2000; Lehrer & Schauble, 2000; Little, Gearhart, Curry, & Kafka, 2003; Richardson, 1990, 1994; Schifter, 1997; Warren & Rosebery, 1995). We recognize that successful professional development that leads to ongoing, generative teacher learning cannot be created by considering only the forms of professional development, such as the use of cases or lesson study or the use of student work. In creating learning opportunities for teachers, we have found that it is not just the form of the professional work that makes a difference, but the content-driven interactions within those forms. So, on the one hand, we can enumerate some key general principals for supporting professional development: (a) Support the development of communities where teachers engage in inquiry; (b) understand that student thinking provides a tool that engages teachers, makes explicit their and their students' participation with the mathematics, and provides a trace of the community's ideas; and (c) understand the histories and cultures of the communities that we enter. On the other hand, what we now realize is how deeply these general ideas interact with the specific mathematical content on which the professional development is based.

Our change in mathematical terrain has highlighted for us the role of specific mathematical content in coming to understand how to create learning opportunities that teachers can take advantage of and make sense of in meaningful ways. Given the variations across content in cultural practices, including the extent to which that content is already represented in standard texts and curricular expectations, in political and social contexts surrounding it, and teachers' understanding and stance toward it, the artifacts and forms of engagement also may need to vary.

This chapter begins to articulate in what particular ways content may make a difference in the design of learning opportunities for elementary school teachers. We recognize that professional development for teachers requires constant attention to and explicit articulation of the mathematical content being addressed. We are more convinced than ever that content matters in designing professional development.

ACKNOWLEDGMENTS

The research reported in this chapter was supported in part by a grant from the National Science Foundation (No. ESI9911679) and a grant from the Department of Education Office of Educational Research and Improvement to the National Center for Improving Student Learning and Achievement in Mathematics and Science (No. R305A60007-98). The opinions expressed in this chapter do not necessarily reflect the position, policy, or endorsement of the National Science Foundation, the Department of Education, OERI, or the National Center.

REFERENCES

Balacheff, N. (1988). Aspects of proof in pupils' practice of school mathematics. In D. Pimm (Ed.), *Mathematics, teachers and children* (pp. 316–230). London: Hodder & Stoughton.

Behr, M., Erlwanger, S., & Nichols, E. (1980). How children view the equal sign. *Mathematics Teaching, 92*, 13–15.

Blanton, M., & Kaput, J. (2002, April). *Developing elementary teachers' algebra "eyes and ears": Understanding characteristics of professional development that promote generative and self-sustaining change in teacher practice*. Paper presented at the AERA annual meeting, New Orleans, LA.

Carpenter, T. P. (1985). Learning to add and subtract: An exercise in problem solving. In E. A. Silver (Ed.), *Teaching and learning mathematical problem solving: Multiple research perspectives* (pp. 17–40). Hillsdale, NJ: Lawrence Erlbaum Associates.

Carpenter, T. P., Fennema, E., & Franke, M. L. (1996). Cognitively guided instruction: A knowledge base for reform in primary mathematics instruction. *Elementary School Journal, 97*(1), 1–20.

Carpenter, T. P., Fennema, E., Franke, M., Levi, L., & Empson, S. (1999). *Children's mathematics: Cognitively guided instruction*. Portsmouth, NH: Heinemann.

Carpenter, T. P., Franke, M. L., & Levi, L. (2003). *Thinking mathematically: Integrating arithmetic and algebra in elementary school*. Portsmouth, NH: Heinemann.

Chailkin, S., & Lesgold, S. (1984, April). *Prealgebra students' knowledge of algebra tasks with arithmetic expressions*. Paper presented at the annual meeting of the American Educational Research Association, New Orleans, LA.

Clement, J. (1982). Algebra word problem solutions: Thought processes underlying common misconception. *Journal for Research in Mathematics Education, 13*, 16–30.

Collis, K. F. (1975). *The development of formal reasoning*. Newcastle, Australia: University of Newcastle.

Davis, R. B. (1964). *Discovery in mathematics: A text for teachers*. Palo Alto, CA: Addison-Wesley.

Driscoll, M. (1999). *Fostering algebraic thinking*. Westport, CT: Heinemann.

Erlwanger, S., & Berlanger, M. (1983). Interpretations of the equal sign among elementary school children. In J. C. Bergeron & N. Herscovics (Eds.), *Proceedings of the North American Chapter of the International Group for the Psychology of Mathematics Education* (Volume 1, pp. 250–259). Montreal, Canada: Program Committee.

Falkner, K. P., Levi, L., & Carpenter, T. P. (1999). Children's understanding of equality: A foundation for algebra. *Teaching Children Mathematics*, *6*(4), 232–236.

Fuson, K. C. (1992). Research on whole number addition and subtraction. In D. Grouws (Ed.), *Handbook of research on mathematics teaching and learning* (pp. 243–275). New York: Macmillan.

Greeno, J. G., & Middle School Mathematics Through Applications Project. (1998). The situativity of knowing, learning, and research. *American Psychologist*, *53*, 5–26.

Grossman, P., Wineburg, S., & Woolworth, (2000). *What makes teacher community different from a gathering of teachers?* Paper published by the Center for the Study of Teaching and Policy, University of Washington.

Kaput, J. (1998). Transforming algebra from an engine of inequity to an engine of mathematical power by "algebrafying" the K–12 curriculum. In the National Council of Teachers of Mathematics & the Mathematical Sciences Education Board (Eds.), *The nature and role of algebra in the K–14 curriculum: Proceedings of a national symposium* (pp. 25–26). Washington, DC: National Research Council, National Academy Press.

Kieran, C. (1981). Concepts associated with the equality symbol. *Educational Studies in Mathematics*, *12*, 317–326.

Kieran, C. (1989). The early learning of algebra: A structural perspective. In S. Wagner & C. Kieran (Eds.), *Research issues in the learning and teaching of algebra* (pp. 33–56). Reston, VA: National Council for Teachers of Mathematics.

Kieran, C. (1992). The learning and teaching of school algebra. In D. Grouws (Ed.), *Handbook of research on mathematics teaching and learning* (pp. 390–419). New York: Macmillan.

Lave, J. (1996). Teaching, as learning, in practice. *Mind, Culture, and Activity*, *3*, 149–164.

Lehrer, R., & Schauble, L. (2000). Model-based reasoning in mathematics and science. In R. Glaser (Ed.), *Advances in instructional psychology* (Vol. 5, pp. 101–159). Mahwah, NJ: Lawrence Erlbaum Associates.

Little, J. W. (1999). Organizing schools for teacher learning. In L. Darling-Hammond & G. Sykes (Eds.), *Teaching as the learning profession: Handbook of policy and practice* (pp. 233–262). San Francisco: Jossey-Bass.

Little, J. W., Gearhart, M., Curry, M., & Kafka, J. (2003). Looking at student work for teacher learning, teacher community, and school reform. *Phi Delta Kappan*, *85*, 185–192.

Loucks-Horsley, S., Hewson, P. W., Love, N., & Stiles, K. E. (1998). *Designing professional development for teachers of science and mathematics.* Thousand Oaks, CA: Corwin.

Matz, M. (1982). Towards a process model for school algebra errors. In D. Sleeman & J. S. Brown (Eds.), *Intelligent tutoring systems* (pp. 25–50). New York: Academic Press.

Richardson, V. (1990). Significant and worthwhile change in teaching practice. *Educational Researcher, 19*(7), 10–18.

Richardson, V. (1994). Conducting research on practice. *Educational Researcher, 23*(5), 5–10.

Rogoff, B. (1997). Evaluating development in the process of participation: Theory, methods, and practice build on each other. In E. Amsel & A. Renninger (Eds.), *Change and development* (pp. 265–285). Hillsdale, NJ: Lawrence Erlbaum Associates.

Saenz-Ludlow, A., & Walgamuth, C. (1998). Third graders' interpretations of equality and the equal symbol. *Educational Studies in Mathematics, 35,* 153–187.

Schifter, D. (1997, April). *Developing operation sense as a foundation for algebra.* Paper presented at the annual meeting of the American Educational Research Association, Chicago, IL.

Schifter, D. (1999). Reasoning about algebra: Early algebraic thinking in grades K-6. In L. V. Stiff & F. R. Curcio (Eds.), *Developing mathematical reasoning in grades K–12* (pp. 62–81). Reston, VA: National Council of Teachers of Mathematics.

Sowder, L., & Harel, G. (1998). Types of students' justifications. *Mathematics Teacher, 91,* 670–674.

Warren, B., & Rosebery, A.S. (1995). Equity in the future tense: Redefining relationships among teachers, students and science in language minority classrooms. In W. Secada, E. Fennema, & L. Adajian (Eds.), *New directions for equity in mathematics education* (pp. 298–328). New York: Cambridge University Press.

Wenger, E. (1998). *Communities of practice: Learning, meaning, and identity.* Cambridge, England: Cambridge University Press.

Wertsch, J. V. (1998). *Mind as action.* New York: Oxford University Press.

14

Building District Capacity for Teacher Development in Algebraic Reasoning

Maria L. Blanton
James J. Kaput
University of Massachusetts, Dartmouth

How can elementary teachers who have been schooled in a way of doing mathematics defined largely by the memorization of facts and procedures emerge from the constraints of practice that this creates to build classrooms that teach a more powerful and general mathematics for understanding? How can this type of change be supported systemically and institutionally so that it becomes deeply embedded in instruction and not limited to cosmetic features that might be discarded if other demands shift teachers' attention? These are some of the questions that have guided our thinking as we engaged in a 5-year, district-wide research and professional development project designed to transform how elementary teachers understand and teach mathematics so that algebraic thinking is at the heart of instruction. This is an especially challenging task when the change under consideration is transformative rather than additive, and when it involves all dimensions of a teacher's practice. Our effort did not involve adding new curriculum on top of existing curricula, nor did it involve substituting a new curriculum for the existing one as occurs when one implements a standards-based curriculum to replace a traditional basal text, for example. Rather, it involved getting teachers to see mathematics in a new way, as a process of generalization and formalization, to reorganize their classroom practice to make teaching to this new vision of mathematics possible and viable, and to change their relationship to their instructional materials base from one of implementer/consumer to one of active transformer.

Our intent in this chapter is to examine the teacher learning that occurred in this project as a process situated within an institutional context (Cobb, McClain, deSilva Lamberg, & Dean, 2003) and how both process and context interact reciprocally to support development. Our findings are not necessarily unique, but contribute to empirical and theoretical arguments on what is entailed in bringing about generative and self-sustaining teacher learning (e.g., Coburn, 2003; Elmore, 1996; Franke, Carpenter, Fennema, Ansell, & Behrend, 1998; Franke, Carpenter, Levi, & Fennema, 2001; Garet, Porter, Desimone, Birman, & Yoon, 2001) with special focus on the sustained integration of algebraic reasoning in elementary mathematics. We draw evidence from a large data corpus gathered over a 5-year period and consisting of teachers' reflections and written work, students' written work, observations of classrooms and seminars with teachers and principals and administrators, and interviews with teachers and principals.

SUPPORTING TEACHERS IN ALGEBRAFYING THEIR PRACTICE

Our professional development project, Generalizing to Extend Arithmetic to Algebraic Reasoning (GEAAR), asks teachers to *algebrafy* their practice by transforming and extending the mathematics they teach (typically, arithmetic) to algebraic thinking and to establish classroom norms of participation so that argumentation, conjecture, and justification are routine acts of discourse. An important intrinsic feature of this *algebrafication* of elementary mathematics is its own generality (see Kaput, Blanton, & Moreno, chap. 2, this volume). It cuts across most domains of mathematics defined by national standards in elementary mathematics (National Council of Teachers of Mathematics, 2000), and it impinges on virtually all aspects of elementary teachers' work. Our approach with teachers has been to exploit this generality across three dimensions by helping teachers transform the ways they (a) algebrafy their existing instructional materials by turning these materials into starting points for generalization and its associated symbolization processes, (b) interpret and build on student thinking, and (c) construct classroom practice and culture, so that algebraic thinking is at the heart of classroom mathematics rather than an easily marginalized enrichment activity (Kaput & Blanton, 2005).

Our content goal is to deepen the nature of elementary school mathematics, shifting its focus from the particulars of number and computational technique to purposeful generalization of mathematical ideas and the expression of generalities with increasingly sophisticated symbol systems (Kaput, 1999, 2001; Kaput & Blanton, 2005). Although such generalizations naturally arise in arithmetic and in the use of arithmetic to model situations and phenomena, they can be developed in any mathematical domain and can result in a richer and more varied mathematical experience for young children (see e.g., Boester & Lehrer, chap. 9, this volume; Dougherty, chap. 15, this volume; J. Smith & Thompson, chap. 4, this volume). At their heart are symbolization processes that are not part of mainstream practice but are at the core of most mathematical thinking.

Our concept of algebrafying classrooms also addresses teachers' classroom practice. Algebraic thinking thrives in an instructional context that both elicits students' thinking and uses it to build a climate of conjecture and argumentation so that conjectures can be established or rejected as valid mathematical claims, especially conjectures regarding the generality of claims. Thus, in algebrafying classrooms, we are also asking teachers to unpack years of instructional practice that likely did not attend seriously to students' mathematical thinking, including the various symbolization processes students use (Kaput et al., chap. 2, this volume).

Because the complexity of change needed in teachers' mathematical and instructional knowledge to algebrafy elementary classrooms is substantial, we need to understand how school and district contexts support and/or constrain teachers' growth. This chapter describes how the school district in which we worked helped build capacity for teacher change so that algebraic thinking was supported, made intelligible to teachers, integrated as a regular part of instruction, and related to other district priorities and initiatives. We organize our discussion around some of the areas that we have come to view as essential in helping teachers algebrafy their classrooms: the development of a *professional community network*; a *distributed* approach to district leadership practice, the development of a school mathematics culture, the integration of district professional development initiatives, and the development of teachers' capacity to algebrafy their own instructional resource base.

First, Change Teachers' Experiences

One perspective on research in teacher development characterizes it as a process of changing teachers' beliefs and practices (Cooney, 2001). In a similar vein, Coburn (2003) describes deep change as going "beyond surface structures or procedures ... to alter teachers' beliefs, norms of social interaction, and pedagogical principles as enacted in the curriculum" (p. 4). We characterize such change as effected by a perturbation in what teachers experience, accompanied by a shift in how they perceive these experiences relative to their existing beliefs and practice. Thus, we have come to interpret the heart of professional development to be about changing teachers' mathematical and pedagogical experiences in their classroom settings and providing a subsequent context in which teachers can collectively reorganize these experiences. Although this may seem overly simple, it illustrates an important shift in our own thinking about our role and that of teachers in professional development. In particular, our intent became to situate teachers' learning within the intellectual space in which they could most authentically think about their beliefs and practices—their own classrooms. Our assumption was that teachers' notions about what constitutes good practice or good mathematics, whether professed or enacted, could be most effectively challenged in the real time of classroom instruction and through the collective negotiation of their classroom experiences within a professional teacher community. As Franke et al. (2001) describe it, professional development should "create opportunities for teacher learning through professional communities whose activities are embedded in teachers' everyday work" (p. 655).

We want to emphasize this point because it clarifies our task as professional development providers: to support teachers in modifying their actions in the classroom so that they will experience content and instruction, and its effects on students' ways of doing mathematics, in ways that respect the many and subtle variations in their personal histories and their school contexts. It also underscores that our task was not to provide a contained program—one that operated from an extant, finalized set of ideas and resources—for teachers to study outside the parameters of their school experiences. In other words, we came to value contextualizing teachers' experiences within their own school communities, and the larger district community, as a way to build the intellectual space for deeper insight, messiness, and ambiguity—out of which could come real growth.

The Role of a Professional Community Network

The Necessity of a Teacher Community for GEAAR. Perturbing teachers' classroom experiences raises the need for a cohesive setting—a community—where teachers can collectively make sense of these experiences. It is widely recognized that the existence of community is crucial for teacher learning to be generative and self-sustaining over time (National Research Council, 1999, 2001). We found that community was critical for building, sustaining, and extending teachers' capacity to algebrafy their classrooms. In contrast, learning is not generally maintained through random isolated professional development, such as the professional development that occurs when teachers take ad hoc courses or attend brief workshops removed from the intellectual space of their school experiences.

Lave and Wenger (1991) argue that teacher learning occurs as teachers participate in a *community of practice*. As teachers negotiate discrepancies in their expertise in order to improve their practice, they arrive at new meanings and understandings that "derive from and create the situated practice in which individuals are co-participants" so that "learning becomes a by-product of participation in joint activities for which teachers have mutually held goals" (Stein, Silver, & Smith, 1998, p. 29). Thus, a community of practice supports collaborative relationships that engage teachers in research and practice (Carpenter & Fennema, 1992) and is constituted by mutual professional trust among its members. It exists as a network by which teachers share ideas and adopt the ideas of others, and reflexively, as teachers engage in this type of collaboration, the community's capacity to sustain growth expands.

Our professional development design used the following iterative cycle of activity based on approximately biweekly after school seminars that consisted of teachers representing multiple grade levels. Teachers were recruited through district channels (which relied heavily on school principals) and compensated according to established district patterns:

- Working in small groups, teachers solved authentic and challenging algebraic reasoning problems (which they sometimes contributed).
- They shared and compared the various groups' mathematical solutions.
- They discussed how to adapt the problems to the various grade levels they taught.
- Outside of seminars, working mostly on their own or with school colleagues, they further adapted the problems and implemented them in their own classrooms, writing brief reflections on the classroom episodes.
- At the next seminar, they collectively analyzed classroom stories based on teachers' records and reflections on students' thinking and artifacts of instruction, including student work. Discussion specifically focused on the grade-to-grade differences in the problems and in the student work, particularly the evolution of symbolizations across the years. It also included examination of differences in teachers' strategies and student classroom responses across grades.
- A subset of the problems and their adaptations, teacher reflections, and student work was selected to become part of a continually evolving district resource that teachers used as teacher leaders with new participants.

A second version of this cycle was based on examining teachers' existing text and other available materials for standard one-answer arithmetic problems, as well as somewhat more complex patterning problems, and systematically modifying these to build opportunities for generalization and expression of generality (we refer to this process as algebrafying). Quite often algebrafying involved varying the givens of the problem so that it became a sequence of problems that served as a base for generalization. For example, in the Handshake Problem, instead of treating it as a count of the number of handshakes for a group of specific size, it was modified to be a problem about a group of any size (Blanton & Kaput, 2004).

This approach necessarily keeps the attention on teachers' own practices, how they understand and are able to transform the mathematics they typically teach, and how students in their classrooms are able to think mathematically and how that thinking evolves across grade levels. In all, it creates a context in which teachers are intellectually vulnerable and that vulnerability requires the development of professional trust in order for collegial sharing to occur and to be viable when it does occur.

We have seen that as teachers experiment with ideas, reflect on classroom experiences, and discuss how students are reasoning algebraically and how their instruction supports this, they forge a community comprised of those who engage, at some level, in this process. Not all teachers participated from within the 15 (district-targeted) participating schools (there are 28 small elementary schools in the district). Some selected not to be part of the professional development project; a small minority (10%) attended at least 1 year of professional development but remained isolated within the group. We maintain that those in the latter case were unable to algebrafy their classrooms in isolation and ultimately suspended their work with the group as other professional development options arose. Some teachers were especially worried regarding their ability to deal with "algebraic" thinking and took a less visible role in the seminars; those who were otherwise active in professional development had different professional obligations that took priority for them, particularly a high-profile literacy initiative that had begun several years earlier. As a result, the teacher communities that developed emerged organically around those teachers who were open to collaboration and inquiry about mathematics and teaching mathematics and who subsequently became leaders in their respective communities.

The Development of Community. Thus, whereas the need for community is not in question, the sticking point is understanding what it takes to support teachers in creating a community of practice. All organizations have idiosyncratic features that affect the ways that learning occurs and renders certain learners (e.g., teachers) potentially unaffected—or affected in different ways—by external interventions. Moreover, as we have found and as other research bears out (Garet et al., 2001), constructing community is a fragile process that requires long-term, sustained collaboration among teachers. We estimate that it requires 3 to 4 years to establish a community of teacher leaders who are at some level of generativity and self-sustainability in algebrafying their classrooms and in leading school-based groups to do so.

We find reasons for this time scale in the notion that communities are constituted by trust relationships, which are in turn constructed through experience as "ideas and concepts migrate throughout the community via mutual appropriation" (Brown, Ellery, & Campione, 1998, p. 349). For example, the development of the GEAAR teacher leader community has involved ideas, practices, and knowledge developing in individual teachers through experimentation, being shared through collaboration, then ultimately traveling across schools in the district as teachers test the ideas of their peers—a process requiring years rather than months.

A Professional Community Network: Extending the Notion of Community. Thus, we have found that a *teacher* community is critical in supporting teachers as they algebrafy their classrooms, and that constructing community is a long-term, delicate enterprise. However, we maintain that even the development of a robust teacher community without a broader network in which it can exist—or better yet—thrive, can isolate teachers and threaten the life of this community. Cobb and McClain (2001) note the difficulties that arise when teachers form a professional community that is locally viable and responsive yet disconnected from district leadership. Our view is that the development of teacher communities must be situated within the context of a larger network that accounts for and is supported by the institutional setting in which teachers learn and grow. As such, the broader intent of GEAAR evolved to deliberately generate distributed capacity at the teacher, school, and district levels by helping to build networked communities of teachers, principals, and other administrators. In other words, not only do we advocate for a teacher community of practice, but also for what we describe as a *professional community network*. By this, we mean an infrastructure consisting of interconnected but distinct communities that have parallel purposes and that can and do support each other toward a common goal. We view it as an ever-widening community system that is strengthened as the complexity of its internal connectivity evolves and that could conceivably extend beyond district borders to include, say, parent–teacher communities.

We observed at least three categories of communities evolving within this district (although they were not all at the same point in their growth after 5 years) with the mutual goal of supporting teachers' capacity to algebrafy their classrooms: teacher communities, principal communities, and communities based on teacher–principal partnerships. We viewed these as distinct communities because they were based on differences in relationship dynamics (e.g., hierarchical vs. peer) and because each group necessarily addressed a unique set of issues and purposes. We use community in plural because of fluid boundaries that allowed participants membership in like, but logistically separate, communities. For example, teachers within a particular school sometimes collected as a community within that school, but also reconnected as part of a district-wide teacher community. The result was that membership in multiple communities layered the complexity and potential effectiveness of relationships within the network.

Bringing District Leaders Into the Professional Development Agenda. To create systemic support for teacher communities, we offered professional development for district principals and administrators. This enabled them to (a) understand and support algebraic reasoning and its implications for teacher practice so that the evaluation and hiring of teachers would be aligned with the teacher professional development agenda; (b) assist in structuring professional development to allow for ongoing teacher collaboration; and (c) promote mathematics literacy school-wide and to integrate it with other district initiatives so that change would be truly systemic. We led a superintendent-sponsored Leadership Academy for principals and administrators, where activities were patterned after those used with our teachers and teacher leaders were invited to share their classroom experiences with district leaders. For example, during one meeting of the Leadership Academy, participants solved an *algebrafied* version of the Handshake Problem (see Blanton & Kaput, 2004, for a detailed treatment of this problem), viewed video of district third graders solving it, and then discussed students' strategies with the presenting teacher who was also the instructor for the third-grade class in the video. In another meeting, a second-grade teacher discussed with participants how her students' work incorporated reading and writing in order to integrate the goals of the district literacy program. In fact, our mathematics effort was renamed as the "Mathematics Literacy Program" to establish it as a parallel program in administrators' and teachers' eyes.

As a result of the Leadership Academy, we observed the emergence of a small group of principals who showed serious (sometimes hands-on) support for the work in which teachers were engaged with GEAAR. This included principals and teachers collaborating in specific schools for the purpose of designing and leading staff development to incorporate the ideas of GEAAR in existing programs, or principals participating in school-wide GEAAR teacher professional development to gain more understanding about the kinds of knowledge teachers need. Moreover, principals in the district began participating in a separate (but ideologically parallel) professional development program, organized and supported by district administrators, whose goal was to think about the kinds of content and instructional knowledge needed to evaluate classrooms that support learning mathematics with understanding.[1]

Our intent was to foster the development of a professional community network. Through joint leadership meetings with district leaders (e.g., the mathematics curriculum coordinator) and the provider of principal professional development subsequent to our Leadership Academy, we engaged in coordinating the respective goals and purposes of both teacher and principal professional development so that they were as mutually supportive as possible. As we will explore in the following section, the existence of communities based in part or whole on *principal* membership and connected to the goals of the teacher community seemed to develop from a particular approach to leadership and emerged as an agent of teacher change in this district.

The Role of District Leadership

Cobb et al. (2003) point out the need to move beyond the dichotomy of thinking that exists in research on teacher change that focuses on either "the role of professional development in supporting teachers' reorganization of their instructional practices and their views of themselves as learners" or "the structural or organizational features of schools and ... how changes in these conditions can lead to changes in classroom instructional practices" (p. 13). Instead, they and other scholars increasingly view teacher learning as a process situated within an institutional context and thus intrinsically constrained or enabled by this context (Cobb et al., 2003; Franke et al., 2001; Gamoran, 2001; Gamoran et al., 2003; Gamoran, Secada, & Marrett, 2000; Nelson, 1999).

This suggests that an interpretation of teacher change should account for how the institutional setting is organized, how its practices are defined, how its human and material resources are allocated, and so forth. Moreover, it draws our thinking to the nature of leadership[2] practice of principals and administrators in the district, because this practice in large part defines how the institutional setting is organized, and consequently,

---

[1]The Lenses on Learning Program (Grant et al., 2003; Nelson, 1998) was targeted toward principals of the schools with lowest scores on the state fourth-grade achievement test. Note that the district was under considerable pressure to raise scores because its scores were among the lowest in the state, especially at the upper grade levels.

[2]By leadership, we refer here to formal leadership positions such as *principal* or *curriculum coordinator* as opposed to its organic, informal forms.

how it constrains or enables teachers in algebrafying their classrooms. Spillane, Halverson, and Diamond (1999) assert that leadership must be *distributed* throughout an organization (e.g., district or school) and should include the use of teachers as leaders. Gamoran (2005) further argues that district leadership can increase schools' capacity for change by providing substantial time for professional development, giving teachers a choice in content and instructional practice, and appropriating resources in ways that respond to teachers' efforts. Moreover, Gamoran found that leadership supports teacher change when it allows teachers the autonomy to develop their own expertise and assume leadership roles that reflect this.

A Brief District History. Primarily, the nature of the decisions and actions of school principals and district-level administrators are what constitutes its formal leadership practice. Like any district, the one in which we worked had its own administrative idiosyncrasies and histories that have shaped its direction. And, like the administrative bodies in most districts, it has had a fluid membership whose collective characteristic is its flux in purpose: as personnel are replaced, district priorities are often recast in subtle or dramatic ways. This district experienced three superintendents during our work in the district, the first and third being district veterans and the second an external hire with a sophisticated understanding of curriculum and instruction. When the first superintendent departed, the long-time, influential assistant superintendent for curriculum and instruction accepted a position at a nearby university. She had been an academic leader in the district for almost 20 years and had more recently promoted the district's literacy program, a strong elementary grades science program, and in cooperation with us, major improvements in K–12 mathematics curricula and instruction. Her leadership style was informal and collegial, and given a relatively free rein on academic matters, she promoted the development of school-based study groups to address issues of curriculum and instruction through a distributed, organic system of leadership (Gamoran, 2005). Soon after her departure, and with her political support, the district hired curriculum coordinators for social studies, language arts, mathematics and science, and—from within the district—a new assistant superintendent for curriculum and instruction. The mathematics coordinator, at one time a high school teacher within the district, was in tune with the reform intentions of the outgoing assistant superintendent for curriculum and instruction.

This significant flux in personnel resulted in varying levels of commitment to mathematics teacher professional development at the district level, underscoring for us the susceptibility of teacher learning to an intrinsic feature of institutional settings: shifts in regimes. The consistency of support for our work was through the Title I director and his chief academic assistant, a former science department chair at the high school with strong interest in mathematics and teaching and excellent rapport with teachers across the district. They regularly attended GEAAR teacher leader seminars and provided critical support in identifying and recruiting teachers. Whereas this led to a disproportionate percentage of Title I teachers as leaders (about 30%), the connection with Title I provided us a critical avenue for recruiting teachers and learning about the district from an inside perspective, as a *lived-in enterprise* (Cobb et al., 2003; Wenger, 1998).

The net effect of the structures, styles, and specific circumstances described earlier was to provide both flexibility and credibility for the elementary mathematics work, as well as specific insider perspectives and pragmatic assistance. GEAAR's capacity-building approach fit within a larger district reform effort that was given varying degrees of support through to the superintendent level. The result is that we had long-term access to a subset of the district's elementary teachers as volunteers. Whereas a few elementary schools adopted a whole-school approach to GEAAR, a number of teachers participated as individuals rather than as members of school-based team. Not surprisingly, whole-school configurations of teacher groups best supported the development of teacher communities, whereas individual teachers participating in GEAAR risked remaining isolated within their schools. However, whole-school efforts seemed to have been largely led by individual principals and not promoted as vigorously at the district level. Our belief was that increased collaboration between principals and district administrators—that is, the strengthening of a principal/administrator community as part of the professional community network—would increase the possibility for whole-school teacher development, which would in turn support the development of teacher communities of practice. As the next section discusses, during our work there were positive indications of an emerging principal learning community being more explicitly supported by priorities set at the district level.

A Principal's Distributed Leadership Practice. Interviews with district principals and administrators revealed cases of principals' distributed approach to leadership and its effect on how teachers participated in GEAAR. We illustrate this first with the case of Julia[3] and Joan, a principal and third-grade instructor for one of the schools in the district. Julia characterized her principal leadership practice as evolving from a top–down perspective to a school team approach. In her school, a school-based leadership team evolved from an existing literacy team that had been formed to discuss issues arising from the school's literacy collaborative:

[3]All names are pseudonyms.

> Initially the group began as a training team to know more about the [literacy] initiative, and slowly we began to discuss other issues in the school, other parts in the curriculum as well. So ... very quickly, within a year, we began to notice that we were becoming more of a leadership team than just a literacy team, um, just by the nature of the needs of the school. And it created a much better collaboration among teachers because for the first time they were able to discuss in an, in almost an informal format, concerns about children and curriculum that we had not had the opportunity to talk about before. (Julia, interview)

As she described it, there were two leadership teams for the school—one for Grades K to 2 and one for Grades 3 to 5. The teams met biweekly outside school hours and all teachers were invited to participate on their grade-level team. Regarding her role, Julia noted:

> "I am part of the team ... and all the concerns and issues and problems are put forth to the team and most of the decisions are made from there. What I bring to that team is the ability to facilitate, um, the decisions because I might see the broader picture or have the connections to carry out the decision."

We see Julia's practice of shared decision making through team collaboration as an example of the type of distributed leadership scholars advocate in order to support teaching for understanding (Gamoran, 2005). Moreover, it underscores why she was able to collaborate with Joan to bring GEAAR into school-wide focus. Prior to this, Joan had been one of only two teachers from her school that had participated in GEAAR. Having been with the project from its inception, she was one of our veteran teacher leaders and had approached Julia about designing a school-based teacher professional development day that focused on algebraic reasoning. After meeting with Julia, Joan wrote to us: "I am doing a professional development day. ... I have put together quite a little packet for the teachers and would like to do a few activities with them. ... I spoke to the principal of the [school] about my idea. She thought that would be great and [would] impress the seriousness of the math program."

As it turned out, Joan's one-day workshop marked the beginning of a collaboration with Julia that grew into a multiyear effort to integrate the mathematics of GEAAR with the school's literacy program. The integration grew out of two very practical needs: (a) classroom instruction time required for the highly structured literacy program consumed the first 3 hours of the school day, leaving very little room for the time-intensive kinds of mathematics we were advocating; and (b) the school had access to large literacy grants and could justify the expenditure of these funds on mathematics if there was a legitimate connection to literacy.

Joan wrote to us, "[Julia] is very interested in professional development for the staff next year. I told her it would be something that we would talk about. It seems as if it is falling into place." In describing their collaboration, Julia noted, "[Joan] and I sat and developed a bibliography of titles that would be good 'read alouds' and good literacy books to work in the math curriculum, to connect with the math curriculum and from there we were able to expand the Math Initiative (as GEAAR was termed by the district) by making that connection. So [Joan] was able to teach the [math] initiatives as well as make that connection with literacy."

We later revisit the significance of integrating GEAAR with other initiatives across the district. Our point for now is that the distributed nature of Julia's leadership practice allowed her to utilize Joan's expertise and interests to support teacher professional development, as Gamoran (2005) advocates. Although Julia was involved in the process, she transferred autonomy to teachers by supporting *teachers* in leading professional development. The result was that Joan was able to integrate her GEAAR expertise into school-wide staff development.

Principal Leadership Practice in the District

As with GEAAR teachers, we observed an emerging community of *principal* leaders that we anticipated could affect the culture of leadership among their peers, provided there was sustained, ongoing support at the district level. Indeed, Julia's case was not unique among principals in this district. Leslie, an experienced principal at a small (two-unit) school described a similar practice of leadership based on forming group consensus among staff. Part of her practice involves immersing herself in what teachers are doing: "I'm involved, I work with the children, I'm in the classrooms, uhm, I go to all the workshops. I'm there for them." She also encourages teachers to identify the programs they are interested in to pursue for further development.

Marcia, a principal in her third year at a three-unit school, is also deeply involved in staff development for her school and tries to give her staff the independence to pursue their own professional development interests. She noted, "I don't ask my teachers to do anything that I wouldn't do myself. So any professional development activities that I have asked them to participate in, I have participated in with them." (In fact, she enthusiastically participated in one of our summer institutes for teacher leaders.)

Moreover, to support her school's implementation of GEAAR, Marcia spent a lot of time in the classroom working with students. As she explained, "I need to figure out how to do these things myself and ... I need to see how the children will react to it." She described that she and her teachers engage deeply with GEAAR activities and are thus able to bring that experience of struggle to bear on classroom practice:

> We felt the same frustrations that the kids feel when we were learning things. So then we took that and went into the classroom. Of course, in addition to the classroom teachers some of the Title I teachers participated, so we'd go in as a group and work with the class. Sometimes there'd be four teachers in the classroom doing the same thing. It was fun. (Marcia, interview)

As a result of her participation in the GEAAR professional development at her school (led by two GEAAR teacher leaders), it seemed that Marcia had come to appreciate the value and complexity of GEAAR mathematics for teachers and students and so was able to share in how teachers were experiencing GEAAR. She also subsequently volunteered to be a teacher leader for her school. As she explained, "I'm doing it for selfish reasons because math is not my strength, so I thought maybe I could win something here."[4]

Principals such as Julia, Leslie, and Marcia formed the core of an emerging principal community that existed as part of the district's professional community network and whose distributed leadership practice supported the development of teachers' understanding of how to integrate algebraic reasoning into their classrooms. These principals also partnered with teachers in their schools, thus bringing about a community based on principal–teacher collaborations. What was common among the practices of these principals was a commitment to working with teachers for the benefit of their own understanding of the professional development in which teachers were engaged, preserving teachers' autonomous role in leading and designing professional development, and sharing decision authority with teachers on issues that affected the school community.

Although these forms of leadership practice existed prior to our work in the district, the Leadership Academy was able to leverage this depth of practice by raising the awareness of principals and administrators to the kinds of knowledge teachers need to support students' algebraic reasoning. Moreover, the academy helped initiate a GEAAR principal community (although this community was still fairly young relative to its potential influence on the district) that specifically supported teachers in the implementation of GEAAR through choices in how resources were allocated, through planning and participating in the professional development with

---

[4]Interestingly, each of the principals claimed that mathematics was a weakness for them personally in their own educations, although each of them proved to be sophisticated mathematical thinkers. Similar comments were made by many of the more successful participating teachers, leading us to believe that algebrafying tapped into intellectual capacities that are not well utilized by a computationally oriented curriculum.

teachers, through joint instruction with teachers in classroom GEAAR activities, and so forth. Leslie noted that her participation in the Leadership Academy had given her an initial understanding of "how important and exciting" the mathematics advocated by GEAAR was for children. This, in turn, gave her the legitimacy with teachers to support it as a school-wide initiative.

Given what we observed with how principals in the district were coming to understand and support GEAAR based on their experiences with it, we maintain that effective *teacher* professional development is enabled by parallel *principal* professional development whose purposes include building a principal community that supports the particular goals of teacher professional development and whose members operate within a distributed approach to leadership. Because new GEAAR teacher participants in the fourth and fifth years of the project were increasingly drawn from a population who were less enthusiastic about professional development in teaching mathematics (those eager to participate were naturally among the first wave of participants), we found that the existence of a cadre of principals who wanted to integrate algebraic thinking into their schools was an essential resource for accessing this more reticent teacher population. We should note that, although participation was always voluntary, if a school principal decided (usually in concert with both district-level administrators and one or more teacher leaders who happened to be in their school) to promote GEAAR, then not volunteering was to resist peer pressure in a conspicuous way. Hence participation at whole-school sites was typically at 90% or higher.

Developing Congruency Across Professional Development Initiatives

Garet et al. (2001) assert that teacher professional development has more impact when teachers perceive the program as relevant to other school activities. We watched this play out firsthand in the district, where increasing congruency between GEAAR and other professional development initiatives helped leverage the influence of GEAAR across the district by bringing access to teachers with a focus on other agendas. The goal of what we describe as *professional congruency* is to increase connectivity within teacher development by identifying and strengthening ways in which teachers' professional obligations and interests can reflexively support each other. The flexibility of the GEAAR algebrafication approach is that it is not constrained by a curriculum, but rather is designed to help teachers transform and extend the mathematics they teach, regardless of their existing instructional resource base. This flexibility allowed us to draw from and connect with other district initiatives to address professional congruency. In our work, this occurred most visibly in GEAAR's integration with (a) MCAS, a statewide, high-stakes assessment on which this district's performance has been among the lowest in the state; (b) the district's highly intensive and structured Literacy Initiative; and (c) in our fifth year, an extension to another mathematics teacher development effort, Developing Mathematical Ideas (Schifter, Bastable, & Russell, 1999), adopted by the district to support entry into GEAAR by the least mathematically prepared teachers and lowest performing schools.[5]

Integrating Competing School Agendas. The high-stakes nature of MCAS, with student scores reported at the district, school, and classroom levels, understandably generates pressure among teachers and administrators across the district to orient all that they do to improving students' performance on the test. This is complicated by the inability of classroom instruction that focuses on memorizing number facts and arithmetic procedures to prepare students for the complexity of problem solving and the diversity of topics (e.g., probability) found on the highly challenging MCAS tests. So, to the extent that rote instruction on arithmetic skills existed across the district, there was an urgent need to change teacher practice so that student performance could improve. In all, this placed MCAS at the center of what teachers perceived as relevant to their daily practice.

To address this, one of our objectives was for teachers to algebrafy MCAS, that is, to identify instances among the existing test items where algebraic reasoning occurred or to extend tasks so that they involved algebraic reasoning, and to adapt these fourth-grade items across grades K–5 for teachers to implement in their daily instruction. We should note also that we utilized released eighth-grade MCAS items as a source of problems for the teacher seminars described earlier. In doing so, we were able to tap into substantial teacher energy and administrative support that could be aligned with the algebrafication agenda. Simultaneously, as teachers were made aware of ways in which MCAS already expected algebraic reasoning, this strengthened the legitimacy of our project in teachers' eyes.

The K–5 Literacy Initiative, begun 3 years prior to GEAAR, was a substantial and well-funded district-wide focus that required significant human and material resources and thus constrained what was available for other professional development initiatives. It began as voluntary,

[5]Not only is Developing Mathematical Ideas compatible with GEAAR, but it is being extended to include the development of algebraic reasoning. The essence of that work is reflected in the two chapters (6 and 16) coauthored by Schifter in this volume. Indeed, the second author is an advisor to that development.

whereby a school could choose to become a "Literacy Collaborative." This committed the school to supporting the development of a school literacy coordinator who, in turn, would train the remaining teachers through an intensive and highly structured program. With extensive summer and academic year training that took coordinator trainees from their regular classrooms for as much as a week per month, and a daily 3-hour commitment of classroom instructional time to literacy, Literacy Collaborative schools tended to be dominated by this initiative, especially during the first 3 years of participation as a Literacy Collaborative school.

Teachers found—not surprisingly—that there was not enough time to account for all the necessary innovations required for literacy, GEAAR, and other professional development programs offered by the district. Because of these felt constraints, we tried to design GEAAR so that it would be congruent with, not in conflict with, the Literacy Initiative. One of the forms this took was supporting GEAAR teacher leaders, through regular meetings with us, to integrate the mathematics of GEAAR into their literacy program and in some cases, to lead school-based programs in doing so.

To get a more concrete look at how this played out in the district, we return first to the case of Julia and Joan. As we described earlier, a one-day professional development workshop integrating GEAAR mathematics and literacy that Joan conducted for her school was well received by teachers. As a result, Julia and Joan collaborated in the design (to which Julia, the school principal, credited mostly Joan) of a multiyear professional development agenda in which teachers could explore algebraic thinking in the context of their ongoing Literacy Collaborative. The core task was to locate children's literature that could be used in conjunction with existing algebraic thinking tasks or for which new tasks could be developed. In the process of examining the literature, teachers would solve the mathematical tasks and thus increase their understanding of algebraic reasoning. Joan wrote to us about the program's design:

> I have talked to Julia and it seems that we are doing a course next year with or without funding. It would run similarly as what you did. Begin with a share. Then what we want to do is integrate a piece of children's literature. We want each question to align with the [Massachusetts Curriculum Frameworks]. Also, I am looking for an MCAS question to accompany the problem. The teachers would walk away with a valuable binder that they could add to. Hopefully, teachers would bring in more problems that they could add.

Joan's plan was to model the teacher meetings after the GEAAR seminar structure, with teachers sharing their experiences and their students' thinking and contributing resources that were or could be algebrafied and adapted across the grade levels. Perhaps more significantly, we viewed Joan's intent to incorporate MCAS questions and to link teachers' work to the state frameworks as recognition of the practical need for professional congruency across efforts.

As Joan described to us, the program was enthusiastically received by teachers and led to new ways of thinking about mathematics. She wrote:

> Our first grade teacher [Carol] has been very involved in the literacy program. . . . She is also very quiet, but always brings samples of her students' work. I really didn't know how she was thinking. Not knowing her very well, I got up the nerve to ask her if she always taught math this way or was this new since we started the class. She told me that she never really taught this way, but knew what she had been doing before really wasn't working. She has admitted that she has changed the way she teaches math (more open-ended and listens more to how the children are thinking). She actually looks forward to the problems and is willing to make them appropriate for her grade level. Some teachers are asking what am I going to do next year to keep this ongoing in the school. A lot of enthusiasm. [sic]

We took Carol's case as an example of how professional congruency could provide teachers access to a diversity of essential innovations while strengthening their awareness of connections across these innovations. That is, Carol was simultaneously thinking about algebraic reasoning, children's literacy, and how the two domains could be connected. In 4 years, Carol had not participated in GEAAR, yet was heavily involved in the school's literacy program. By integrating GEAAR with the literacy collaborative, Joan provided a setting that would overlap with Carol's interests and open her thinking to new ways of teaching mathematics. And perhaps most importantly, aligning the professional development would provide coherence to students' academic experience.

Joan's arrangement was special because she was working within a *literacy* professional development agenda. Other teachers successfully used the GEAAR project as a platform to think about and plan ways to integrate mathematics and literacy. Gina, a first-grade teacher leader from a different school in the district, worked creatively in her own practice to integrate mathematics into the literacy portion of her school day by selecting literature for "read alouds" that connected to mathematical concepts. In connection with this, she used what she called a "Math Board" as an interactive forum for students to write math stories connected to their readings, usually during the literacy portion of the day. The Math Board has a prescribed structure for story writing that includes identifying number, subject, setting (i.e., Where is the subject found?), and action (i.e., What does the subject do?). For example, a poem Gina selected for a read aloud described "measuring yards and yards of string." Students studied measurement concepts, measured string in order to make a kite tail, then wrote stories for which the subject was *kite*, the setting was *beach*, and the action was *twirl*. Whereas this particular activity did not involve algebraic reasoning per se, it is a reflection of an important level of mathematical awareness that has grown out of GEAAR.

What teachers such as Gina and Joan were doing in their individual classrooms and schools was subsequently shared with teacher leaders as a resource for the seminars they led throughout the district so that, ultimately, these ideas traveled in some form across the district. The result was that interest in integrating math and literacy spread across teacher leaders, who by the fifth year of GEAAR were actively identifying children's literature to be used in connection with mathematics. For example, teacher leaders selected a book whose story line involved doubling quantities as a context for in/out problems building functional thinking. Mathematically, this genre of problems became significant for teachers as a basis for student explorations into more complex and formalized notions of function (for examples of this, see Blanton & Kaput, 2004, 2005a, 2005b. Situating these problems within children's literature conceptualized the mathematics and thereby gives students an additional way to access the problem semantics.

As a final example of professional congruency, we describe a collaboration with Developing Mathematical Ideas (DMI), a national mathematics professional development project adopted by the district in our fifth year. DMI is consistent with the goals of GEAAR not only because it emphasizes student thinking and understanding mathematics conceptually, but also because it supports real growth in content knowledge necessary for GEAAR mathematics (Bastable & Schifter, chap. 6, this volume; Schifter, Monk, Russell, & Bastable, chap. 16, this volume). One of the challenges of GEAAR was the difficulty in enlisting teachers to think about algebraic notions when more fundamental mathematical concepts (e.g., place value and the properties of the operations) were at issue and when the teachers had no experience in trying to systematically examine student thinking. DMI addressed an important, foundational content need for the district and we, in turn, worked with district teacher leaders to think about the role of algebraic thinking in DMI mathematics.

In all, we have found that increasing the congruency between our own and other professional development and district initiatives supported teachers in algebrafying their classrooms for the 110 teachers and 15 schools in which GEAAR had a major presence. At a minimum, it increased the range of contexts in which algebraic reasoning could occur and so has increased its potential frequency of integration. Perhaps more significantly, we maintain that the process of integrating algebraic reasoning in diverse subject areas and settings involves a reorganization of conceptual structures that can fundamentally alter teachers' views of mathematics and how it could be taught. Moreover, we conjecture that building algebraic reasoning within the context of other subject areas (e.g., reading and writing) is fundamentally different than teaching it within the context of a mathematics class. In part, it draws on a different set of teacher skills and knowledge that requires teachers to find mathematics in a nonmathematical context. But also, the subject areas themselves draw on a different *personality of instruction*. By this, we mean that teachers who feel more connected to a particular subject area because of inherent interests, knowledge of the subject, or other biographical factors, can bring a certain personality to bear on how they teach that particular subject. For example, it is widely recognized that weak content knowledge and diffidence in a particular subject evokes a tendency to control and direct classroom conversation (if there is any) so that dialogue does not veer into territory unfamiliar to the teacher. Conversely, a teacher who feels particularly connected to and confident in a subject is more likely to be open and flexible in how she teaches it. Thus, situating mathematics within a subject area more comfortable for a teacher can leverage her personality of instruction for that subject. It was not uncommon for us to find teachers such as Carol who were less comfortable with mathematics but were deeply involved in literacy instruction. We claim that integrating mathematics with literacy supported their mathematics teaching because it accessed their personality of instruction for teaching literacy.

Thus, we maintain that teachers can develop a deeper and more compelling view of how to teach in ways that support students' algebraic reasoning when they have occasion to think about it across the breadth of their daily and professional experiences. This has required us to look inclusively at the district's professional development initiatives in order to build connections that increased how teachers perceived GEAAR as relevant to their daily practice. Finally, we strongly emphasize that the coordination of professional development initiatives by and for teachers has required the support of a professional community network that included the type of (distributed) leadership practice that would encourage teachers in designing and leading school-wide professional development.

Developing a School-Wide Mathematics Culture

> I wanted to tell you also that we now have a small study group working on making a math day a month. We are looking at the end of the math book to see what kinds of activities are back there that no one gets to. We are considering having Fraction Day, where all classes will work on fraction activities appropriate to their grade level. Probability will be on another

> day. . . . It is almost like "back of the math book awareness." We are also thinking about giving all classes in the school certain problems to work on and have an answer on a given day. All of the details will be worked out as soon as the group is in place. (Joan, e-mail)

This account might seem somewhat removed from the idea of algebrafying classrooms, but we maintain that it reflects what became an important part of reaching out to new participants and supporting them in transforming their classrooms. Joan's excerpt is about creating a school-wide network that can support a culture of mathematics in general and the ideas of GEAAR in particular. The idea of a school mathematics culture is connected to the existence of a professional community network; it is difficult, if not unlikely, that one teacher can impact the culture of a school without the resources and support of administrative leadership or the intellectual availability of teachers. Indeed, one of the challenges teachers faced early in GEAAR's implementation was being the only teacher or one of only a few teachers from a particular school who were participating in GEAAR. Thus, although these teachers became part of a district-wide community of teachers, they did not have frequent, daily interaction with this community and were essentially isolated within their schools.

"Monthly Math Day"

What Joan described, a tentative idea for getting an entire school to focus on math concepts that were often neglected due to other curricular demands, grew into a significant event that embedded cross-grade investigations into fractions, symmetry, measurement, triangles, and geometric solids into shared school interests. For example, as part of Fraction Day, each class had to design a class flag whose components could be fractionally described as parts of the whole. The flags were displayed on the school bulletin board and were used as part of the Olympic ceremonies in the "Math Olympics" organized by the gym teacher (this ran in parallel with the 2002 Winter Olympics).

"Our school wide initiative is still building up steam. Teachers are looking forward to it and we try to get all the supportive people involved as well" (Joan, e-mail). As the school year progressed, increasing attention was given to making math a more significant and visible part of the school culture. Students' work on common projects across the elementary grades was posted throughout the school building; nonmathematical components of mathematical tasks were distributed to those with the expertise or facilities to accommodate the task. For example, the gym teacher helped students collect basketball data during gym class so that these data could be analyzed in a subsequent math class, whereas the art teacher worked with students on the artistic development of a Geometric Circus, in which students constructed three-dimensional geometric shapes to create a circus with its various components.

There was a reflexive process at work here by which school culture supported the development of and was simultaneously sustained by a professional community network, particularly school-based communities of practice. This process helped build connectivity among teachers, leaders, and ultimately schools via the principals' network. Because they are based on shared experiences, school-wide activities such as Monthly Math Day can help diminish the isolation teachers feel and serve as an incubator in which community can be nurtured and maintained. At the same time, the community itself can support the extension of a mathematics culture so that it becomes more deeply embedded in the fabric of school activity. What seemed clear in our work is that the development of a school-wide mathematics culture required the existence of a professional community network. As we observed this play out, GEAAR teachers who were able to impact school culture did so by leveraging congruency across professional development initiatives with the support of a distributed, rather than a centralized top-down, leadership practice, drawing support from the professional community network as it existed within their own schools or the district at large.

Enabling Teachers to Construct Their Own Resource Base

> Just an interesting aside so early in the year [sic]. I started right out with prime and composite numbers, using the number of days we have been in school. We talked a bit about what would make a number a prime number and a composite. When we got to the number 5, Anthony started to think that all odd numbers would be prime. I asked him to give me another number hoping it would be 7. It was. Jackson really started getting angry with me because I wouldn't call on him, but I wanted this to evolve. I called on Jackson and he told Anthony that he was wrong. Anthony started to argue with him, trying to prove his point explaining that all of the numbers were odd up to this point. Jackson told him that 9 was odd, but there were different ways to make 9, so it had to be composite. Anthony just shook his head, and agreed with him. Then he said I should have gone further. We took the opportunity to talk about how you have to try many examples before you can prove your point. I congratulated Anthony for his conjectures. He looked at me very strangely. I told him that was a good thing, then explained what conjectures were. Not bad for the first week of school. (Joan, e-mail)

The vignette described here, sent to us as Joan, a teacher leader, began her fifth year as a GEAAR participant, illustrates one of our central strategies for professional development: to help teachers algebrafy their instructional resource base so that their ability to identify and extend opportunities for algebraic reasoning would not depend on professional development providers or resources. As we have detailed elsewhere (Kaput & Blanton, 2005), this involved teachers developing new activities or adapting existing materials to use in classroom instruction. Perhaps more significantly, it required learning to respond flexibly to students' thinking in the course of daily instruction so that algebraic thinking could be spontaneously integrated into instruction in meaningful ways. The mining and algebrafying of instructional materials is designed to build teacher capacity to exploit their instructional resource base, independent of its curricular particulars; GEAAR teachers were encouraged to find and act on their local resources to serve the algebrafication objectives.

Practically speaking, teachers learned to algebrafy their resource base as they worked in small groups within GEAAR seminars to modify problems for use in their own classes, with the intent that tasks were to be integrated within their regularly defined curricular responsibilities and not marginalized as enrichment activities. After teachers implemented their modified tasks in their own classrooms, they discussed their experiences, including students' strategies and representational systems, in GEAAR seminars. As different grade- level approaches to the problem were compared, teachers came to see their students' work within the longitudinal context of the grades that surrounded their grade level, thus situating their own activity within the broader curriculum.

In our work, the heart of algebrafying existing arithmetic tasks is to transform them from single-numerical-answer problems into opportunities for pattern building, conjecturing, generalizing, and justifying mathematical facts and relationships. This might occur through varying a given problem parameter (see Blanton & Kaput, 2004), making known quantities unknown so that the task involves the analysis of relative differences (Carraher, Brizuela, & Ernest, 2001), or describing and applying generalized properties of whole numbers (see e.g., Bastable & Schifter, chap. 6, this volume; Kaput et al., chap. 2, this volume; Schifter et al., chap. 16, this volume). In the vignette from Joan's classroom, Andrew's idea that all odd numbers are prime presented an opportunity for algebraic reasoning, for examining whether his conjecture was a valid mathematical generalization for a class of numbers. Joan was able to spontaneously extend the conversation so that through peer argumentation, students disproved the conjecture and subsequently began to examine what counts as mathematical proof.

We maintain that the type of mathematical autonomy exhibited here by Joan is essential in order for teachers to operate independently of a limited professional development program. Teachers need a mathematical knowledge that is self-sustaining so that they are not dependent on extant, and sometimes inaccessible, materials as their sole source of algebraic reasoning. Our objective was to help teachers see opportunities for algebraic reasoning in their available resources, or as these opportunities arose in daily classroom instruction.

Gamoran (2005) notes that forgoing a prepared curriculum, at least initially, "helped teachers to match their curricular interests to their students' emergent thinking" (p. 310). We maintain that teachers' capacity to algebrafy their classrooms is supported when teachers are able to generate their own resource base by rethinking the mathematics they are currently teaching so that it can be extended in powerful, algebraic ways. As Coburn (2003) suggests, teacher change requires going beyond surface characteristics such as changing materials or adding specific activities. When teachers are helped to understand content so that mathematical resources are treated as a dynamic and flexible body of knowledge that can be adapted for particular classroom mathematical purposes, they are able to function in more generative and self-sustaining ways.

GEAAR teachers seemed to be learning, at different levels of proficiency, to algebrafy their resources and develop new tasks that incorporated algebraic reasoning (Blanton & Kaput, 2005a; Kaput & Blanton, 2005). Gina was typical of our teacher leaders in that, as she reported, for the first year of her participation in GEAAR, she focused exclusively on the tasks that we provided, but grew to regularly identify tasks from her own resources that could develop students' algebraic reasoning skills. We describe this as the development of teachers' algebra "eyes and ears" (Blanton & Kaput, 2003). As a teacher leader, Gina observed her "client" teachers become increasingly able to connect their resources with the mathematics of GEAAR. It became an important activity not only because it generated mathematical autonomy, but because it also increased teachers' perception of professional congruency. Moreover, the development of teachers' capacity to algebrafy their own instructional resource base evolved as a shared process situated within a professional community network. Whereas some teachers showed greater individual resourcefulness in how they retooled their curricular resources, their continual sharing of task ideas in community seemed to embed teacher growth across a larger teacher population so that the end result was a more deeply instantiated notion of how algebraic thinking fit into the curriculum.

CONCLUSIONS

Our intent in this chapter was to detail ways in which we observed a district build its capacity to support teachers in algebrafying their classrooms.

What seemed to be common in our observations was the need to build connectivity so that teacher change occurred within a complex network of mutually supportive relationships and ideas. This included building connections among people and groups through the development of a professional community network, building connections across teachers' professional activities so that distinct teacher development agendas did not occur in isolation but would make professional congruency visible and explicit for teachers, building connections across content domains so that teachers developed a broader, deeper way of thinking about the content they taught and were able to help students appreciate these relationships, and building connections across the curriculum so that teachers saw how the content (in our case, algebraic thinking) could be embedded across a broad expanse of topics. Practically speaking, it involved the coordinated professional development of teachers and principals and administrators, being flexible in our purposes in order to increase professional congruency for teachers, and giving teachers autonomy in developing mathematical insight, making curricular decisions, and providing peer leadership. What emerged from this was the development of communities of learning where teachers and principals worked to effect change in how their schools thought about mathematics.

We were never far from the recognition that our work was a fragile process subject to constraints and events over which we sometimes had no control. State budgets and mandates, district priorities, changes in leadership, teacher interests and availability, all converged to define how GEAAR would affect the district. In all, it was a fluid process that required flexibility in our curriculum, goals, and implementation. Our intent was to help teachers become generative and self-sustaining in their instructional knowledge and content knowledge so that the principles of GEAAR would survive the inevitable vagaries of educational circumstance, outlast our presence in the schools, and continue to evolve more deeply in teachers' daily and highly situated mathematical practices. This will be a measure of the viability of the approaches we have taken. Ultimately, as we have come to more deeply appreciate, how teacher learning develops apart from our intervention will be in part a reflection of the institutional setting in which teachers live their professional lives.

ACKNOWLEDGMENTS

We wish to thank Tona Williams and Adam Gamoran for sharing their expertise in educational leadership research on the work reported in this chapter.

The research reported here was supported in part by a grant from the U.S. Department of Education, Office of Educational Research and Improvement, to the National Center for Improving Student Learning and Achievement in Mathematics and Science (R305A600007-98). The opinions expressed herein do not necessarily reflect the position, policy, or endorsement of the supporting agencies.

REFERENCES

Blanton, M., & Kaput, J. (2003). Developing elementary teachers' algebra eyes and ears. *Teaching Children Mathematics, 10*(2), 70–77.

Blanton, M., & Kaput, J. (2004). Design principles for instructional contexts that support students' transition from arithmetic to algebraic reasoning: Elements of task and culture. In R. Nemirovsky, B. Warren, A. Rosebery, & J. Solomon (Eds.), *Everyday matters in science and mathematics* (pp. 211–234). Mahwah, NJ: Lawrence Erlbaum Associates.

Blanton, M., & Kaput, J. (2005a). Characterizing a classroom practice that promotes algebraic reasoning. *Journal for Research in Mathematics Education, 36*(5), 412–446.

Blanton, M., & Kaput, J. (2005b). Helping elementary teachers build mathematical generality into curriculum and instruction. In J. Cai & E. Knuth (Eds.), *Zentralblatt Didaktik der Mathematik, 37*(1), 34–42. (International Reviews on Mathematical Education)[Special issue].

Brown, A. L., Ellery, S., & Campione, J. C. (1998). Creating zones of proximal development electronically. In J. G. Greeno & S. V. Goldman (Eds.), *Thinking practices in mathematics and science learning* (pp. 341–367). Mahwah, NJ: Lawrence Erlbaum Associates.

Carpenter, T. P., & Fennema, E. (1992). Cognitively guided instruction: Building on the knowledge of students and teachers. In W. Secada (Ed.), Curriculum reform: The case of mathematics education in the U.S. [Special issue]. *International Journal of Educational Research, 17*(5), 457–470.

Carraher, D., Brizuela, B., & Earnest, D. (2001). The reification of additive differences in early algebra: Viva La Différence! In H. Chick, K. Stacey, J. Vincent, & J. Vincent (Eds.), *The future of the teaching and learning of algebra* (Proceedings of the 12th ICMI study conference, pp. 163–170). Melbourne, Australia: University of Melbourne.

Cobb, P., & McClain, K. (2001). An approach for supporting teachers' learning in social context. In F.-L. Lin & T. Cooney (Eds.), *Making sense of mathematics teacher education* (pp. 207–232). Dordrecht, the Netherlands: Kluwer Academic.

Cobb, P., McClain, K., deSilva Lamberg, T., & Dean, C. (2003). Situating teachers' instructional practices in the institutional setting of the school and district. *Educational Researcher, 32*(6), 13–24.

Coburn, C. (2003). Rethinking scale: Moving beyond numbers to deep and lasting change. *Educational Researcher, 32*(6), 3–12.

Cooney, T. (2001). Considering the paradoxes, perils and purposes of conceptualizing teacher development. In F. Lin & T. Cooney (Eds.), *Making sense of mathematics teacher education* (pp. 9–31). Dordrecht, the Netherlands: Kluwer Academic.

Elmore, R. F. (1996). Getting to scale with successful educational practice. In S. Furhman & J. O'Day (Eds.), *Rewards and reform: Creating educational incentives that work* (pp. 294–329). San Francisco: Jossey-Bass.

Franke, M. L., Carpenter, T. P., Fennema, E., Ansell, E., & Behrend, J. (1998). Understanding teachers' self-sustaining, generative change in the context of professional development. *International Journal of Teaching and Teacher Education, 14*, 67–80.

Franke, M. L., Carpenter, T. P., Levi, L., & Fennema, E. (2001). Capturing teachers' generative growth: A follow-up study of professional development in mathematics. *American Educational Research Journal, 38*(3), 653–690.

Gamoran, A. (2001). American schooling and educational inequality: A forecast for the 21st century. *Sociology of Education*, pp. 135–153.

Gamoran, A. (2005). Organizational support for teaching for understanding in mathematics and science. In T. A. Romberg, T. P. Carpenter, & F. Dremock (Eds.), *Understanding mathematics and science matters*. Mahwah, (pp. 307–321). NJ: Lawrence Erlbaum Associates.

Gamoran, A., Anderson, C. W., Quiroz, P. A., Secada, W. G., Williams, T., & Ashman, S. (2003). *Transforming teaching in math and science: How schools and districts can support change*. New York: Teachers College Press.

Gamoran, A., Secada, W. G., & Marrett, C. (2000). The organizational context of teaching and learning: Changing theoretical perspectives. In M. T. Hallinan (Ed.), *Handbook of the sociology of education* (pp. 37–63). New York: Kluwer Academic/Plenum.

Garet, M., Porter, A., Desimone, L., Birman, B., & Yoon, K. (2001). What makes professional development effective? Results from a national sample of teachers. *American Educational Research Journal, 38*(4), 915–945.

Grant, C. M., Nelson, B. S., Davidson, E., Sassi, A., Weinberg, A., & Bleiman, J. (2003). *Lenses on Learning, Module 1: Instructional leadership in mathematics*. Parsippany, NJ: Dale Seymour.

Kaput, J. (1999). Teaching and learning a new algebra. In E. Fennema & T. A. Romberg (Eds.), *Mathematics classrooms that promote understanding* (pp. 133–155). Mahwah, NJ: Lawrence Erlbaum Associates.

Kaput, J. (2001, October). Learning algebra using dynamic simulations and visually editable graphs of rate and totals quantities. In J. Abramsky (Ed.), *Reasoning, explanation and proof in school mathematics and their place in the intended curriculum: Proceedings of the QCA international seminar* (pp. 88–100). London, England: Qualifications and Curriculum Authority.

Kaput, J., & Blanton, M. (2005). Algebrafying the elementary mathematics experience in a teacher-centered, systemic way. In T. A. Romberg, T. P. Carpenter, & F. Dremock (Eds.), *Understanding mathematics and science matters*, (pp. 99–125). Mahwah, NJ: Lawrence Erlbaum Associates.

Lave, J., & Wenger, E. (1991). *Situated learning: Legitimate peripheral participation*. New York: Cambridge University Press.

National Council of Teachers of Mathematics. (2000). *Principle and standards for school mathematics*. Washington, DC: Author.

National Research Council. (1999). *How people learn: Brain, mind, experience, and school*. Washington, DC: National Academy Press.

National Research Council. (2001). *Adding it up: Helping children learn mathematics.* Washington, DC: National Academy Press.

Nelson, B. (1998). Lenses on learning: Administrators' views on reform and the professional development of teachers. *Journal of Mathematics Teacher Education, 1,* 191–215.

Nelson, B. (1999). *Building new knowledge by thinking: How administrators can learn what they need to know about mathematics education reform.* Paper series published by the Educational Development Center.

Schifter, D., Bastable, V., & Russell, S. J. (1999). *Developing mathematical ideas. Casebooks, facilitators guides, and videotapes for two modules: Building a system of tens and making meaning for operations.* Parsippany, NJ: Dale Seymour.

Spillane, J. P., Halverson, R., & Diamond, J. B. (2001). Towards a theory of leadership practice: Implications of a distributed perspective. *Educational Researcher, 30*(3), 23–28.

Stein, M. K., Silver, E. A., & Smith, M. S. (1998). Mathematics reform and teacher development: A community of practice perspective. In J. G. Greeno & S. V. Goldman (Eds.), *Thinking practices in mathematics and science learning* (pp. 17–52). Mahwah, NJ: Lawrence Erlbaum Associates.

Wenger, E. (1998). *Communities of practice: Learning, meaning, and identity.* New York: Cambridge University Press.

15

Measure Up: A Quantitative View of Early Algebra

Barbara Dougherty
University of Mississippi

Imagine the following dialogue in first grade as Caylie and Wendy compare three volumes, *D*, *K*, and *P*:

> "I think that volume *D* is greater than volume *K*," said Caylie.
> "How do you know that, Caylie? We didn't directly compare those two volumes," said Mrs. M.
> "Well," said Caylie, "we found out that volume *D* is equal to volume *P* and volume *P* is greater than volume *K*, so volume *D* must be greater than volume *K*."
> "I agree with Caylie," said Wendy. "Because volume *D* and volume *P* are really the same amount so if volume *P* is greater than volume *K*, then volume *D* also has to be greater than volume *K*."

The previous dialogue comes from a classroom that is part of the Measure Up (MU) project at the Curriculum Research & Development Group (CRDG), University of Hawaii. The MU approach derives from work originally conducted by a group of mathematicians, psychologists, and mathematics educators in Russia (Davydov, 1975a, 1975b). Their work stemmed from a need to improve student achievement so that students entering secondary education could successfully deal with more complex mathematics. Rather than focus on changing mathematics in later grades, they decided that perhaps enhancing the mathematics in earlier grades was the key to improvement.

In their view, beginning in the early grades was not indicative of rearranging the sequence of conventional elementary topics. Instead, the Russian group thought more about what children naturally do at early ages—they compare things. Young children continually ask "who has more" and create ways in which to measure such as laying objects on top of each other or putting them side by side. Combining this notion with Piaget's work on child development and Vygotsky's ideas on how to let instruction lead development, the Russian team began a series of a studies focused on changing elementary mathematics from a number-centered approach to a mathematical structure approach.

Traditionally, early elementary students would begin with learning to count and move from that to whole number addition and subtraction. The Russian team wanted students to develop an understanding of the structure of mathematical systems but this could not be accomplished by solely working with whole or natural numbers. If students understood *structure*, Davydov and his colleagues believed that they would be able to apply properties and underlying foundations to any number system, rather than only to natural numbers. They used that perspective as the basis for their mathematical content development.

Vygotsky's reference to scientific concepts (1978) was used to further support this perspective. In Vygotsky's view, there are two basic means by which students learn, through either spontaneous or scientific concepts. Spontaneous or empirical concepts are developed when children can abstract properties from experiences with specific cases, in this instance number. In particular, these concepts in conventional elementary programs progress from natural numbers to whole, rational, irrational, and finally real numbers, in a very specific sequence. Topics are taught within each number system, and often not connected across the different systems. Vygotsky distinguishes scientific concepts from spontaneous concepts. Scientific concepts, as he describes them, develop from experiences that focus on big ideas or conceptual foundations and then lead to identifying, applying, and analyzing those generalized big ideas in more specific instances. If developing scientific concepts is the basis of instruction, then teaching would focus on properties of real numbers by using tasks that embody the conceptual side of number and operations, with specific cases found in natural, whole, rational, and irrational numbers at the same time. Students do not see each number system as a special case but rather see that the mathematical structure applies to all types of numbers.

Davydov (1975a) conjectured that a general to specific approach in the case of the scientific concept was much more conducive to student understanding than using the spontaneous concept approach. To do this, the Russian group use nonspecified, continuous quantities, such as volume or length, to model real number properties and operations. These quantities are represented by literal symbols, or letters, so that relationships of and actions on those quantities can be described. Because no numbers are used during this period of instruction, Davydov refers to this as the prenumeric stage. It constitutes more than half of a traditional school year. Having young children working exclusively with quantities and literal symbols for the initial part of the school year is not typical in an elementary mathematics program.

The use of scientific concept development to promote an understanding of mathematical structures led the Russian team to recommend changes in elementary mathematics teaching from specific instruction about number to a more generalized approach involving measurement contexts. This work has inspired MU and has led to collaboration between the Institute of Developmental Psychology and Pedagogy in Moscow and their Russian partner institutions.

Using Davydov's and his colleagues' early research work, MU has adapted the measurement context in which to develop most mathematics topics, especially number, operations, measurement, geometry, and algebra. This chapter describes how MU approaches algebraic thinking through measurement and the implications of such an approach to classroom implementation.

MU incorporates the research findings and recommendations from the Davydov team. However, the Russian research was first carried out with 8-year-old children. Because MU begins with first grade, it is important to consider other developmental factors.

Another general premise of MU is that students need to have the physical models linked simultaneously with diagrams (part–whole and line segments) and symbolism. Instead of first using a sequence of physical models, then using diagrams, and finally moving to symbols, students experience each of these representations at the same time. This helps students include a mental picture of the physical model as part of a more robust and substantive understanding of the symbolic representations.

HOW IS ALGEBRAIC THINKING DEVELOPED THROUGH MEASUREMENT?

MU, like Davydov's team, assumes that children enter first grade with a view of quantities that centers around comparisons. First graders are concerned with who has more than they do, who has less, and who has the same amount. Their intuitive and spontaneous approach to measurement is the base for mathematical development that promotes first graders' building, recognizing, and using properties of real numbers before they deal with whole or natural numbers. This is called the *prenumeric*

FIGURE 15.1. Demonstration of area.

stage in which students work with nonspecified, continuous quantities such as area, length, volume, and mass, rather than discrete numbers.

Children in grade one first identify what attributes of an object can be compared. They come to realize that comparing colors, textures, and overall shapes does not tell "how much" of something there is. They focus instead on four measurable attributes: length, area, volume, and mass. These measures are developed, not in the formal sense, but in terms of working definitions that are usable by children at this level. For example, first graders treat area as the region within the boundaries of a shape or object by moving their hand over that region as they discuss area. This allows them to compare areas, as well as other measures, of different objects without being confined by a more formal definition (see Fig. 15.1). These working definitions are important because students can use the measures to perform actions that model many types of relationships, as described later in this section.

Within the first week of school, children can rather quickly learn to compare and describe measurements along the above four dimensions in terms of *equal to*, *not equal to*, *greater than*, and *less than*. As they perform measurements on several objects, the teacher presents a problem in which

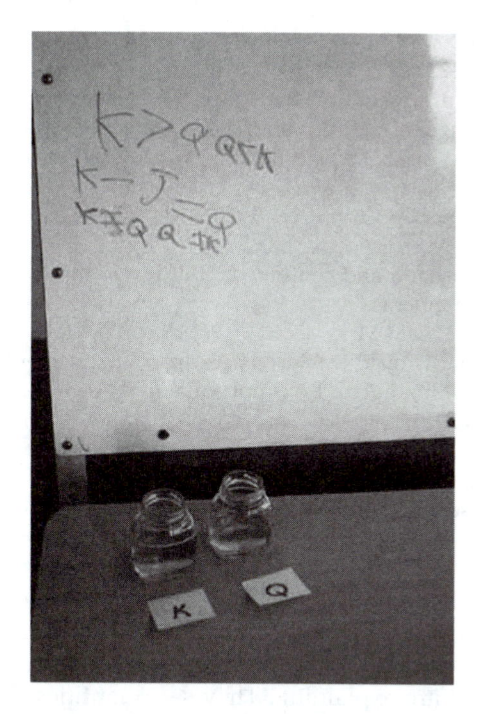

FIGURE 15.2. Measurement and statements.

the physical objects disappear so that students must communicate in writing or orally to someone else about the results of a comparison. They find it difficult to do so because they do not have the linguistic tools for quantifying measures.

This problematic situation creates a need for students to find names for the quantities they compare (see Fig. 15.2). They may suggest using colors, but soon realize that this is not sufficient as the color of the object or the attribute could be the same in both comparisons. For example, if two containers hold volumes of green liquid, there is no way to describe which volume is larger by using color. Thus, the teacher guides students to move to naming the quantities with a letter. The use of the literal symbols is a precursor to variables and comes early in the elementary grades, contrary to others' thoughts (see Carraher, Schliemann, & Schwartz, chap. 10, this volume).

The fundamental properties of equality—reflexive, symmetric, and transitive—are then easily introduced. Because the children are modeling the properties with physical quantities, they can clearly see and describe how these properties work. The following dialogue taken from a first-grade class illustrates how they describe them in the context of mass:

"I see something," said Justin.
"What do you see?" asked Mrs. M.
"Look," said Justin. "Every time two things are equal we can only write two statements. Like when mass K and mass B are equal. We can say mass K equals mass B and mass B equals mass K. Mass K and mass B are the same amount."
Mrs. M wrote $K = B$ and $B = K$ on the board. "That's a good observation, Justin," said Mrs. M.
Wendy raised her hand and continued, "But if two things are unequal we can write four statements."
"Four statements?" said Mrs. M.
"I agree with Wendy," said David. "There are four statements. See, volume F is not equal to volume A and A is not equal to F. And F is less than A and A is greater than F."
Mrs. M wrote F/A, A/F, $F \neq A$, and $A \neq F$ on the board.
"Hey," said Mia, "We can even write more because $A = A$."
(Taken from observation notes, Education Laboratory School, September 2003)

It is important to note the significance of writing statements in the way in which these first graders are doing. The use of the equal sign in the statement $K = B$ is the symbolic representation of an experience that Justin, Wendy, and David are explaining. They see quantities that are the same amount and use the equal sign to record their observation. This is an instance where no operation is used in the equation; it is merely a recording that indicates two quantities of the same amount. Yet, the use of the equal sign in instances like these help students to not confuse the equal sign as an operator, as studies (Kieran, 1981) have shown.

The introduction to the reflexive and symmetric properties moves to the transitive property. Students are asked to compare two objects by some attribute, say length. The students are told that the objects cannot be moved, they have to think through how the comparison can be made. They decide that they can create an intermediate measure that is compared to the objects' attributes. By comparing the outcomes of these measures, students can then infer the comparison of the two attributes.

In Figure 15.3, lengths G and X were first compared. Length X is now the intermediate measure, used to measure another object's length, length B.

From the comparison statements students can infer that length G is less than length B without directly comparing those two lengths. As David described it: "We use an intermediate measure when we have to compare two things that we can't move" (Taken from observation notes, Education Laboratory School, February, 2004).

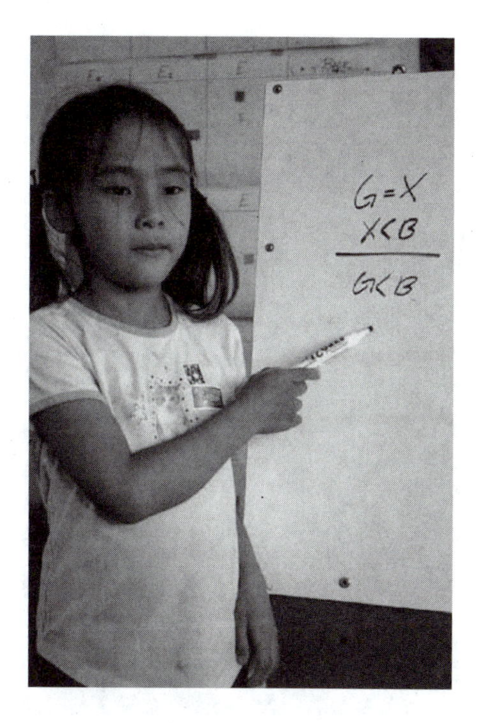

FIGURE 15.3. Student work with transitive property.

From the equality properties and comparison experiences, a variety of number and operation concepts emerge. First, students explore how to make unequal amounts equal. For example, if they have two lengths, length A and length B, that are unequal such that $B > A$, they explain, for example, "you can make the two lengths equal if you do one of two actions. You could subtract a quantity, say length H, from B or you could add the same length H to length A." Length H is given a special name, *difference*. Students explain that the difference can be added to the amount that is less or subtracted from the amount that is greater to make the two quantities equal (Fig. 15.4).

This quantitative modeling enables students to see at least two meanings of subtraction. One is the action of take away and the other is comparison. The comparison model helps students see that if two quantities are unequal, the amount by which they are unequal is the difference. Consequently, if the quantities are equal, then the difference is zero. The difference is the amount that is added to or subtracted from the quantities to make them equal.

FIGURE 15.4. Physical and symbolic representations with length.

Second, students explore the effects of changing two equal quantities with the constraint that they must maintain the equal relationship by adding or subtracting. By modeling the actions of adding or subtracting with length, area, volume, or mass, they make the generalization that it is possible to keep two quantities equal if you add (or subtract) the same quantity from both.

Third, first graders must change two quantities while maintaining a given unequal relationship of those two quantities, which is somewhat more difficult. As Richard noted:

> To keep them unequal, you can add or subtract but you have to be careful. If you add too much, they'll be equal. Or if you subtract too much, they'll be equal. You can't add or subtract the difference. It has to be more or less than the difference—it depends on what you're trying to do. That's what I think. (Taken from observation notes, Education Laboratory School, November, 2001)

Finally, numbers are introduced when situations arise that require the quantification of differences in comparisons. How much larger is length *K* than length *Q*? At this point, students decide that a unit must be found to measure the lengths of both objects. Without a unit, it is not possible to quantify the differences or to make definitive comparisons between the two lengths.

This is a significant point in children's development of mathematical thinking because the generalized approach now shifts to specific quantities and the counting of units that make up those quantities. In a traditional classroom, children would count a collection of discrete objects, assuming that each object is *one*. In essence, they see each object as a unit in and of itself. This is not the case in the progression found in MU, where ad hoc units play an important role at the outset. Students first identify what is meant by a unit—it could be an area unit that is made up of three squares or it might be a volume unit that can be used to measure two volumes in noncongruent containers that could not be compared visually. The multiplicity of ad hoc units, the recognition that units may be different depending on the context, creates a more flexible approach to thinking about relationships among quantities.

The conjectures that are made at this point by the students are fundamental to understanding relationships between units and quantities. Physical measurements help students see that the smaller the unit, the more times it must be iterated to measure a given quantity. However, students have developed residual mental images of the process of measuring a quantity with a unit, that is, mental pictures of the actions they have done. This helps them to make inferences from only the symbolic representations. For example, given $\frac{B}{E} = 5$ and $\frac{B}{Y} = 8$, (read as quantity B measured by unit E is 5 and quantity B measured by unit Y is 8), they conclude that because the quantities B are the same, then unit Y must be smaller than unit E. They justify this with the explanation that unit Y had to be used more times to make B than unit E was used. This means that unit Y is smaller.

In a similar way, if two different quantities are measured by the same unit and represented only symbolically, a comparison can be made without seeing the actual quantities. For example, if $\frac{W}{E} = 10$ and $\frac{P}{E} = 7$, then $W > P$ because it took more unit Es to measure W. Being able to draw conclusions with symbolic representations like these is typically not possible for first graders in a traditional curriculum that has focused solely on whole or natural numbers. And, it focuses on a different view of quantitative reasoning than offered by Smith and Thompson (chap. 4, this volume). The consistent use of letters to name physical quantities gives students a feeling of confidence with literal symbolic representations. Quite naturally, they begin to manipulate the letters in ways characteristic of older and sophisticated mathematics students. For example (see Fig. 15.5 where Caylie is measuring volume units), as students are finding the relationship between the volume unit E and volume H, they write the following equations:

FIGURE 15.5. Using a volume unit to measure volume.

$E + E + E + E + E + E = H$	$H - 3E = 3E$
$6E = H$	$E = H - E - E - E - E - E$
$H = E6$	$\frac{H}{E} = 6$

(Taken from observation notes, Education Laboratory School, February 2003)

By thinking of different ways that a quantity can be separated into parts, students can describe those ways with multiple symbolic representations. The previous representations show that quantities can be separated into parts, which, when (re)combined, make up the whole. Each of the equations expresses a different way to think about the relationship of the unit E to the quantity H.

Diagrams are also used to illustrate these relationships. These diagrams are consistent throughout MU and are explicitly called part–whole diagrams. If the aforementioned example is put in diagram form, students may write something as shown in Figure 15.6. The diagrams in Figure 15.6 show that the whole, quantity H, is made up of six equal parts of unit E, which can be broken up in different ways. This expresses an additive relationship even though the symbolism that children used may be more formally thought of as multiplicative.

Given any of the three diagrams, children could construct quantity H by iterating unit E the number of times indicated in the diagram. The example used leads to questions about the relationship of the parts to the

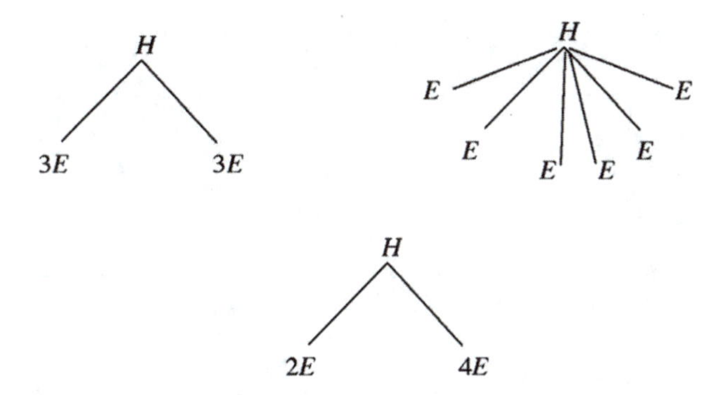

FIGURE 15.6. Part–whole diagrams.

whole. Does every whole have to be broken down into equal parts? Or, conversely, do equal parts have to be used to make a whole? Students' generalizations about these relationships usually include statements that the parts do not have to be equal, but the unit needs to be clearly identified so that it is possible to determine the relative size of the parts.

That the sum of the parts is to equal the whole is well established with first graders even as they move to the more specific cases of number. A natural link between the general cases shown in the measurement context is the number line, thought of as a length constructed with equal-sized units. It can be constructed using any size unit and allows students the opportunity to create multiple number lines (or rays or segments) to fit a specific purpose. For example, if volume is the context, and a volume unit *L* has been identified, a number segment can be created by placing a piece of tape from the bottom of a container to its top. Each time a volume unit is added to the container, a unit is marked on the tape. When the container is completely filled, students have constructed a number segment that shows how many volume units were used to fill it. In Figure 15.7, Justin is creating the number segment by adding a volume unit and marking its height on the tape.

The number segment that Justin created is part of a number line. Rather than thinking of a number line as something that is just given to students without any meaning or context, students in MU create one using a given unit length (see Fig. 15.8). In doing this, they realize that a number line must use a consistent length unit. If the units are not consistent, then quantities cannot be compared.

Representing quantities on a number line requires that students link one type of measurement—say volume—to a different representation—length. To use a number line with understanding, students have to associate two measurements that do not "look" the same. In Figure 15.8,

FIGURE 15.7. Creating a number segment using volume units.

Michael has shown multiple representations of volume using a picture of the physical model and then representing it with a number ray.

A number line is used to represent any quantity that is found in length, area, volume, or mass. Calvin is creating a number ray using a Cuisenaire™ rod as his length unit in Figure 15.9. This ray will be used to represent the mass units that Stephanie measured.

Once students begin working with units and number lines, they are careful not to jump to conclusions about number relationships. For example, in an interview with a first grader, the following dialogue took place:

> "Which is larger, 3 or 8?" asked Mrs. W.
> "It could be 8," said Caitlin. "But it could be 3."
> "Why could it be either one?" asked Mrs. W.
> "Because it could be a small 8 or a big 3. See, if you have 3 really, really, really big units, then 3 could be greater than 8. Or you could have 8 really, really, really, really small units. Then 8 would be less than 3. So it's hard to tell if you don't know the unit," said Caitlin.
> "What if it's on a number line?" asked Mrs. W.
> Caitlin thought and then said, "Oh, then that's ... uh, different. You know that 8 has to be larger than 3 because the units are the same."

(Taken from interview dialogue, Education Laboratory School, February, 2002)

The impact of this child's experience with units clearly shows in her response. Rather than assuming that three and eight have a static relationship

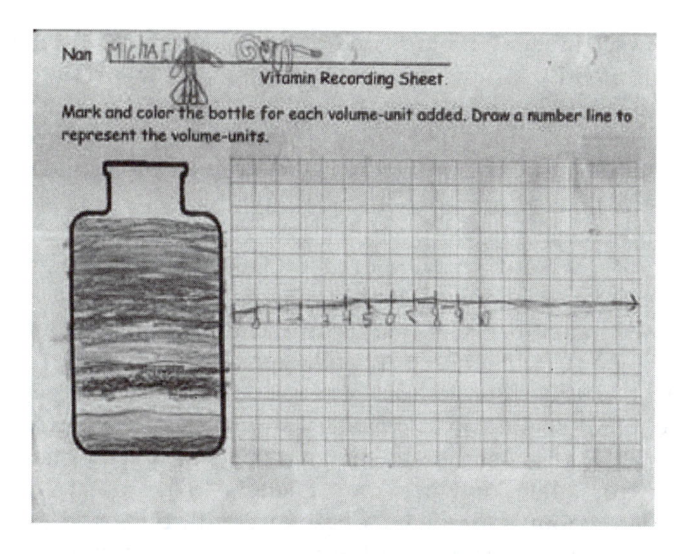

FIGURE 15.8. Multiple representations of volume.

FIGURE 15.9. Mass represented by length.

based on a common unit taken implicitly to be one, she placed a caveat on their relationship that is dependent on the unit used to create the quantity represented by the two numbers. Minskaya (1975) pointed to the importance of unit in these comments:

Many first graders who are good at counting (by the ordinary standards) still identify a number (a set of units) with an actual aggregate. They make no distinction between what they are counting and the method of recording the result and ... they do not understand that number depends on the base, which is chosen. As a result, these children do not acquire a full-fledged concept of number, and this has a negative effect on all their subsequent study of arithmetic. (p. 211)

MU findings agree that, without an understanding of unit, the values or magnitude of number has little meaning for students. It is only after a unit has been identified that one can make assumptions about relationships among quantities.

Building on the work of units and relationships within generalized, and now more specific, quantifications, students begin to actually use a literal variable to represent an unknown number or specific quantity. This is a shift from their previous use of letters to represent continuous, nonspecified amounts. Students' experience with units and part–whole relationships enables them to deal with known and unknown quantities in a more sophisticated way. Equations that are written to represent the part–whole relationships are fluid—that is, students can think of the parts as addends or as the difference and the subtrahend.

If quantities are known, equations (and inequalities) can be written, such as:

$$7 = 3 + 4 \qquad 4 + 3 = 7 \qquad 7 - 3 = 4 \qquad 7 - 4 = 3$$

These four equations represent a fact team that is useful as students move to the next phase of combining known and unknown quantities. For example, if students know that 3 and x are the parts, and 7 is the whole, then they can show the relationships among the parts and whole with the following equations

$$3 + x = 7 \qquad x + 3 = 7 \qquad 7 - 3 = x \qquad 7 - x = 3$$

To find the unknown quantity, some students recognize that the third equation ($7 - 3 = x$) is helpful in finding the missing part. Other students describe the process of finding the unknown quantity as the amount that when added to 3 gives 7 as shown by the first two equations.

Students generalize that the fact team relationship can be used across a variety of numbers. It is not limited to whole numbers because, as they have come to know, numbers are only descriptions of quantities. If a part–whole relationship is known, a fact team can be used to represent the relationships. See Figure 15.10 as an example of the kind of generalization that occurs regardless of the measurement context. Jensen has shown that

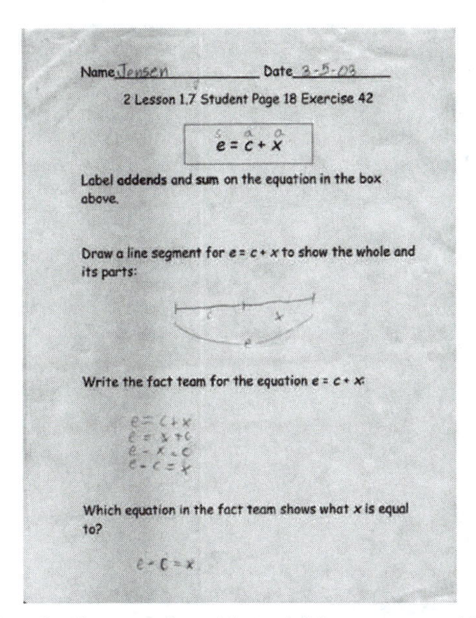

Name Jensen Date 3-5-03

2 Lesson 1.7 Student Page 18 Exercise 42

$e = c + x$

Label addends and sum on the equation in the box above.

Draw a line segment for $e = c + x$ to show the whole and its parts:

Write the fact team for the equation $e = c + x$:

e = c + x
e = x + c
e - x = c
e - c = x

Which equation in the fact team shows what x is equal to?

e - c = x

FIGURE 15.10. Student works with variables or unspecified quantities.

nonspecified quantities can be used to write equivalent statements without knowing the quantities that any of the variables represent.

Measurement contexts provide opportunities for students to develop an understanding of place value that is quite different from the traditional models using the base 10 blocks for whole numbers. Because an understanding of place value is the foundation for working with all computational algorithms and for creating a strong number sense, students should understand how and why place value processes (e.g., regrouping and exchanging) work the way they do. It is difficult to model physically the building of the different places in base 10 because each successive place to the left of the ones increases exponentially. Thus, MU begins with smaller bases, like base 3 or 4.

In a base 4 system, students can build a three or more place quantity and model it with any of the four measures that might include length, area, volume, or mass. If they use length, they realize that four length units make up the Place II position (represented with 10_4) and four of the Place II units make the third place value to the left (100_4). With each successive place to the left of the Place III, the pattern continues. In Figure 15.11, a portion of a wall chart created by second graders is shown. They used the chart to find similarities and differences across bases and to make generalizations about how quantities compare.

Patterns within and across bases promote a deeper understanding of when and why exchanging (or regrouping) is necessary. By beginning

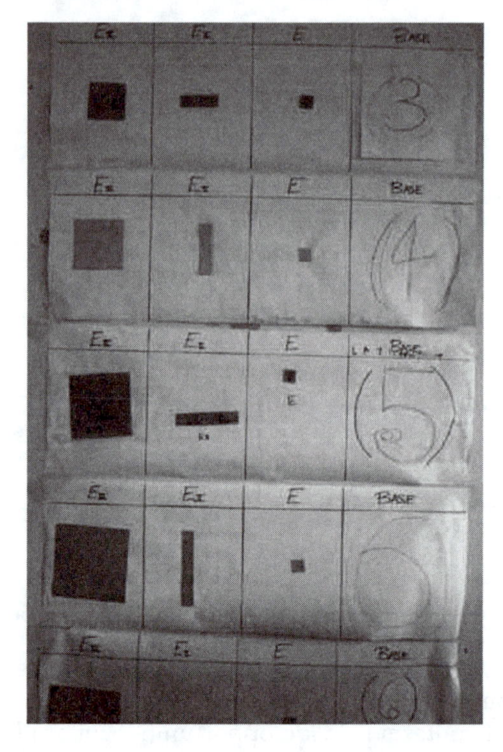

FIGURE 15.11. Wall chart depicting base notations.

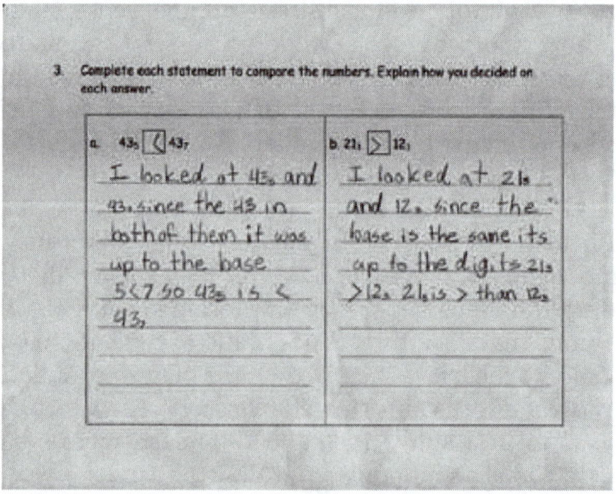

3. Complete each statement to compare the numbers. Explain how you decided on each answer.

a. 43_5 [<] 43_7	b. 21_3 [>] 12_3
I loked at 43_5 and 43_7 since the 43 in both of them it was up to the base 5<7 so 43_5 is < 43_7	I looked at 21_3 and 12_3 since the base is the same its up to the digits 21_3 > 12_3 21_3 is > than 12_3

FIGURE 15.12. Student process for comparing numbers in different bases.

with a specified unit of measure, students build quantities that are based on the way in which number is developed, second graders become fluid in dealing with bases from 2 to 10. They can use the patterns they notice and their experiences with the physical models to create multiple algorithms for a variety of processes, including comparing numbers in different bases, comparing numbers in the same base, and adding or subtracting multidigit numbers (see Fig. 15.12).

At this point, students develop a stronger ability to make and defend generalizations from patterns observed in both specific and general cases. For example, students may be asked to compare the following 2 two-digit numbers in the base 10 system:

$$5\gamma \quad 4\gamma$$

It is made clear to students that the unknown quantity, represented by γ, is a digit of a two-digit number. In this example, they establish that the base is the same and it is a base greater than base 5, as indicated by the size of the digits in the Place II position. From there, they reason that the quantity on the left must be greater than the quantity on the right because the value of Place II is greater. That is, the Place II digit 5 represents a greater quantity than the Place II digit 4.

Students are also confronted with a problem such as

$$\lambda\phi \quad \lambda\alpha.$$

This is a generalized representation similar to the previous problem and students are advised to note that the symbols represent digits of a two-digit number. Again, they are asked how the two quantities compare. Although some students would like to illustrate their reasoning with specific examples in which they substitute numbers for the symbols, others write that they cannot compare them because the Place II value is the same in both quantities so their relationship is dependent on the values in Place I.

Because place value is developed from a conceptual basis, students in the MU project see addition and subtraction in terms of units (e.g., ones, tens, hundreds, and thousands in the decimal system). Thus, each place value has a specific relationship to the initial unit (or ones), called a main measure. In the decimal system, students explain that it takes 10 main measures to create a Place II or 10s unit. To make a Place III or 100s unit, you need 10 of the 10s units. One generalization that they make as a class is that, in base 10, "any time you have 10 of any unit, it creates the next place." The idea is generalized to any base so that exchanging or regrouping is viewed as an integral part of adding or subtracting (combining, taking away, or comparing).

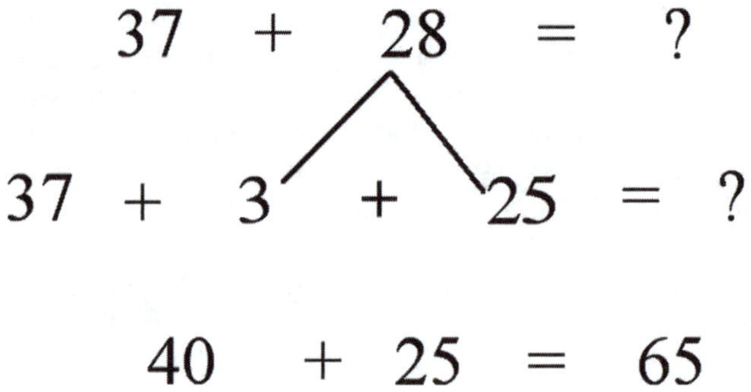

$$37 + 28 = ?$$

$$37 + 3 + 25 = ?$$

$$40 + 25 = 65$$

FIGURE 15.13. Using a round-number strategy and part–whole relationships.

This concept supports the construction of algorithms that use 10 as their basis. For example, in the "round number strategy," numbers are broken apart so that they can make a 10 (Fig. 15.13). In this case, 28 can be thought of as 3 and 25. Thus, 37 and 3 can be combined to make a round number, 40. It is now much simpler to add 40 and 25 to get the final sum.

Having the ability to compose and decompose numbers based on the parts and whole gives students the opportunity to construct specific quantities in multiple ways. They realize that combining 12 and 17 results in 29 regardless of whether you add 12 to 17 or 17 to 12. This enhances their understanding of properties, including the Commutative Property of Addition, Identity Property of Addition, and the Associative Property of Addition. These properties are made explicit by name and linked to the actions that show why they "work."

Almost seamlessly, students move from the models of addition and subtraction to multiplication and division. The concept of unit permeates the foundation of these operations in that the introduction of multiplication and multiplicative comparisons stems from the need to create an intermediate unit.

The scenario used is that the teacher needs to find how many volume unit *E*s make up volume *Q*, which is in a rather large container. Volume unit *E* is quite small. As two third-grade students begin measuring *Q* with *E*, they quickly realize that it will take a very long time to find the number if *E*s in *Q*. To speed up the process, they suggest that an intermediate unit *C* could be created. Then the intermediate unit could be used to measure volume *Q*. Their reasoning is that if they know how many volume unit *E*s it takes to make intermediate unit *C*, and they know how many intermediate unit *C*s it takes to make quantity *Q*, they are able to find the number of volume unit *E*s needed to make the quantity *Q* by, in essence, repeated addition. The diagram of this process shows that an initial unit *E* is used nine times to make the intermediate unit *C*. The intermediate unit is then used eight times to make quantity *Q*. That would mean that *Q* is equal to the sum of nine 8s or the product of 9 and 8 as shown in Figure 15.14.

Although this example illustrates the repeated addition model for multiplication, the model also allows a division concept to develop naturally, by focusing on the relationship among the quantity, unit, and intermediate unit. If, for example, the quantity is known and the unit is known, students can use their understanding of the relationship among the unit, the intermediate unit, and the quantity to find the missing amount. In this case, the missing amount is the intermediate unit. In a similar fashion, if the quantity and the intermediate unit are known, students can find the missing amount, the unit. Consequently, multiplication and division can be simultaneously developed rather than sequentially—analogous to addition and subtraction being treated simultaneously. This supports the need for students to be able not only to build quantities but to be able to decompose them as well.

The fluidity that comes from the introduction of all number operations in MU promotes the development of properties often neglected until students reach much higher grade levels. Properties that students found when they worked with addition (e.g., commutative, associative, and identity) apply to multiplication and division as well and are facilely modeled in a similar fashion. For example, the Commutative Property of Multiplication is modeled by selecting a unit (unit *E*) and iterating five times to produce an intermediate unit *D*. Intermediate unit *D* is used four times to create quantity *J*. This illustrates that quantity *J* is comprised of 20 unit *E*s. Likewise, unit *E* could be iterated four times producing intermediate unit *Q*. Unit *Q* is then iterated five times to make quantity *J*. Again, unit *E* was used 20 times to make quantity *J*. It is illustrated in the two arrow notations in Figure 15.15.

Other, more complex, properties are accessible by students through a generalized arithmetic, or patterning, approach. For example, in a property that students call the Division with Multiplication Property, they made this conjecture:

> If you have a number that is divided by a product, it gives the same result if you divide the number by the first factor of the product and then divide that quotient by the second factor.

With prompting from students, this property is written in the generalized form of a $a \div (b \times c) = (a \div b) \div c$. Detecting and then representing the property in this generalized form gives students a tool to help them in computational settings where they can apply it as needed. Some students

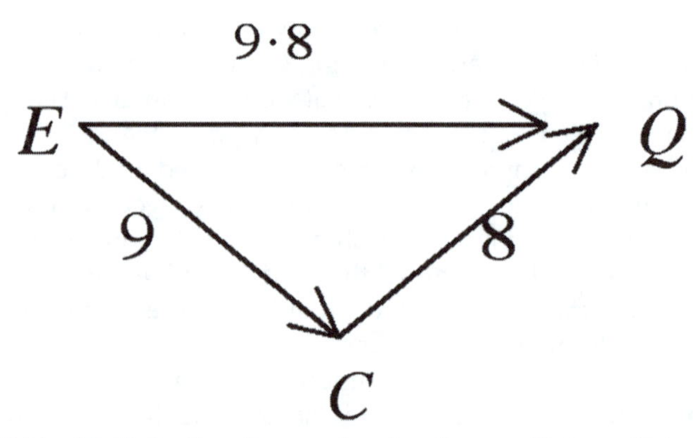

FIGURE 15.14. Multiplication diagram showing the use of intermediate measure C.

find that it is easier for them to do a series of divisions rather than find a product first and then divide by that product, especially if they are not sure about the value of the product. Other students may find that they prefer to find the product first, and then do the division as the left side of the generalization is stated. Regardless, being able to state properties in a generalized form gives students more fluency with numbers and operations.

The multiple representations used in MU (physical, diagrammatic, and symbolic) provide a structure for solving computational word problems. Problems like the following can be seen as a description of relationships of quantities rather than an unsolvable problem:

> Jason has v mass-units of rice. Jon has w mass units less than Jason. How many mass units of rice do Jason and Jon have together?

One model that could be used to solve the above problem is with a line segment diagram (see Fig. 15.16). Students see that it is the combining of v and $v - w$ that gives the amount of mass that Jason and Jon have altogether. As one student noted, it ends up being "like two Vs put together with w taken out" (Macy, Education Laboratory School, December 2003). Writing an equation that represents the quantities in the problem becomes a literal translation of the diagram— $V + (v - w) = A$, where A represents the amount they have altogether.

Similarly, given problems that show a multiplicative relationship, students are able to think about the way the quantities interact to solve the problem. For example, the following problem was given to third-grade students:

Jessica has 7 oranges. Raul has 4 times as many oranges as Jessica. How many oranges do they have altogether? Students use the relationship of the unit to the quantity to create a diagrammatic representation.

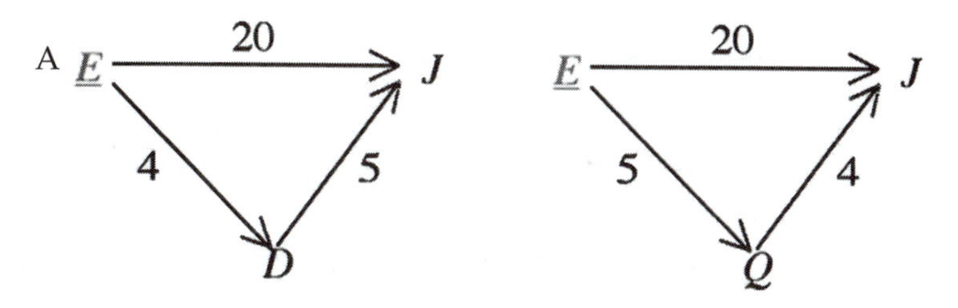

FIGURE 15.15. Diagrams showing the commutative property of multiplication.

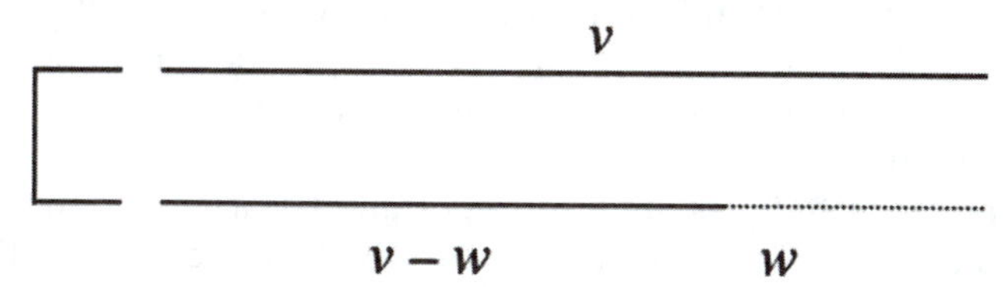

FIGURE 15.16. Solving a word problem with a diagram.

The relationships can be expressed in multiple ways:

$$E + (E + E + E + E) = 5E \text{ (}E\text{ represents seven oranges as a unit).}$$
$$5E = 35 \text{ oranges.}$$

In this problem, Richard's diagram (observation notes, Education Laboratory School, third-grade) shows that two quantities. In the first line, the 7-unit represents Jessica's quantity. The 7-unit repeated four times in the line beneath indicates Raul's quantity. Figure 15.17 shows by the end notation that the total amount is the addition of the two quantities.

WHAT ARE THE IMPLICATIONS FOR IMPLEMENTING A CURRICULUM SUCH AS MEASURE UP?

Developing mathematical understanding through measurement and with an algebraic reasoning foundation requires teachers to understand mathematics in a different way. They must be able to see the quantitative, prenumerical relationships in grade one and understand how they lead to the specific numerical cases that will follow. In a traditional curriculum where new topics are introduced with sometimes little or no connections to previous topics, there is no undercurrent that provides the cohesion for a full curriculum. In MU, the teacher has to be aware of how the topic was

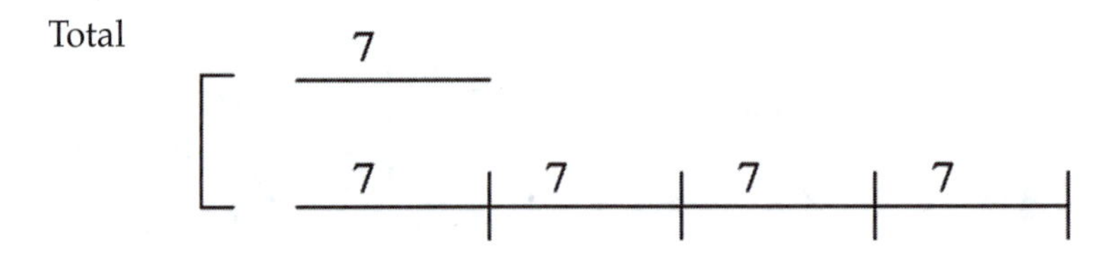

FIGURE 15.17. Diagram representing a multiplicative relationship.

developed, what topics are approaching, and how the bridges are made between and among them.

Most important, teachers who are using MU must be open to thinking about mathematics in a different way. They need to have multiple ways of thinking about relationships and be able to apply those relationships to a more formal mathematical structure. Even in early grades, this content knowledge must be well developed.

Pedagogical techniques are also important. Clearly, if young children are to be able to access a curriculum rich in algebraic ideas in the early grades, the pedagogy aligned with that mathematics must be different than what would be expected in a conventional mathematics classroom. The teacher has to expect twists and turns within a lesson and be ready to handle them. The unpredictability of what students will say makes the daily lesson a challenge.

The orchestration of the physical actions with objects, intermediary representations, and symbolism is challenging. Teachers must constantly be thinking about how the action is linked to the symbolism and then encourage and model appropriate languaging about the students' observations of such actions.

Aligned with Vygotsky's (1978) suggestion, the pedagogy must include a strong language component so that children have opportunities to describe the relationships they find, make conjectures, justify their thinking, and use multiple representations to describe what they see and think. Without this, young children cannot make sense of the mathematics in a meaningful way. It is this sense making that helps children apply the mathematics appropriately across a wide variety of situations. This does, however, demand a student-centered classroom that has an inviting environment to motivate students sharing ideas.

In preliminary professional development institutes associated with MU, a focus on mathematical content has taken precedence. With institute instructors modeling the pedagogy, teachers are given authentic learning tasks that allow them to experience the mathematics in a different way than they may have originally learned it. The tasks are derived directly from student tasks so that teachers too have the opportunity to learn mathematics through measurement. Such experiences challenge teachers' conceptions of what they have believed mathematics to be and what they have thought to be true.

Because MU is still in its infancy stage, curriculum materials from the project are not available to teachers in the professional development institute. Instead, the institute focuses on helping teachers enhance the content in their programs by adapting tasks and creating higher order questions. The adaptation process and the creation of the questions are based on teachers' new and altered understandings of the mathematics found in elementary curricula.

Thus, the implications for instruction and the level of teacher content knowledge may be complex. In the early stages of this project, it appears that teachers will have to learn more and different mathematics, as well as become comfortable in using a wide array of instructional techniques even with young children.

CONCLUSIONS

An approach to elementary mathematics that focuses on nonspecified, generalized quantities is often thought to be too abstract and thus, not accessible by young children. MU preliminary results, however, support Davydov's claims that it offers young children a meaningful foundation on which to build sophisticated and complex mathematics. Understanding the structure and properties of mathematics creates a way for children to construct solid underpinnings that lead to substantive mathematics (this sentence seems to be a tautology). It builds confidence so that even within nonroutine or unfamiliar situations, children can reason through the relationships expressed in the problem.

The combination of physical, diagrammatic, and symbolic representations in a measurement context that underlies all of the mathematics appears, in an early analysis in MU, to have a strong positive influence on children's abilities to deal with more complex mathematics at an earlier age. Davydov (1975b) contended that "there is nothing about the intellectual capabilities of primary schoolchildren to hinder the algebraization of elementary mathematics. In fact, such an approach helps to bring out and to increase these very capabilities children have for learning mathematics" (p. 202).

Davydov's work has caused our group to rethink what might be possible if we step out of the box of conventional elementary mathematics. Using continuous measures rather than focusing on counting discrete objects provides a context in which children can explore and describe mathematical relationships that are fundamental yet substantial mathematics. We have seen that young children can grapple with and come to some resolution about mathematical ideas that are typically found in later years such as the relationships of quantities measured by different-sized units. They are capable of using symbolic notation as a means of conveying their understanding and interpretation of problem situations. Students see relationships among operations, noting that addition and subtraction or multiplication and division have commonalities that go beyond the way you get an answer. Equations are viewed as statements of relationships with the equal sign functioning as an indicator of that relationship and not an operator. They can flexibly use multiple representations, interchanging how they express their ideas. However, the Davydovian approach has also raised issues about what are the trade-offs when an elementary program is constructed around these premises.

How are data analysis and probability developed with young children when measurement is the underlying context? Because children use multiple strategies and decompose numbers so flexibly, how do they develop algorithms that support an efficient way to solve number problems? Is this important? What if mathematics for very young children started in pre-kindergarten (before age 5) with a focus in measurement? What is the potential?

By using specific diagrams to represent the mathematics, we are concerned that students focus on structure rather than on the mathematics that drives the structure. This will be an issue that we continue to monitor as the project matures.

Preliminary results seen in students participating in MU indicate that Davydov's recommendations are worthy of further exploration in our journey to improve students' achievement in mathematics.

REFERENCES

Davydov, V. V. (1975a). Logical and psychological problems of elementary mathematics as an academic subject. In L. P. Steffe (Ed.), *Children's capacity for learning mathematics. Soviet studies in the psychology of learning and teaching mathematics* (Vol. 7, pp. 55–107). Chicago: University of Chicago.

Davydov, V. V. (1975b). The psychological characteristics of the "prenumerical" period of mathematics instruction. In L. P. Steffe (Ed.), *Children's capacity for learning mathematics. Soviet studies in the psychology of learning and teaching mathematics* (Vol. 7, pp. 109–205). Chicago: University of Chicago.

Kieran, C. (1981). Concepts associated with the equality symbol. *Educational Studies in Mathematics, 12*, 317–326.

Minskaya, G. I. (1975). Developing the concept of number by means of the relationship of quantities. In L. P. Steffe (Ed.), *Children's capacity for learning mathematics. Soviet studies in the psychology of learning and teaching mathematics* (Vol. 7, pp. 207–261). Chicago: University of Chicago.

Vygotsky, L. S. (1978). *Mind in society. The development of higher psychological processes*. Cambridge, MA: Harvard University Press.

16

Early Algebra: What Does Understanding the Laws of Arithmetic Mean in the Elementary Grades?

Deborah Schifter
Education Development Center, Inc.

Stephen Monk
University of Washington

Susan Jo Russell
TERC

Virginia Bastable
Mount Holyoke College

A *New York Times* article from December 19, 2002 ("Cutting Jargon, Klein Offers a Report Card Johnny Can Read," Abby Goodnough), reports on a controversy sparked by the introduction of a new report card for New York City's elementary school children. Parents, teachers, and Schools Chancellor, Joel Klein, complained the report card was too long and too dense. Offered as an example of the *bewildering* descriptions of the skills fourth graders were expected to acquire was the rubric, "understands the commutative, associative, and distributive properties."

The properties cited in New York City's report card fracas help to define the basic operations. The methods students use when operating on two or more numbers can be defined in terms of these fundamental properties or laws. From the standpoint of mathematical analysis, then, inclusion of "understands the commutative, associative, and distributive properties" among the standards prescribed for the elementary curriculum would seem inarguable, however unfamiliar to nonmathematician adults the terms in which they are described.

This chapter holds that curriculum appropriate to children in the elementary grades cannot be determined through mathematical analysis alone, but must be developed in conjunction with systematic work exploring how children's mathematical thinking develops. The view that underlies this thesis grows out of many years of collective experience, ours and that of many others in mathematics education, spent trying to answer questions like, "What does it mean to say of a fourth grader that she or he 'understands the commutative, associative, and distributive properties'?" What we, the authors of this chapter, have seen is that, as children learn about addition, subtraction, multiplication, and division—developing an understanding of the kinds of situations that can be modeled by the operations, sorting out various representations for them, and figuring out how to compute—they observe and comment on regularities in the number system. They may notice, for example, that the calculations 72 – 38 and 74 – 40 produce the same result, or that successive answers to a series of problems (10 + 1, 10 + 2, 10 + 3, . . .) increase by 1. In our view, such regularities, emerging naturally from children's work, become the foundation not only for exploration of generalizations about number and operations, but also of the practices of formulating, testing, and justifying such generalizations—and it is these practices that are at the heart of what we mean by *early algebra*.

To repeat, when making decisions about content of the elementary mathematics curriculum, one must consider both the perspective of the mathematician who identifies fundamental concepts of the discipline as well as that of the child first learning how to explore and negotiate numbers and operations on them. This chapter examines the three fundamental properties (also called the laws of arithmetic) from both points of view. We find that, at times, there is considerable overlap and, at times, there is significant difference.

BACKGROUND

Recent calls for the improvement of mathematics education in the United States have set, alongside demands for computational proficiency, ambitious goals for conceptual understanding (Cohen, McLaughlin, & Talbert, 1993; Hiebert et al., 1996; Kilpatrick et al., 2001; National Council of Teachers of Mathematics, 2000). These calls have also urged elimination of many of the most serious barriers to a full and enabling mathematics education for all students. Among these barriers (foremost among them, some have argued; Moses & Cobb, 2001; Moses, Kamii, Swap, & Howard, 1989) is school algebra, which continues to baffle large numbers of middle and high school students, including many who have been generally successful in their first 6 or 7 years of mathematics study. To the extent that middle- and high school algebra courses act as social filters for a variety of future career opportunities, children who do not succeed in algebra are largely shut out of such mathematics-related careers as medicine, engineering, and business (Steen, 1995). A number of researchers have argued that to overcome this barrier, K–5 students should be exposed to the ways of thinking and communicating that form the foundation out of which algebra develops and is formalized (Kaput, 1995; National Council of Teachers of Mathematics, 2000; RAND Mathematics Study Group, 2003).

But, in the history of mathematics education reform in the United States, these recent calls for the inclusion of algebra at the elementary level are not unprecedented. On the contrary, as those who remember the "New Math" of the Sputnik era, the late 1950s and the 1960s, can attest. "The new emphasis on operations was largely a means of providing a systematic, logical rationale for the operations of algebra. For example, $7a + 2a = 9a$ not because of the fact that 7 apples together with 2 apples amounts to 9 apples but because of the distributive property of multiplication over addition" (Osborne & Crosswhite, 1970, p. 284). The Cambridge Conference report of 1963 includes in its recommendations for grades K–2: "Questions that lead the children to 'discover' the commutative nature of addition and multiplication," and for Grades 3–6, simply, "Commutative, associative, and distributive laws" (DeVault & Weaver, 1970). Readers may have encountered the remnants of this reform in their school textbooks, which encouraged committing definitions of these terms to memory, but offered little opportunity to understand their significance.

As evidenced by the *Times* article, current efforts to improve the teaching and learning of mathematics urge renewed attention to the laws of arithmetic. The focus on integrating the building blocks of algebra throughout the grades, rather than waiting for the single course in high school, is one reason educators and curriculum designers are thinking about what it means for elementary school children to study these properties. For example, in the Grades 3–5 section of the *Principles and Standards for School Mathematics* (National Council of Teachers of Mathematics, 2000), among the expectations listed under algebra is that all students should be able to "identify such properties as commutativity, associativity, and distributivity and use them to compute with whole numbers" (p. 158).

But what does such a statement mean? Readers may have seen, and may have even been victims of, attempts to incorporate study of these properties in the elementary curriculum in ways that lead only to struggles with strange-sounding terms—and memories of these struggles may be responsible for the reactions of the New York City parents and teachers. How does the demand for "identify[ing] such properties as commutativity, associativity, and distributivity and us[ing] them to compute with whole numbers" translate into the classrooms of 6- to 12-year olds who are making meaning for the basic operations? How do we build on students' emerging knowledge of number and operations to help them engage with the ideas of algebra?

In our work, the authors of this chapter have seen that the regularities children are prepared to notice are not usually of the same order of abstraction as the mathematicians' axioms. In the elementary classroom, exploration of ideas that may seem trivial to the mathematician can prove immensely fertile, and some distinctions important to mathematicians can make no sense to children. Therefore, as designers of K–5 mathematics curriculum, we have chosen the path of identifying regularities implicit in students' work on arithmetic, investigating which of those regularities can be explored productively, and collaborating with teachers to focus classroom activity on those same regularities. We hypothesize that significant foundational work can occur during these years, work that will prepare students for later encounters with algebra. We hypothesize further that this approach to early algebra can help students meet more confidently the challenges already part of every K–5 program—to understand number and operations and develop computational proficiency.

In this chapter, we step into the elementary classroom to see how ideas related to the commutative, associative, and distributive properties can be engaged. First, we view young children's varied and textured considerations of the significance of the order of terms in an arithmetic expression—the substance of the commutative property. We next inquire into their work on propositions related to the associative property, establishing that students' proofs can flow quite naturally from their visual representation of the operations. And, finally, we examine two episodes in which children discuss, first, strategies for multiplication and, second, a proof that the sum of two even numbers is even—two consequences of the distributive property. These examples are preceded by a description of the context of our work and the data from which we draw. We close with reflections on implications for curriculum design.

THE CONTEXT OF OUR WORK

Our observations come out of the work of the algebra team (the four authors of this chapter) responsible for revising a component of the K–5 curriculum, *Investigations in Number, Data, and Space.*[1] In the context of this project, our central tasks are two-fold: to bring out, for teachers and for students, the generalizations that underlie students' work in number and operations; and to design a unit at each grade level for a K–5 strand on functions and the mathematics of change. In the last 2 years, we have worked together to investigate how early algebra can arise quite naturally out of the work of the elementary classroom, how teachers can further encourage and develop it, and how a curriculum can help guide teachers as they do so.

Although we have been working on the *Investigations* curriculum revisions for just 2 years, members of our team have been pursuing questions about children's algebraic thinking for many more (Bastable & Schifter, chap. 6, this volume; Monk, 2003; Schifter, 1999; Tierney & Monk, chap. 7, this volume). For example, in the context of previous professional development projects, teacher participants had written cases based on mathematics discussions that had taken place in their classrooms, some of which touched on ideas of early algebra.

Now, in our current project, a group of teachers[2] has been collaborating with us, bringing examples of early algebra from their classrooms and helping us understand what happens when they structure their lessons in particular ways or include certain questions as part of their classroom routine. For example, what happens when children build the habit of addressing such questions as: Why does it work out that way? Will it always work that way? How do you know? A teacher might say, I'm not convinced; prove it to me. Can you convince your classmates? Can you convince a younger child?

During our first year of work together, monthly meetings of project staff with collaborating teachers were organized around both an initial outline of the main mathematical ideas to be treated in the curriculum and a set of mathematical tasks developed to help teachers explore those ideas. In addition, teachers read and discussed some of the cases, reflective and usually detailed descriptions of classroom episodes, produced in the earlier projects.

As teachers engaged together on the mathematics and on the cases, they refined their understanding of the mathematical ideas involved,

---

[1]The authors have also created a professional development module in the *Developing Mathematical Ideas* series called *Reasoning Algebraically About Operations* (RAO; Schifter et al., 2008). Many of the classroom episodes used as illustration in this chapter are included as cases in the RAO casebook.

[2]The collaborating teacher group includes representatives from urban, suburban, and rural school systems, as well as monolingual, bilingual, and ESL classrooms.

analyzed children's responses, and considered which classroom activities would best support children across the grades as they take up these ideas. The teachers introduced these or related mathematics activities to their students, then wrote their own cases (meant to be shared with the group) documenting the resulting classroom process. They reported on their students' thinking as this was reflected in classroom conversations, the ways in which representations were used by their students, and the questions this episode brought up for them about their own teaching practice. Staff members read and responded to each case, highlighting in particular the mathematical-conceptual issues in play.

These cases, together with videotapes from the classrooms of a small subset of teachers in the group, provided data we used to refine our own thinking about the key algebraic ideas to be addressed in the curriculum, the development of those ideas across the grades, and the classroom tasks that could draw out those ideas.

At the time of this writing, in the second year of the project, the teachers are field testing the curriculum materials so far produced. Written and oral feedback from the teachers, their own case writing, field notes by classroom observers, and video footage are analyzed as we seek to further refine the curriculum.

EXPLORATIONS IN THE PROPERTIES OF THE OPERATIONS

In the context of our project, teachers have been explicitly addressing generalizations about number systems with their students. As we begin to attend to student thinking about, for example, the consequences of changing the order of the numbers in a calculation, we find their ideas to be much more nuanced and complex than we had previously imagined. Some of the questions this raises for us are: What does it mean for students to understand the properties of the operations? What kinds of representations and arguments can students of different ages use? What are the steps along the way?

The Complexity of Apparently Simple Ideas: Approaching the Commutative Law of Addition

> I asked [my third-grade students] if 0 + 10 was the same as 10 + 0. As a group, they decided the answer was YES *and* NO: "YES, because they are the same numbers and they equal the same, 10; but NO, because they're in a different order. They're not in the same spots."

In this short vignette, teacher Jan Szymaszek captures what it means to find regularity in our number system—to identify what stays the same among things that are changing. In the context of a discussion about addition pairs that make 10, Szymaszek's students had written $0 + 10, 1 + 9, 2 + 8, 3 + 7, 4 + 6$, and $5 + 5$, and then concluded, "You can switch them around for the other ones." Upon questioning by their teacher, they specified that (now paraphrasing) even though the order of the addends changed, the addends, themselves, stayed the same, and so did the value of the sum.

When given opportunities to articulate their observations in a mathematics classroom, this idea—that changing the order of two addends does not affect the sum—arises frequently. However, probing more deeply, this simple idea turns out to be a more textured notion. To illustrate this point, we visit three classrooms.

Ana Vaisenstein's First-Grade Class, November: Localized Generalizations. In the first few months of Ana Vaisenstein's first grade, students periodically worked on a type of problem called "How Many of Each?" Students are told they have a given number of objects (e.g., seven vegetables) comprised of two categories (e.g., some are peas and some are carrots). Students answer the question, how many of each? Early in the year, they might find just one answer. Later, they find several. Still later, they might address the question: How do you know if you have found all possible solutions?

In November, Vaisenstein wrote about her students' thinking about "How Many of Each?":

> In conversations we have had about the different ways one could have, for example, seven peas and carrots, children brought up the idea of "opposites." By that, they meant that 2 peas and 5 carrots is the opposite of 5 peas and 2 carrots. The "opposites" became an idea that many children began to adopt. They brought it up in observations of other children's work, or as a strategy to solving this type of problem.... As these first graders are holding onto [the idea of "opposites"] very strongly, I think it is worthwhile exploring it as much as possible.... The most important thing right now is to watch out for keeping the expression "opposites" alive, and not just a word that the children say repeatedly which loses meaning in the process.

The children in this first-grade classroom are intrigued by a regularity they see in their "How Many of Each?" problems. Within this class of problems, they can find what stays constant within the differences: Even though the total number of objects may change from one problem to the next, whenever x of an object of Type 1 and y of an object of Type 2 is a solution, so is y of an object of Type 1 and x of an object of Type 2. The very fact that children have given a name to this phenomenon—"opposites"—indicates that they have formulated a generalization. At first glance, the students seem to have discovered $x + y = y + x$.

However, we must be careful not to impose on these children a generalization immediately obvious to us. First, their observation may still be very local, confined to the set of problems characterized in their program as "How Many of Each?" Second, perhaps more importantly, these children may not yet have a notion of addition as an operation. At this point in first grade, they are simply listing pairs of numbers. As of now, *opposites* is a property of "How Many of Each?" and not necessarily a property of addition.

Carol Walker's Second-Grade Class, April: Justification and the "=" Sign.[3] By April in second grade, the children in Carol Walker's class certainly do have an idea of addition as an operation, and the generalizations they make extend beyond the context of particular problem types. Walker wrote:

> *Turn arounds* came up first when we were generating ways to make ten early in the year. The children made a list like 5 + 5, 4 + 6, 3 + 7—and then would suggest 6 + 4, 7 + 3, etc. and referred to them as turn arounds. Soon everyone was calling 4 + 6 and 6 + 4 turn arounds and it became almost a vocabulary term without ever really discussing its implications. So I decided to ask them to think about turn arounds and see if they might define it or describe it or illuminate something about it for me. These are some snippets of what I watched and heard.
>
> Natalie[4]: Turn arounds always work. I just know they do.
> Me: How do you know?
> Natalie: Well, look. 27 + 4 = 31 and 4 + 27 = 31.
> Me: But does this always work, for any number, no matter how big it gets?
> Natalie: Well, let me try it.
>
> So Natalie tried numbers in the hundreds and added them together both ways and felt convinced that it always worked. Her reasoning seemed to be based on her having done many of them and having had them always work out to be the same answer.
>
> I did ask her if something like 13 + 23 = 23 + 13 is true. Her immediate response was that, no, it didn't work. Several other children confirmed that it didn't work either, saying, "There's no answer here." Even after they felt sure that 13 + 23 = 36 and that 23 + 13 = 36, no one felt like the original statement could be true.
>
> Other children spent their time making up additional problems and solving them both ways. Ingrid's [written] work [which starts with single- and

[3]Part of this vignette is described in Schifter (1999).
[4]Pseudonyms are used for the children appearing in teachers' cases.

two-digit numbers, extending into hundreds, thousands, and ten thousands] is an example. She was using a calculator and expressed real satisfaction with this work.

Early in the year, when Walker's students first made the observation about turn arounds, one could have concluded that they had discovered additive commutativity. However, now in April, once Walker probes more deeply, she discovers that there are issues yet to be pursued. Although Natalie is convinced that turn arounds "always work; I just know they do," other children in the class find satisfaction testing pairs of numbers, extending to quantities larger than those they are familiar with—providing a form of self-motivated, mindful computational practice as well as development of number sense.

When asked about the equation 13 + 23 = 23 + 13, children say "it doesn't work," although they don't doubt that 13 + 23 and 23 + 13 result in the same sum. Most likely at issue here is their interpretation of the equal sign. Many children seem to believe that "=" is a directive to write down the answer to the calculation on the left (Carpenter, Franke, & Levi, 2003; Carpenter & Levi, 2000; Kieren, 1981, 1992). "There's no answer here," Walker's children say in explanation for why the equation doesn't work.

The issue of the equal sign aside, many of the children in this class seem convinced that changing the order of two addends does not change the sum, and they are convinced because they have tried it for many pairs of numbers. Yet, not all children in the class take that position. A colleague from Walker's school, Lisa Seyferth, observed the same lesson and wrote about the whole group discussion that took place.

> At the end of class, when the children shared what they worked on, Emily showed how she worked with adding 70 and 35. She had 70 cubes in stacks of 10, and this was separated by a wooden block from 35 cubes in stacks of 10 and one 5 (Fig. 16.1):
>
> She said she added 35 to 70 by counting on: 80, 90, 100, 105. Then she moved the two groups of cubes so that the 35 was to the left of the block and the 70 was to the right. She counted on again: 45, 55, 65, 75, 85, 95, 105. This pair worked because they added up to 105 in either order. Walker asked if Emily thought it would still work for different numbers. Emily said she didn't know because she only did these ones.
>
> A few more children shared their thoughts and then Emily raised her hand again. She said that she could use the same cubes but divide them up differently and it would still work. When asked to demonstrate, she moved the block to another spot and said that the two new parts would also add up to 105, no matter what order she added them. She said it would work no matter how she divided the cubes because there would always be 105. ... Nathan [added] she will always get the same answer because she is always starting with the same number of cubes.

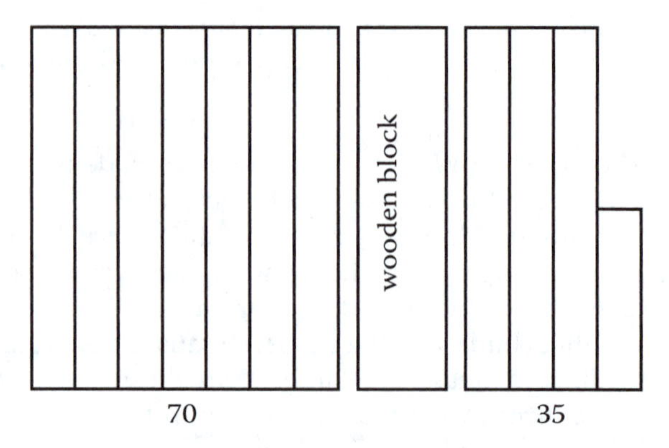

FIGURE 16.1. Emily used an arrangement of cubes to demonstrate that 70 + 35 and 35 + 70 both equal 105.

Emily has taken a different approach to studying turn arounds, modeling the operation of addition with cubes. Having 70 cubes on the left side of a wooden block and 35 cubes on the right, addition is accomplished by counting the number of cubes in the joined collection. She shows how, if 35 is moved to the left and 70 to the right, the total stays the same, and she demonstrates by counting them all again. Initially, when asked if this would work for other numbers, Emily says she doesn't know because she only tried these.

However, after thinking about the question for a bit, she offers some new thoughts. No matter how the 105 is decomposed, you can switch the order and the parts would still add up to 105. Although still limited to the sum of 105, Emily has extended the application of her representation to any two whole numbers that total 105. In making this observation, Emily has also shifted her interpretation of addition. Whereas before she started with two quantities that were joined, now she begins with a total and decomposes it into two parts.

We do not see Emily or Nathan taking this idea further. However, the physical representation they have built could provide the basis for a proof—an argument for the original proposition that does not rely on "trying it lots of time." That is, given any number of blocks, that number could be decomposed to represent an addition statement, and the parts rearranged while preserving the number of blocks one started with. The representation embodies the operation of addition, and (in imagination) can accommodate any whole number of blocks because it does not depend on the actual number of blocks at hand. It depends on a concretely based realization of the invariance of the total across rearrangements of its parts. The students are leveraging their experience of invariance in the physical world to serve their understanding of addition. Also note that students are learning to pay attention to *what counts* for two expressions to be considered equivalent. As they learn to identify what stays the same among things that are changing, Emily and Nathan understand that when determining equality, quantity does count, while location does not.

Jan Szymaszek's Third-Grade Class, September: Changing the Order of Terms—for Which Operations Over Which Domain?[5] It was in Jan Szymaszek's third-grade class that students articulated what is the same and what is different about 0 + 10 and 10 + 0, as recounted at the beginning of this section. Given the initial discussion, Szymaszek chose to "push them on this a bit more" and so, with their input, she wrote out the "Switch-Around Rule" she thought they were expressing: When you add two numbers together, you can change the order and still get the same total. Yet, after they discussed it further, Szymaszek decided she needed more information about what each child believed. She wrote:

> The next day, I asked them to write the "Switch-Around Rule" in their own words, and give examples to show if it's always, sometimes, or rarely true. Their comments helped me to see that the term "switch-around" meant different things to different students. [Among their comments were:]
>
> - If you switch around the numbers in a math problem, you will get the same answer. I think it's true, but I'm not sure about division.
> - The switch-around rule says 2 – 1 = 1 and 1 – 2 = 1.
> - The switch-around rule is you put two numbers together and you switch the numbers and it equals the same thing, like 7 + 3 = 10 and 3 + 7 = 10, 100 + 700 = 800 and 700 + 100 = 800, 30 + 40 = 70 and 40 + 30 = 70, 3 + 27 = 30 and 27 + 3 = 30, 6 + 5 = 11 and 5 + 6 = 11, 1000 + 8000 = 9000 and 8000 + 1000 = 9000.
> - The switch-around rule is an example that 3 + 7 = 10 is the same as 7 + 3 = 10 and 5 – 3 = 2 but 3 – 5 = ?, so it does not work for some number sentences.
> - The switch-around rule is when you take two numbers and add them together and try switching them around. Say you had 5 + 3 = 8. You would switch them around to 3 + 5 = 8. For 5 – 3 = 2, then switch them around to 3 – 5 = –2.
> - The switch-around rule is if you have two numbers and you put the numbers in the other one's place, it will equal the same thing, but, you can't use subtraction otherwise it will not work, like 4 + 5 = 9 and 5 + 4 = 9, but 8 – 7 = 1 and 7 – 8 = –1.

---

[5]This episode is excerpted from a case that appears in Schifter et al. (2008).

- The switch-around rule is if I have, say, three apples and four plums, if instead I said I have 4 plums and 3 apples it will still equal seven pieces of fruit.
- When you add or subtract two numbers and you switch them around you get the same total. Most of the time true 8 + 7 = 15 and 7 + 8 = 15, 8 – 7 = 1 but 7 – 8 = 0. I think the switch around rule works only when you add.
- The switch-around rule says that no matter which way they're put, they equal the same.
- The switch-around rule says you can change the order of the numbers. It works every time you add two or more different numbers. 5 + 7 + 3: 7 + 3 = 10, 5 + 10 = 15. 4 + 8 + 1: 8 + 1 = 9. 9 + 4 = 13. 8 + 9 + 4: 8 + 4 =12 and 12 + 9 = 21. 7 + 3 + 4 + 1: 4 + 1 = 5, 7 + 3 = 10, 5 + 10 = 15.

What is particularly striking in the entire list, of which this is an excerpt, is how many of the children interpret the switch-around rule to apply to other operations, even though the rule was written "when two numbers are added together." Some children consider addition, subtraction, multiplication, and division; others addition and subtraction; and some specify that the switch-around rule is about addition. Some children do not identify whether the rule applies to particular operations: "The switch-around rule says that no matter which way they're put, they equal the same."

Of those students who consider switch-arounds for subtraction, some say it works, others say it doesn't. Of those who say it does not work, the evidence they offer includes "8 – 7 = 1 but 7 – 8 = 0" (this child hasn't yet learned to think in terms of integers) and "5 – 6 = –1 and 6 – 5 = 1" (this child has). Among those who say switch-arounds do work for subtraction, one child writes, "The switch-around rule says 2 – 1 = 1 and 1 – 2 = 1" (misunderstanding what is involved in subtracting a larger whole number from a smaller), and another offers, "For 5 – 3 = 2, then switch them around to 3 – 5 = –2." In this last example, the student seems to differ from her classmates over what counts as the same and what counts as different—an interpretation of sameness that departs from what is considered relevant in the commutative law.

One student wrote, "The switch-around rule is if I have say three apples and four plums, if instead I said I have four plums and three apples it will still equal 7 pieces of fruit." Although the rest of the students all wrote in terms of "naked numbers," devoid of context, this child's thinking harkens back to ideas about order encountered by Ana Vaisenstein's first-grade students.

Another child extends the idea of switch-arounds to multiple addends, offering examples that involve a combination of the commutative and associative properties. This child sees that, when given three or four addends, you can group them and order them in any way, yet preserve the sum. Here, commutativity and associativity are not seen as distinct properties—all of the examples illustrate joining addends in various orders without affecting the total. (Although this student may understand that, no matter how many addends are involved, order doesn't change the sum, in other classrooms this is an open question; Bastable & Schifter, chap. 6, this volume.)

The following day, Szymaszek began with a discussion in which the class clarified for itself that switch-arounds work differently for addition and subtraction; she then asked the class to think only about addition:

Teacher: How many people would agree that if we're talking about addition, the switch-around rule works all the time? ... Raise your hand if you're sure that it's always going to work for addition.
Mark: Do you mean for every number in the entire world?
Teacher: For every number in the entire world.
Chris: Two different numbers?
Teacher: Two different numbers, two same numbers ... Add them together, switch them around, they're still going to equal the same amount.

Many hands go up:

Teacher: Raise your hand if you're not sure.
Steve: If you're not positive?
Teacher: Yup.

Two hands go up. One is Steve and the other is Marina:

Teacher: So, Marina, what do you think would convince you? What are you still not sure of that would make a difference to you?
Marina: Well, I'm not too sure it will work for *every single* number because ...
Steve: Because we haven't tried every single number.

Perhaps it seems as if Steve and Marina are behind their third-grade classmates and even those second graders who are convinced that switch-arounds, or turn arounds, will always work. However, we have seen that by third or fourth grade, many students become aware of the infinite nature of the number system and now realize checking particular pairs of numbers is not enough to establish the general rule (Ball & Bass, 2003). They may still be thinking about whole numbers, which they now know extend forever and reach quantities they cannot imagine. Or perhaps they are becoming aware of other classes of numbers—

integers, for example—that may have a different character. Indeed, in other third-grade classrooms, the question of order arises again once work with integers is begun: If you add 4 and –6 (move from zero on the number line up four and back six), do you land at the same place if you add –6 and 4 (move from zero on the number line back six and forward four)? And will this always be true?

Thus, even as we see that, early in first grade, children recognize a regularity that some educators might call the commutative law of addition, *how* children make sense of this law involves considerable complexity. These vignettes reveal several issues.

When faced with the question about whether changing the order of addends always leaves their sum unaffected, many children interpret *always* to refer both to all numbers and to all operations. (Although not depicted here, the children in Carol Walker's class also considered turn arounds for subtraction.) Thus, they must sort out that it does apply to addition, but not to subtraction. In fact, deliberately contrasting these two operations is not only a particularly effective way to deepen students' understanding of commutativity, but also, as well, of the operations of addition and subtraction themselves. Once they move into their studies of multiplication and division, they will have to think through why turn arounds work for multiplication, but not for division. (See Schifter, Bastable, & Russell, 1999, for cases about the commutative law for multiplication.)

Furthermore, even when the discussion is restricted to addition, the issue of whether one can *always* change the order of addends is, in itself, complex. Can we ever say *always* when we can't test all numbers? What constitutes proof at the elementary level? Even after *always* is established within a domain (e.g., whole numbers), the question must be revisited when that domain is extended to other kinds of numbers (e.g., integers).

Finally, for children, questions about the effect of reordering two addends can extend to the reordering of any number of addends. From the students' point of view, the properties of commutativity and associativity for addition are not separable, but are implicated in the same question, "Does the order of addends matter?" Therefore, as the number of addends increases, students may again need to reconsider this question.

What Is Basic for Whom? Considering the Associative Law of Multiplication

The previous section presented vignettes from classrooms in which children's observations hit directly on the basic concern of additive commutativity—the order of addends. In contrast, we find that questions about associativity do not arise with the same immediacy in elementary

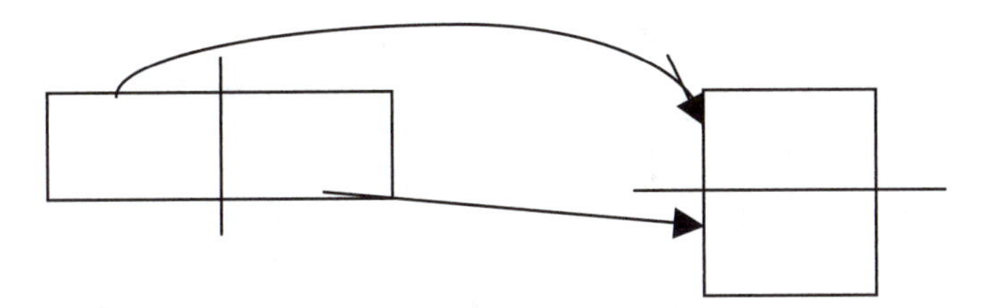

FIGURE 16.2. Doubling one factor and halving the other keep the product constant.

classrooms. In fact, especially with regard to multiplication, the law expressed algebraically as $(xy)z = x(yz)$, elementary students rarely see problems with three or more factors and so have no occasion to notice what happens when factors are regrouped. However, we have found considerable interest in related propositions, which, in fact, follow from the associative property for multiplication.

For example, some students realize that you can double one factor in a multiplication statement if you halve the other, keeping the product constant (see, e.g., Russell, 1999). This proposition follows from the associative law: $(x2)z = x(2z)$, the factor $(x2)$ being halved, the factor z, doubled.[6] For children, it is their representations of multiplication that they call on as the basis of their justifications (Fig. 16.2). "If I had half as many groups, I've got to have twice as many things in them to equal the same amount." Or, "If I cut this rectangle in half and rearrange the parts, I still have the same area."

Another observation that engages elementary children is that the factors of a number are also factors of that number's multiples, a proposition that follows directly from the associative law of multiplication: If a is a factor of b ($b = ma$) and c is a multiple of b ($c = nb = n(ma)$), by associativity, $c = n(ma) = (nm)a$, which means that a is a factor of c. But again, for children, it is their representations of multiplication that are the basis of their proofs.

To illustrate this point, we will visit two classrooms where this idea is explored.

[6]As we have shared this problem with mathematically sophisticated adults, some see it as a direct application of the associative law of multiplication; others see it as a combination of commutativity, associativity, the product of 2 with its multiplicative inverse, and the multiplicative identity. Students' observations and their use of representations of the operations to justify them at once give meaning for what the operation does and provide students with the ability to maneuver about the number system flexibly.

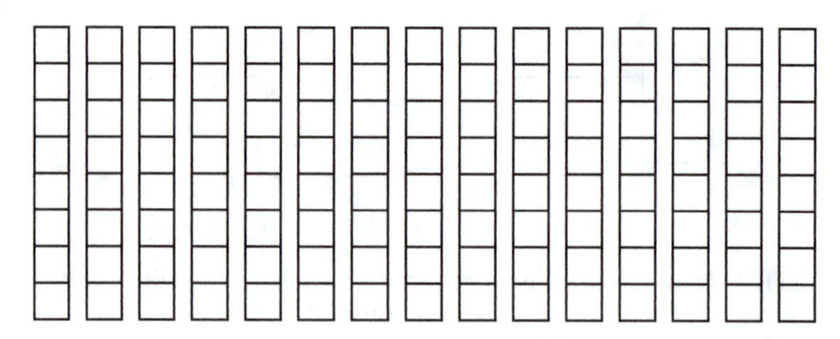

FIGURE 16.3. 8 is a factor of 120.

Jan Szymaszek's Third-Grade Class, April: Factors of Factors and Physically Based Justification.[7] In this third-grade classroom, students had been working on what numbers one can count by to land on a particular number. For example, to land on 30, one can count by 2s, 3s, 5s, 6s, and so on. They began to notice patterns and formulated questions about them. One such question had engaged their interest: "Is a factor of a factor also a factor of the number?" Szymaszek described the following lesson:

> The next day, when my class was finding factors of 120, I also asked them to find out whether all the factors of the factors of 120 were also factors of 120. As I checked in with them while they were working on this problem, I overheard one partner pair, Allan and Ben, talking about a short cut way of checking for factors of factors. When the class gathered for a whole group discussion about what they had learned, I asked Allan and Ben to share their idea first.
>
> Teacher: When I saw what Ben and Allan did to show how the factors of the factors were also factors of 120, I wondered if this might help us see how we could show it could work for all numbers and all factors.
>
> Ben: We had ours in "eight-sticks," the way we do now.
>
> The array he had in front of him looked like this that depicted in Figure 16.3:
>
> Ben: When you came over to us you asked a question if all the factors of eight are also factors of 120. I said, "Sure!" and I just took one stick and I just took the factors out.
>
> Teacher: Could you show us how you did that?
>
> Ben broke one of the eight-sticks into two 4s, then four 2s, then eight 1s (Figure 16.4):

[7]This episode is excerpted from a case that appears in Schifter et al. (2008).

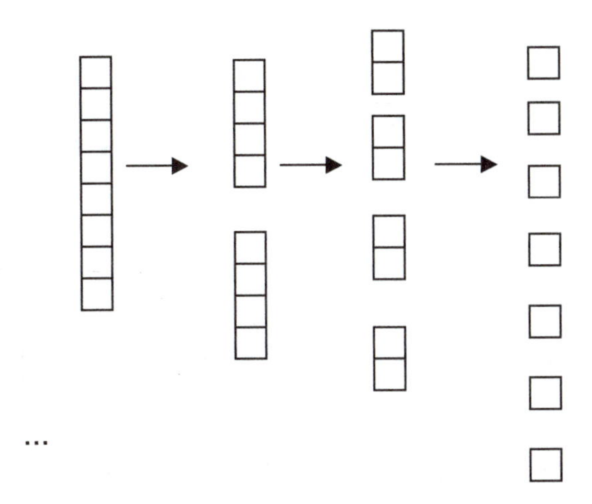

FIGURE 16.4. 4, 2, and 1 are factors of 8.

Teacher: Allan, how did you two know that all those factors were also factors of 120?

Allan: Because all the multiples ... Well, every stick is the exact same thing, because it's made with the same number of cubes, so if there are certain factors of 8, they have to be the same factors of all the other 8s.

At this point, many students sighed with "Oh!" as a sign of acknowledgment that this idea could work for them, too:

Teacher: Let's let that sit with us all for a few seconds, and then we'll see if anyone else can find a way of saying that idea in another way.

After a short silent "think time," I asked Allan to call on someone who wanted to restate his idea. He picked Sharon:

Sharon: What they were saying, I think, is that since you could split *one* of the 8s into 1s or 2s or 4s, you could split *all* of the 8s into 1s or 2s or 4s, and that would mean that the 120, you can split into 1s or 2s or 4s.

Allan agreed that Sharon's paraphrase matched what he was saying.

When Ben and Allan presented their proof that all factors of 8 must also be factors of 120, some of their classmates recognized that their method offers proof for the general claim. The children explain what the cubes show: If a number is broken into groups of equal size, then if one group is broken into its factors, all of the groups can also be broken into smaller groups of that same size—demonstrating that the smaller number is also a factor of the original number.

Table 16.1

Factors of Hundreds Numbers

100	*200*	*300*	*400*
1	1	1	1
2	2	2	2
4	4	3	4
5	5	4	5
10	8	5	8
20	10	6	10
25	20	10	16
50	25	12	20
100	40	15	25
	50	20	40
	100	25	50
	200	30	80
		50	100
		60	200
		75	400
		100	
		150	
		300	

Nancy Buell's Fourth-Grade Class, January: Factors of Factors—More Representations, More Justification Strategies.[8] In Nancy Buell's fourth-grade class, children worked on a similar idea, calling on different representations to prove their claims. The students' explorations arose following a lesson in which the class had listed the factors of 100, 200, 300, and 400 (Table 16.1).

Using the phrase "hundreds numbers" to mean multiples of 100, Buell posed the following questions: "What patterns do you see in looking at the factors of different hundreds numbers? How might the patterns help you figure out the factors of a hundreds number that is not yet on our chart?" By the end of the period, the class had come up with 25 observations. The following day, Buell handed out compilations of those observations and then asked each student to pick one they wanted, determine if it is always or sometimes true, and offer a proof.

As they got to work, different children selected observations at different levels of generality. Their proofs relied on their representations of multiplication or division, highlighting different interpretations of what these operations do.

---

[8]This episode is excerpted from a case that appears in Schifter et al. (2008).

The first observation on the list was, "1, 2, 4, 5, 10, 20, 25, 50 and 100 are always or sometimes factors of all the other hundreds numbers." Some children chose to prove the claim of *always* for each factor separately. For example, Betsy suggested, "All hundreds numbers are even and all even numbers can be divided by 2, so 2 is a factor of all hundreds numbers."

Shavon said that she thought of dollars and quarters when considering 25 as a factor. Buell asked the class, How many quarters in $2? $3? $4? $5?:

Teacher: So what does that tell you about 25?
Joey: That 25 goes into 100 four times and, so, however many dollars, there are, umm, there will be four quarters for each one—or four 25s.
Teacher: So what does that prove?
Joey: That there's 25 in any hundred.

Still other children, also working from observation 1, interpreted it as "All factors of 100 are factors of other hundreds numbers." Thus, they offered a single proof for all 9 factors.

Chang, Ivan, and Khalid, working as a small group, thought about this claim in terms of skip counting. Although they *spoke* in terms of specific numbers, they explained that they were talking about *all* factors of 100. Khalid used 25 as his first example—you go up four for each additional 100.

Chang interrupted, excited by his own insights:

> Because 100 has those numbers as factors. Say I'm skipping by 2. I need, umm, 50 skips to get to 100. Umm. Just add 50 more, then you got a factor of 200. Just keep adding by 50s, you get to higher numbers, the other hundred numbers.

When asked if there are any hundred numbers that you can't get to that way, Ivan spoke up. "700 has 7 hundreds in it. So all those numbers are factors. You just have to take more jumps."

Jon, Williamson, and James worked on the observation, "If you double the numbers in the 100 column you get numbers in the 200 column." (If you double a factor of 100, you get a factor of 200.) Because this rule is about a finite set, it could be checked for each case. However, these boys chose to think about it in terms of arrays. First they wrote:

I think this works because 100 is half of 200, and therefore the factors of 100 are half of the factors of 200. Then they drew the following picture, as shown in Figure 16.5.

When asked if there was any way to see factors in the picture, Jon filled in some more numbers. "See, 20 × 10. And inside here there is 10 × 10."

Jon's picture (Fig. 16.6) illustrates a rectangle whose area is 200, with vertical sides of 10 and horizontal sides of 20. Half the rectangle, whose area is

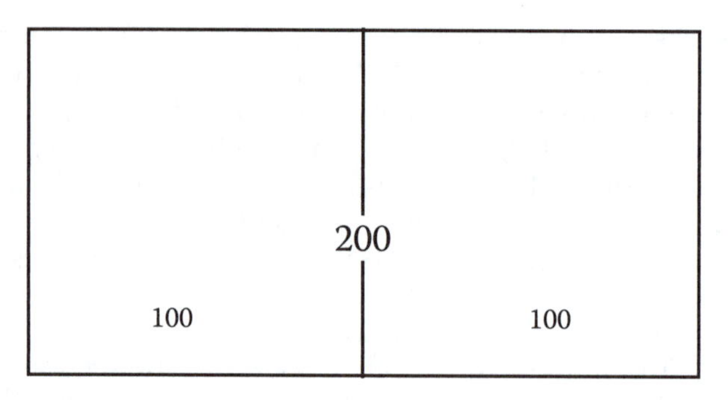

FIGURE 16.5. Jon argues that the factors of 100 are half the factors of 200.

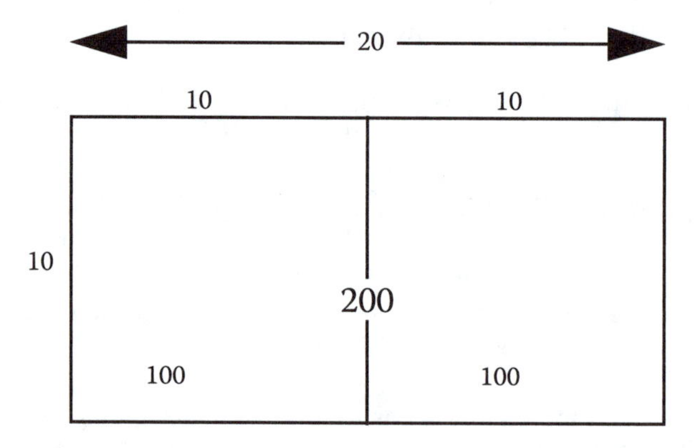

FIGURE 16.6. 20 is a factor of 200, and half of 20 is a factor of 100.

100, has vertical sides of the same length (10), and horizontal sides half the length (10). However, Jon did not stop to explain this. Instead, he went ahead and drew a picture to demonstrate the same notion with another pair of factors. Starting with 50 × 4 as factors of 200, he cut his rectangle in half to show that half of 50, 25, is the corresponding factor of 100 (Fig. 16.7).[9]

These boys also made a general claim, but, as in Chang's group, relied on specific examples to justify it. And again, as in Chang's group, their justifications were not merely a matter of testing the claim by checking particular number facts. In this case, the children were using representations of multiplication to illustrate how the quantities involved are related. As had Emily in Carol Walker's second-grade class, or Ben and Allan in Jan Szymaszek's case, they, too, employed a representation with particular numbers. However, unlike Emily, they seemed to be saying that their representations demonstrated the general claim. They were seeing the

[9]It is not clear to us that Jon was excluding the case of an odd factor of 200, where his approach would not give new whole-number factors of 100.

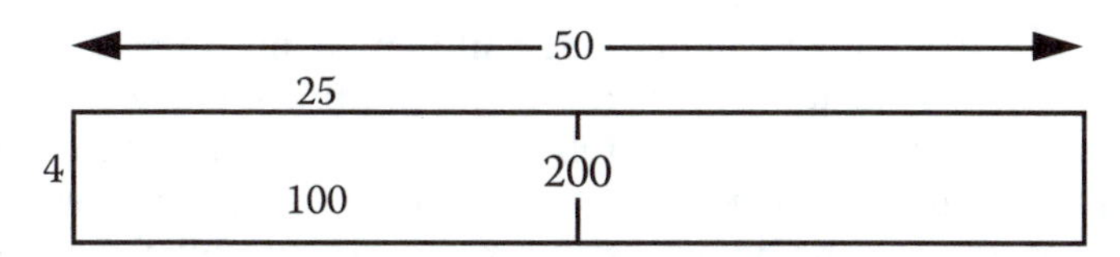

FIGURE 16.7. 50 is a factor of 200, and half of 50 is a factor of 100.

general in the particular, or perhaps more accurately, treating the particular as the general (Mason, chap. 3, this volume).

Tyrone, who wrote about the observation, "300 has the most factors,"[10] ended up working at a higher level of generality: "I don't think 300 has the most factors. Here's my proof. Because 600 is a multiple of 300 it has all the factors of 300 plus itself and 8."

An adult observer who came by as Tyrone was writing out his explanation asked about the claim that 600 has all the factors of 300. Here is Tyrone's written response:

> The reason for this is that any factor of a lower number will be a factor of any given multiple of that number. This is because you simply multiply the number the factor is being multiplied by as many times as the lower number goes into the multiple.

Based on their observations of lists of factors, the children in this class explore particular cases of the proposition that a factor of a number is also a factor of that number's multiples. The general proposition is closely associated with multiplicative associativity: $x(yz) = (xy)z$. At Tyrone's level of abstraction, he has articulated an idea that is close to the following formal algebraic statement, where a, b, c, m and n are taken to be whole numbers: If a is a factor of b (which can be written as $ma = b$) and c is a multiple of b (which can be written as $c = nb$), then a is a factor of c because $c = n(ma) = (nm)a$.

But Tyrone doesn't think in terms of grouping and regrouping factors and the sort of algebraic manipulation illustrated previously. "You simply multiply the number the factor is being multiplied by as many times as the lower number goes into the multiple." He is, instead, thinking about the roles the different factors play in the relationships he is noticing. The children in both these classrooms rely on various representations of multiplication to think about the generalizations they have articulated. They are not thinking about grouping factors, but are thinking in terms of factors and multiples and various visual representations of their relationships.

[10]The class' observations came from lists of the factors of 100, 200, 300, and 400. Indeed, of those four lists, the one for 300 was longest. Tyrone set out to prove that the claim was false once you consider multiples of 100 more generally.

The particular generalization these children are making—the factors of a number are also factors of that number's multiples—arises naturally in the course of their work on multiplication. Its power stems from the fact that it stimulates the curiosity of these children who are engaged in building their understanding of factors and multiples, multiplication and division. By raising these observations for explicit consideration, the teachers create opportunities for their students to learn that it is possible to make—and prove—claims about infinite classes of numbers. There is no reason to explore or articulate explicitly the associative property of multiplication at this point, and when, in later grades, these children are introduced to its formal algebraic expression, they will be able to attach meaning to it. Nevertheless, they are connecting ideas about regularities of the operations to their underlying properties. As they investigate the generalizations that emerge from their work, they deepen their understanding of multiplication and learn to more fluently use their representations to show multiplicative relationships. The children's proofs rely on their specific representations, for these are what are most accessible to them.

What's the Same Generalization to Me May Not Be the Same to You: What Does It Mean to "Understand" Distributivity?

When students begin their study of multiplication of whole numbers, they think of the operation of multiplication as adding up the number of objects in a set of groups of fixed size—repeated addition. And, as they figure out how to calculate, implicit in their work is the idea of distributivity. For example, in order to find the number of sodas in five 6-packs, they might find the number in three 6-packs added to the number in two 6-packs. At a later point, children may act with a generalized notion of how one factor can be decomposed, each part multiplied by the other factor, and the subproducts totaled to find the answer to the original problem. Yet when children encounter what is—from a formal mathematical perspective—the same idea, but is now placed in a different context, they fail to recognize it.

To illustrate this point, consider two episodes, 6 weeks apart, from the same third-grade classroom.

Susan Smith's Third-Grade Class, May: Regrouping and the Distributive Property.[11] In this lesson, Susan Smith posed the following problem:

> On the weekend, I found many flowers in my garden. In the morning, I picked 4 bunches of flowers to give to my family. That afternoon I picked 3 more bunches to give to some friends. Each bunch had 8 flowers. How many flowers did I pick?

[11]This episode is excerpted from a case that appears in Schifter et al. (2008).

After the children had a chance to solve the problem individually, they spent the rest of the lesson sharing their solution methods. Later that day, with the distributive property still on her mind, Smith met with some students who had volunteered to spend recess thinking further about this problem. She directed their attention to Laura's solution strategy:

Teacher: So I've got another problem for you to think about. It seems like many people did what Laura did. She thought of the answer as 7×8, but she didn't know how much that was. So she counted 2 eights and then 2 more eights, until she had 7 eights.

Smith then asked the class to think about another problem: How might they "pull apart" 12×6 to solve that multiplication?:

Linda: That means you have to have 12 six times. I'd take 2 of the 12s and add them. So then I'd have 12×2 and I'd take that answer and add it 2 more times.
Teacher: Could you do $(12 \times 2) + (12 \times 2) + (12 \times 2)$? Can we write it this way? Is this a true statement? $12 \times 6 = (12 \times 2) + (12 \times 2) + (12 \times 2)$.

Linda said yes, and the rest agreed:

Teacher: Why can we write it that way?
Louisa: Because it is still six 12s, just broken up into 2 at a time and added together.
Teacher: Is there another way we could break it up?
Elizabeth: We can do $(12 \times 3) + (12 \times 3)$.
Catherine: We could also do $(12 \times 4) + (12 \times 2)$.
Laura: $(12 \times 5) + (12 \times 1)$
Teacher: Why can we do it all these ways?
Elizabeth: It's like before, when we did $(8 \times 4) + (8 \times 3)$. It's still the same answer.
Laura: It's because they all equal the same number. They are all 12×6.
Teacher: Are you saying all of these give you 12 six times?

They all nodded. Since they had been only breaking up the six, I wanted to see if they also thought you could break up the 12:

Teacher: You've found lots of ways to break up the six. Can we break up ...
Louisa: ... break up the 12?
Teacher: Yes, can we break up the 12 so you would have six 12 times?
Louisa: $(6 \times 6) + (6 \times 6)$.
Teacher: Does that work? How many groups of six do we have?
Elizabeth: It's six groups of six and another six groups of six, so it's still 12 sixes.
Teacher: Is there another way to break it up into sixes?

I (the teacher) decided to put up my own answer. I wrote $(10 \times 6) + (2 \times 6)$.

Teacher:	What do you think of this?
Burt:	It's just another way to make 12. It doesn't matter how we do it. It's still the same.
Louisa:	It's like all those other times, when we break numbers up and switch them around. The answer doesn't change because we didn't change what we had.

When given multiplication problems to solve, these children, calling on their sense of what it means to multiply, know that they can break the problem apart and multiply in chunks, in effect, applying the distributive property. Although they don't have a name for the rule, they work with the principle fluently and explain why it works: You can calculate 12×6 by solving $(6 \times 6) + (6 \times 6)$ because "it's six groups of six and another six groups of six, so it's still 12 sixes." Their sense of the invariance of the overall collection seems to be the basis of their certainty.

Burt and Louisa explain that it doesn't matter "how we do it," "the answer doesn't change because we didn't change what we had." Their understanding that these multiplication problems entail counting the total number in a given number of groups of a fixed size enables them to conclude that the groups can, themselves, be grouped in any way, without changing that total. The children have articulated a general rule which, written symbolically, is $(x + y)z = (xz) + (yz)$. That is, they seem to be close to articulating the distributive law for multiplication over addition.

From a mathematical perspective, one can consider a simple application of this law: The sum of two even numbers is even. That is, representing an even number as any integer multiplied by 2: $2n + 2m = 2(n+m)$. Dispensing with algebraic notation, we can say that when one number represented as a "bunch of pairs," and another number represented as a "bunch of pairs," are brought together, they make still another "bunch of pairs." However, returning to Susan Smith's classroom in June (next section), it appears that the children do not see this as the same generalization they had hit upon in April.

Susan Smith's Third-Grade Class, June: Sums of Evens—An Entirely Different Experience of Distributivity.[12] The class has spent time investigating even and odd numbers: defining terms, identifying particular numbers as even or odd, and predicting whether sums of given pairs of numbers are even or odd. Based on this work, many of the students suggest generalizations—the sum of two evens is even; the sum of two odds is even—but none of the children offers a proof. That is, nobody offers a proof until Amanda speaks up in whole group discussion:

[12]This episode is excerpted from a case that appears in Schifter et al. (2008).

Amanda: Two evens, no matter what they are, have to equal an even.
Teacher: Why?
Amanda: Um. I just figured out something. . . . If you counted something by 2s, and 2s always work on an even number, they can't work on an odd number, and um every even number you count by 2s with it and if you added the 2s of both even numbers on top of each other, they both count by 2s, so they would have to equal an even.
Teacher: Somebody know what she's talking about? Ellen?
Ellen: She's kind of talking about—No, I'm confused.
Teacher: Somebody want to hear it again? Elizabeth wants to hear it again, Amanda.
Amanda: Well, if you have 2 even numbers, 2s work on both of them, so if you put them on top of each other. Umm. Can I have some cubes?

(The reader is invited to reread what Amanda said and try to determine her reasoning and then compare with what is revealed later.) As Amanda builds two sticks of cubes representing even numbers, she explains her idea again, showing the two sticks of cubes that can be counted by 2s, and then joining them to make one stick:

Amanda: I have this, both of them [the two sticks] count by 2s, so if I put them on top of each other, you keep counting by 2s, and then you get to an even number.
Teacher: Ohhhh. That's not what I thought you meant when you said you put them on top of each other. Ohh. What do people think about what she just said? Elizabeth?
Elizabeth: I think she said that if you have two even numbers and they're counting by 2s, then you put them on top of each other.
Teacher: You stick them together, I think that's what she meant.
Elizabeth: Yeah. But then I don't know what she said after that.
Teacher [Checking with Amanda to make sure she's correctly paraphrasing]: So you've got an even number over here, and an even number over here, and you stick them together, it has to give you an even number.
Amanda: Because you can count by 2s up to 6 and if I add the four on, you can just keep counting up by 2s, and that would have to equal an even number because 2s only get you to even numbers.

At this point, most of the children seem lost. Only Elizabeth is still working to follow Amanda's proof. At first, she seems to agree that the generalization feels right. But then she reconsiders:

Elizabeth: Well, it just, well um because, it just, um, you just, well, I don't know, because in some cases, well, um, I can't really think of it now but like, if you had one that was an even plus an even, if like, I haven't figured this out, but sometime maybe it could equal an odd.

Teacher: And Amanda's saying it couldn't ever equal odd. Is that what you're saying Amanda? That an even plus an even could never equal odd? And Elizabeth is wondering if it sometimes could be an odd, but you're saying it could never be an odd? Do you want to say more about why that is?

Amanda: Because 2s don't get to odds. And if they're two even numbers, they're both counting by 2s, and if you put them on top of each other you keep counting by 2s and that always equals an even number.

Elizabeth: So she's saying she already knows that it always equals it.

Teacher: She thinks she knows that it's always going to be even.

In whole-group discussion, Amanda says she can prove that the sum of any two even numbers is even. Relying on a definition of even as a number you land on when counting by 2s, she builds two stacks of cubes, each stack representing an even number. If she puts them together, that represents the sum. If you start counting by 2s, you get to the end of the first number and start counting by 2s on the second number, and you have to land on the last cube. Amanda says she knows it will work for any two even numbers. However, the class has trouble following her. Elizabeth tries, but as she works to paraphrase Amanda's argument, she gets confused and can't follow through. After Amanda repeats it, Elizabeth can't take on the idea of "all" even numbers. "If you had one that was an even plus an even ... sometime maybe it could equal an odd."

In discussion the previous May, a group of children, Elizabeth among them, articulated what we would call distributivity and provided an explanation for why it works. However, the discussion of the sum of even numbers has a very different character. From the mathematician's perspective, it might seem strange that the children cannot see a direct application of a property they seem to have clearly understood. However, viewed from the child's perspective, we might ask, why should they see the two discussions as related? After all, the discussion in May was about multiplying (with addition used as part of the calculation strategy); the discussion in June was about adding (with multiplication implied in the definition of even numbers). In the first discussion, children were thinking in terms of multiple groups; in the second, they were counting by 2s. In the first discussion, they were working on a strategy for calculation; in the second, they were considering a general claim for a class of numbers.

How does one answer the question: Do these children understand distributivity? On the one hand, they do not see how to apply this principle to prove that the sum of two even numbers is even. However, as children are working to develop their capacities to calculate and maneuver about the number system, they are not thinking in terms of applying principles. Rather, they are developing multifaceted "local" meanings for the operations, becoming fluent with a variety of representations, and learning about how to make and justify general claims about the numbers and operations (while, of course, enriching their familiarity with the characteristics of specific numbers). In light of this, we might define "understanding distributivity" for a third grader to mean breaking apart one factor to perform multiplication and successfully explaining why it works.

Summary

If curriculum were to be determined exclusively through mathematical analysis, an effort to bring algebra into the early grades would involve the expectation that students be able to state and apply the commutative, associative, and distributive laws. However, as we have seen, taking account of how children learn to negotiate the number system leads to different conclusions.

Early in their mathematical learning, children notice regularities that close in on the commutative law of addition. At first, however, they seem to be thinking about *numbers* rather than *operations*. Thus, when they notice that changing the order of numbers in some problems does not change the result, they might miss the fact that the regularity applies to a particular operation (or operations). Examination of the regularity, considering when it works and when it does not, can bring them to begin thinking about the operations as entities, each with its own properties.

In contrast to additive commutativity, associativity does not arise naturally as children explore number and operations. Grouping and regrouping addends is simply a feature of the extension of the commutative law of addition—when given multiple addends, they can be added in any order. And although children become interested in and prove generalizations that can be seen as consequences of the associative law of multiplication, they rely on representations of the operation as the basis of their justifications. Their insights do not derive from grouping and regrouping factors (which describes the associative law of multiplication), but to rearrangements of visual representations of quantities.

When working with whole-number multiplication (finding the total number of elements in a set of equal-sized groups), children will employ, articulate, and explain a strategy that approaches the distributive property of multiplication over addition. In fact, the idea of additively decomposing 1 factor, multiplying each of the parts, and summing the subproducts is at the basis of almost all children's multidigit multiplication strategies.

IMPLICATIONS FOR CURRICULUM DESIGN AND IMPLEMENTATION

The work the authors are pursuing with collaborating teachers is in the service of revising a K–5 curriculum. In this section, we articulate some of the conclusions we have drawn that inform our approach to curriculum design.

The generalizations written into a curriculum cannot be determined from mathematical analysis alone, but must be selected in conjunction with systematic inquiry into the ideas most salient to children of different ages. For example, from the standpoint of mathematical analysis, the commutative and associative properties would be among the first to be included in an elementary curriculum, and they would likely be given equal weight. However, as we have shown, as children begin to explore number and operations, additive commutativity and associativity fall under the topic of order and are not necessarily distinguishable.

Rather than investigate associativity in itself, it is more fruitful for children to explore related ideas, ideas children pursue with energy (e.g., that the factors of a number are also factors of that number's multiples). And some generalizations that can seem powerful to young children—if you add 1 to an addend, your total increases by 1, is an example that second graders discuss—would appear, at best, uninteresting and trivial from the standpoint of a purely mathematical analysis.

Among the generalizations we are introducing into the Investigations curriculum are the following:[13]

- The order of addends does not affect the sum, but the order of the terms in a subtraction problem does affect the difference.
- The order of factors does not affect the product.
- Subtraction "undoes" addition, as in $22 + 8 - 8 = 22$.
- Any missing addend problem can be solved by subtraction and vice versa.
- In an addition problem, if you subtract a certain amount from 1 addend and add it to the other, the sum remains constant.
- In a subtraction problem, if you subtract (add) the same amount from (to) both numbers, the difference remains constant.
- The less you subtract, the larger the result.
- Adding or subtracting 0 does not change the amount you start with.
- Multiplying or dividing by 1 does not change the amount you start with.

---

[13]Precisely because some generalizations do not lend themselves to algebraic notation, we write them here in English.

- If you add 1 to an addend, the sum increases by 1. (And, for older students, if you add [or subtract] any number to an addend, the sum increases [or decreases] by that number.)
- If you double one factor and halve the other, the product remains constant.
- The factor of a number is also a factor of that number's multiples.
- In a multiplication problem, you can decompose one factor, multiply the parts by the other factor, and add the subproducts.
- The more you divide by (the larger the divisor), the smaller the result (the smaller the quotient).

These generalizations initially arise in the context of students' work in whole numbers. As the domain of number expands—to rational numbers and integers—many of these generalizations must be revisited and arguments must be developed with representations that apply to the larger domain. In addition, students refer to these generalizations as new issues arise, for example, determining which of two fractions is larger.

Although this chapter focuses on early algebra as generalization about number systems, the habits of generalizing and justifying in mathematics are not restricted to explorations of number and operations. Thus, in their study of geometry, students apply the same habits of mind to find (e.g., to find the formula for the area of a triangle). Similarly, in their study of functions, function rules must be developed, an activity that involves generalizations about situations in context.

Ideas about generalization are implicit in the current elementary curriculum, but are not necessarily addressed either explicitly or in a sustained manner. Our work does so. The tendency to generalize is a natural human one (Mason, chap. 3, this volume), and young children do make generalizations about the regularities they see in the number system. This chapter has shown how first graders notice regularities that arise in the set of problems called "How Many of Each?" and how second graders name the pattern they see as turn arounds. Furthermore, as children devise procedures for calculation, they employ methods in which certain laws are implicit. (Almost all strategies, including the standard algorithms, for adding multidigit numbers involve decomposing addends and summing the parts in a different order.) An early algebra curriculum can build on children's natural tendencies and draw from their work with calculation, turning these observations and strategies into objects of study in themselves, encouraging children to articulate and prove generalizations.

For example, when Jan Szymaszek asked her students to consider their question, "Is a factor of a factor of a number always a factor of that number?" their discussion led to the development of a proof based on representations of multiplication applicable to all whole numbers. In many of the examples in this chapter, students had noticed regularities for numbers and operations, some of which are true for all numbers and some of which are not. However, without the classroom work that raised these ideas for collective reflection, extraordinary opportunities for learning would have gone unnoticed.

Generalizations should be revisited throughout the grades, with each encounter enriched by deepened understanding of number and operations and of the notion of proof. As the example of "How Many of Each?" shows, before young children actually construct a notion of addition as an operation, they can encounter an activity in which additive commutativity is foreshadowed. Later, as they work on addition of small numbers, they notice an intriguing regularity and some, working with an image of addition as joining two quantities, can explain why it works. But, as they become aware of numbers larger than they can quantitatively imagine, or when they consider adding multiple addends, students must revisit the idea of switch-arounds. And, later, when confronted with positive and negative integers, children need to think through the issue once more.

Inquiry into the properties of the other operations will prove to be another avenue leading to revision of their ideas about order. Working on order in addition, even in the early grades, children often raise questions about subtraction. What happens when two numbers in a subtraction problem are interchanged? Which regularities become apparent? In later grades, students consider which representations can be used to demonstrate that reordering factors in a multiplication problem does not affect the product and examine the regularities that become apparent when the order is changed for division.

Students' understanding of commutativity grows in power as the domain of number expands and as they consider all four operations. But developing simultaneously is their understanding of what it means to prove that something is always true. We find that young children, frequently as early as in first grade, make general claims and speak with assurance that something always works. However, in the middle elementary years (most frequently in third grade), students are more likely to reject claims of always. No longer satisfied with testing several (or many) examples, they explain that because numbers go on forever, you cannot test all cases. There might be some number out there that doesn't work the same way—recall Elizabeth's statement that sometime maybe the sum of two even numbers could be odd.

Still later, students make general claims, but offer different kinds of arguments to defend them. In Nancy Buell's class, children made such claims (e.g., that 25 is a factor of *all* multiples of 100, that all of the factors of 100 are factors of *all* multiples of 100), relying on representations of multiplication that can accommodate a class of numbers. Because any number of dollars can be converted to quarters, 25 is a factor of any multiple of 100. Because counting by a factor of 100 lands you on 100, as you repeat centuries, you land on all the multiples of 100.

This progression is not necessarily developmental. For example, first and second graders can argue that whenever you add a certain number and then subtract it, you get the number you started with, (algebraically, $x + a - a = x$). Their justifications are not based on simply trying it lots of times, but on what it means to add and subtract whole numbers. Certain generalizations and the arguments to prove them may be accessible to younger children, whereas others must be revisited and refined as knowledge of the number system, of representations of operations, and of methods of justification deepens.

Justification in the elementary grades should be based in visual representations (diagrams, manipulatives, etc.) of operations. We have offered a number of examples to illustrate our view that for children in the elementary grades, it is the meaning of the operations, as represented visually in diagrams, manipulatives, and so on, that forms the basis of justification of claims of generality. Thus, proofs involving addition or subtraction might be demonstrated by joining or separating stacks of cubes, comparing amounts, or tracking movement along a number line. Arguments involving multiplication or division might rely on images of groups of equal size, or on arrays, or counting by a fixed number.

To be useful in validating a general claim, representations must accommodate a class of numbers. However, it is sometimes difficult to distinguish, from students' words and actions, when a representation is being used to argue for a particular instance rather than to defend a claim of generality. In several examples in this chapter, the authors interpret students' use of particular quantities as placeholders for a general number (a variable). For example, when Amanda demonstrates her proof that the sum of two even numbers is even, she holds up cube sticks that necessarily contain a certain (small) number of cubes. To her, they stand for any two even numbers, and she makes her argument for the general case. Similarly, Allan seems to be making a general claim—that factors of a factor of a number are factors of that number—even though he is arguing the case that factors of 8 are also factors of 120: "Well, every stick [of 8 cubes] is the exact same thing, because it's made with the same number of cubes...."

In these examples, the role of representation is central to the articulation, investigation, justification, and communication of generalizations. Allan's explanation, based on his representation, gives his classmates access to an understanding of why this general idea is true.

Given these four conclusions, we believe curricular activities must provide opportunities for children to puzzle over regularities of the number system. It is one thing for a teacher to pause for exploration when students happen to notice a generalization in the course of their work with number. It is another to write a curriculum that consistently draws students' attention to significant generalizations, to stimulate puzzlement and the desire to explore. To do this, curriculum writers must identify which generalizations catch the interest of children of different ages and then formulate questions about those generalizations. For example, one might present two related problems and ask: Can you use your answer from the previous problem to help you think about this problem? Or, one might describe two fictional characters presenting different, contradictory arguments and ask: Which character do you think is right, and why? Or again, one might present two different ways to solve the same problem and ask: If you apply these two methods to other problems, will you always get the same answer?

We must emphasize, however, that curriculum alone cannot ensure development of a strong early algebra strand in the elementary grades. As other chapters of this volume attest (Bastable & Schifter, chap. 6, this volume; Franke, Carpenter, & Battey, chap. 13, this volume), engaging students in the formulation, testing, and justification of generalizations involves a teaching practice based on inquiry, and doubly so—students inquiring into mathematical ideas, their teachers into students' thinking. The vignettes included in this chapter were chosen to illustrate how children's discoveries of the behavior of the basic operations are related to the laws of arithmetic. However, these same vignettes can be read as evidence of the classroom cultures the teachers have established, the stances they take as their students offer their own ideas, and the pedagogical moves they make in response to those ideas. In these classrooms, students are encouraged to express their thoughts for the class to reflect on and develop further. Teachers ask questions to elicit ideas or to move their students' thinking in particular directions. For example, after her second graders coined the term *turn arounds,* and after Carol Walker asked them to define or describe turn arounds for her, the question naturally arose: Do turn arounds work for any pair of numbers? Similarly, Jan Szymaszek wrote out the rule for switch-arounds, which she thought her students had articulated. But, only when they wrote it in their own words did she realize that many of her students had a different rule in mind. And as Susan Smith's third graders discussed the equation $12 \times 6 = (12 \times 2) + (12 \times 2) + (12 \times 2)$, Smith asked "Why can we write [12 × 6] that way?" "Is there another way we can break it up?" "You've found lots of ways to break up the 6. Can we break up the 12?" The disposition to inquiry illustrated by these teachers, norms of the kind they have established for their classrooms, and the capacity for such minute-to-minute responsiveness that they evidence are all requisite to the effective enactment of curriculum. Indeed, enacted curriculum is only possible as a partnership between the authors of curriculum and the teachers who employ them.

CONCLUSION

This chapter has argued that some distinctions and principles that follow from mathematical analysis are meaningless to children; and some distinctions and principles significant to children seem trivial in the context of formal mathematical analysis. For this reason, we argue, curriculum designers must make decisions about content drawing both on mathematical analysis and on systematic inquiry into how children's mathematical thinking develops.

Yet there is a fortunate irony here: By organizing lessons around ideas most salient to children, we create a classroom environment in which children are able to engage in activity that is much closer to the mathematician's own practice—formulating, testing, and proving claims of generality. As the *Principles and Standards for School Mathematics* (National Council of Teachers of Mathematics, 2000) comments:

> Discussion about the properties [of operations] themselves, as well as how they serve as tools for solving a range of problems, is important if students are to add strength to their intuitive notions and advance their understanding. . . . Analyzing the properties of the basic operations gives students opportunities to extend their thinking and to build a foundation for applying these understandings to other situations. (p. 161)

At the end of a year of work focusing on generalization, third-grade teacher and project participant Jan Szymaszek wrote about its implications for her teaching practice.

> One thing I learned about generalizations is how powerfully they can operate within a mathematics classroom community. They seem to help students develop a habit of mind of looking beyond the activity to search for something more, some broader mathematical context to fit the experience into. When students work with the generalizations that they have created, they come to appreciate what feels to me (and them?) more like the real process of doing mathematics. By explicitly stating the generalizations, and then finding examples and counterexamples, they are thinking more about the principles underlying their work with number and operation. Generalizations help students see relationships among and between numbers, and among and between operations. Generalizations keep expanding the confines and broadening the ideas with which students are working.

When algebra is viewed in terms of what might take place in the elementary classroom, the possibility emerges of students making their own generalizations based on their own actions and justifying them in terms of their experience with numbers, physical materials, or visual displays.

We have tried to suggest by our examples, that when children are engaged in activities of this kind, they are in a position to develop, in Szymaszek's words, a "habit of mind of looking beyond the activity to search for something more, some broader mathematical context to fit the experience into." It is our belief that such experiences and such habits of mind are not only rewarding in themselves, but serve to prepare children to encounter school algebra not as an impenetrable barrier, but as a height from which realizable life chances can be surveyed.

ACKNOWLEDGMENTS

This work was supported by the National Science Foundation under Grant No. ESI-0095450 awarded to Susan Jo Russell at TERC and Grant Nos. ESI-9254393 and ESI-0242609 awarded to Deborah Schifter at the Education Development Center. Any opinions, findings, conclusions, or recommendations expressed in this chapter are those of the authors and do not necessarily reflect the views of the National Science Foundation.

REFERENCES

Ball, D. L., & Bass, H. (2003). Making mathematics reasonable in school. In J. Kilpatrick, W. G. Martin, & D. Schifter (Eds.), *A research companion to principles and standards for school mathematics* (pp. 27–44). Reston, VA: National Council of Teachers of Mathematics.

Carpenter, T., Franke, M. L., & Levi, L. (2003). *Thinking mathematically: Integrating arithmetic and algebra in elementary school.* Portsmouth, NH: Heinemann.

Carpenter, T. P., & Levi, L. (2000). *Developing conceptions of algebraic reasoning in the primary grades* (Research Report). Madison, WI: National Center for Improving Student Learning and Achievement in Mathematics and Science. Retrieved December 1, 2006 from http://www.wcer.wisc.edu/ncislapublications/index.html

Cohen, D. K., McLaughlin, M. W., & Talbert, J. E. (Eds.). (1993). *Teaching for understanding: Challenges for policy and practice.* San Francisco: Jossey-Bass.

DeVault, M.V., & Weaver, J.F. (1970). Forces and issues related to curriculum and instruction, K–6. In P. S. Jones et al. (Eds.), *A history of mathematics education in the United States and Canada: 1970 NCTM yearbook* (pp. 91–152). Reston, VA: National Council of Teachers of Mathematics.

Goodnough, A. (2002, December 19). *Cutting jargon: Klein offers a report card Johnny can read.* Retrieved December 19, 2002, from http://www.nytimes.com/ 2002/-12/19/nyregion/19SCHO.html?ex=1041298795&ei=1&en=1472a722163c655d

Hiebert, J., Carpenter, T. P., Fennema, E., Fuson, K., Human, P., Murray, H., Olivier, A., & Wearne, D. (1996). Problem solving as a basis for reform in curriculum and instruction: The case of mathematics. *Educational Researcher, 25,* 12–21.

Kaput, J. (1995, April). *Transforming algebra from an engine of inequity to an engine of mathematical power by "algebrafying" the K–12 curriculum.* Paper presented at the annual meeting of the National Council for Teachers of Mathematics, San Francisco, CA.

Kieran, C. (1981). Concepts associated with the equality symbol. *Educational Studies in Mathematics, 12,* 317–326.

Kieren, C. (1992). The learning and teaching of school algebra. In D. Grouws (Ed.), *Handbook of research on mathematics teaching and learning* (pp. 390–419). New York: Macmillan.

Kilpatrick, J., Swafford, J., Findell, B., National Research Council Mathematics Learning Study, Mathematics Learning Study Committee, & National Research Council (Eds.). (2001). *Adding it up: Helping children learn mathematics.* Washington, DC: National Academies Press.

Monk, S. (2003). Representations in school mathematics: Learning to graph and graphing to learn. In J. Kilpatrick, W. G. Martin, & D. Schifter (Eds.), *A research companion to the NCTM standards* (pp. 250–262). Reston, VA: National Council of Teachers of Mathematics.

Moses, R., & Cobb, C., Jr. (2001). *Radical equations: Math literacy and civil rights.* Boston: Beacon.

Moses, R., Kamii, M., Swap, S., & Howard, J. (1989). The algebra project: Organizing in the spirit of Ella. *Harvard Educational Review, 59*(4), 423–443.

National Council of Teachers of Mathematics. (2000). *Principles and standards for school mathematics.* Reston, VA: Author.

Osborne, A. R., & Crosswhite, F. U. (1970). Forces and issues related to curriculum and instruction, 7–12. In P. S. Jones et al. (Eds.), *A history of mathematics education in the United States and Canada: 1970 NCTM yearbook* (pp. 153–297). Reston, VA: National Council of Teachers of Mathematics.

RAND Mathematics Study Panel. (2003). *Mathematical proficiency for all students: Toward a strategic research and development program in mathematics education.* Santa Monica, CA: Rand Corporation.

Russell, S. J. (1999). Mathematical reasoning in the elementary grades. In L. Stiff & F. Curio (Eds.), *Developing mathematical reasoning in grades k–12,* (pp. 1–12). Reston, VA: National Council of Teachers of Mathematics.

Schifter, D. (1999). Reasoning about algebra: Early algebraic thinking in grades K–6. In L. V. Stiff & F. R. Curcio (Eds.), *Developing mathematical reasoning in grades K–12: 1999 NCTM yearbook* (pp. 62–81). Reston, VA: National Council of Teachers of Mathematics.

Schifter, D., Bastable, V., & Russell, S. J. (2008). *Developing mathematical ideas: Casebook for reasoning algebraically about operations.* Parsippany, NJ: Pearson Learning Group.

Steen, L. A. (1995). *How school mathematics can prepare students for work, not just for college* (Harvard education letter). Cambridge, MA: Harvard University Press.

17

Early Algebra: The Math Workshop[1] *Perspective*

E. Paul Goldenberg
Nina Shteingold
Education Development Center, Inc.

When talking about something as messy as teaching and learning, we—the authors of this chapter—are always a bit uncomfortable presenting theoretical frameworks. A theoretical framework, like any abstraction, must simplify from the chaos of life, must ignore parts of the data in order to be truly useful. In education, the almost inevitable danger is oversimplification. When we are being teachers, we find that theory helps us think about the complex events of classrooms, and organize the jumble of facts into a coherent story about learning. When we are being theoreticians, we try to find more theory, or find (or do) other research, to clarify the story, or perhaps to modify or even reject it for a better story. We don't mean to use *story* in any demeaning sense at all. A theory, in science, is a way to organize and explain events in a way that allows us to predict new ones. Unlike a mathematical theorem, whose truth rests on logic alone and is absolute (modulo a set of assumptions that are already taken as "given"), needing no connections with a physical world, a scientific theory is essentially a story whose "truth" lies entirely in its usefulness in explaining the reality we experience and in guiding our practical handling of that reality.

[1]The original materials we describe were named *Math Workshop* (Wirtz et al., 1964). With support form Harcourt School Publishers, the authors and their colleagues at EDC have since developed new materials, published by Harcourt as *Think Math!*, based on that original design. Where we use *Math Workshop* in this chapter, we are referring exclusively to the old materials. Where we use MW, we refer to both the original work and to the *Think Math!* program.

Educational theory is therefore a tricky thing. When we are being teachers, theory does guide us some, but not completely. We find ourselves quite often doing things that don't fully accord with what we believe, not just because we're human and can't always act in accord with our theory, but because sometimes the situation frankly doesn't seem to fit the theory and yet we must act anyway. In these latter cases, real science would deem the theory inadequate—it failed to account for the events—but any clinical practice (teaching and psychotherapy being two good examples) requires considerable art and craft skill along with scientific principles. Because we cannot reject educational theories just for failure to accord with all the data—they *can't* accord with all the data—we live in a fuzzy world in which, depending on which data we care to ignore, we have competing theories or, worse yet, loose or inconsistent standards for judging even the theory we choose to accept. Perhaps we (the authors of this chapter) are doing our own credibility as theoreticians damage by making such a claim, but it seems the only responsible preamble to the presentation that follows, in which we will tell a story (one that we completely believe) as if it is an established truth about children's algebraic thinking.[2] Like its competitors, it is not, but it's interesting to see the practical consequences—in the form of instructional ideas and materials—that follow from this theory and appear to be highly successful.

There are two more disclaimers we'd like to make:

1. While talking about ideas, we often do not cite research works where an idea has been studied in depth as we feel that it would be distracting, and would also "elevate" the claims in an inappropriate way, making this chapter look like a research result though it is not.
2. We do not claim that all the features of the *Math Workshop* and the *MW* that are described here are unique and never occur in other curricula; it is the combination that we believe is unique.

HOW CHILDREN LEARN: AN UNORTHODOX POINT OF VIEW

Sawyer (1964), in *Vision in Elementary Mathematics*, explains:

> It is a defect of most algebra books that they begin by developing a lot of machinery, and it is a long time before the learner sees what he can do with all this machinery. For example, he may learn to simplify $5(x + 3) - 4(2 - x)$ without seeing in just what circumstances he would feel a need to perform this calculation.

[2]According to the NCTM *Standards* (NCTM, 2000) major skills related to algebra are "(1) understanding patterns, relations, and functions, (2) using algebraic symbols to describe and analyze mathematical situations, (3) using mathematical models to represent and understand quantitative relationships, and (4) analyzing change in various contexts." The first two reflect the most everyday image of algebra; we focus on these.

> It is quite possible to use simultaneous equations as an introduction to algebra. Within a single lesson, pupils who previously did not know what x meant, can come, not merely to see what simultaneous equations are, but to have some competence in solving them. No rules need to be learnt; the work proceeds on a basis of common sense. The problems the pupils solve in such a first lesson will not be of any practical value. They will be in the nature of puzzles. Fortunately, nature has so arranged things that until the age of twelve years or so children are more interested in puzzles than in realistic problems. So the puzzle flavor of the work is, if anything, an advantage. (p. 40)

Then follows a whimsical story about a father with same-height twin boys (Fig. 17.1) (Sawyer, 1964).

The *Math Workshop* implementation of Sawyer's idea in the fourth-grade uses two more pictures, and very few words, to present a puzzle, leaving the children to figure out how tall the three people are (Fig. 17.2) (Wirtz et al., 1964).

Letters—like x, n, a, and b—have been used since first grade to stand for numbers, but in this puzzle presentation "no symbols are used." Of course, that is not true at all. The pictures are symbols and, for this purpose at this time, they are deemed the appropriate ones, but we will say more later about the way the use of letters as symbols is developed and why alternative choices are made even after that development is well under way. What is important to note for now is that the children easily figure out the heights, and love the puzzles.

Two important assumptions buried in this statement are at significant variance with some of the traditional thinking about early mathematical learning:

1. The assumption that, in some ways, algebra is quite natural for children—implied in Sawyer's claim that algebra can start with what is often deemed as an advanced topic (his example names simultaneous equations) and that "no rules need to be learnt; the work proceeds on a basis of common sense."
2. The assumption that young children "are more interested in puzzles than in realistic problems" (Sawyer, 1964).

This chapter explores the background and meaning of these two assumptions[3] and, with concrete examples, their consequences for teaching.

---

[3]Both of these assumptions follow from the fact that children live in a messy world and have less control of the environment around them than do adults. Therefore, they need to be adapted to make sense of and learn in that messy world, and to find pattern and order in whatever fragmentary and disorganized data they get. Children are constantly solving puzzles in their attempt to makes sense of the real world. This involves seeking structure, while ignoring some details. It is an act of abstraction that children naturally start with (see e.g., chap. 3, this volume, or the ideas described by Pinker, 1984, a body of ideas deriving principally from research in cognitive science and interpreted in the light of evolutionary psychology).

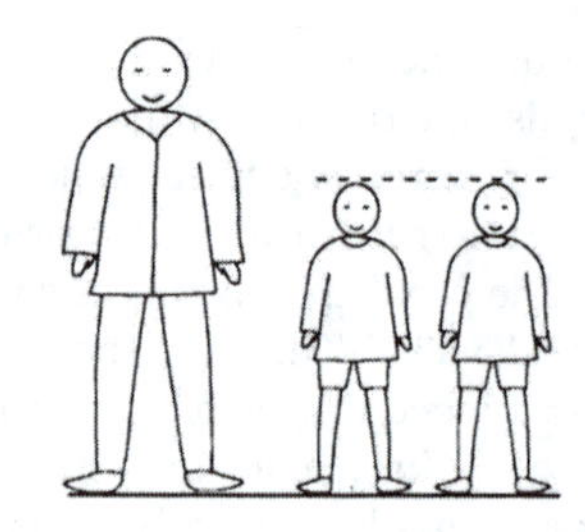

FIGURE 17.1. A father and twins.

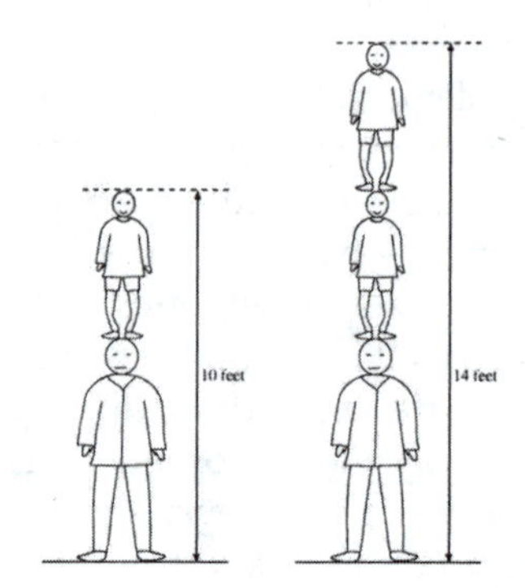

FIGURE 17.2. A puzzle: How tall are the people?

CHILDREN START WITH ALGEBRAIC THINKING AND USE IT TO ACQUIRE ARITHMETIC KNOWLEDGE

When children are very young, before they "conserve number,"[4] formal arithmetic cannot make much sense at all. For many 4-year-olds, even those who can reliably count beyond seven, seven objects spread out seem like "more" than the same objects bunched together. Although that is still the way their logic works, we cannot sensibly ask them to make sense out of the claim that 4 + 3 is the same amount as 7. Until their logic develops further, one seven does not always seem to be the same amount as another *seven*. Some children are really

[4]Here, conservation of number refers to solidity of quantities greater than three or four. For small enough numbers of objects, babies at 11 months seem to have not only stability of number but essentially addition as well (see e.g., Feigenson, Carey, & Spelke, 2002). But 11-month-olds' abilities to do arithmetic are not so readily observed by the casual onlooker and require advanced methods to be detected. The algebraic ideas and learning described in this chapter are all based on observations anyone can make, and reproduce, with school-age children.

ready when they first encounter 4 + 3 = 7, but the nonconserving child, faced with the requirement to assert that 4 + 3 = 7 has only two options: either to play along and divorce common sense from mathematics—after all, what the child sees[5] is that the two quantities are not the same—or to give up altogether and simply fail to "get it." But here are two remarkable phenomena. First, conservation of number will develop normally, regardless of school, and a brief delay in the requirement to "know that 4 + 3 = 7" would, therefore, put fewer children in this difficult position. Quite possibly, the delay could be more than compensated[6] by the advantages of working with children who have not already suffered a pretty serious assault to their logic, being forced, in effect, to abandon what makes sense to them in favor of what the adult says is correct. One doesn't sit by idly in the meantime waiting for maturation, of course—it is always possible to nourish the child's intellect—but the best leverage point for intellectual growth is the place where the child's logic is. Second, the development of what we call "conservation" is essentially an algebraic idea, arriving without our intervention before the child has learned the arithmetic facts (see also E. Smith, chap. 5, this volume).

Here is a concrete example. Shown two hands like this 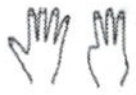 , a conserving child may not know how many fingers there are without counting. But if the hands are moved like this  or in this  , the child will be sure that the number of fingers is the same. Put another way, if the child is not yet sure, then the whole notion of a "total" makes no sense: There is no stability to the quantity. The conserving child knows that 3 + 5 is the same as 5 + 3, an algebraic idea, before knowing what number 3 + 5 is, a fact of arithmetic.

We are inclined to extrapolate from this and raise a question about development. Piaget's theory included a kind of graduality—new forms of logic (e.g., conservation) do not arrive suddenly, whole, and rock solid, and may be seen, especially at first, to come and go depending on the context or situation the child faces. But, we wonder if there are not some ideas that do arrive in a nongradual way, and are fairly complete when one first sees them. We'll play this out by extending the 3 + 5 example to a

[5]This is not evidence of children being enslaved by their senses. It is not that appearance is winning out over counting; it is that counting isn't yet logic. The child is trying to make sense of the world and has not yet built the abstraction we call number, at least not for such a high number. Logic requires that the child use what is available.

[6]See, for example, the bold move of Benezet (1935), who, as superintendent of schools in Manchester, New Hampshire, eliminated the teaching of mathematics in the early years, with no untoward effects on student learning after it was later introduced.

situation involving larger numbers. Bastable and Schifter (chap., 6, this volume) describe a classroom event in which a teacher asked a group of third graders if they thought that "the amount of money may be a different amount" if they added some coins in a different order. Some students seemed uncertain, and the interpretation was that they "hadn't internalized that you can add numbers in any order and maintain the same sum." In the spirit of the revisiting of Piaget's experiments (see e.g., Donaldson, 1978), imagine a variant on the problem in which you draw a fistful of change out of your pocket and, without showing any of the coins, ask students how much money you have. Of course, they can't know without seeing the coins, and a suitable presentation of the question would have to allow children to feel comfortable saying that, perhaps acknowledging the joke of the situation, and not imagining that the adult thinks they could know or, equally bad, is just asking them to guess. But, given such a situation, doesn't the question, itself, carry the assumption that there is an amount—one fixed quantity—and that the only thing left to know is what that amount is? It seems hard to believe that any children capable of understanding the question would think that the amount, itself, could change depending on how they looked at it. So, what could have been behind the children's uncertainty, reported in Bastable and Schifter, when they were asked "Will you get the same answer [if you add the coins in a different order]?" One possibility worth investigating, if a suitable method can be devised, is that the very posing of the question by the teacher implied to the children that their logic—their certainty of commutativity and associativity with small numbers and their certainty that there is a unique amount—is not shared by the adult teacher. If this is the case, then rather than the discussion having revealed an unanticipated uncertainty, it might well have *created* the uncertainty. Alternatively, it may seem that the teacher is hinting that the procedure by which children find sums (the process we call addition) might be unreliable. Any deliberate classroom intervention (including deliberate inaction) has associated hopes and risks. In this case, the hope in asking such a question was presumably to nudge children into thinking more deeply about the underlying mathematical issues; handled sensitively, that may indeed be the common outcome. But a risk of questioning a child's emerging logic in the context of arithmetic is that the child learns (along with whatever else) that one's own way of seeing things is not necessarily to be trusted in this subject.[7]

[7]As a side note, the common use of "can you be certain this will always be true" as a nudge toward proof carries the same risk—it implies that what the child has arrived at by whatever (even immature) logic the child used is not to be trusted without proof. By contrast, proof that explains a mechanism (i.e., proof as answering "why does this work" rather then "is it really true") engages the child's logic in taking a step further, rather than questioning the validity of the results it has already produced.

We might go so far as to suggest that prior to the age at which the question "How much money might I have" becomes reasonable, the question "Will you get the same answer [if you add the coins in a different order]?" is not reasonable and we should not ask it. After such an age, we should not assault children's logic by asking them to verify that different ways of counting the money would give the same result. They may need experiences of adding up coins in various orders in order to learn that some orders make the addition more convenient than others, but they cannot even understand the question "how much money" without the full conviction that there *is* an amount.

Life, of course, is a bit more nuanced than we've just portrayed it, especially when one is quite near the frontier of a child's thinking. The *Math Workshop* approach certainly did not avoid questions that pushed at the edges of children's thinking (nor does *MW*). It also greatly values surprise: results that the child's logic/intuition did not predict. But it always aims to pose activities that use (and thereby affirm the usefulness of) the logic a child already has in the development of new knowledge. When that logic is genuinely inappropriate for a task that the child must (eventually) learn to perform—as the nonconserving child's logic is inappropriate for learning addition—we must be thoughtful about ways to feed and exercise the growing edge of the child's logic without risking circumventing or ignoring it.

The point of this is that learning the arithmetic facts and the procedures for adding both rely on some algebraic ideas (about the behavior of aggregation and partition that we encode with the operation of addition and its inverse subtraction) that must already be in place. A curriculum is rarely in a good position to teach those ideas, but is excellently placed to use, hone, and extend them. It can also give those ideas names, when children are ready to take yet new steps of abstraction that require talking about (and therefore having names for) the ideas, but early on, the use, not the names, is generally most important.

MW's approach makes great use of children's natural and productive ways of abstraction: principally their keenness at pattern finding, which provides interesting ideas worth talking about, their eagerness to show these patterns, and their facility at learning language (including some work-saving mathematical shorthand) to describe what they've found.

YOUNG CHILDREN ARE MORE INTERESTED IN PUZZLES THAN IN REALISTIC PROBLEMS

Judd (1928) explicitly questioned the assumptions being made about so-called real-world contextualization. Nothing in the 1960s *Math Workshop* materials explicitly raises this as an issue but, whereas *MW* does help children learn to decode word problem formats, and does use some conventional application contexts, these are both done very sparingly compared to most contemporary (or old) curricula. The effect, however, is not at all a naked arithmetic approach. Everything needs a context to help give it meaning, but, for Sawyer (1964), children's enjoyment of puzzling things out was seen as, itself, a context—a genuine and motivating one, and one that is especially well suited to mathematical learning. Sawyer was not egocentrically taking the mathematician's love of mathematics and assuming that everyone would put up with any kind of dry activity just to learn the subject. In fact, his thought-experiment proposal for teacher training suggests how strongly he valued watching what truly grabbed students' interest: "Go into a street or park or public place where there are children over whom you have no disciplinary powers, and start doing something to see how many children come round you, how long they stay, and what questions they ask."

The real world of the young child is play and fantasy; if connecting with a child's reality really does improve motivation,[8] then children, especially young ones, are likely to be best involved by puzzles, inherently mathematical games, and appeals to their imagination. Learning to apply logic, organization, and focused attention to the child's own real world—a world of pretend and make believe in which children suppose, much as we do in mathematics, and then follow the consequences—might well have a better chance of surviving the transition to adulthood than learning to apply mathematics in problems that derive from the real world of adults.

Children's verbal tricks, riddles, and puzzles are not mathematics, but have a character that is similar enough to be a good leverage point. As they play their various games (e.g., sports or cards) they discuss the rules (think postulates) and the consequences of the rules (think theorems) in much the way we'd want to hear (and refine, of course) in a mathematics class. The real world of children is, in this regard, more classically mathematical than the (applied) real world of adults in which the pragmatic utility of mathematical results may actually work against the development of mathematical sensibility. (If the value of some mathematical

[8]It seems hard to believe that connecting with a child's out-of-school reality does not improve motivation, but is there actual scientific research evidence for this? Because this seems such an unquestionable theory, it may not need research, but we should be asking what evidence shows that the real-world applications we currently employ actually motivate students, or are responsible for any significant component of the successes we see. New programs that excite teachers help to excite the students, but it is hard to tease out what factors are responsible. Of course, perhaps it is enough that these adult applications excite the teacher: We may not care, then, whether they directly or only indirectly motivate the student.

result is only in its utility, one hardly needs to understand why it works or go through the thinking required to prove that it works, as long as a recognized authority has approved the result.) Being "more interested in puzzles than in realistic problems" is an intellectual stance that fits well with Sawyer's assertion that algebra can start with advanced topics. Why would anyone care about the heights of the father and his twin sons in that silly story *Math Workshop* tells? The problem is completely silly, but discovering the unknown, as long as it is not dreary, appeals to people.

THE CURRICULAR CONSEQUENCES

The word *algebra* does not appear anywhere in the tabular scope and sequence of the *Math Workshop* curriculum, and no unit is devoted to a focused development of algebraic ideas before Grade 4.[9] As with all big ideas in *MW*, algebraic thinking becomes a pervasive theme, rather than a topic, and is brought in without fanfare and used everywhere, as befits anything worthy of the name *big idea*.

Fostering a Puzzling-Out Frame of Mind. *MW* does this in several ways. An obvious way is to use puzzles explicitly and often, but there are other ways as well. It provides a variety of clues instead of explicit instructions, with sufficient redundancy of information to allow students to make conjectures about what to do, and to check those conjectures. This allows for multiple strategies early on. In fact, this approach also helps students learn to read mathematics. Unlike English text, mathematical information on a page does not unfold left-to-right-and-top-to-bottom. An entry in a table gets its meaning from the row and column in which it is found; the vertical bar on a bar-graph has meaning only if both its vertical size and horizontal placement are both understood; a point on a coordinate system has a similar two-dimensional meaning; the meaning of an algebraic statement that describes the pattern of growth shown in a sequence of pictures is clear only when one looks both at the statement and at the pictures; and so on. Figure 17.3 gives an example from the first day of Grade 2. Rather than telling students precisely what to do, they are asked to look over the entire page. What do they see? There is a question, but where should the answers go? How can they check?

[9]In fact, unitlike structure hardly exists at all before the last third of second grade of the original material, as the development of all mathematical ideas tends to be so interconnected.

FIGURE 17.3. The first day of second grade in *MW*.

The page can be completed just by counting, but children are rewarded for finding shortcuts, for noticing structure that helps them count, or noticing that *J* and *P* and *G* are built of the same pieces, and so must have the same number of dots. The page even provides gentle clues that these shortcuts exist. There are ideas worth discussing even after students have completed the page. To begin with, is there any significance to the way the answer boxes are grouped?

Presenting Problems With Different Parts Missing. *MW* also generally presents problems with missing results (e.g., 3 + 2 = —) and problems with missing parts (e.g., 3 + — = 5) mixed together. Thus, students develop a habit of paying more attention to the part–part–whole structure of a problem, and not use appearance of the symbol "+" as a signal to add.

(Students also encounter, very early, problems like 8 + 4 = — + 9.) Our experience observing students is that when their experience is not one of problems segregated by type, but encountered together in this fashion from the start, students learn relatively quickly that the position of the missing number is important. The result is that whatever differences in difficulty the two forms present are essentially invisible in the classroom, and students seem equally comfortable with all forms. In fact, we would argue that difficulties are minimized because problems are not segregated "by type," thus subordinating form to structure. Rather than learning addition and subtraction, each as a topic, students are learning the relationship between a whole and its parts.

A particularly interesting example of a problem with missing parts shows up as students are learning to perform multidigit multiplication. Consider the kind of thinking you must do—and how different your thinking is in each case—in order to solve the problems in Figure 17.4. When puzzles like this are presented in a considerate way, so that the very first problems are not quite as challenging as these, children love them. Each problem is just different enough to be fun and maintain children's interest while they are quite literally learning their multiplication inside out and backward. The fact that the problems are different means that students cannot mindlessly and repetitively use the same technique; their attention is focused on choosing the technique, not applying it, and they get their needed practice in the algorithm without monotony.

Even if one agrees that it is educationally valuable for children to puzzle things out, it is certainly reasonable to ask what connection it has to algebra. The process of solving puzzles requires considering a variety of different clues at the same time; it develops one's ability to focus on connections, relationships, and patterns—exactly the object of algebra. The language of algebra might or might not be used, but the mental disposition is essentially equivalent to solving a system of equations. To develop their algebra, students need both the set of mental tools and inclinations, and the appropriate language(s) in which to express their thinking.

MW Treatment of Written Language. *MW* regularly presents a kind of "word problem without the question," a storylike format where students are to fill in missing parts, or even invent the question, rather than answer an explicit, single, closed question posed at the end of the problem. For example: "We have 22 children in our class. There are not many more girls than boys." What questions come to mind? Learning what questions *can* be asked and answered is part of learning the modeling part of mathematics. Also, the enterprise of "solving for the unknown" seems more

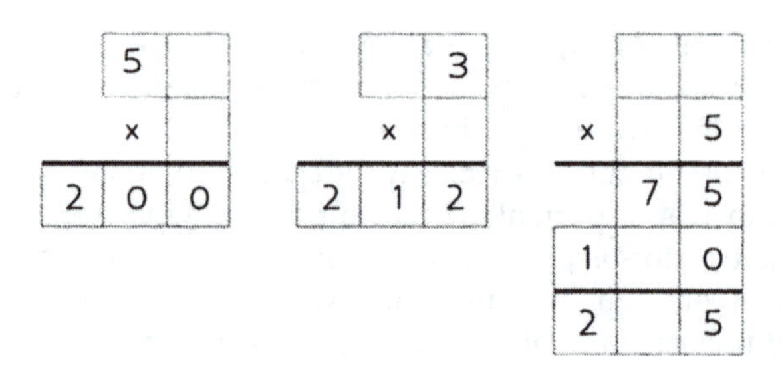

FIGURE 17.4. Fourth grade: learning multiplication inside out and backward.

meaningful if one gets experience asking what *is* unknown and knowable. To some children, this problem might suggest making a table. Over time, children learn to ask many kinds of questions, like: Could there be one more girl than boys? Alternatively, the situation might be presented this way: "There are two more girls in our class than boys." Again, the first question is "What could we figure out about this situation?"

A feature of *MW*'s algebraic approach that requires a bit more explanation is its specific treatment of using language.

It is clear enough why curriculum materials should sometimes present mathematical problems verbally—as story problems, for example. The most obvious, if least glamorous, is that tests do so and, because they do, children need to be protected from them by being prepared. But there are some solid educational reasons, too. Real life (except the real life of academia) does not present us with word problems; it presents us with situations in which we find problems, which we often express in words. To focus the exercises used in teaching, educators simplify the situations, narrow the choice of possible problems one might find in them, and write up the verbal descriptions: Word problems are a printable substitute for real-life problems, but only a very approximate, thin, and typically stilted one. Calls for yet closer approximations date far back. In the long history of education, Dewey[10] is recent, but we tend, these days, not to look back even that far. The various integrated math–science–technology programs

[10]Dewey's take on real-life problems talks about the real life *of the child*. "School must represent present life—life as real and vital to the child as that which he carries on in the home, in the neighborhood, or on the playground ... much of present education fails because it ... conceives the school as a place where certain information is to be given, where certain lessons are to be learned, or where certain habits are to be formed [the value of which] is conceived as lying largely in the remote future" (Dewey, 1897, pp. 77–80).

and investigation-style curricula are modern efforts in a similar direction—to let mathematical problems arise for students in much the way they do in real life, without being encountered only as canned word problems. And even the canned problems can muster some plausible defense. Although they are their own genre of writing, the ones that are not formulaic caricatures of the genre do present a kind of exercise in translating between verbal symbols (words) and mathematical symbols (letters, numbers, operations) that is plausibly of value in mathematical learning.

There are also many reasons why we might want to present a large part of mathematics to children *without* embedding it in words, written or even spoken. Perhaps the most obvious reason is that there are many children for whom the written word is yet one more barrier to get through, a cognitive load that takes attention away from the mathematics. These include, of course, children for whom English, itself, is a challenge, but also those with reading disabilities, those with learning disabilities arising in disordered language processing, and those whose optimal learning mode is not verbal.

But there are also reasons that tie directly to the nature of the algebraic process. At the surface level, algebraic symbolism is, itself, a language designed to convey certain essential mathematical ideas better than words do.

So, one message that students should receive is that information can be obtained not only through the text. The puzzlelike presentation helps here, too, allowing for an informal and natural introduction of new formats and ideas. Unlike most classic puzzles, *MW* student pages typically provide more information than is logically required; this redundancy of information allows students to check their initial guesses to see if they are correct, and it also allows some formats and ideas to appear as something of a cross between additional clues and new objects to explore.

Consider the exercises in Figure 17.5, drawn from Grades 1 and 2. Over time, algebraic notation gradually becomes a more and more valuable clue, but not yet the sole clue, required for completing a problem successfully.

This way of introducing an algebraic notation has several additional advantages: allowing different students to learn on different levels, demonstrating the usefulness of algebraic notation while introducing it, and saving time on teaching just the notation.

So far, we've discussed only MW's treatment of written words while introducing a problem (using words as just one of possible clues, not even the major one). *MW* also uses a pedagogical technique that does not use even spoken words.[11]

---

[11]This technique is supported by Hendrix (1961), who presented situations where avoiding spoken words is beneficial for learning.

Find a rule.

0	3	6	1	8	9		18		29
5	8	11				19		32	

Find a Rule

7	10	8	13	19			33	40		100	n
2	5	3			19	0			37		n-5

Find a Rule

a	8	6	17	15	18		26	12		27	21
b	2	3	5	10	6	8			18	25	
a + b	10	9	22			17	35	15	50		36

FIGURE 17.5. Find-a-rule puzzles.

This first stage is a game whose rules do not allow speaking—either by the teacher or by the students. For example, while the class watches, the teacher silently writes on the board:

$$7 \rightarrow 11$$
$$9 \rightarrow 13$$
$$12 \rightarrow$$

and then offers the marker to the class. A volunteer takes the marker, silently records a number, and the teacher nods yes or no. The teacher writes another number and again offers the marker to the class. This goes on until many students have provided responses. The teacher can vary the difficulty of entries to provide more support for some students or more challenge for others, depending on where the students are. Then, the game is declared to be over, and the students discuss the ways they thought of responses.

We have observed this technique in many classrooms where *MW* was piloted, and the attitude to it of both teachers and students is unanimous: They all love it. Students, we think, love it because of its gamelike appearance, because the pressure of necessity to figure it out right away is lifted, but the possibility to participate in the game remains. Teachers love it because while playing the silent game students are highly interested and attentive, because it is accessible for students of different levels, because the game gives teachers a good chance to informally assess students' understanding, and because it prepares students for a better, more focused discussion afterward. One of the teachers said: "This class isn't usually quiet, and listening is hard for them. It's great to see them all captivated, paying attention, thinking hard [during a silent game.]" We hear similar remarks from other teachers.

We think (although it is only our opinion, supported by observing over 10 classrooms directly and getting feedback from over 60 classrooms that pilot *MW*) that in addition to being highly motivating, silent games are very effective in developing students' thinking in general, and algebraic thinking in particular. Initial absence of explanations by students or the teacher allows the majority of students to find the rule all by themselves, using their own methods. Students can take their time thinking; each new entry in the table increases the data available to other students; the whole class is involved in studying, analyzing, and applying a pattern, and all students can participate.

Asking students to explain their thinking can serve several purposes: to help the teacher understand a child's correct (or incorrect) thinking; to show the class that there are multiple ways to solve a problem; and to sometimes help the student who explains the thinking to clarify that thinking. Despite this, *MW* places less emphasis on having students explain their thinking than is currently fashionable: Some of the goals of having students explain their thinking can be achieved in other ways, and having students explain their thinking is not entirely without risk. For example, virtually all *MW* work invites multiple strategies (or, in the silent teaching example, different inputs to find an output), affording the teacher many opportunities other than a student's explanation to "see" and diagnose a student's thinking. Moreover, the puzzlelike structure of the material lets students self-correct without having to mediate the analysis of their own thinking through words. (Of course, we do not suggest abandoning language as a medium for self-correction, we merely suggest that perhaps it should not be the only or the first way.) Articulate verbalization, not to mention analyzing their own thinking well enough to know what to say, is very hard for young children—often harder than having the mathematical idea and just demonstrating understanding—and can sometimes get in the way of mathematical learning. Explanations that are not articulate, even if they are correct, can increase confusion instead of reducing it, even for the child who is making the explanation, and certainly for others whose understanding is shaky. Explanation, whether by the teacher or by other children, steals the chance for others to make their own discovery. A heavy reliance on words creates barriers for many children, clearly for those whose English is limited, but for others as well. Learning to *express* mathematical ideas is worthwhile, but not if it competes with *developing* the mathematical ideas; for young children it is best when their thinking is already pretty clear, and the major purpose of trying to express their ideas is learning to do so. One might argue that explanations could also facilitate discovery; but much more effectively when the discussion takes place after a silent game, when all students had enough time for independent thinking; even incorrect ideas had time, if not to be self corrected, then to get condensed into a more accessible form.

MW'S APPROACH TO USING AND EXTENDING ALGEBRAIC IDEAS

Building on logic already available to children, ideas of commutativity and associativity are combined in the "any order, any grouping" principle. Given a collection of small objects like these, young children spontaneously sort, making little piles by color or size or both, or perhaps deliberately mixing large and small to make "families," but the inclination and cognitive apparatus for finding pattern in (or ascribing meaning to) the randomness is already natural. As mathematics teachers, we help them extend the usefulness of the logic they already have by adding organization and by appealing to their ability to shift flexibly from one classification scheme (e.g., color) to another (e.g., size) and to summarize, perhaps numerically, what they know about each (Fig. 17.6).

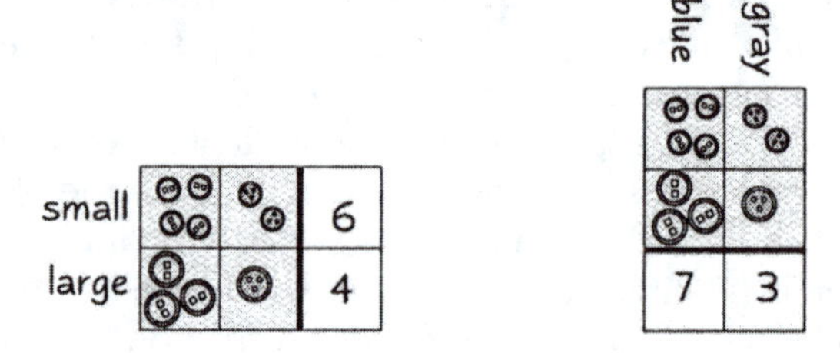

FIGURE 17.6. Buttons: data from a summary of the sorting.

Substituting numbers for the original objects doesn't change the ultimate logic. Adding either pair of subtotals in the white boxes—the ones that had previously recorded the summary of small and large buttons, or the ones that had recorded the numbers of blue and gray—must give the same grand total (Fig. 17.7).

Nine numbers in this chart are connected by six addition sentences—three vertical and three horizontal—and these sentences are also connected among themselves. Later, students consider these sentences without the

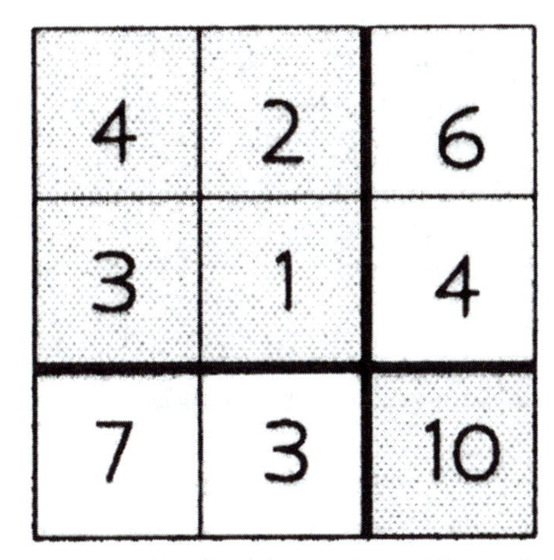

FIGURE 17.7. Recording the sorting with numbers only.

frame that used to surround all the numbers $\begin{smallmatrix} 4+2=6 \\ 3+1=4 \\ 7\;\;3\;\;10 \end{smallmatrix}$. For students who have been sorting buttons and working with the chart format for a long time, the idea that adding two addition sentences produces a new true sentence, 7 + 3 = 10, comes as a delightful surprise, but a surprise that they can explain easily and naturally. A consequence of this presentation is that when students later in their education encounter simultaneous equations, they can add or subtract those just like they added simple addition sentences. The idea comes equally naturally and with *no* surprise $\begin{smallmatrix} 5x+3y=23 \\ 2x+3y=11 \\ 3x+0=12 \end{smallmatrix}$.

The point is not merely that the algebra is there, but that it is founded directly in the children's logic. As Sawyer wrote, "No rules need to be learnt; the work proceeds on a basis of common sense."

From "Any Order, Any Grouping" to the Distributive Law and Beyond

By making a highly structured collection, and splitting the collection in a way that preserves that structure, we can extend the notion that one may safely split a collection into several smaller collections to help one number the elements. For example, in trying to figure out how many items are in a 5×9 array, a child might split it into two arrays—5×4 and 5×5—whose counts are already known, and add the $20 + 25$ to get the total: $5 \times 9 = 45$.

Extending that logic to a 16×27 array, we first partition the dots into four regions to make the counting of each region easier (see Fig. 17.8).

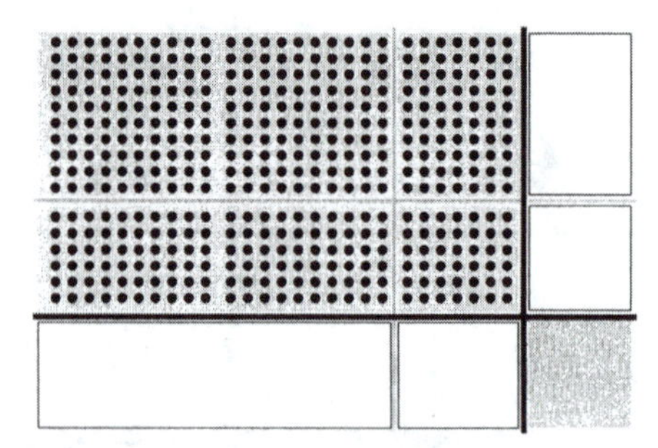

FIGURE 17.8. Instead of a few buttons, many dots, arranged to help us count them.

Rewriting the results—200, 70, 120, and 42—in the familiar format, students can summarize the number of dots in the rows and columns, as they summarized the numbers of buttons (Fig. 17.9). This pushes their algebraic logic in a different direction. Instead of arriving at the reasoning behind certain manipulations of equations, this route uses the child's algebraic reasoning—an intuitive understanding of the distributive law—to build the standard American multiplication algorithm.

Children naturally apply the distributive law in concrete situations. Given $36 in the form of three $10 bills and six $1 bills, and asked to share the money equally among three people, most children will not simply hand each person three bills regardless of their denomination: Without hesitation, they will distribute the tens evenly and the ones evenly, and conclude that each person gets $12. In other words, they already intuitively feel, although cannot express it symbolically yet, that $\frac{36}{3} = \frac{30}{3} + \frac{6}{3}$. Building the symbolic from the intuitive takes work, but this intuitive use of distributive law in another guise is the foundation both for division (we divide a large number by dividing portions of it and then combining the results) and for a sensible understanding of the addition of fractions with like denominators. Many children, when faced with structures like $\frac{1}{3} + \frac{1}{3}$, even when interpreting it as a sum of two quantities, are lured by the plus sign to add everything in sight to make false statements like $\frac{1}{3} + \frac{1}{3} = \frac{2}{6}$. But children, in effect, already know that $\frac{30}{3} + \frac{6}{3}$ does not equal $\frac{36}{6}$, and *MW* children have been taught to notice what they know, and use it. Their ideas about fractions are not built on new rules that the children might mix up or forget, but on old logic, their own understanding of related situations, which they cannot forget or, at least, can easily re-derive.[12] As Sawyer wrote: "No rules need to be learnt; the work proceeds on a basis of common sense."

[12]This is a major point of situated cognition research.

200	70	270
120	42	162
320	112	432

```
  16      27
 ×27     ×16
 112     162
 320     270
 432     432
```

FIGURE 17.9. The numbers in the summary of the array match those in the algorithm.

Pattern-Seeking in Numerical Contexts

If true mastery requires high-speed random-access to the facts, to borrow a metaphor from computer science, that still does not mean the process of *learning* the facts should be random. For example, even though we want 4×7 to be its own independently known fact, curricula often try to teach children how to use the knowledge of 4×6 to compute 4×7 because the same logic applies equally to deriving 41×25 mentally from easier problems like 40×25, which we want them to be able to derive from 4×25. Do children need to perform such problems mentally? No, but the process is an algebraic one that we want to be well rooted in familiar logic, and well exercised by the time children encounter it in the form $a(b + 1) = ab + a$. In the fourth grade, when a large part of children's job is to master multiplication facts, *MW* helps children see what is already familiar about $a(b + 1)$ and generalize it, record it with generic notation, connect it with prior knowledge, and use it to aid computation. *MW* also makes use of other patterns. Children who already know the facts for the perfect squares love seeing that 6×8 is one less than 7×7, and that the pattern holds for any square and the product of its nearest neighbors. While trying many cases in order to find the pattern, students practice, but not mindlessly. After becoming convinced that the pattern holds, students use the familiar arrays of dots to show why it is so. And then they generalize the rule, seeing that the product of the next-nearest neighbors is always 4 less than the perfect square, and, in general, $(a + b)(a - b) = a^2 - b^2$. Again, the discovery entails considerable multiplication practice; the knowledge—apart from the fluency it generates through practice—is itself a tool to make certain new facts easier to remember; and the algebraic idea has a practical application. For a child, practical means that it can be used for fun. "Mommy, give me any two-digit number." "OK, um, 53." "Well, I can multiply that by 47 before you can!" As mommy reaches for paper and pencil, the child thinks: "Mommy's number was 3 more than 50. I picked 47 because it is 3 less than 50. So 47×53 is 50×50 minus 3×3. That's $2500 - 9$, or 2491." Of course, the child already has the standard algorithm for doing completely random problems on paper, which pleases parents, but this extra mental trick leaves children feeling brilliant. And the few who take it a step further and combine it with the $a(b + 1) = ab + a$ idea can then multiply 48×53 in their heads, too. It is, after all, 47×53 plus another 53. Does this have any practical value? Absolutely. Not for getting the answers, of course; for that purpose, the standard methods or a calculator will do. But, for feeling smart, which in education is of *real* practical value. *MW* also has students look for patterns in combinatorics. Young students see how many ways they can arrange n objects in two boxes (there are $n + 1$ ways to make n as $a + b$) or three boxes (there are $n(n + 1)/2$ ways to make n as $a + b + c$). The particular patterns are of some interest, but the real value is in the structured searching, the systematic thinking that is involved,[13] and the way that the physical act of rearranging the objects leads to an understanding of why the patterns are what they are. Imagine, for example, placing all objects in one box, and then moving one object at a time from that box to the other, each time recording the results. How many moves can you make? Each move results in a new arrangement, but the very first arrangement was recorded before you started moving: $n + 1$. Now picture what you do with three boxes. Place all the objects in one box. Each time, as you diminish that box's contents by one, count and record all the arrangements in the other two boxes. But that is just the previous problem all over again, summing the results of several investigations. So, $1 + 2 + 3 + 4 + \ldots$.

Function Composition

Function machine imagery gives children a good notation for figuring out the effects of combining operations. The assembly line diagram (see Fig. 17.10) provides another useful image. Because the three paths in the figure (red, green, and blue) must all have the same result (adding 6), the two operations along the green path (adding 1 and adding 5) must simplify to just +6, as must the two operations along the blue path. If, in a new diagram, the blue path is –3, –4, and the green path is + 3, –10, the seeds of a new arithmetic on a new kind of (signed) number are being planted, but children do not see new numbers here or think of themselves as performing a new kind of arithmetic. They are still simply combining operations in a familiar and fully understood way. "No rules need to be learnt." The activity is algebraic in two aspects. First, it focuses on performing arithmetic on

[13]English (2004) provides a strong modern case for combinatorics in elementary school and the research background that supports it, and refers explicitly to the value of the systematic thinking.

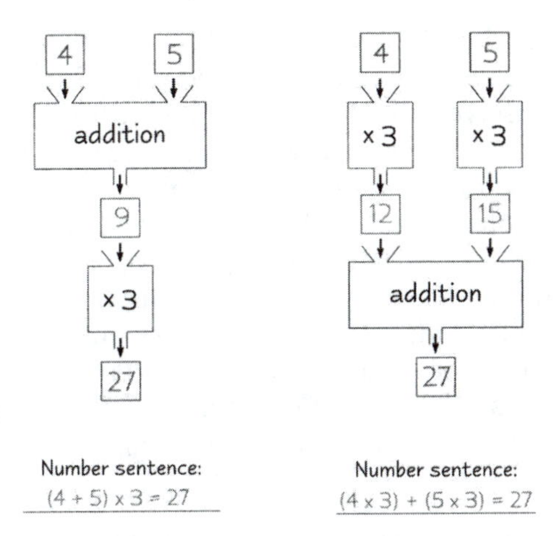

FIGURE 17.10. Function machines showing that $(4 + 5) \times 3 = 4 \times 3 + 5 \times 3$.

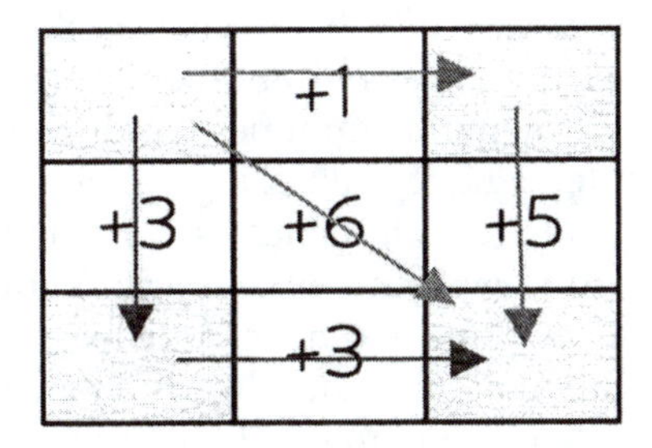

FIGURE 17.11. Function machines showing that $(4 + 5) \times 3 = 4 \times 3 + 5 \times 3$.

numbers you don't know (those numbers even are not showing up in the colored boxes). Second, it introduces a new kind of objects and invites generalizations on how to work with them.

There is yet another way to introduce variables and algebraic expressions. In fourth grade, children learn a number trick:

Think of a number.
Add 3.
Double the result.
Subtract 4.
Cut that in half.
Subtract the number you first thought of.
Your answer is 1!

They love the trick, but there is more to do with it than merely memorize it to play on friends. Students soon learn how a trick works, and make up other tricks themselves. To do so, they have to use notation. At first students use pictorial notation: a bag with an unshown number of marbles in it represents the number "you thought of" (see Table 17.1).

The left-hand column shows the steps of a trick, the middle column shows an image of the result of that step: the algebraic expression on the right (children do not see this expression at first), records the history of the steps taken so far. One cannot readily see from $\frac{2(x+3)-4}{2}-x$ that "Your answer is 1!", but the bag notation, being an icon (see Kaput, Blanton, & Moreno, chap. 2, this volume), naturally suggests simplifying on every step. Children very readily learn this, and have no difficulty with the transition from to x.

There are two things related to the use of algebraic notation that are interesting to point out. First is the fact that it appears only when it is badly needed, to simplify an activity, not as an additional thing to learn. To check whether a trick works on all numbers, the child must keep in mind a generic amount—the picture of a bag. When tricks become longer and drawing bags becomes more of a nuisance, omitting the top and the bottom of the bag while drawing it simplifies the drawing to look like an x. Second, although students have been using variables and simple expressions since the first grade (e.g., in tables as was shown earlier), they are absolutely not expected to make use of that prior knowledge; they have a fresh start. In addition to giving another chance to some students, this also makes the use of notation simple and already familiar to most students, letting them concentrate on thinking and having notation support thinking and communication, rather than being the centerpiece of learning.

TABLE 17.1
A Number Trick

Think of a number.		x
Add 3.	•••	$x+3$
Double the result.	••• •••	$2(x+3)$
Subtract 4.	• •	$2(x+3)-4$
Cut that in half.	•	$\frac{2(x+3)-4}{2}$
Subtract the number you first thought of.	•	$\frac{2(x+3)-4}{2}-x$

AN OVERVIEW OF MW'S APPROACH TO BUILDING ALGEBRAIC LANGUAGE

Algebra uses letters to stand for numbers; it is remarkable how many ideas are embodied in that notation. From first through fourth grade, *MW* introduces these various ideas:

- Letters as labels for problems (beginning of Grade 1)
- Letters as labels for numbers (middle of Grade 1)
- Literal expressions to record or describe processes, patterns, or properties (middle of Grade 1–4)
- Abstract arithmetic, summarized and encoded (Grade 4)
- Variables (late in Grade 4)
- Unknowns (late in Grade 4)
- Identities (late in Grade 4)
- Algebra as abstract arithmetic (late in Grade 4)

Letters as Labels for Problems

This, of course, can happen in any text with problems. In *MW*, however, these labels are used to pair problems with answers. Figure 17.8, which shows the first worksheet of Grade 2, is an example of such pairing. Numbers derived from the problem are written in the appropriate place on the table, as indicated by the corresponding letter. This happens very often starting with the first grade.

Letters as Labels for Numbers

This step is not sharply distinct from the previous one, nor does it occur at a different stage in the development of algebraic thinking in *MW*, but there are subtle differences. Not only is there a labeled space in which to write an answer, but that label is used again, to let that answer (output) be the input to a new problem, as shown on Figure 17.12, a worksheet from about the middle of second grade.

Literal Expressions to Record or Describe Processes, Patterns, or Properties

From first grade on, students will see function tables from which they are to induce a simple rule. Most of the time, these also have a pattern indicator—the n and $n + 10$ in the example on Figure 17.13—to which *no* attention is called. At this stage, students don't need to decode these captions in order to do the work. Students find the pattern exactly as one might

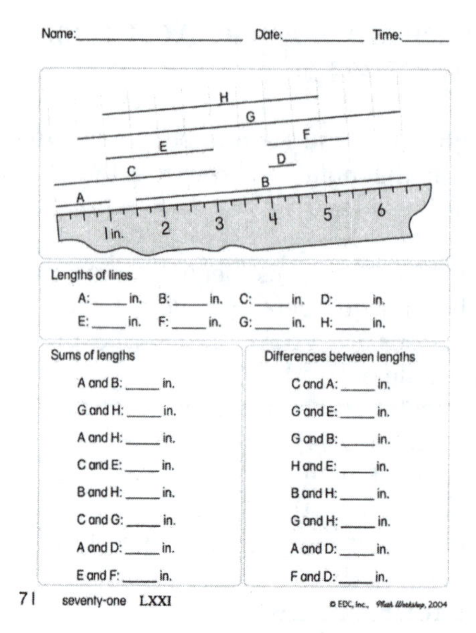

Name:____________ Date:________ Time:______

Lengths of lines

A:____in. B:____in. C:____in. D:____in.

E:____in. F:____in. G:____in. H:____in.

Sums of lengths	Differences between lengths
A and B:____ in.	C and A:____ in.
G and H:____ in.	G and E:____ in.
A and H:____ in.	G and B:____ in.
C and E:____ in.	H and E:____ in.
B and H:____ in.	B and H:____ in.
C and G:____ in.	G and H:____ in.
A and D:____ in.	A and D:____ in.
E and F:____ in.	F and D:____ in.

71 seventy-one LXXI

© EDC, Inc., Math Workshop, 2004

FIGURE 17.12. Letters as labels for problems (upper part) and letters as labels for numbers (lower part).

n	7	13	42	5		28			8	
n+10	17	23	52		33		16	10		51

FIGURE 17.13. Children invest algebraic symbols with meaning derived from context, just as they do with unfamiliar words in their native language.

imagine: They look at the numbers, figure out what's going on, and fill in the table.

But they do decode them! Their natural inclination to learn language in context causes many of them to invest meaning in the few marks that have been given no meaning by the teacher or the instructions. One little second grader blushed as if she wasn't supposed to know what those symbols meant—as if knowing was somehow cheating, because it gave away the pattern—and said that she'd figured out the pattern from the numbers "but, really, it even said the answer right there!" (pointing at the column with pattern indicators). Notice, as always, that the table is filled in so that the function must be computed both ways: sometimes we know n and must figure out $n + 10$; at other times we must work from $n + 10$ back to n. Other tables have three with captions like c, d, and $c - d$ (see Fig. 17.10).

Algebraic notation plays its usual role, but it is not the focus of the activity: the challenge of the activity is pattern finding, the drill is addition and subtraction, the context and image is function, and the notation is the algebra of structure. This use of symbols is also familiar from various guess-my-number games, in which students ask questions like, "Is your number greater than 4?", and the teacher notates the question first with all the words, and then, explicitly calling attention to the nuisance of writing it all out, as "number > 4?" and finally (all in one session) "$n > 4$?" followed by the answer, which is always Y or N. Soon, children also need to write their questions, and greatly appreciate the abbreviated forms. Starting in Grade 2, children also play various clue games in which they are trying to determine another player's secret number from clues about it. The secret keeper only writes clues (doesn't speak them). The clues may be fully spelled out in words (e.g., "My hundreds digit is greater than my tens digit" or "the sum of my digits is less than 10"), but they are strongly motivated to use shorter ways of writing. As a result, by Grade 3 they quickly pick up models provided by the puzzles in the curriculum (and the recommendations to the teacher) that present the same information in more abbreviated ways, including $h > t$ and $h + t + u < 10$ (initially with enough context to make the meanings clear).

Abstract Arithmetic, Summarized and Encoded

Experiments with magic squares allow students to look at properties of operations without being "prejudiced" by their expectations about these operations in standard arithmetic. Their results are "labeled" in ways similar to the "pattern indicator" captions described earlier, but still without fanfare. See a page from the beginning of Grade 4 on Figure 17.14 as an example.

Variables

The idea that a particular symbol may represent many values is introduced before the idea that one can assign a single assured value (solve for an unknown) to it. Interestingly, the notion of variables is first introduced in a non-numeric context. In fourth grade, children play a word game in which they are to find words that match a given pattern. The original *Math Workshop* materials introduce this puzzle through a story about two children who invented it. The girl, Numa, explains, "let's hunt for patterns in the letters of words.... I pick a Pattern Indicator *abbc*. That means I want to find 4-letter words with the two middle letters alike." They list *meet*, *room*, and *boot*, and leave the students to list others. Then Numo asks if he can use *deed*. "That's all right, too!" says Numa, "Pattern

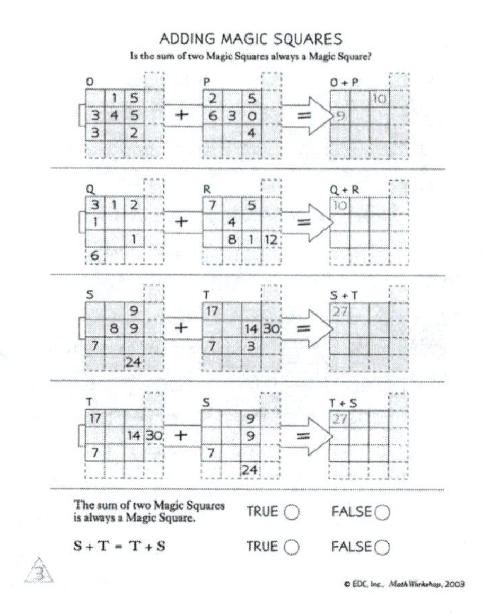

FIGURE 17.14. A page from the beginning of grade 4: The sum of two magic squares is also a magic square.

Indicators only tell you the number of letters and which ones must be alike." In a page full of puzzles with six different pattern indicators and a small amount of additional dialogue, the whole mystery of $a + a + b = 9$ is solved. Yes, not only are (0, 9), (1, 7), (2, 5), and (4, 1) solutions, but so is (3, 3). Something about this context makes the ideas quite nonmysterious.

Unknowns

All of these uses of letters to stand for numbers (or other letters) treat them as variables. The think-of-a-number tricks described earlier introduce two other notions simultaneously. One is that we can "do arithmetic on a number" without knowing what the number is. The other is that we can figure out what a number must have been by looking at how the arithmetic turned out.

Identities

Different structures can represent the same values. Think-of-a-number tricks compose several arithmetic steps that might be recorded algebraically as, say, $\frac{2(x+3)-4}{2}$. The four steps, recorded in terms of their results at each step, end with $x + 1$. So $\frac{2(x+3)-4}{2} = x+1$. Children in fourth grade are not in any way thinking through the algebraic steps that convert one expression to another, partly because they are never even faced with the more complex of those expressions. But they do record their experiments with 7×7 and 6×8 by filling in tables that are captioned with n, $n \times n$, $n \times n - 1$, $(n - 1)$, $(n + 1)$, and $(n - 1) \times (n + 1)$, and they do conclude, experimentally, that $(n - 1) \times (n + 1) = n \times n - 1$, and they do illustrate that geometrically (see Fig. 17.15) (and many students see the proof in this illustration).

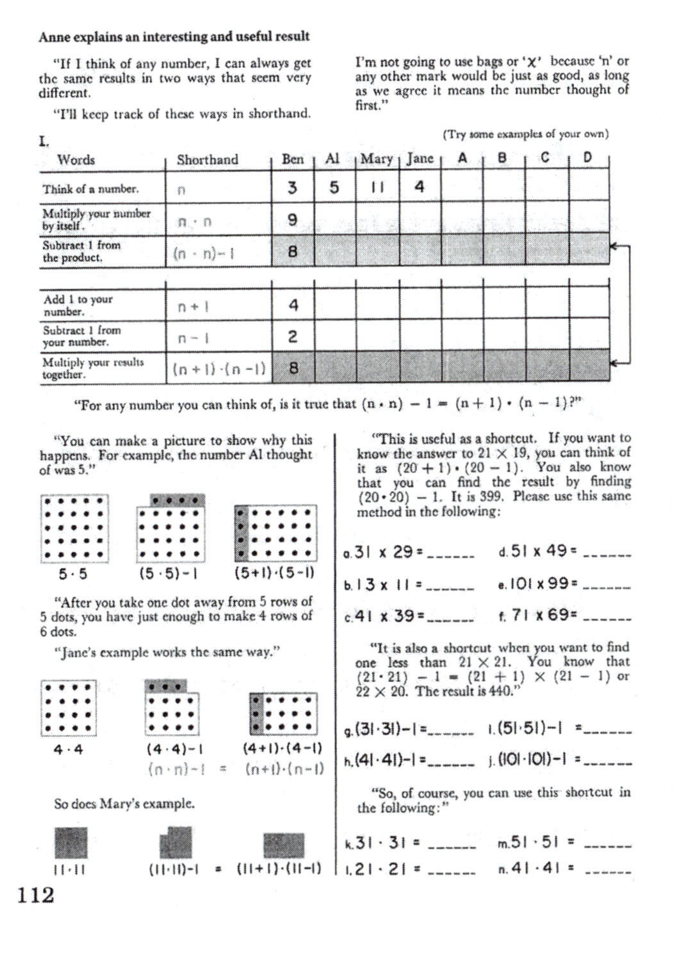

Anne explains an interesting and useful result

"If I think of any number, I can always get the same results in two ways that seem very different.

"I'll keep track of these ways in shorthand.

I'm not going to use bags or 'x' because 'n' or any other mark would be just as good, as long as we agree it means the number thought of first."

I. (Try some examples of your own)

Words	Shorthand	Ben	Al	Mary	Jane	A	B	C	D
Think of a number.	n	3	5	11	4				
Multiply your number by itself.	n · n	9							
Subtract 1 from the product.	(n · n) − 1	8							
Add 1 to your number.	n + 1	4							
Subtract 1 from your number.	n − 1	2							
Multiply your results together.	(n + 1) · (n − 1)	8							

"For any number you can think of, is it true that (n · n) − 1 = (n + 1) · (n − 1)?"

"You can make a picture to show why this happens. For example, the number Al thought of was 5."

5 · 5 (5 · 5) − 1 (5+1) · (5−1)

"After you take one dot away from 5 rows of 5 dots, you have just enough to make 4 rows of 6 dots.

"Jane's example works the same way."

4 · 4 (4 · 4) − 1 (4+1) · (4−1)

(n · n) − 1 = (n+1) · (n−1)

So does Mary's example.

11 · 11 (11 · 11) − 1 = (11+1) · (11−1)

"This is useful as a shortcut. If you want to know the answer to 21 × 19, you can think of it as (20 + 1) · (20 − 1). You also know that you can find the result by finding (20 · 20) − 1. It is 399. Please use this same method in the following:

a. 31 x 29 = ______ d. 51 x 49 = ______

b. 13 x 11 = ______ e. 101 x 99 = ______

c. 41 x 39 = ______ f. 71 x 69 = ______

"It is also a shortcut when you want to find one less than 21 × 21. You know that (21 · 21) − 1 = (21 + 1) × (21 − 1) or 22 × 20. The result is 440."

g. (31 · 31) − 1 = ______ i. (51 · 51) − 1 = ______

h. (41 · 41) − 1 = ______ j. (101 · 101) − 1 = ______

"So, of course, you can use this shortcut in the following:"

k. 31 · 31 = ______ m. 51 · 51 = ______

l. 21 · 21 = ______ n. 41 · 41 = ______

112

FIGURE 17.15. A page from the fourth grade of Wirtz et al. (1964).

Algebra as Abstract Arithmetic: Proof

In fact, in fourth grade, students generalize that result to $(a + b)(a - b) = a^2 - b^2$ by examining a tabular evaluation of the several functions of *a* and/or *b* that are combined in the expressions that make up that equation. The geometric proof involves creating an $a \cdot a$ array of dots, shaving off one column so that it becomes an $a \times (a - 1)$ array, and replacing that column as a bottom row of the resulting array. The column is too long, by one dot, to fit the new array, but without that dot, the result is an $(a + 1) \times (a - 1)$ array. For all a, $a^2 - 1 = (a + 1) \times (a - 1)$. The children cannot use algebra to prove this syntactically—the algebra serves no function for them yet except to record what they know—but for that limited purpose, it serves well. Characteristic of this very integrated approach, the pattern also supports their learning of multiplication facts. Students who know 7×7 and 8×8 therefore know 6×8 and 7×9.

CONCLUSIONS

The view that children's acts of abstraction (including their play) and their skill at pattern finding and natural language learning provide the ideal foundation for algebra leads *MW* to a particular approach to the development of algebraic ideas and algebraic language. It also leads to particular pedagogical consequences: a generally less-is-more view of instructions and explanations, in the print materials, by the teacher, and even by children, prior to and during the first stages of activities; frequent use of puzzles; simultaneous presentation of problems with different parts of information missing (e.g., a sum, and then an addend; an output, and then an input); early introduction of algebraic notation; and learning language (including the symbolic written language of algebra) by encountering it in context.

ACKNOWLEDGMENTS

The writing of this chapter was supported, in part, by a grant from the National Science Foundation. Opinions expressed here are not necessarily those of the NSF. We gratefully acknowledge the help of our colleagues in preparing this chapter, especially Wayne Harvey, who—as always—is a valued critical reader, discussant, and friend.

REFERENCES

Benezet, L. P. (1935). The teaching of Arithmetic I: The story of an experiment. *Journal of the National Education Association, 24*(8), 241–244.

Dewey, J. (1897). My pedagogic creed by Dewey. *The School Journal, 54*(3), 77–80.

Donaldson, M. (1978). *Children's minds*. New York: Norton.

English, L. (2005). Combinatorics and the development of children's combinatorial reasoning. In Jones, Graham A. (ed.). *Exploring probability in school: Challenges for teaching and learning*. pp. 121–141. New York: Springer.

Feigenson, L., Carey, S., & Spelke, E. (2002). Infants' discrimination of number vs. continuous extent. *Cognitive Psychology, 44*, 33–66.

Hendrix, G. (1961). Learning by discovery. *The Mathematics Teacher, 54*, 290–299.

Judd, C. (1928). The fallacy of treating school subjects as "tool subjects." In J. Clarke & W. Reeve (Eds.), *The National Council of Teachers of Mathematics third yearbook: Selected topics in the teaching of mathematics* (pp. 1–10). New York: Bureau of publications of Teachers College, Columbia University.

National Council of Teachers of Mathematics. (2000). *Principles and standards for school mathematics*. Reston, VA: Author.

Pinker, S. (1994). *The language instinct: The new science of language and mind*. New York: Penguin.

Sawyer W. W. (1964). *Vision in elementary mathematics*. New York: Penguin.

Wirtz, R., Botel, M., Beberman, M., & Sawyer. W. W. (1965). *Math workshop*. Chicago: Encyclopaedia Britannica Press.

18

Early Algebra as Mathematical Sense Making

Alan H. Schoenfeld
University of California, Berkeley

The explorations of new intellectual territory in this book reminded me of the evolution of cartography. Early maps, dating back to the 7th century, divided the (flat) earth into three major parts corresponding to Asia, Europe, and Africa. Such maps, which persisted through the 14th century, served more as metaphorical depictions than as cartographic representations (see e.g., a version printed in 1472 at http://www.en.wikipedia.org/wiki/Etymologiae).

As knowledge of the terrain increased, increasingly functionally representational maps were produced. The navigational maps produced by the mid-14th century provided reasonable guidance about known terrain. However, the big picture left something to be desired: One could barely recognize the earth as we know it from the maps of the 1400s (see http://www.bl.uk/onlinegallery/features/ptolemylge.html).

Once Mercator projection was introduced in the mid-1500s and its conventions were understood, the maps that resulted were as good or bad as the data that generated them. Mercator introduced a representational framework that had its own systematic distortions, but which offered a unifying big picture perspective. An interactive version of Mercator's atlas of Europe, compiled in the 1570s, can be accessed by clicking on the link "First Atlas of Europe" that appears in the right-hand column of the online gallery located at http://www.bl.uk/onlinegallery/ttp/ttpbooks.html. When Mercator's data were good, the maps in his atlas look very much like contemporary maps. When his data were not (as in the case of early maps of the Americas), the maps left a great deal to be desired. Nonetheless, the big picture he offered helped put things in relation to one another. Indeed, by 1699, large parts of the known world were mapped with some accuracy and maps of the world looked very much like today's maps (see e.g., http://www.library.yale.edu/MapColl/wrld1699.htm).

Fast-forward to the present, and maps are commonplace. Some maps are interactive: On http://www.maps.google.com/ I was just able to zoom in from a (Mercator's) view of the earth to Europe, to England, and all the way in to the street address of a colleague in Kegworth. If I wanted, I could get driving directions online. Moreover, representations of maps continue to evolve: On Google Earth one can now view maps, or photographs, or maps superimposed atop of photographs, of many parts of the world.

This book captures pioneering efforts to map the interior of early algebra, much of which is unknown territory. Part I provides a structural decomposition of algebra, descriptions of precursors to and pathways into algebra, and elaborations of aspects of (pre)algebraic thinking. Parts II and III provide essential existence proofs along multiple dimensions, showing that some aspects of (pre)algebraic thinking are indeed accessible to children in grades K–6. They also offer a clear and concise statement of what early algebra is not ("Early Algebra Is Not the Same as Algebra Early," the title of chap. 10). Chapters in Part III discuss necessary conditions for the successful implementation of early algebra instruction. However, early algebra is not yet a coherent entity. The foci of different chapters are, at times, so different that if one read them in isolation, one might wonder whether they were leading into the same or different terrain. The question, then, is whether there is a Mercatorlike perspective for early algebra—one that, although undoubtedly introducing distortions of its own, nonetheless allows one to see all of the parts as fitting into one coherent whole. Is there an underlying coherence to the chapters in this book, although different chapters deal with representational thinking, with "making and expressing generalizations (about arithmetic patterns)," and with "reasoning about situations of change over time using number patterns, symbols, graphs and tables"? I believe the answer is yes. The balance of this chapter is devoted to presenting and explicating that perspective.

ALGEBRAIC THINKING REVISITED

In chapter 1, Kaput stipulates the following: "We take the essence of algebraic reasoning to be symbolization activities that serve purposeful generalization and reasoning with symbolized generalizations." I concur, but to prevent a narrow reading of this sentence, I would like to expand on my understanding of purposeful symbolization activities. To do so I start with a classic in cognitive science, Greeno's (1983) piece entitled "Conceptual Entities." Greeno re-interpreted data from D. D. Simon and H. A. Simon's (1979) "A Tale of Two Protocols." That paper compared the physics problem solving of a novice physics problem solver with that of an expert. The novice was a good problem solver, but relatively new to the domain of physics being studied.

D. D. Simon and H. A. Simon (1979) showed that both problem solvers were effective at context-free symbol manipulation. To find the answer to a kinematics problem, the novice determined the variable(s) she needed to solve for, and then worked through the equations at hand to find the value(s) she needed. In contrast, the expert was more structured in his approach, using means–end analysis (a problem-solving heuristic) to structure his work with the equations. The expert's and the novice's approaches were both consistent with one aspect of algebra: the manipulation of formal symbols independent of their semantic referents. However, one (the expert's) was more purposeful and sophisticated than the other.

Greeno's re-analysis of Simon and Simon reveals another dimension to the expert's competence. Yes, the expert is capable of operating on symbols qua symbols. But, he also knows that in the case at hand, the symbols have meanings (they refer to theoretical or conceptual entities in the system being analyzed) and is comfortable using those meanings to guide his symbolic operations when it is appropriate. Thus, for example, when asked to find the total distance that an object under constant acceleration has traveled, the expert can reason this way: "The distance traveled is equal to the object's average velocity multiplied by the amount of time spent traveling. Because the change in velocity is linear, the average velocity is $1/2\ (v_o + v_f)$, which makes the total distance traveled equal to $D = 1/2\ (v_o + v_f)(t_f - t_o)$." As Greeno noted, the term *average velocity* was a *conceptual entity* for the problem solver: It was a functional object in the problem solver's mental model of the situation. Moreover, the expression for average velocity, $1/2\ (v_o + v_f)$, could be derived from an understanding of the situation. In sum, the expert's competence was derived both from an understanding of the semantics of the symbols (what they meant and how they related to each other) and the ability to operate on the symbols in purely syntactic terms when such manipulation was called for.

This example suffices to motivate the definitions of *algebraic thinking* and *the purpose of early algebra* that follow. After the definitions, there is a detailed discussion of two examples of early algebraic thinking in order to illustrate what is intended by the definitions. Then the discussion indicates that the apparently disparate approaches to early algebra in this book are, in fact, entirely consistent with them—and the definitions can serve as useful heuristic guides for future efforts in early algebra.

Definition: Algebraic Thinking

A core aspect of algebraic thinking consists of the ability to make effective and purposeful use of symbols in ways that are inherently sensible and meaningful. At times, this means operating on symbols in ways that are purely syntactical. At times, this means making meaningful use of the contexts and relations that gave rise to the symbols. The act of symbolizing may itself serve purposes of generalization, classification, or abstraction. Thinking algebraically means having access to various forms of representation, including symbolic representation; being able to move flexibly from one representation to another when one representation or another provides better affordances for the task at hand; being able to operate on the symbols meaningfully in context when called for, and according to the relevant syntactic rules when called for. One sees these manifestations of algebraic thinking in the three strands identified by Kaput in chapter 1: the study of structures and systems, often abstracted from computations and relations; the study of functions, relations, and joint variation; and in modeling. In sum, algebraic thinking is a particular form of mathematical sense making related to symbolization. It involves meaningful symbol use, whether the meaning is simply guided by the syntactical rules of the symbol system being used or the meaning is related to the properties of a situation that has been represented by those symbols, or related to other representations of it.

Definition: The Purpose of Early Algebra

The purpose of early algebra is to provide students with the kinds of sense-making experiences that will enable then to engage appropriately in algebraic thinking.

A First, Extended Example

In what follows, I explore aspects of sense making in a traditional (and traditionally difficult) area of the upper elementary mathematics curriculum. The topic came up as a group of sixth-grade teachers with whom I was working were engaged in re-thinking their curricular program for the following year.

The issue at hand was fractions and rates. We had decided to create assessment tasks that represented important ideas, and then to explore both the mathematical thinking, and the student thinking, that such tasks might entail. Here is the assessment item the teachers suggested.

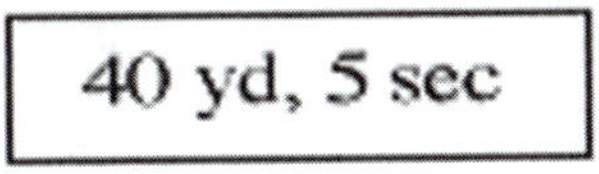

40 yd, 5 sec

FIGURE 18.1. A building block representing John's rate of travel.

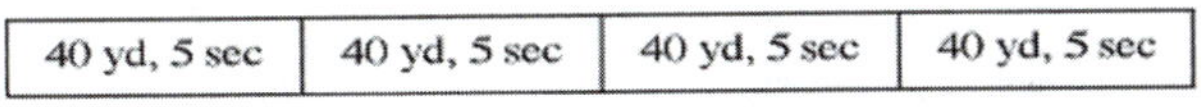

40 yd, 5 sec	40 yd, 5 sec	40 yd, 5 sec	40 yd, 5 sec

FIGURE 18.2. A train of building blocks representing John's travel over time.

Problem 1: John can run 40 yards in 5 seconds. Mary can run 50 yards in 6 seconds. Who is faster, John or Mary?

What the teachers had in mind was that the students should ultimately be able to see John's rate as 8 yards per second and Mary's rate as 8 1/3 yards per second. Comparing the two rates indicates that Mary is faster. Having established this as one plausible goal, we then decided to explore the ways in which students might think about the mathematics, and about directions in which the mathematics involved in this problem might lead.

Students with different backgrounds or inclinations might approach such a task in diverse ways. For example, I have seen students create a "building block" representation, in which a block represents one unit of John's travel (Fig. 18.1). The students went on to build a train of such blocks, representing travel over time (Fig. 18.2).

In this manner, they could see that (subject to the assumption that his speed was constant, and that he could keep running at that spend forever—a not-quite-legitimate assumption, but one commonly made in such problems) John traveled 160 yards in 20 seconds, 200 yards in 25 seconds, 240 yards in 30 seconds, and so on. A similar train for Mary indicates that she travels 150 yards in 18 seconds, 200 yards in 24 seconds, 250 yards in 30 seconds, and so on.

This kind of chunking actually allows for two comparisons. Students look for either common times or common distances. Some observe that John covers 240 yards in 30 seconds, whereas Mary covers 250 yards in the same time—hence Mary is faster, because she goes a greater distance in the same time. Other students notice the common distance of 200 yards. Whereas John took 25 seconds to get that far, Mary took 24 seconds to travel the same distance.

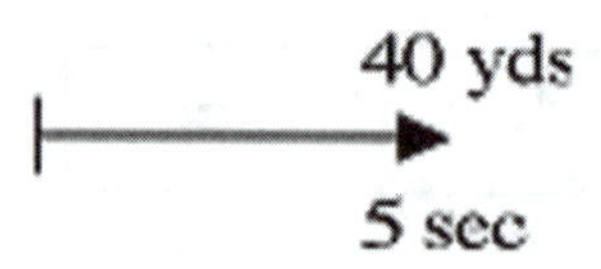

FIGURE 18.3. An iconic representation of Mary's rate of travel.

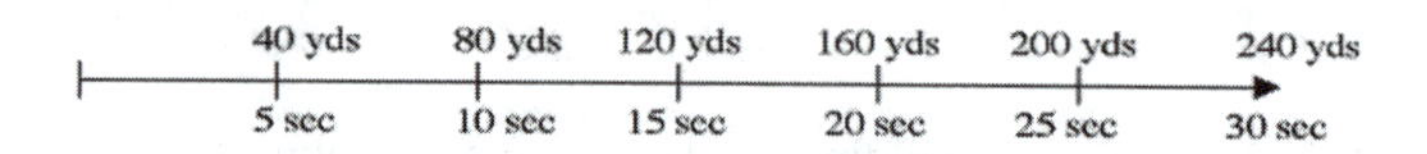

FIGURE 18.4. An iconic representation of Mary's travel over time.

Other students build equally idiosyncratic—or conventional—representations. I have seen arrow diagrams, starting with Figure 18.3, to describe the first sentence in the problem statement, and others such as Figure 18.4 for extensions; I have also seen students (with and without encouragement) make tables that recorded the combined data for the two racers (Table 18.1).

The teachers and I discussed ways to move forward from this. We noted first that both kinds of comparisons, on the basis of distance traveled per fixed time (how far away are the runners 30 seconds after they start running?) and time to reach a fixed distance (who gets to the 200 yard finish line first?), are legitimate ways to conceptualize the problem, and students should come to understand both. We then looked for simplifications, and ways to standardize things. What if a third runner were introduced? You could spend a long time looking for common times or distances, and the numbers could get pretty messy. This ultimately led to the idea that we could always (assuming proportionality) find out how far each of John and Mary could run in 1 second, and simply compare those distances. That comparison would tell us who was faster. And that comparison could be arrived at sensibly for the children: "If John covers 40 yards in 5 seconds, how far does he get in 1 second?" is a problem the students can solve using knowledge they already have (and quite possibly the representations they have just used). Given this, they can make sense of Mary's speed (stated as "how far she would get in 1 second"), and compare the two. We seemed to have tied up the sense-making issue cleanly.

At that point I asked, "What if we graphed the two students' trajectories?" This suggestion was met with incredulity. In essence, the reaction

Table 18.1

A List of Times and Distances for Mary and John

Time (seconds)	*John's Distance (yds)*	*Mary's Distance (yds)*
5	40	
6		50
10	80	
12		100
15	120	
18		150
20	160	
24		200
25	200	
30	240	250

was: "Why bother? We already have all the information we need." The teachers saw no need to pursue the issue, in terms of the sixth-grade curriculum. From their perspective, the graph lacked the concreteness and the affordances of the other representations, and it did not yield the practice with fractions or the reduction to rates that the other representations did.

Graphs are both an extension of the sense making that has been discussed thus far and as a precursor of the algebraic work that sixth-grade students would be undertaking in the next year or two. Questions of the type "You know how far John runs in 1 second. How far does he run in 7 seconds? In 22 seconds? Can you give me instructions that tell me how to figure out how far John has run, mo matter how many second he has run?" lay the foundation for thinking about the distance John has run as a function of time, and expressing it as a rule ("to get the distance John has traveled, multiply the number of seconds John has been running by 8") and then as a function ($D = 8t$). Plotting John's distance-versus-time graph allows one to see that it is linear; that he goes the same number of units vertically for every horizontal unit; and that the line itself is an extension of the table and the function (you can read off how far he has traveled after 15 or 12.5 seconds), and of the inverse function (you can see graphically how long it would take John to run 200 yards, by dropping vertically from where the line $D = 200$ crosses the line representing John's trajectory). The same applies for the graph of Mary's trajectory by itself. When both trajectories are graphed on the same set of axes, there is yet more information. One can learn to see how far along Mary and John are at any given time. Tracing distance as a function of time enables one to see that Mary will reach any given distance (any hypothetical finish line) before John. This is related to the fact that the line representing Mary's trajectory is above the line representing

John's trajectory—and this fact, in turn, raises questions about what is formally known as the slopes of the two lines.

Why do all this? Among other things, operating in the graphical domain provides opportunities to reinforce the work on rate and ratio that is considered core to sixth grade. It provides new ways to think about the same content, introducing the ideas of functional relationships in contexts that are meaningful to the students. It introduces representations they will have to be come fluent with, such as graphs, and ties the interpretations to meaningful contexts. It also prepares the students for problems that they will encounter when they do study algebra, for example:

> Problem 2: John can run 40 yards in 5 seconds. Mary can run 50 yards in 6 seconds. They are going to run a 200-yard race. How much of a head start should John be given in order for them to tie at the finish line?

The point I want to stress in the extended discussion of this example is that the more that students see mathematics as being coherent, the more they see themselves as being able to make sense of it, and the more they have multiple perspectives and representations of any mathematical idea, then the more robust their understandings will be. Seeing mathematics as a sense-making activity should be at the core of early algebra, because it lays the foundation for the meaningful use of algebraic tools and techniques when students encounter them. I return to this theme in discussing the mathematics examples that appear throughout the book.

A Second (Set of) Example(s)

The preceding example focused on the use of symbols with meaningful referents (specifically, time and distance). Sense making seems natural in that context, and in the physics example discussed previously. After all, the symbols refer to entities that have certain properties. Thus, the known properties of, and relationships among, the objects represented might well shape the way one chooses to operate on the symbols that represent them. But what about situations when there are no such referents, that is, when one is faced with what has been called *naked symbol manipulation*? It is a grave mistake to think that meaning and sense making are not involved in such situations. My intention here is to give a few quick examples to show that is the case.

Kaput (1979) pointed out that there often is a context-driven semantics that belies the syntactics of algebraic symbolism. For example, the relationship $A = B$ is presumed, in formal terms, to be symmetric, transitive, and reflective. The symmetry condition says that $A = B$ implies $B = A$, and vice versa. This is taken to mean that positioning on the left or right of the equals sign is irrelevant, and the two expressions $A = B$ and $B = A$ convey identical information. Drawing on Kaput's work, I wrote the following in 1985:

> Yet in practice [$A = B$ and $B = A$] may carry radically (semantically) different meanings, Consider, for example, the following two mathematical statements.
>
> $$\frac{2}{x+3} + \frac{5}{x-2} + \frac{3}{x^2-1} = \frac{7x^3 + 14x^2 + 10x - 7}{x^4 + x^3 - 5x^2 + x - 6} \quad \text{(Equation 1)}$$
>
> $$\frac{7x^3 + 14x^2 + 10x - 7}{x^4 + x^3 - 5x^2 + x - 6} = \frac{2}{x+3} + \frac{5}{x-2} + \frac{3}{x^2-1} \quad \text{(Equation 2)}$$
>
> Despite the fact that they are formally equivalent, Equation 1 will generally be interpreted as representing the simple addition of algebraic fractions, while Equation 2 will be taken to represent the result of a complicated process, the decomposition of a complex rational function by means of partial fractions. In both cases the equals sign is read as yields and suggests the operation that provides the result. At a more elementary level,
>
> $$2 \times 3 = 6$$
>
> represents a simple multiplication,
>
> $$6 = 3 \times 2$$
>
> a factorization into primes, and
>
> $$2 \times 3 = 3 \times 2$$
>
> an embodiment of commutativity—to the cognoscenti. All three equations are formally equivalent, and students are generally presented solely with the formal meanings of mathematical statements. Thus formal mathematics statements carry, in context, semantic meanings that contradict (or at least qualify) the formal meanings of the statements. Students are generally expected to pick up this semantic meaning on their own. ... It should not be surprising that students have difficulty with the semantics of mathematics. (Schoenfeld, 1985)

Those difficulties are still prevalent today. As Kaput, Blanton, and Armell write in chapter 2, syntax-driven actions on symbols:

> ... require attention to the form, the configuration, of the symbols—the sense in Frege's terms. It is here where adjustments to approaches to arithmetic are especially important because, as pointed out by Carpenter and Levi (2000), Fujii (2003), Livneh and Linchevski (1999), in our own work (Blanton & Kaput, in press), and by Smith and Thompson (chap. 4, this volume), most arithmetic statements are read as instructions to compute and are indeed executed as such, typically leading to a numerical result. ... In algebra they need to be thought about in a fundamentally different way.

Consider, for example, the example discussed by Franke, Carpenter, and Battey: "When asked to solve a problem like 8 + 4 = □ + 5 students want to put 12 in the box. Some want to include the 5 in their total so they put 17 in the box. Others create a running total by putting a 12 in the box and an' = 17' following the 5" (Franke et al., chap.13, this volume). That is because, on the basis of their prior experience, students read the equal sign as an instruction to compute. Such a reading is impossible, of course, if one understands the formal meaning of equality—the syntactics of the equation. But a solution is facilitated when one understands the semantics. One possible approach to this problem is as follows:

> What I face is an equation, with a number I don't know. I am supposed to find the number in the box. The two sides of the equation must be equal. I do know how to find the sum on the left-hand-side of the equation: 8 + 4 = 12. So, I can rewrite the equation as
>
> $$12 = \square + 5,$$
>
> or maybe more comfortably as
>
> $$\square + 5 = 12.$$
>
> So, now I'm looking for the number that has the property that when I add five to it, I get 12. I know how to do that. The answer is 7, so 7 goes in the box. And, I can check: 8 + 4 = 12 and 7 + 5 = 12, so 8 + 4 = 7 + 5.

What Franke et al. (chap. 13, this volume) call *relational thinking* is, in essence, an understanding of the semantics underlying the equation. One learns the semantics, I would argue, by treating actions on equations as sense-making activities. That is, one respects the semantics of what are ostensibly purely syntactic operations.[1]

The point is that meaning and sense making are central to all mathematics learning. To the degree that they permeate the mathematics leading up to algebra, they prepare students for algebra. I now turn to the task of showing that "early algebra as sense making" is indeed a theme that permeates this volume, and can be seen as the theme that unifies it.

---

[1]This is not the place for an extended discussion, but I do note that even an expression as simple as "2 + 3 = 5" is not purely syntactic. The sentence has a meaning akin to "a collection of two objects, together with a collection of three objects, has five objects altogether." Thus the symbols have referents—and it could be argued that even purely syntactic operations are meaningful in that sense. See Lakoff and Nunez (2000) for the extreme version of this argument.

Table 18.2
Table 1.1 From Chapter 1

The Two Core Aspects

(A) Algebra as Systematically Symbolizing Generalizations of Regularities & Constraints.
(B) Algebra as Syntactically-Guided Reasoning and Actions On Generalizations Expressed in Conventional Symbol Systems.

Core Aspects A & B Are Embodied in Three Strands

1. Algebra as the Study of Structures and Systems Abstracted from Computations and Relations, Including Those Arising in Arithmetic (Algebra as Generalized Arithmetic) and in Quantitative Reasoning.
2. Algebra as the Study of Functions, Relations, and Joint Variation.
3. Algebra as the Application of a Cluster of Modeling Languages both Inside and Outside of Mathematics.

COMMENTS ON THE CHAPTERS COMPRISING PART I

Table 18.2 (Table 1.1), reproduced here, provides the taxonomy of algebra that shapes the volume. In chapter 2, Kaput, Blanton, and Armella focus on symbolization as a process. They rightly demonstrate its complexity, that is, the kind of analytic detail in the chapter is necessary to develop a theoretical understanding of how people come to symbolize meaningfully.

I agree with the broad characterization in Table 1.1, and with the need to develop a deep understanding of the process of symbolization. Although I might use somewhat different theoretical language to address the issues addressed in those two chapters, our bottom lines are similar. For example, Kaput, Blanton, and Armella focus on the transition from arithmetic to algebra and the disruption that it can cause: "Most arithmetic statements are read as instructions to compute and are indeed executed as such, typically leading to a numerical result. Because in algebra they need to be thought about in a fundamentally different way... students have particular difficulty making the transition unless their work with arithmetic treats arithmetic statements in a more algebraic way."

I agree, and I would go further. I believe that "most arithmetic statements are read as instructions to compute" because they are taught as such, without an eye toward the future use of the symbols. On the basis of repeated experience with arithmetic examples in which arithmetic statements are treated as instructions to compute, it is only natural that students will come to understand them that way. The issue here is one of beliefs and understanding. It is very much akin to cases where students come to believe that "all problems can be solved in five minutes or less" because that has been their experience, over and over, in classrooms (see e.g., Schoenfeld, 1992). If students are taught mathematics as a set of procedures, they are likely to conceptualize it that way—and act accordingly. If they are taught mathematics as a sense-making activity, then there is a chance that they will engage in it that way. In this case, as Kaput and colleagues make clear, understanding the broader set of meanings associated with arithmetic statements is a precursor to—better, a foundation for—algebraic thinking.

Mason (chap. 3, this volume) makes the case eloquently:

> Th[is] chapter describes various powers such as imagining and expressing (in various modes), focusing and de-focusing, specializing and generalizing, conjecturing and convincing, and classifying and characterizing.... What is needed at school is not direct instruction in what to do but rather the invocation of those powers through suitable challenges so that learners experience the development and use of their own powers rather than having texts and teachers try to do the work for them.

In other words, students who experience arithmetic from the very beginning as a mathematical domain in which they can perceive structure, make and test hypotheses, and operate meaningfully on (arithmetic) symbols, will be prepared to do the same when they encounter algebra.

In many ways, Smith and Thompson's view (chap. 4, this volume) is consistent with this. They argue that "for too many students and teachers, mathematics bears little useful relationship to their world. It is first a world of numbers and numerical procedures (arithmetic), and later a world of symbols and symbolic procedures (algebra). What is often missing is any linkage between numbers and symbols and the situations, problems, and ideas that they help us think about." They argue—in contradiction to much of common practice but with good reason—that it is a mistake to accelerate through the curriculum, importing parts of an algebra I course into the earlier grades. Rather, they say, preparation for algebra "should involve changing elementary and middle school curricula and teaching so that students come to use symbolic notation to represent, communicate, and generalize their reasoning."

Smith and Thompson's discussion of Krutetski's problem—"I walk from home to school in 30 minutes, and my brother takes 40 minutes. My brother left 6 minutes before I did. In how many minutes will I overtake him?" illustrates a number of important points. First, multiple solutions are a "good thing": People with flexible understandings of a domain have various ways to think about a problem situation. Second, not all solutions to algebraic problems are algebraic: Smith and Thompson offer a nonalgebraic solution that depends on making sense of the relationships between various real and imagined quantities in the problem. In this case, Smith and Thompson reason about the hypothetical distance between the problem solver and his brother. In Greeno's (1983) terms, Smith and Thompson are working with a *conceptual entity* in precisely the way that the expert problem solver in D. P. Simon and H. A. Simon (1979) did. Third, understandings of the properties of that conceptual entity drive the ways Smith and Thompson solve the problem.

Let us pursue this line of reasoning a bit further, using the Krutetski example. As it happens, I also solved the problem without using algebraic formalism. I noted that it takes the problem poser's brother 4 minutes to cover the distance covered by the problem poser in 3 minutes. Thus, the problem poser gains a minute on his brother every 3 minutes. Because the brother had a 6-minute head start, it would take 18 minutes to catch up. This checked: After 18 minutes, the poser would be 18/30 = 3/5 of the way to school, and his brother, with a 6-minute head start, would be 24/40 = 3/5 of the way to school.

Here is the connection I want to make. As Smith and Thompson note, there are various ways to solve this problem algebraically. Indeed, this problem—although stated in a slightly nonstandard way—is one of an equivalence class of problems, in which two linear functions are given and the task is to find out when they have the same value. (Recall Problem 2 earlier.) These can be "head start" problems, as in Problem 2. In similar fashion, students may be asked to compare of the costs of two cell phone plans, automobile rentals, and so on.

Typically, one function is expressed as $y_1 = mx + b$, the second $y_2 = Mx + B$. Students will find out where the functions intersect by setting $mx + b = Mx + B$ and determining that they have the same y-value when $(x + B - b/m - M)$. But what does this mean? The way that I think of it is as follows. At $x = 0$, the functions have values of b and B respectively—so (assuming $B > b$), y_2 has a "head start" of $(B - b)$. But (assuming $M > m$), y_1 "catches up" $(m - M)$ units each time x increases by 1. Hence it takes $(B - b/m - M)$ units for y_1 to catch up with y_2. I call this my catching up schema, and it is part of the knowledge I bring to any problem that deals with the interpretation of simultaneous linear equations. That knowledge is sometimes generative in that it guides a solution; it sometimes serves as a constraint in that it helps me to recognize that a potential solution violates a constraint it imposes and must therefore be incorrect. It is one of many tightly connected pieces of information I have regarding pairs of linear equations. All of this information—about slope, about how changes in x relate to changes in y, about the x- and y-intercepts of a line, and more—works in ways that guides my thinking about problems and shapes what I do as I try to solve them (Schoenfeld, Smith, & Arcavi, 1993). The richer the connections, the deeper my understanding, and the more flexible I can be in approaching problems. Smith and Thompson's point is that the development of such understandings can and should begin long before a formal algebra course. Smith argues, in chapter 5, that a form of mathematical certainty can emerge from students' working at making meaning, exploring connections, representing them, and explaining them. The examples given in chapter 5 and the other chapters differ to varying degrees, but the underlying concept is the same.

COMMENTS ON THE CHAPTERS COMPRISING PART II

I begin with comments about the chapters that focus significantly on aspects of algebra as generalized arithmetic and the symbolic codification of patterns (chaps. 6, 7, 8, and 12). I then turn to a discussion of representational issues, including functional notation and multiple representations of geometric and other objects.

Part II offers rich exemplifications of theme of early algebra as sense making. Consider, for example, the discussion of the commutativity of whole number multiplication discussed by Bastable and Schifter in Episode 1 of chapter 6. Lauren creates a (7×3) array and a (3×7) array, and superimposes one on the other to show that they yield the same product; Jeremy rotates the (7×3) array by 90° to show that it *is* the (3×7) array; Anna acknowledges the generality, because the demonstrations by Lauren and Jeremy could be done for all numbers. This is a hands-on representational practice attaching meaning to symbols—with significant generality.

The extra dimension here, and common to part II, is that there is a *communal* dimension to the practice of sense making. Part of what makes the instruction powerful is that the students, with careful orchestration by the teacher, work together to explore the mathematical issues at hand. With this kind of reasoned discourse one can see (or at least imagine) the development of productive habits of mind—that students expect to reason things through and to explain them; that explanations must take relevant information into account; that explanations must stand up to scrutiny; that there is more than one way to understand something (e.g., that a problem can be conceived of as an addition problem and as a subtraction problem); and more.

Beyond that, a main focus of chapter 6 is generalization. I think the chapter makes it clear that students in the elementary grades can and do produce mathematical generalizations—perhaps not in the language of algebra, but correct nonetheless. Bastable and Schifter note that "many adults ... who read Adam's 'amazing' discovery—If you take two consecutive numbers, add the lower number and its square to the higher number, you get the higher number's square'—feel the need to translate it into terms more familiar to them: $n + n^2 + (n + 1) = (n + 1)^2$."

Indeed, the algebraic notation is clean. It provides a clear proof, in a formal language we have come to depend on. The proof is general. But I am

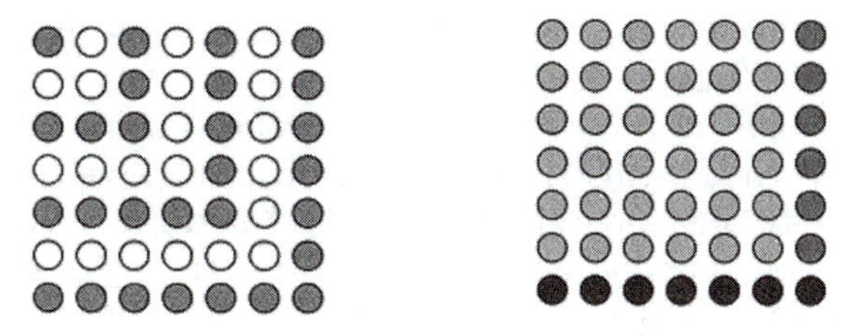

The sum of consecutive odd numbers, starting with 1, is always a square

A number squared, plus itself, plus the next number, is always a square

FIGURE 18.5. Two proofs without words.

willing to bet that students who have developed the kinds of habits of mind discussed in this chapter will be ready for the algebraic symbolism when they encounter it; and, if the symbolism comes as the codification of something they already understand, then it will have meaning and be seen as a powerful tool. That is precisely the purpose of early algebra. I would also note (cf. chap. 17, as well) nonverbal arguments can provide compelling evidence that leads to proof. One of the proofs without words in Nelsen's (1993) book by the same name—on the cover, in fact!—is a pictorial argument that the sum of the first n odd numbers is n^2. A slight variation on this picture provides a proof without words of Adam's claim (see Fig. 18.5).

One may still feel the need to symbolize algebraically—but there is no question that even for those who already know the algebra, the pictures can add an extra dimension to one's understanding; and for those who have not yet learned it, the picture(s) may serve as a means of bootstrapping the desired algebraic understandings.

One sees similar underpinnings for algebraic thinking in chapter 7. In simplest terms, the backward problems in the elevator scenarios described by Tierney and Monk provide a nonstandard model of number sentences in which the first term is the unknown. For example, "This time I don't know the starting floor. I go on the elevator. I push the plus two and the minus three and I end on floor one" can be expressed as:

$$\square + 2 - 3 = 1.$$

There is more to the examples than this, however. Mathematically speaking, each of the hypothetical buttons on the elevator represents a function. The "plus two" button moves the elevator up two floors no matter where it starts, the "minus three" moves it down three floors, and so on. Call the first function $f(x)$, the second function $g(x)$. When Kadisha says, "We're supposed to end on floor 1. If I were to go +2 and –3, I would end up one below where I started," she is, in effect, computing the composite function

$h(x) = f(g(x))$; when she reasons "so I must be one above the ending floor, so that's plus two" she is, in effect, arguing that one "reverses" the action of the "minus one" function with the "plus one" function. When Sylvia says "I just switched them around. I made the plus two a minus two and the minus three a plus three and then I did it," and generalizes by saying "If I'm starting on floor one and I don't know my ending place, I'd have to switch the whole way," she is, in essence, making a general claim about the structure of inverse functions in this domain. I say "in essence" because the students are not using that language; at this point they might well be confused by it. Nonetheless, they are clearly showing evidence of thinking in algebraic ways. One can argue similarly about Rose's discussion of the maximum value of a sequence of addition and subtraction functions on p. She does not compute what happens to the value of 4, but rather discusses the impact of the "+" and "–" functions, individually and collectively. In Episode 3, one sees students grappling with fundamental issues of representation, with tables and graphs. Episode 4 adds representational complexity, with the students working to understand the properties of verbal and tabular descriptions and to see if both tell the same story.

There are direct parallels to the stories of Kadisha, Sylvia, and Rose in the examples discussed by Peled and Carraher in chapter 12. In the section of chapter 12 entitled, "Equivalent Transformations: The Long Way and Shortcuts," Peled and Carraher introduced functions as transformations, independent of their origin: Carolina and James performed +2 and –1 transformations on the number line. Peled and Carraher then posed inverse questions: If Carolina and James ended at 3 and 5, respectively, where did they start? The students, asked to think about "shortcuts" relating the +2 and –1 transformations, ultimately wound up producing composite functions (which corresponded to the addition of the signed numbers +2 and –1). Interestingly, as the authors note, taking a more advanced (essentially algebraic) approach to displacement produces a more robust understanding of the arithmetic topic of the addition of signed numbers than traditional arithmetic approaches.

At the risk of flogging a dead horse, I want to be especially clear about the claims I am making before moving on. I am not arguing that students are doing algebra when they treat "+3" as a function and compute its inverse as "–3," or when they discuss the properties of graphs and tables. I claim that they are engaging in sense-making activities that help them to develop the habits of mind that will serve them well when they engage in the formalization of similar kinds of ideas in an algebra course.

In chapter 8, Mark-Zigdon and Tirosh examine kindergarteners' and first graders' capacity to handle the syntax of number sentences. The issue they raise is: At what point will students be able to symbolically represent numbers and number sentences? There are, I think, at least three dimensions to this kind of question. One is, when will students master the syntax involved—can they recognize the symbols for objects and operators and combine them correctly? This is the issue examined in the current study. A second is, under what conditions and at what age will such number sentences make sense semantically to young children? This depends on experience, of course. For example, it is reasonable to conjecture that young students who have experienced an instructional program such as cognitively guided instruction (Carpenter, Fennema, & Franke, 1996; Carpenter, Fennema, Franke, Levi, & Empson, 1999; Carpenter, Franke, & Levi, 2003) would get the semantics—understanding how stories generated (or could be represented by) number sentences—whereas other students who have not had the relevant symbolizing experience would have a harder time with such sentences and the problem solving they entail. A third question is, how are the answers to the first and second related? Would there be an interaction between semantics and syntactics? That is, do the students who have a meaningful understanding of the ways in which number sentences represent concrete situations develop increased mastery of, and fluency with, number sentences because the referents are meaningful to them? These are interesting issues to be explored as we map out the territory of early algebra.

I now turn to the issues of functional relationships and multiple representations. In chapter 9, Boester and Lehrer explore relationships between algebraic and geometric reasoning: "Rather than place these two strands in competition for curricular space and time, we propose synergy: Visualization bootstraps algebraic reasoning and algebraic generalization promotes 'seeing' new spatial structure." That sounds like sense making to me! Think of the five process standards in NCTM's (2000) *Principles and Standards:* problem solving, reasoning and proof, communication, connections, representation.

"Problem solving" is not immediately apparent in chapter 9, at least as most people envision it. Indeed, the authors do not use the term. Yet consider the task implicitly described in this sentence: "For example, when students first attempted to determine a symbolic expression for rectangles in the ratio of 1 to 4, they were stymied because it was one of the first groups to contain a rectangle with non-whole number dimensions." The challenge to the students was to use what they knew to come up with a meaningful representation of a new class of objects. Not only is that a form of problem solving, it is a form of mathematical thinking in which mathematicians often engage. (Its use in this case also illustrates an important pedagogical principle: Problematic situations can be used as the contexts within which students can develop mathematical understandings.) The unit has a major focus on reasoning and communication: "The teacher always elicited students' thinking, nearly always insisted on justifications for that thinking, and generally conducted a classroom emphasizing mathematical conversations. She also promoted mathematics as a form of literacy." Connections are what the unit is all about, and the mechanism for making those connections is the development of meaningful representations of the same objects. "The four different representational forms explored were: rules in the form of verbal descriptions (e.g., long side is twice short side), re-expressed and re-interpreted as symbolic equations (e.g., $LS = 2 \times SS$), Cartesian graphs, tables, and quotients representing ratios." In short, the content of the unit is material that grows into the content of a standard algebra course, and the unit involves varied processes called for by major reform documents.

Although in a very different domain, similar themes are announced by the authors of chapter 10. There is, from the beginning, a focus on meaning ("If algebra is meaningless at adolescence, why should it be meaningful several years earlier?"). Then there are assertions I take as being axiomatic for the authors:

1. "Early algebra builds on background contexts of problems." The issue, in fact, is more general. Two paragraphs ago, I noted that grappling with meaningful problems is often a powerful mechanism for students' development of mathematical ideas. More important, powerful mathematical ideas are always grounded in examples (see e.g., Rissland, 1978).
2. "In early algebra formal notation is introduced only gradually." Note the parallels between the introduction and use of letters in this chapter ("Letters Can Name Indeterminate Amounts") and the ways that the authors of chapter 10 introduced mnemonic variable names ("long side is twice short side" is reexpressed as the symbolic equation "$LS = 2 \times SS$.")
3. "Early algebra tightly interweaves existing topics of early mathematics." It may be worth recalling that teaching algebra as a separate course, divorced from the rest of mathematics, is an American curricular anomaly: In much of the world (including nations that have consistently trumped the United States on international comparison tests such as TIMSS), many of the ideas of algebra are developed in the context of pure and applied mathematics in a wide range of domains. Algebraic symbolism often serves to capture patterns, whether those patterns are numerical, geometric, or probabilistic; and, the "unreasonable effectiveness" of mathematics (Wigner, 1960) is due to the meaningful mathematical representation of objects and relations from outside of mathematics. Why not start early? The key, as exemplified here, is that the processes involved must be meaningful to the students.

Sense making via the productive use of multiple representations comes to the fore in the discussion of the "Best Deal" problem in chapter 11. As in the case of chapter 9, linkages across representations are seen as a powerful way of enhancing understanding: Different representations of the same phenomenon can provide affordances for different kinds of insights into that phenomenon. Let me address that issue in algebra, then return to the early version.

Some years ago, my research group worked on the development of curricular materials (including computer software and hands-on activities) designed to have students learn about the properties of linear functions. Our explicit concern was that students be able to represent functions in multiple ways, and think about them in multiple ways—precisely because using one perspective,[2] or one representation, might make it much easier to solve a problem than any other. Moschkovich, Arcavi, and I (1993) wrote the following:

> Our first major goal in writing this paper was to introduce and elaborate the framework for understanding functions that was outlined in schematic form in Table 1. There we pointed to two ways of viewing functions (the process and object perspectives) and the three most prominent representations of functions (in tabular, graphical, and algebraic form) We hope to have indicated that competence in the domain consists of being able to move flexibly across representations and perspectives, where warranted: to be able to see lines in the plane, in their algebraic form, or in tabular form, as objects when any of those representations is useful, but also to switch to the process perspective (in which an x-value of the function produces a y-value), where that perspective is appropriate. ... For us, now augmented by columns representing real-world contexts and verbal representations, it serves as both a heuristic guide to curriculum development (Does any curriculum we propose make adequate connections across representations and perspectives? If not, it had better be revised) and for assessment of student learning (Can the student move flexibly across representations and perspectives when the task warrants it?). (p. 97)

If *early algebra* is to have the meaning discussed in this book, then the processes of developing such meanings and understandings should indeed start early. Chapters 9 through 11 provide insights into how to do so.

---

[2]There is a large literature on the dual nature of real-valued functions, as *processes* (instructions to compute) and as *objects* (e.g., the whole graph of $f(x)$ is raised by one unit when one graphs $g(x) = 1 + f(x)$.) It is sometimes advantageous to conceptualize a function using one perspective, sometimes the other (see e.g., Harel & Dubinsky, 1992.)

COMMENTS ON THE CHAPTERS COMPRISING PART III

Part III deals with issues of implementation. All told, chapters 13–17 raise three sets of issues: the nature of the mathematics to be taught, the kinds of mathematical knowledge it takes to teach early algebra effectively, and systemic issues related to supporting the effective implementation. I deal with each in turn, drawing from the body of chapters as I go.

The Mathematics of Early Algebra (Continued)

The arguments made in parts I and II apply here as well. In fundamental terms, each of the chapters is concerned with having students encounter early algebra as a sense-making activity. Earlier in this chapter, for example, I discussed the semantics of the equation

$$8 + 4 = \square + 5,$$

which plays a central role in the narrative of chapter 13. Franke, Carpenter, and Battey demonstrate clearly that being able to see that equation relationally is a key pathway to learning to think about equations in algebraic terms. The same is the case for being able to articulate the fundamental properties of number and operations, and for being able to justify one's answers. And (see Issue 2, later) they note that thinking in such terms is hardly natural for the teachers with which they have worked.

The same issue of semantics—of thinking about equations as capturing *relationships*—comes front and center in chapter 15. Consider the following classroom dialogue:

> Mrs. M wrote F, A, A, F, F $<$ A, and A $>$ F on the board
>
> "I see something," said Justin.
> "What do you see?" asked Mrs. M.
> "Look," said Justin. "Every time two things are equal we can only write two statements. Like when mass *K* and mass *B* are equal. We can say mass *K* equals mass *B* and mass *B* equals mass *K*. Mass *K* and mass *B* are the same amount."
> Mrs. M wrote $K = B$ and $B = K$ on the board. "That's a good observation, Justin," said Mrs. M.
> Wendy raised her hand and continued, "But if two things are unequal we can write four statements."
> "Four statements?" said Mrs. M.
> "I agree with Wendy," said David. "There are four statements. See, volume *F* is not equal to volume *A* and *A* is not equal to *F*. And *F* is less than *A* and *A* is greater than *F*."
> Mrs. M wrote *F* ? *A*, *A* ? *F*, $F < A$, and $A > F$ on the board.
> "Hey," said Mia, "We can even write more because $A = A$."

It provides powerful evidence that first graders can think in relational terms. Likewise, the fact that young students can work in ways that depend on the symmetric and transitive properties of equality, and can express multiplicative relationships in proto-symbolic ways (see the discussion of the problem "Jessica has 7 oranges. Raul has 4 times as many oranges as Jessica. How many oranges do they have altogether?") provides an important existence proof. With the right kinds of support, young students can learn to think about the relationships embodied in ostensibly concrete situations, and thus be prepared for the use of such relationships when they encounter them in more formal terms in an algebra course.

One sees the same theme in chapter 16, with an additional emphasis on the kinds of reasoning that provide the underpinnings of mathematical generalization. Not only are students encouraged to explore turnarounds ($4 + 6 = 10 = 6 + 4$), but they are encouraged to think about whether turnarounds always work:

> Turnarounds always work. I just know they do.
>
> Me: How do you know?
> Natalie: Well, look. $27 + 4 = 31$ and $4 + 27 = 31$.
> Me: But does this always work, for any number, no matter how big it gets?
> Natalie: Well, let me try it.
>
> So Natalie tried numbers in the hundreds and added them together both ways and felt convinced that it always worked. Her reasoning seemed to be based on her having done many of them and having had them always work out to be the same answer.
>
> I did ask her if something like $13 + 23 = 23 + 13$ is true. Her immediate response was that, no, it didn't work. Several other children confirmed that it didn't work either, saying, "There's no answer here." Even after they felt sure that $13 + 23 = 36$ and that $23 + 13 = 36$, no one felt like the original statement could be true. . . .
>
> Other children spent their time making up additional problems and solving them both ways. Ingrid's [written] work [which starts with single- and two-digit numbers, extending into hundreds, thousands, and ten thousands] is an example. She was using a calculator and expressed real satisfaction with this work.

Generalization is a natural process but, as we know, one that does not always produce correct results. For example, all too many students generalize the distributive law $(a + b)(c) = ac + bc$ to the case of exponents, distributing the exponent 2 in the expression $(a + b)^2$ to obtain the incorrect expression $a^2 + b^2$ (see e.g., Matz, 1982). The question is, how does one deal with such things? In the excerpt provided, students are encouraged to observe patterns and then challenge them. That strikes me as exactly right. In particular, the following comment by the authors resonated with me completely:

> Now, in our current project, a group of teachers has been collaborating with us, bringing examples of early algebra from their classrooms and helping us understand what happens when they structure their lessons in particular ways or include certain questions as part of their classroom routine. For example, what happens when children build the habit of addressing such questions as, Why does it work out that way? Will it always work that way? How do you know? A teacher might say, I'm not convinced; prove it to me. Can you convince your classmates? Can you convince a younger child?

This set of questions bears a remarkably close resemblance to the three stages of making compelling arguments (i.e., proofs) suggested by Mason, Burton, and Stacey (1982):

> Convince yourself;
> Convince a friend;
> Convince an enemy.

As Tall (1991) indicated:

> The idea is first to get a good idea how and why the result works, sufficient to believe its truth. Convincing oneself is, regrettably, all too easy. So pleased is the average mortal when the "Aha!" strikes that, even if shouting "Eureka" and running down the street in a bath towel is de rigeur, it is very difficult to believe that the blinding stroke of insight might be wrong. So the next stage is to convince a friend—another student, perhaps—which has the advantage that, to explain something to someone else at least makes one sort out the ideas into some kind of coherent argument. The final stage in preparing a convincing argument, according to "Thinking Mathematically" is to convince an enemy—a mythical arbiter of good logic who subjects every stage of an argument with a fine toothcomb to seek out weak links. (p. 24)

This sequence of arguments is, I believe, a habit of mind possessed by mathematicians, who will trust their insights to a degree, but also understand that a general claim must withstand any and all counterarguments—attacks by an "enemy." It seems exactly right to begin to develop this and related habits of mind in meaningful contexts, early in students' school careers. Of course, one does not expect formal reasoning or formal proof; but, the general enterprise of mathematical sense making can be made accessible. Schifter, Monk, Russell, and Bastable offer a set of guidelines, grounded in their empirical experience, for doing so.

Chapter 17 is a fitting end, in two ways. First, the approach covers a huge amount of territory: The K–6 curriculum entitled *Math Workshop* (Wirtz, Botel, Beberman, & Sawyer, 1965) is pretty much an entire early algebra curriculum. In it we see a version of relational thinking as discussed in previous

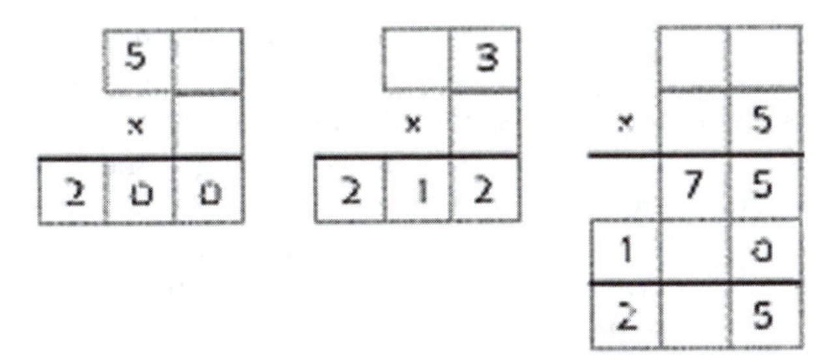

FIGURE 18.6. Fourth grade: learning multiplication inside out and backward.

chapters ("presenting problems with different parts missing," such as 3 + _ = 5). What the authors call "puzzle solving" with regard to the multiplication problem in Figure 18.6 (Fig. 17.9, reproduced here) is very much an exercise in working through base 10 structure.

In this chapter, as in some of the others, we see discussions of arithmetic foundations for the distributive law, find a rule (i.e., function) problems, the use of icons to represent what will be identities (the explanations for numerical magic tricks), what is essentially matrix addition (the sum of magic squares), and the geometric representations of algebraic proofs. The very scope of the approach shows that it is possible to unify the various strands of early algebra over the years. The curriculum appears to include (to at least some degree) antecedents for all of the algebraic activities described in Figure 1.1.

Second, Sawyer's (1964) description of his intentions anticipates this commentary in remarkable ways:

> It is quite possible to use simultaneous equations as an introduction to algebra. Within a single lesson, pupils who previously did not know what x meant, can come, not merely to see what simultaneous equations are, but to have some competence in solving them. No rules need to be learnt; the work proceeds on a basis of common sense. The problems the pupils solve in such a first lesson will not be of any practical value. They will be in the nature of puzzles. Fortunately, nature has so arranged things that until the age of twelve years or so children are more interested in puzzles than in realistic problems. So the puzzle flavour of the work is, if anything, an advantage. (p. 30)

To paraphrase using my language, Sawyer is arguing that mathematics can and should be a sense-making activity. Moreover, he argues (and I agree) that students are naturally curious, and that if one can harness the energy unleashed by their curiosity, one can begin to do some real mathematics. Here, as in the other chapters, the idea is to foster the development of productive habits of mind consistent with the formalization that, when it comes, will officially label what the students are doing as the study of algebra. The hope and expectation of all the chapters in this volume is that when such formalization comes it will serve as the codification of coherent understandings, a natural evolution of the students' earlier experiences in mathematics. That would stand in stark (and welcome) contrast to most students' current experience.

Issues of Teacher Knowledge

Teaching early algebra effectively will, at minimum, demand the following from teachers: a solid understanding of whichever aspects of early algebra are to be taught, and a comparably solid command of the relevant pedagogical techniques. The challenges with regard to both are substantial.

In recent years, various such programs of research on teacher knowledge (e.g., Ball & Bass, 2000, 2003; Lampert, 2001; Ma, 1999; Sherin, 2001) have begun to unravel the kinds of mathematical and other knowledge that are necessary to provide the support structure for rich mathematics instruction. Such research suggests that there is little likelihood that even a significant proportion of the current K–6 teaching force has anywhere near the mathematical or the pedagogical knowledge required. This is for two reasons. First, teachers who are asked to teach early algebra are being asked to teach unfamiliar content. The content of early algebra as envisioned in this volume is conceptually challenging, because it is focused on the semantics of the mathematics at hand—and few students of mathematics (current teachers included) have ever been taught to grapple with such issues. Discussions of relational thinking with regard to number sentences, or of the power and use of multiple representations, are not exactly common. Thus, teachers will have to learn the new content.

Second, one cannot conceptualize early algebra in meaningful ways unless one has a reasonable grasp of the semantics of algebra. There is not much reason to believe that many K–6 teachers have such knowledge. In most states, elementary school teachers earn a multiple subject credential, which certifies them to teach all elementary school subjects. Teaching the whole child is critically important in elementary school. Thus, credential programs are often strong on developmental issues.

But, because such programs have to prepare teachers to teach everything, they are necessarily thin on any particular subject matter. There is often a focus on literacy, which is typically high priority. There is not, in general, that much of a priority on mathematics in elementary teacher preparation programs. Mathophiles (and mathematics majors) tend to go into single-subject credential programs, and to teach middle and secondary school mathematics. In some states, it is possible for elementary school teachers to be credentialed after having studied only one or two postsecondary courses in mathematics.

The sixth-grade teachers in the first extended example discussed in this chapter illustrate the issues to be confronted. These are good teachers who care about their students, know a lot about developmental issues, and work hard to implement carefully chosen reform curricula as best they can. But, these teachers were not at all sure about the value of exploring multiple connections to the simple rates involved in the straightforward solution to Problem 1 ("John can run 40 yards in 5 seconds. Mary can run 50 yards in 6 seconds. Who is faster, John or Mary?"). They did not see the value of graphing the trajectories of the two students, because they did not know that their students would be expected to make use of such ideas the following year. To put things simply, it is hard to prepare one's students for what is to come when one does not know what is coming! Moreover, it is nearly impossible for someone who does not understand algebra to grasp the semantics of early algebra in meaningful ways. I would like to believe that the story described in the first extended example is anomalous, but the fact is that the teachers I was working with are comparatively well-prepared. The evidence of teachers' mathematical knowledge in the books by Ma (1999) and Cohen (2004) does not provide grounds for optimism.

The situation with regard to pedagogy is similar. As Part II of this book makes clear, the success of early algebra instruction will depend in large measure on the implementation of a discourse-oriented pedagogy, in which students and teacher grapple with issues of sense making, and in which attempts to understand the mathematics are made public and reflected on. Unfortunately, most of the current generation of teachers was taught via the traditional didactic model of instruction. Hence, they do not have models of the appropriate pedagogy to implement; so, what they do implement can be far from what one would like. This was made clear in chapter 13:

> After leaving the professional development, teachers relied on existing practices to help them. They created worksheets of true/false and open number sentences so they could practice getting the correct answers to problems like 8 + 4 = _ + 5. The teachers used the worksheets for continued practice. They were collected, graded and returned to the students. The worksheets were not used to promote discussion, to challenge existing ideas, or to figure out why 12 would not go in the box. The worksheet was an artifact of existing practice. The teachers used the sequences we discussed in class in the context of an existing practice and changed the purpose of the number sentences themselves. Here the teachers appropriated

a practice that was not helpful in relation to developing algebraic thinking, particularly relational thinking.

In short, to learn to teach early algebra effectively will require unlearning, as well as learning: Many teachers will have to divest themselves of established habits and to learn both new content and pedagogy. This is a decidedly nontrivial matter, and it takes time. One might well ask where the time for this will come from (see the discussion of systemic issues later).

For early algebra to become a reality on a large scale, the content and pedagogy in teacher preparation programs must ultimately change, to increase the likelihood that prospective teachers will experience mathematics in a manner more consistent with the ways we hope they will teach it. And, professional development must be sufficiently intensive, over extended enough periods of time, to allow teachers to develop both the content knowledge and the pedagogical orientation and skills to be able to carry out the kinds of early algebra instruction discussed in this volume. The chapters in Part III of this volume offer existence proofs. They show that, under certain circumstances, teachers can learn the requisite skills; they show that skilled teachers can make early algebra happen in classrooms. The challenge is to move from existence proofs to large-scale reality. When one tries to do that, one confronts issues at the systemic level.

Systemic Issues

The knowledge and inclinations of individual teachers will only account for part of the story of the implementation of early algebra over the years to come. Even though teachers often have some degree of autonomy in the isolation of their "egg crate" classrooms (Lortie, 2002), the embedded contexts of their departments, school sites, districts, and professional affiliations will support and constrain what is possible for them to do within those classrooms (McLaughlin & Talbert, 2001; McLaughlin, Talbert, & Bascia, 1990). As a framework for what follows, I point to a set of conditions (Table 18.3) that are catalytic for the implementation of successful mathematics instruction.

There is evidence that when all of these conditions are met, there can be steady improvement in a system—including the reduction of racial performance gaps (see e.g., Schoenfeld, 2002, in press). There is also, alas, evidence that when one or more of the five conditions listed in Table 18.3 are missing, attempts at the implementation of novel instruction can be undermined to the point where they are ineffective. This is not the place for an extended discussion, but here are some of the reasons, numbered to correspond to the conditions mentioned previously: (a) The need for well thought out curricular goals (a.k.a. standards) is obvious. (b) Research dating

Table 18.3
Systemic Conditions That Support Improved Student Performance and a Reduction of Racial Performance Gaps

1. A well-designed, mathematically rich set of standards for instruction
2. A well-designed curriculum aligned with the standards
3. Well-designed assessments aligned with the standards
4. Well-designed professional development aligned with the standards
5. Enough time and stability in the system for all of the above to take hold

Note: Reproduced with permission from Schoenfeld (in press).

back more than a quarter century (e.g., Stake & Easley, 1978) documents the dependence of teachers on textbooks as sources of authority regarding mathematics and the appropriate sequence of learning activities. "Ownership of mathematics rests with the textbook authors and not with the classroom teacher. Departures are rarely made" (Romberg & Carpenter, 1986, pp. 867–868). (c) Pressures to "teach to the test" are well known; when tests focus on skills to the exclusion of conceptual understanding, instruction suffers (see e.g., Shepard, 2001). (d) Effective professional development, like all effective instruction, must be well designed. (e) Mixed and conflicting messages leave teachers unsure about where to focus their efforts, and undermine the coherence of instruction. (Borko et al., 1992). If a district does not deliver a consistent message over a long enough time period for it to take effect, then the impact of short-term professional development efforts are likely to wash out.

This broad frame helps to situate the attempts at professional development discussed in this volume, and the challenges the field faces if it wants to see larger impact. These issues lurk beneath the surface in chapter 13. The professional development discussed there has as its goal significant changes in both the beliefs and practices of the teachers involved. Such change takes time (see e.g., Cooney, 2001), and few districts make much time available for professional development. (For example, the Diversity in Mathematics Education project negotiated a monthly 2-hour professional development session with middle school teachers with one of its partner districts. When I mentioned to one of our cooperating principals that the 16 hours we met over the course of the year was equivalent to just one third of a one semester course at the university, she pointed out how serious conditions really are: The 16 hours of DiME time represented half of the professional development time those teachers had that year.)

Systemic issues are the central focus of chapter 14. In describing their professional development work, Blanton and Kaput write:

> The chapter is organized around five areas that we came to view as essential in supporting teachers as they incorporated algebraic thinking into their classrooms: (1) the development of a *professional community network*; (2) a *distributed* approach to district leadership practice; (3) the development of a school mathematics culture; (4) the integration of district professional development initiatives; and (5) the development of teachers' capacity to algebrafy their own instructional resource base.

Research shows that to survive in the long run, teachers need the support of peers engaging in similar activities; the efforts of teachers working in isolation tend to falter (McLaughlin & Talbert, 2001; McLaughlin et al., 1990). Thus, one sees the necessity of Blanton and Kaput's focus on the development of a professional community network. Because so much of the context of a teacher's professional work is at the school or department level, the development of a school mathematics culture is also essential. Because district mandates (and skill) can come and go, there is a need for distributed commitment to professional development. Moreover, distributed expertise both reduces the risk of having reform dependent on just one person, and "spreads the wealth" in obvious ways. And, teachers are often bombarded with multiple and conflicting messages about what is important; hence the integration of district professional development initiatives is a mechanism for coherence and focus. In sum, the field will face significant systemic challenges in taking the ideas of early algebra to scale.

CONCLUSIONS

When done properly (i.e., with understanding), engaging in mathematics is a coherent, sense-making activity. Actions are not arbitrary; one does what one does for good reason. If one understands those reasons, then everything fits together.

Algebra represents one of humankind's great intellectual achievements—the use of symbols to capture abstractions and generalizations, and to provide analytic power over a wide range of situations, both pure and applied.

The fundamental purpose of early algebra should be to provide students with a set of experiences that enables them to see mathematics—sometimes called *the science of patterns*—as something they can make sense of, and to provide them with the habits of mind that will support the use of the specific mathematical tools they will encounter when they study algebra. With the right kinds of experiences in early algebra, students will no longer find algebra to be a new and alien body of subject matter. Rather, they will find it to be the extension and codification of powerful modes of sense making that they have already encountered in their study of mathematics.

The efforts in this volume are all of a piece in that regard. All ask students to see how and why the mathematics fits together. In doing so, they help to lay a foundation for the experience of mathematics as a sense-making activity. As the preceding section indicates, one should not underestimate the work that it will take to make this vision a reality. At the same time, there is reason for guarded optimism. The current political context, with an emphasis on high-stakes testing (often of basic skills) is not an easy one to work with, but such things come and go. (I have been around long enough to see President Reagan "zero out" the education budget at NSF, and the subsequent recovery of the field.) Real change takes time. For example, the research on mathematical thinking and problem solving done in the 1970s and 1980s, including a number of small-scale existence proofs, provided the intellectual underpinnings of the 1989 NCTM *Standards*. Twenty-five years after the original research, standards-based curricula occupy a nontrivial segment of the marketplace, and what evidence there is (e.g., Senk & Thompson, 2003) indicates that students using such curricula tend to outperform those who do not. Early algebra is at the conceptualization, beginning research, and existence proof stage. This volume represents some important first steps. It will be interesting to see where the next steps take the field.

REFERENCES

Ball, D. L., & Bass, H. (2000). Interweaving content and pedagogy in teaching and learning to teach: Knowing and using mathematics. In J. Boaler (Ed.), *Multiple perspectives on the teaching and learning of mathematics* (pp. 83–104). Westport, CT: Ablex.

Ball, D. L., & Bass, H. (2003). Making mathematics reasonable in school. In J. Kilpatrick, W. G. Martin, & D. Schifter, D. (Eds.), *A research companion to* Principles and Standards for School Mathematics (pp. 27–44). Reston, VA: National Council of Teachers of Mathematics.

Blanton, M., & Kaput, J. (2005). Design principles for instructional contexts that support students' transition from arithmetic to algebraic reasoning: Elements of task and culture. In R. Nemirovsky, B. Warren, A. Rosebery, & J. Solomon (Eds.), *Everyday matters in science and mathematics* (pp. 211–234). Mahwah, NJ: Lawrence Erlbaum Associates.

Borko, H., Eisenhart, M., Brown, C., Underhill, R., Jones, D., & Agard, P. (1992). Learning to teach hard mathematics: Do novice teachers and their instructors give up too easily? *Journal for Research in Mathematics Education, 23*(3), 194–222.

Carpenter, T. P., Fennema, E., & Franke, M. L. (1996). Cognitively guided instruction: A knowledge base for reform in primary mathematics instruction. *Elementary School Journal, 97*(1), 1–20.

Carpenter, T. P., Fennema, E., Franke, M., Levi, L., & Empson, S. (1999). *Children's mathematics: Cognitively guided instruction*. Portsmouth, NH: Heinemann.

Carpenter, T, P., Franke, M. L., & Levi, L. (2003). *Thinking mathematically: Integrating arithmetic and algebra in elementary school*. Portsmouth, NH: Heinemann.

Carpenter, T. P., & Levi, L. (2000). *Developing conceptions of algebraic reasoning in the primary grades* (Research Report). Madison, WI: National Center for Improving Student Learning and Achievement in Mathematics and Science. Available at http://www.wcer.wisc.edu/ncislapublications/index.html

Cohen, S. (2004). *Teachers' professional development and the elementary mathematics classroom*. Mahwah, NJ: Lawrence Erlbaum Associates.

Cooney, T. (2001). Considering the paradoxes, perils and purposes of conceptualizing teacher development. In F. Lin & T. Cooney (Eds.), *Making sense of mathematics teacher education* (pp. 9–31). Dordrecht, the Netherlands: Kluwer Academic.

Fujii, T. (2003). Probing students' understanding of variables through cognitive conflict problems: Is the concept of a variable so difficult for students to understand? In N. Pateman, B. Dougherty, & J. Zilliox (Eds.), *Proceedings of the 27th conference of the International Group for the Psychology of Mathematics Education* (Vol. 1, pp. 49–66). Honolulu, HI: University of Hawaii.

Greeno, J. G. (1983). Conceptual entities. In D. Gentner & A. Stevens (Eds.), *Mental models* (pp. 227–252). Hillsdale, NJ: Lawrence Erlbaum Associates.

Harel, G., & Dubinsky, E. (Eds.). (1992). *The concept of function: Aspects of epistemology and pedagogy* (MAA Notes No. 25). Washington, DC: Mathematical Association of America.

Kaput, J. (1979). Mathematics and learning: Roots of epistemological status. In J. Lochhead & J. Clement (Eds.), *Cognitive process instruction* (pp. 289–303). Philadelphia: Franklin Institute Press.

Lakoff, G., & Nuñez, R. (2000). *Where mathematics comes from: How the embodied mind brings mathematics into being*. New York: Basic Books.

Lampert, M. (2001). *Teaching problems and the problem of teaching*. New Haven, CT: Yale University Press.

Livneh, D., & Linchevski, L. (1999). Structure sense: The relationship between algebraic and numerical contexts. *Educational Studies in Mathematics, 40*, 173–196.

Lortie, D. (2002). *Schoolteacher: A sociological study* (2nd ed.). Chicago: University of Chicago Press.

Ma, L. (1999). *Knowing and teaching elementary mathematics*. Mahwah, NJ: Lawrence Erlbaum Associates.

Mason, J., Burton, L., & Stacey, K. (1982). *Thinking mathematically*. London: Addison-Wesley.

Matz, M. (1982). Towards a process model for high school algebra errors. In D. H. Sleeman & J. S. Brown (Eds.), *Intelligent tutoring systems* (pp. 25–50). London: Academic Press.

McLaughlin, M., & Talbert, J. (2001). *Professional communities and the work of high school teaching*. Chicago: University of Chicago Press.

McLaughlin, M., Talbert, J., & Bascia, N. (Eds.). (1990). *The context of teaching in secondary schools: Teachers realities*. New York: Teachers College Press.

Moschkovich, J., Schoenfeld, A. H., & Arcavi, A. A. (1993). Aspects of understanding: On multiple perspectives and representations of linear relations, and connections among them. In T. Romberg, E. Fennema, & T. Carpenter (Eds.), *Integrating research on the graphical representation of function* (pp. 69–100). Hillsdale, NJ: Lawrence Erlbaum Associates.

National Council of Teachers of Mathematics. (1989). *Curriculum and evaluation standards for school mathematics*. Reston, VA: Author.

National Council of Teachers of Mathematics. (2000). *Principles and standards for school mathematics*. Reston, VA: Author.

Nelsen, R. (1993). *Proofs without words: Exercises in visual thinking*. Washington, DC: Mathematical Association of America.

Rissland, E. (1978). Understanding understanding mathematics. *Cognitive Science, 2*(4), 361–383.

Romberg, T., & Carpenter, T. (1986). Research in teaching and learning mathematics: Two disciplines of scientific inquiry. In M. Wittrock (Ed.), *Handbook of research on teaching* (3rd ed., pp. 850–873). New York: Macmillan.

Schoenfeld, A. H. (1985). *Mathematical problem solving*. Orlando, FL: Academic Press.

Schoenfeld, A. H. (1992). Learning to think mathematically: Problem solving, metacognition, and sense-making in mathematics. In D. Grouws (Ed.), *Handbook for research on mathematics teaching and learning* (pp. 334–370). New York: Macmillan.

Schoenfeld, A. H. (2002). Making mathematics work for all children: Issues of standards, testing, and equity. *Educational Researcher, 31*(1), 13–25.

Schoenfeld, A. H. (in press). Theory meets practice: What happens when a mathematics education researcher gets involved in professional development? In *Proceedings of the 10th International Congress on Mathematics Education*. Copenhagen: International Congress on Mathematics Education.

Schoenfeld, A. H., Smith, J. P., & Arcavi, A. A. (1993). Learning. In R. Glaser (Ed.), *Advances in instructional psychology* (Vol. 4, pp. 55–175). Hillsdale, NJ: Lawrence Erlbaum Associates.

Senk, S., & Thompson, D. (Eds.). (2003). *Standards-oriented school mathematics curricula: What does the research say about student outcomes?* Mahwah, NJ: Lawrence Erlbaum Associates.

Shepard, L. (2001, April). *Protecting learning from the effects of high-stakes testing*. Paper presented at the annual meeting of the American Educational research Association, Seattle.

Sherin, M. G. (2001). Developing a professional vision of classroom events. In T. Wood, B. S. Nelson, & J. Warfield (Eds.), *Beyond classical pedagogy: Teaching elementary school mathematics* (pp. 75–93). Hillsdale, NJ: Lawrence Erlbaum Associates.

Simon, D. P., & Simon, H. A. (1979). A tale of two protocols. In J. Lochhead & J. Clement (Eds.), *Cognitive process instruction* (pp. 119–132). Philadelphia: Franklin Institute.

Stake, R. E., & Easley, J. (1978). *Case studies in science education* (Vols. 1–2). Urbana, IL: Center for Instructional Research and Curriculum Evaluation, University of Illinois at Urbana-Champaign.

Tall, D. (1991). To prove or not to prove. *Mathematics Review, 13,* 21–32.

Wigner, E. (1960). The unreasonable effectiveness of mathematics. *Communications in Pure and Applied Mathematics, 13*(1), 1–14.

Wirtz, R., Botel, M., Beberman, M., & Sawyer, W. W. (1965). *Math workshop.* Chicago: Encyclopaedia Britannica Press.

Author Index

G

H

I

J

K

N

Subject Index